HARCOURT BRACE COLLEGE OUTLINE SERIES

INTERMEDIATE ALGEBRA

Alan Wise

Department of Mathematics, University of San Diego

Richard Nation and Peter Crampton

Department of Mathematics, Palomar College

Harcourt Brace College Publishers

Fort Worth   Philadelphia   San Diego
New York   Orlando   Austin   San Antonio
Toronto   Montreal   London   Sydney   Tokyo

To our wives,
Carol, Barbara, and Susana

PREFACE

The purpose of this book is to present a complete course in intermediate algebra in the clear, concise form of an outline. This outline provides an in-depth review of the principles of intermediate algebra for independent study, and contains essential supplementary material for mastering intermediate algebra. Or, this outline can simply be used as a valuable, self-contained refresher course on the practical application of intermediate algebra.

Although comprehensive enough to be used by itself for independent study, this outline is specifically designed to be used as a supplement to college courses and textbooks on the subject. The **Textbook Correlation Table** located on the inside front cover of this book shows how the pages of this outline correspond by topic to the pages of eight leading textbooks on intermediate algebra currently in use. So, should the sequence of topics in this outline differ at all from the sequence of topics in your textbook, you can easily locate the material you want by consulting the table.

Notice, too that the topics in this outline are more narrowly defined than the topics in many textbooks on the same subject. For instance, whereas inequalities are included in the discussion of equations and inequalities in some books, in this outline they are covered in their own separate chapters. This isolation not only helps you to find the specific topics that you need to study but also enables you to bypass the topics that you are already familiar with.

Regular features at the end of each chapter are also specially designed to supplement your textbook and course work in intermediate algebra.

RAISE YOUR GRADES This feature consists of a checkmarked list of open-ended thought questions to help you assimilate the material you have just studied. These thought questions invite you to compare concepts, interpret ideas, and examine the whys and wherefores of chapter material.

SOLVED PROBLEMS Each chapter of this outline contains a set of exercises and word problems and their step-by-step solutions. Undoubtedly the most valuable feature of this outline, these problems allow you to become proficient with the fundamental skills and applications of intermediate algebra. Along with the sample midterm and final examinations, they also give you ample exposure to the kinds of questions that you are likely to encounter on a typical college exam. To make the most of these solved problems, try writing your own solutions first. Then compare your answers to the detailed solutions provided in the book. If you have trouble understanding a particular problem, or set of problems, use the example reference numbers to locate and review the appropriate instruction and parallel step-by-step examples that were developed earlier in the chapter.

SUPPLEMENTARY EXERCISES Each chapter of this outline concludes with a set of drill exercises and practical applications and their answers. The supplementary exercises are designed to help you master and retain all the newly discussed skills and concepts presented in the given chapter, and also contain example references.

Of course there are other features of this outline that you will find very helpful, too. One is the format itself, which serves both as a clear guide to important ideas and as a convenient structure upon which to organize your knowledge. A second is the attention devoted to methodology and the practical applications of intermediate algebra. Yet a third is the careful listing of learning objectives in each section denoted by the uppercase letters A, B, C, D, or E.

We wish to thank Mel Freidman for his comments and suggestions during the development of this text.

San Diego, California

ALAN WISE
RICHARD NATION
PETER CRAMPTON

CONTENTS

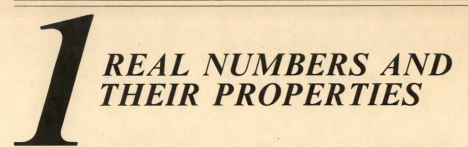

1 REAL NUMBERS AND THEIR PROPERTIES

THIS CHAPTER IS ABOUT

☑ Types of Real Numbers
☑ Computing with Signed Numbers
☑ Properties of Real Numbers
☑ Sets
☑ Translating Words into Symbols

1-1. Types of Real Numbers

Real numbers are the type of number you have used in mathematics all your life. They are sometimes called the **signed numbers,** because they include all the positive and negative numbers, and zero. There are other types of numbers, such as imaginary or complex numbers, but these will not concern us until Chapter 10.

A. Integers

The **integers** are the most familiar types of real numbers, and the most familiar type of integer is the numbers used for counting, 1, 2, 3, . . . , called the **counting numbers** or the **natural numbers.**

Note: The ellipsis symbol (. . .) indicates that the list continues, in the manner indicated by the pattern of the listed elements, until the final listed element is reached. If there is no final element listed, the ellipsis indicates that the list continues indefinitely.

If we include zero with the list of natural numbers, 0, 1, 2, 3, . . . , we call the resulting list the **whole numbers.**

If we include with the list of whole numbers their negative opposites, . . . , -3, -2, -1, 0, 1, 2, 3, . . . , we call the resulting list the **integers.**

Note: We always designate negative numbers with a negative or "minus" sign ($-$). If there is no sign before a number, a positive or "plus" sign is understood (e.g., $4 = +4$). Zero is neither positive nor negative.

B. Rational numbers

Another type of real number is the **rational number.** Any number that can be written in the form m/n, where m and n are integers, and n is not equal to 0, is a rational number. For example, $\frac{1}{2}$, $-\frac{1}{2}$, and $\frac{321}{4321}$ are all rational numbers. All the integers are also rational numbers, because they may be written in the form $m/1$, where m is any integer.

Note: When we need to represent a real number for which no specific value is assigned, we usually use a letter, called a **variable.** In the paragraph above, m and n are variables. A variable can represent a *negative* real number ($x = -3$). When we use a variable to designate a negative number we do not need a minus sign.

C. Irrational numbers

There are real numbers that cannot be written in the form m/n, where m and n are integers. These are the **irrational** numbers. The ratio of the circumference of a circle to its diameter (represented by π, the greek letter "pi"), and the number which when multiplied by itself equals 2 (represented by $\sqrt{2}$), are examples of irrational numbers. Because these numbers cannot be written as the ratio of two integers (that is, in the form m/n), they are frequently written using different types of symbols, such as letters (e), greek letters (π), or radical signs ($\sqrt{2}$).

D. Decimal representation of real numbers

All real numbers may be written in **decimal form.** For example, the integer 3 may be written as 3.0, and the rational number $\frac{3}{4}$ may be written as 0.75. A decimal that ends in a specific place is called a **terminating decimal;** 0.75 is a terminating decimal because it ends in the hundredths place. A decimal that repeats one or more digits without end is called a **repeating decimal;** $\frac{39}{11}$, which has the decimal form 3.54545454 . . . , is a repeating decimal, because the digits 5 and 4 are repeated without end.

Note: The digit or digits that repeat may be indicated by a bar; $3.\overline{54}$.

When irrational numbers are written as decimals, they neither terminate nor repeat. The decimal representation of π, for instance, begins 3.141592653 . . . , but there is no termination point, nor does the pattern repeat.

· Every rational number can be written either as a terminating or a repeating decimal.
· Irrational numbers cannot be written as terminating or repeating decimals.

EXAMPLE 1-1: Determine whether each of the following numbers is rational or irrational:

(a) $\dfrac{3}{8}$ (b) $-\dfrac{5}{8}$ (c) 0.4   (d) 1.47   (e) 0.25225222522225 . . .

Solution: Recall that a rational number can be written in the form m/n, where m and n are integers, and $n \neq 0$. A rational number can also be written as a terminating or a repeating decimal; and irrational number is a nonterminating, nonrepeating decimal.

(a) $\dfrac{3}{8}$ is a rational number because it is in the form $\dfrac{m}{n}$, 3 and 8 are integers, and $8 \neq 0$.

(b) $-\dfrac{5}{8}$ is a rational number.

(c) $0.\overline{4}$ is a rational number because it is a repeating decimal.

(d) 1.47 is a rational number because it is a terminating decimal.

(e) 0.25225222522225 . . . is an irrational number because it is a nonterminating, nonrepeating decimal.

E. The real number line

Every real number can be represented as a point on a line. To construct a **real number line,** we draw a horizontal line and choose an arbitrary point to be associated with the number zero. This point is called the **origin** of the number line. An arrow on the right end of the line indicates the direction of positive increase, so those points to the right of the origin are positive, and those to the left are negative. For each point on the line we can associate one and only one real number. We call that number the **coordinate** of the point. Conversely, for each real number we can associate one and only one point on the line.

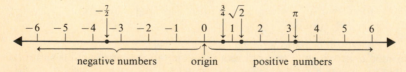

We can use the real number line to show the relative value of signed numbers. To do this we also use the following symbols, called the **relational operators:**

>   greater than

≥   greater than or equal to

=   equals

<   less than

≤   less than or equal to

If we consider any two real numbers a and b, we can say that

- if $a > b$ (a is greater than b), then a is located to the right of b on the real number line.
- if $a \geq b$ (a is greater than or equal to b), then a is located either to the right of b, or precisely at b. (E.g., we can say that $\frac{8}{4} \geq \frac{4}{2}$.)
- if $a = b$ (a is equal to b), then a is located precisely at b.
- if $a < b$ (a is less than b), then a is located to the left of b.
- if $a \leq b$ (a is less than or equal to b), then a is located either to the left of b, or precisely at b.

Conversely, if we are given two points a and b on the real number line, we can say that if a is to the right of b, then $a > b$, and so on.

EXAMPLE 1-2: Determine whether each statement is true or false:

(a) $3 < 4$ (b) $0 > 5$ (c) $-2 < -1$ (d) $4 \geq -1$ (e) $6 \geq \frac{36}{6}$ (f) $-2 \leq 3$
(g) $-2 \leq -\frac{4}{2}$

Solution: Refer to each number's corresponding point on the real number line.

    (a) $3 < 4$ is a true statement; 3 is to the left of 4 on the line.
    (b) $0 > 5$ is a false statement; 0 is not to the right of 5.
    (c) $-2 < -1$ is a true statement; -2 is to the left of -1.
    (d) $4 \geq -1$ is a true statement; 4 is to the right of -1.
    (e) $6 \geq \frac{36}{6}$ is a true statement because $6 = \frac{36}{6}$.
    (f) $-2 \leq 3$ is a true statement; -2 is to the left of 3.
    (g) $-2 < -\frac{4}{2}$ is a false statement because both numbers correspond precisely the same point.

1-2. Computing with Signed Numbers

A. Finding absolute values

The **absolute value** of a real number is its distance from the origin on the real number line. We always express this distance as a positive value, so absolute values are always nonnegative. The symbol for the absolute value of any real number x is $|x|$. To find the absolute value, we use the following definitions:

$$\text{For any real number } x, \; |x| = \begin{cases} x & \text{if } x \text{ is positive.} \\ 0 & \text{if } x \text{ is 0.} \\ -x & \text{if } x \text{ is negative.} \end{cases}$$

Note: The value of a number $-x$ is not necessarily negative. For example, if $x = -3$, then $-x = -(-3)$, or 3.

EXAMPLE 1-3: Find the absolute value of each of the following numbers:

(a) $|5|$ (b) $|0|$ (c) $|-5|$

Solution: Recall the definition of absolute value.

    (a) $|5| = 5$ The absolute value of a positive number is the number itself.
    (b) $|0| = 0$ The absolute value of zero is zero.
    (c) $|-5| = -(-5) = 5$ The absolute value of a negative number is the opposite of that negative number.

B. Adding and subtracting signed numbers

1. Adding signed numbers. When we add numbers, each number to be added is called an **addened,** and the result of the addition operation is called the **sum.** In algebra, the addends are frequently signed numbers, and we must follow certain procedures to insure that the sum has the proper value.

To add two signed numbers with like signs

1. Find the sum of the absolute values of the addends.
2. Write the same sign on the sum.

Note: The sum of two positive numbers is always positive, and the sum of two negative numbers is always negative.

EXAMPLE 1-4: Add **(a)** 7 and 9, and **(b)** -4 and -3.

Solution:

(a) $7 + 9 = ?(|7| + |9|)$ **(b)** $-4 + (-3) = ?(|-4| + |-3|)$ Find the sum of the absolute

$\qquad = ?(7 + 9)$ $\qquad\qquad\quad = ?(4 + 3)$ values.

$\qquad = ?(16)$ $\qquad\qquad\quad = ?(7)$

$\qquad = +16$ $\qquad\qquad\quad = -7$ Write the same sign on the sum.

To add two signed numbers with opposite signs

1. Find the difference of the absolute values of the addends.
2. Write the sign of the addend with the largest absolute value on the difference.

EXAMPLE 1-5: Add **(a)** 7 and -3, and **(b)** -8 and 2.

Solution:

(a) $7 + (-3) = ?(|7| - |-3|)$ **(b)** $-8 + 2 = ?(|-8| - |2|)$ Find the difference between

$\qquad = ?(7 - 3)$ $\qquad\qquad = ?(8 - 2)$ the absolute values.

$\qquad = ?(4)$ $\qquad\qquad = ?(6)$

$\qquad = +4$ $\qquad\qquad = -6$ Write the sign of the largest
absolute value on the difference.

The sum of two numbers with opposite signs but the same absolute value is always zero; $4 + (-4) = 0$, and $-6 + 6 = 0$. These numbers are called **additive inverses.**

2. Subtracting signed numbers. When we subtract two numbers, for example $a - b = c$, we call a the **minuend,** b the **subtrahend,** and c the **difference.** Subtraction is defined as the addition of the additive inverse of the subtrahend. In other words, to subtract a number, add its opposite.

To subtract two signed numbers

1. Change the subtraction operation to one of addition.
2. Write the additive inverse of the second addend (formerly the subtrahend).
3. Follow the rules for adding two signed numbers.

EXAMPLE 1-6: Subtract **(a)** -3 from 12, and **(b)** 8 from -6.

Solution:

(a) $12 - (-3) = 12 + (?)$ **(b)** $-6 - 8 = -6 + (?)$ Change subtraction to addition.

$\qquad = 12 + (+3)$ $\qquad\quad = -6 + (-8)$ Write the opposite of the second
addend.

$\qquad = 15$ $\qquad\quad = -14$ Add as before.

C. Multiplying signed numbers

There are several ways to write the multiplication problem "*r* times *s* equals *t*."

$$\left. \begin{array}{l} r \times s = t \\ r \cdot s = t \end{array} \right\} \text{using multiplication symbols}$$

$$\left. \begin{array}{l} r(s) = t \\ (r)s = t \\ (r)(s) = t \end{array} \right\} \text{using parentheses}$$

$$\text{or simply } rs = t$$

In this problem *r* and *s* are called the **factors,** and *t* is called the **product.**

To multiply two signed numbers

1. Find the product of the absolute values of the factors.
2. Make the product (a) **positive** if the original factors have **like signs.**
   (b) **negative** if the original factors have **opposite signs.**
   (c) zero if either factor is zero.

EXAMPLE 1-7: Find the products of **(a)** -5 times -3, and **(b)** 7 times -4.

Solution:

(a) $-5 \cdot (-3) = ? (|-5| \cdot |-3|)$ **(b)** $7 \cdot (-4) = ? (|7| \cdot |-4|)$ Find the product of the absolute values.

$$= ? (5 \cdot 3) \qquad\qquad\qquad = ? (7 \cdot 4)$$

$$= ? \, 15 \qquad\qquad\qquad\quad = ? \, 28$$

$$= 15 \qquad\qquad\qquad\quad = -28 \qquad \text{Prefix the proper sign.}$$

D. Dividing signed numbers

There are also several ways of writing the division problem "*r* divided by *s* equals *t*."

$$r \div s = t \qquad\qquad r/s = t \qquad\qquad s \overline{)\, r}^{\,t}$$

In this problem *r* is called the **dividend,** *s* is called the **divisor,** and *t* is called the **quotient.**

To divide two signed numbers

1. Find the quotient of the absolute values of the numbers.
2. Make the quotient (a) **positive** if the divisor and the dividend have **like signs.**
   (b) **negative** if the divisor and dividend have **opposite signs.**

EXAMPLE 1-8: Find the quotient of **(a)** -24 divided by 3, and **(b)** -35 divided by -7.

Solution:

(a) $-24 \div 3 = ? (|-24| \div |3|)$ **(b)** $-35 \div (-7) = ? (|-35| \div |-7|)$ Find the quotient of the absolute values.

$$= ? (24 \div 3) \qquad\qquad\qquad = ? (35 \div 7)$$

$$= ? \, 8 \qquad\qquad\qquad\qquad = ? \, 5$$

$$= -8 \qquad\qquad\qquad\qquad = 5 \qquad \text{Prefix the proper sign.}$$

Multiplication and division are called **inverse operations** because one "undoes" the other. For example,

$$12 \div (-3) = -4 \qquad \text{because} \qquad (-4) \cdot (-3) = 12$$

Division involving the number zero has a special role in mathematics.

· Zero divided by any nonzero number equals zero.

$$0 \div (-3) = 0 \qquad \text{because} \qquad 0 \cdot (-3) = 0$$

· Division by zero is undefined. If, for example, we assume that 5 divided by 0 equals the real number n, then $5 \div 0 = n$, which would mean that $n \cdot 0 = 5$. But we know that $n \cdot 0$ cannot equal 5, because the product of every real number and zero is zero. So division by zero must be undefined.
· Zero divided by zero is also undefined, but for a different reason: $0 \div 0 = m$ would require that $m \cdot 0 = 0$. But the statement $m \cdot 0 = 0$ is true for every real number m, and is meaningless because any real number could be a solution. Therefore, zero divided by zero is also undefined.

EXAMPLE 1-9: Find the quotient in each of the following division problems:

(a) $0 \div 7$ **(b)** $0 \div (-7)$ **(c)** $(-2) \div 0$ **(d)** $0 \div 0$

Solution: Recall the rules for division by zero.

       (a) $0 \div 7 = 0$
       (b) $0 \div (-7) = 0$
       (c) $(-2) \div 0$ is undefined.
       (d) $0 \div 0$ is undefined.

E. Evaluating exponents

A short way of writing the product of repeated factors is called **exponential notation.** For example, the exponential notation for $3 \cdot 3 \cdot 3 \cdot 3$ is 3^4. To write the product of repeated factors in exponential notation, we use the value of the repeated factor as the **base,** and the number of repetitions as the **exponent.** So, if n is any natural number greater than 1,

$$a^n = \underbrace{a \cdot a \cdot a \ldots \cdot a}_{n \text{ factors}}$$

where a is the base and n is the exponent.

Note 1: The exponential notation x^2 may be read "x to the second power," or "x squared."
        The exponential notation x^3 may be read "x to the third power," or "x cubed."
        The exponential notation for any n greater than 3 is simply read "x to the nth power."

Note 2: For any real number a, $a^1 = a$. For example, $3^1 = 3$, $(-5)^1 = -5$, and $(2/5)^1 = 2/5$.

Caution: The placement of parentheses is crucial when we are considering negative numbers in exponential notation. For example, $(-3)^4 = (-3)(-3)(-3)(-3)$, or 81. But -3^4 is the same as $-(3^4)$, which is the additive inverse of 3^4, or -81.

EXAMPLE 1-10: Evaluate **(a)** 2^3 and **(b)** $(-4)^3$.

Solution:

(a) $2^3 = 2 \cdot 2 \cdot 2$ **(b)** $(-4)^3 = (-4)(-4)(-4)$ Write as a product of repeated factors.

    $= 4 \cdot 2$ $= 16(-4)$ Compute the product.

    $= 8$ $= -64$

F. The order of operations agreement

A number or variable—or the product or quotient of numbers and/or variables—is called a **term.** Some examples of terms are $2x$, $4y^2$, $-ab$, and xy^5. The variables in a term are called its **literal part,** and the constants (excluding any exponents) are called its **numerical coefficient.** In the term $4xy$, the literal part is xy, and the numerical coefficient is 4.

Note: When no numerical coefficient appears in a term, it is understood that its numerical coefficient is 1 or -1: $xy^2 = 1xy^2$, and $-w = -1w$.

A mathematical **expression** is a single term, or the sum or difference of terms. The term $2m$ may be called an expression, but $2m + 3$ may only be called an expression, because it is the sum of two terms, $2m$ and 3.

When an expression involves more than one arithmetic operation, we agree to perform the operations in a certain order. This is called the **Order of Operations Agreement,** and is a convention used in all mathematics. For example, we can find two solutions for the expression $3 + 6 \cdot 5 = ?$, depending on which operation we perform first.

$$3 + 6 \cdot 5 = ? \qquad\qquad 3 + 6 \cdot 5 = ?$$
$$9 \cdot 5 = 45 \quad \text{or} \quad 3 + 30 = 33$$

Both these solutions are arithmetically correct, but only one is the answer we seek. In order to avoid this kind of ambiguity, all mathematicians follow the steps of the Order of Operations Agreement, which enables us to properly evaluate all expressions.

Step 1. Perform all operations inside grouping symbols. The most common grouping symbols are parentheses ( ), brackets [ ], and braces { }.
Step 2. Evaluate all exponential expressions in order from left to right.
Step 3. Multiply and divide in order from left to right.
Step 4. Add and subtract in order from left to right.

Following these steps, we see that of the two solutions obtained above, 33 is the conventionally correct one.

The Order of Operations Agreement is sometimes referred to by use of the acronym **PEMDAS,** which is compounded as follows:

P	evaluate expressions in **P**arentheses (or other grouping symbols),
E	evaluate **E**xponential expressions,
M	**M**ultiply and
D	**D**ivide in order from left to right,
A	**A**dd and
S	**S**ubtract in order from left to right.

Memorizing the word PEMDAS may make it easier for you to memorize the Order of Operations Agreement.

EXAMPLE 1-11: Evaluate the following expression: $-3 + 6^2 \div 2 + 3 \cdot 5 - (1 + 2 \cdot 4)$

Solution: Use the Order of Operations Agreement:

$$-3 + 6^2 \div 2 + 3 \cdot 5 - (1 + 2 \cdot 4) = -3 + 6^2 \div 2 + 3 \cdot 5 - (1 + 8) \qquad \text{Do operations inside grouping symbols.}$$
$$= -3 + 6^2 \div 2 + 3 \cdot 5 - 9$$
$$= -3 + 36 \div 2 + 3 \cdot 5 - 9 \qquad \text{Evaluate exponential expressions.}$$
$$= -3 + 18 + 3 \cdot 5 - 9 \qquad \text{Multiply or divide from left to right.}$$
$$= -3 + 18 + 15 - 9$$
$$= 15 + 15 - 9 \qquad \text{Add or subtract from left to right.}$$
$$= 30 - 9$$
$$= 21$$

If an expression contains grouping symbols within other grouping symbols, convention requires us to perform the operations within the innermost grouping symbols first.

Note: It is conventional to use parentheses for the first or innermost level of operations, brackets for the next level, and braces for the next. If another level is needed we use parentheses a second time, and continue the series as many times as necessary.

EXAMPLE 1-12: Evaluate the following expression: $20 - [(30 - 4) \div 2]^2$

Solution: Perform the operations in the innermost grouping symbols first, then follow the Order of Operations Agreement.

$$20 - [(30 - 4) \div 2]^2 = 20 - [26 \div 2]^2 \qquad \text{Do operations in the innermost}$$
$$= 20 - [13]^2 \qquad \qquad \text{grouping symbols first.}$$
$$= 20 - 169 \qquad \qquad \text{Evaluate exponential expressions.}$$
$$= -149$$

To evaluate a variable expression containing more than one operation, we substitute given numerical values for the appropriate variables, and perform the indicated operations using the Order of Operations Agreement.

EXAMPLE 1-13: Evaluate $x[2x - (3y + 5)] + xy^2$, where $x = 2$ and $y = -3$.

Solution: $x[2x - (3y + 5)] + xy^2 = 2\{2 \cdot 2 - [3 \cdot (-3) + 5]\} + 2(-3)^2 \qquad$ Substitute 2 for x and -3 for y.

$$= 2[2 \cdot 2 - (-9 + 5)] + 2(-3)^2 \qquad \text{Use the Order of}$$
$$= 2[2 \cdot 2 - (-4)] + 2(-3)^2 \qquad \text{Operations}$$
$$= 2[4 - (-4)] + 2(-3)^2 \qquad \text{Agreement.}$$
$$= 2[8] + 2(-3)^2$$
$$= 2[8] + 2(9)$$
$$= 16 + 2(9)$$
$$= 16 + 18$$
$$= 34$$

1-3. Properties of Real Numbers

Real numbers have certain properties, some of which are listed below, that can often be used to simplify expressions.

The Commutative Property of Addition

$$a + b = b + a$$

Changing the order of the addends does not change the sum.

Example: $4 + 3 = 3 + 4$

$$7 = 7$$

The Commutative Property of Multiplication

$$a \cdot b = b \cdot a$$

Changing the order of the factors does not change the product.

Example: $2(-5) = (-5)2$

$$-10 = -10$$

The Associative Property of Addition

$$(a + b) + c = a + (b + c)$$

Changing the grouping of addends does not change the sum.

Example: $(3 + 2) + 7 = 3 + (2 + 7)$

$$5 + 7 = 3 + 9$$

$$12 = 12$$

The Associative Property of Multiplication	Changing the grouping of factors does not change the product.

$$(a \cdot b) \cdot c = a \cdot (b \cdot c)$$

Example: $(2 \cdot 5) \cdot 7 = 2 \cdot (5 \cdot 7)$

$$10 \cdot 7 = 2 \cdot 35$$

$$70 = 70$$

The Distributive Property of Multiplication over Addition

To multiply a sum by a number, you can first add and then multiply, or first multiply and then add.

$$a(b + c) = ab + ac$$

Example: $2(3 + 7) = 2(3) + 2(7)$

$$2(10) = 6 + 14$$

$$20 = 20$$

$$(b + c)a = ba + ca$$

Example: $(8 + 3)5 = 8(5) + 3(5)$

$$11(5) = 40 + 15$$

$$55 = 55$$

The Addition Property of Zero

The sum of any real number and zero is that real number.

$$a + 0 = 0 + a = a$$

Example: $5 + 0 = 0 + 5 = 5$

The Multiplication Property of Zero

The product of any real number and zero is zero.

$$a \cdot 0 = 0 \cdot a = 0$$

Example: $3 \cdot 0 = 0 \cdot 3 = 0$

The Multiplication Property of One

The product of any real number and one is that real number.

$$a \cdot 1 = 1 \cdot a = a$$

Example: $7 \cdot 1 = 1 \cdot 7 = 7$

The Additive Inverse Property

The sum of any real number and its additive inverse is zero.

$$a + (-a) = (-a) + a = 0$$

Example: $4 + (-4) = (-4) + 4 = 0$

The Multiplication Property of Reciprocals

The product of any nonzero real number and its multiplicative inverse (reciprocal) is one.

$$a \cdot (1/a) = (1/a) \cdot a = 1$$

Example: $4 \cdot \frac{1}{4} = \frac{1}{4} \cdot 4 = 1$

The Zero Product Property

If the product of two real numbers is zero, then at least one of those real numbers must be zero.

If $a \cdot b = 0$, then $a = 0$ or $b = 0$

Example: If $x(x + 2) = 0$, then $x = 0$ or $x + 2 = 0$
$$x = 0 \text{ or } \qquad x = -2$$

A. Identifying properties

The properties introduced above are used frequently in algebra, and it is important not only to be able to use them to simplify expressions, but to be able to recognize when they have been used to rename expressions.

EXAMPLE 1-14: Which properties are being used in the following expressions?

(a) $(2n + 5) + m = 2n + (5 + m)$ (b) $(5 + p)3 = 3(5 + p)$

Solution: (a) The associative property of addition is used here—the order of the terms is the same, but the grouping is different.

(b) The commutative property of multiplication is used here—the grouping of the terms is the same, but the order is different.

B. Simplifying variable expressions

We use the various properties of real numbers to simplify variable expressions. The first step in simplifying variable expressions is to clear any parentheses. We do this by using the distributive property.

EXAMPLE 1-15: Clear the parentheses in the expression $3(2x + 7)$.

Solution: Apply the distributive property of multiplication over addition:

$$3(2x + 7) = 3 \cdot 2x + 3 \cdot 7$$
$$= 6x + 21$$

When parentheses are preceded by a positive sign, we can clear them by simply writing the expression inside without the parentheses (once it has been simplified). When parentheses are preceded by a negative sign, we can clear them by writing the opposite of each term inside. For example, $+(4x - 7)$ simplifies to $4x - 7$, and $-(3v + 4w - 5x)$ simplifies to $-3v - 4w + 5x$.

Terms that have exactly the same literal part (i.e., the same variables raised to the same powers) are called **like terms.** To simplify an expression with like terms, we combine (add or subtract) the like terms using the distributive property. To do this, add or subtract the numerical coefficients of the terms, then write the same literal part on the sum or difference. For example, $2x + 5x = (2 + 5)x = 7x$.

EXAMPLE 1-16: Simplify $4x + 5wy + 7x - 2wy$.

Solution: First use the commutative and associative properties to rearrange and regroup the terms, then combine like terms.

$$4x + 5wy + 7x - 2wy = 4x + 7x + 5wy - 2wy \qquad \text{Rearrange and regroup}$$
$$= (4x + 7x) + (5wy - 2wy) \qquad \text{the terms.}$$
$$= 11x + 3wy \qquad \text{Combine like terms.}$$

Recall: If a variable expression contains grouping symbols inside other grouping symbols, simplify the expression inside the innermost parentheses first, then clear the innermost parentheses.

EXAMPLE 1-17: Simplify $-4\{x + 3[2(x + 5 + 3x) - 12] + 1\}$.

Solution: $-4\{x + 3[2(x + 5 + 3x) - 12] + 1\} = -4\{x + 3[2(4x + 5) - 12] + 1\}$
$$= -4\{x + 3[8x + 10 - 12] + 1\}$$
$$= -4\{x + 3[8x - 2] + 1\}$$
$$= -4\{x + 24x - 6 + 1\}$$
$$= -4\{25x - 5\}$$
$$= -100x + 20$$

1-4. Sets

A. Set notation

In algebra we often classify numbers into groups or collections called **sets.** Each number in a set is called a **member** or **element** of that set. We usually assign a capital letter to name each different set.

Note: The symbol $\in$ indicates that a number is a member of a set; e.g., $3 \in$ the set of natural numbers.

Particular elements of a set may often be denoted by a lowercase letter. Such a letter is called a **constant** when it is used to represent a **specific element** of a set, and is called a **variable** when it is used to represent **any typical element** of a set.

The set with no members is called the **empty set** or **null set,** and is denoted by the symbol $\varnothing$, or by empty braces, { }.

1. The roster method. One way of writing sets is the **roster method,** in which we list all the members of the set, separate them with commas, and enclose them in braces. If, for instance, we wished to make a set of all the lowercase letters we might use as variables, we could denote it with a capital V, and write it as

$$\text{Variable set:} \quad V = \{a, b, c, \ldots, z\}$$

2. Set-builder notation. Another method of writing sets is called **set-builder notation.** With this notation, we use a mathematical expression to state a characteristic or unique property about the members of the set, and a variable to designate the set. For example, set-builder notation for the set of all numbers greater than zero and less than or equal to ten is written and read as

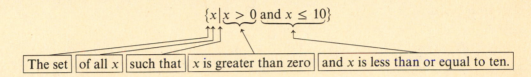

$$\{x \mid x > 0 \text{ and } x \leq 10\}$$

| The set | of all x | such that | x is greater than zero | and x is less than or equal to ten. |

B. Sets of real numbers

We can now write as members of sets the various types of real numbers that we introduced in Section 1-1.

- The set of natural numbers can be denoted by N:
  $$N = \{1, 2, 3, \ldots\}$$

- The set of whole numbers can be denoted by W:
  $$W = \{0, 1, 2, 3, \ldots\}$$

- The set of integers can be denoted by Z:
  $$Z = \{\ldots, -3, -2, -1, 0, 1, 2, 3, \ldots\}$$

Set-builder notation is better suited to writing the sets of both rational and irrational numbers.

- The set of rational numbers can be denoted by Q:
  $$Q = \{m/n \mid m, n \in Z, \text{ and } n \neq 0\}$$

- The set of irrational numbers can be denoted by I:
  $$I = \{x \mid x \neq m/n, \text{ where } m, n \in Z, \text{ and } n \neq 0\}$$

When all the elements of one set are also elements of a second set, we say that the first set is a **subset** of the second. For example, all of the natural numbers are elements of the set of whole numbers, so we can say that N is a subset of W. If we apply this idea of subsets to all the sets of real numbers, we find that

- N is a subset of W
- W (and therefore N) is a subset of Z
- Z (and therefore W and N) is a subset of Q

However, none of the rational numbers Q is a member of the irrational numbers I, or vice versa, so Q is not a subset of I, nor is I a subset of Q.

We can now write the set of all real numbers, of which all the other sets are subsets.

- The set of all real numbers can be denoted by R:
  $$R = \{x \mid x \in Q, \text{ or } x \in I\}$$

EXAMPLE 1-18: Which of the following are true statements?

(a) $-3 \in W$ **(b)** $17 \in Q$ **(c)** $\pi \in R$ **(d)** $-\frac{8}{2} \in Z$ **(e)** $0 \in I$

Solution: **(a)** False. The whole numbers do not include any of the negative integers.
   (b) True. All integers are rational numbers.
   (c) True. Π may be an irrational number, but it is still part of the real number system.
   (d) True. Since this fraction will reduce to an integer, even while it is in factional form it is considered to be an integer.
   (e) False. Zero is an integer, and all integers are rational numbers.

C. Graphing sets

As stated earlier, the point on the real number line associated with a number is called its coordinate. To **graph a number,** we place a dot on the number line at that number's coordinate. To graph a set, we graph each number in that set.

EXAMPLE 1-19: Graph the following sets on the real number line:

(a) $\{-3, 0, 1\}$ **(b)** $\{x \mid x \in N \text{ and } x < 3\}$ **(c)** $\{x \mid x \in R \text{ and } x > -1\}$
(d) $\{x \mid x \in R \text{ and } x \leq 3\}$

Solution:

(a)

Draw a dot on the coordinates -3, 0, and 1.

(b)

The only natural numbers less than 3 are 1 and 2.

(c)

The hollow circle means -1 is not a member of the set. The shaded region to the right of -1 means every real number greater than -1 is included in the graph.

(d)

The solid circle means that 3 is a member of the set, and is included in the graph. The shaded region to the left of 3 means every real number less than 3 is included in the graph.

D. Finding the intersection of sets

The **intersection** of any two sets A and B is the set of elements of A that are also elements of B. We use the symbol $\cap$ to represent intersection; $A \cap B$. For example, $Q \cap I = \varnothing$.

EXAMPLE 1-20: Find the intersections of the following sets:

(a) $\{-3, 0, 1, 5, 15\} \cap \{-3, -2, -1, 0, 1, 4, 7\}$ **(b)** $N \cap Z$ **(c)** $N \cap \{-3, -2\}$

Solution: The intersection of two sets is the set of all numbers common to both sets.

   (a) $\{-3, 0, 1, 5, 15\} \cap \{-3, -2, -1, 0, 1, 4, 7\} = \{-3, 0, 1\}$

The only numbers common to both sets are -3, 0, and 1.

   (b) $N \cap Z = N$

Every natural number is also an integer, so the intersection is the set of natural numbers.

   (c) $N \cap \{-3, -2\} = \varnothing$

There are no numbers common to both sets.

E. Finding the union of sets

The **union** of any two sets A and B is the set of elements in A OR B, or in both A AND B. We use the symbol $\cup$ to represent union; $A \cup B$. For example, $Q \cup I = R$.

EXAMPLE 1-21: Find the unions of the following sets:

(a) $\{1, 2, 5, 7\} \cup \{-3, 6, 7\}$ **(b)** $\{0, 2, 4, 6, 8, \ldots\} \cup \{1, 3, 5, 7, \ldots\}$. **(c)** $N \cup Z$

Solution: The union of two sets is a set whose members consists of all the numbers that are in the first *or* the second set, or in *both* sets.

(a) $\{1, 2, 5, 7\} \cup \{-3, 6, 7\} = \{-3, 1, 2, 5, 6, 7\}$	Note that the common member 7 is listed only once.
(b) $\{0, 2, 4, 6, 8, \ldots\} \cup \{1, 3, 5, 7, \ldots\} = W$	All the even and odd counting numbers and 0, make the whole numbers.
(c) $N \cup Z = Z$	Every natural number is also an integer, so the union of the two sets consists of all the integers.

1-5. Translating Words into Symbols

To help translate words into mathematical symbols you can substitute the appropriate symbols (shown in Table 1-1) for the corresponding key words.

Table 1-1. Key words used to translate a word problem into an equation.

Key Words	Equivalent Math Symbol
the sum of / plus / added to / joined with / increased by / more than / more / and / combined with	$+$
the difference between / minus / subtracted from / take away / decreased by / less than / less / reduced by / diminished by / exceeds	$-$
the product of / times / multiplied by / equal amounts of / of / times as much	$\times$ ·
the quotient of / divides / divided by / ratio / separated into equal amounts of / goes into / over	$\div$
is the same as / is equal to / equals / is / was / earns / are / makes / gives / the result is / leaves / will be	$=$
twice / double / two times / twice as much as	$2\cdot$
half / one-half of / one-half times / half as much as	$\frac{1}{2}\cdot$
what number / what part / a number / the number / what amount / what percent / what price	n (any letter can be used)
twice a (the) number / double a (the) number	$2n$
half a (the) number / one-half a (the) number	$\frac{1}{2}n$

EXAMPLE 1-22: Translate into symbols using key words:

(a) Three times a number is eighteen.
(b) Four less than a number is the same as negative one.
(c) Six subtracted from two-thirds of a number gives two.
(d) Seven times the quantity x plus two is equal to thirty-five.

Solution: Identify the key words, then translate into symbols.

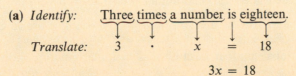

(a) *Identify:*   Three times a number is eighteen.

Translate: 3   · x =   18

$$3x = 18$$

(b) *Identify:* Four less than a number is the same as negative one.

Translate: $n - 4$ $=$ -1

$$n - 4 = -1$$

(c) *Identify:* Six subtracted from two-thirds of a number gives two.

Translate: $\frac{2}{3}n - 6$ $=$ 2

$$\frac{2}{3}n - 6 = 2$$

(d) *Identify:* Seven times the quantity x plus two is equal to thirty-five.

Translate: 7 $\cdot$ $(x + 2)$ $=$ 35

$$7(x + 2) = 35$$

RAISE YOUR GRADES

Can you . . . ?

☑ identify the various types of integers
☑ identify a rational number
☑ identify an irrational number
☑ identify a terminating decimal
☑ identify a repeating decimal
☑ construct a real number line
☑ find the absolute value of a real number
☑ add, subtract, multiply, and divide with signed numbers
☑ write the product of repeated factors using exponential notation
☑ use the Order of Operations Agreement to simplify expressions
☑ identify a term, an expression, and like terms
☑ identify the properties of real numbers
☑ use the properties of real numbers to simplify expressions
☑ write sets using both the roster and set-builder notations
☑ graph sets
☑ find the intersection and union of sets
☑ translate words into symbols using key words

SOLVED PROBLEMS

PROBLEM 1-1 Determine whether each of the following numbers is a rational or irrational number:

(a) $\frac{3}{8}$ **(b)** $0.717717771\ldots$ **(c)** $8.\overline{34}$ **(d)** -4.073

Solution: Recall that a rational number can be written in the form m/n, where m and n are integers, and $n \neq 0$. All terminating decimals and repeating decimals are rational numbers. All non-terminating, nonrepeating decimals are irrational numbers [see Example 1-1].

(a) $\frac{3}{8}$ is a rational number because it is in the form m/n, where m and n are integers, and $n \neq 0$.

(b) $0.717717771\ldots$ is an irrational number because it is a nonterminating, nonrepeating decimal.

(c) 8.34 is a rational number because it is a repeating decimal.

(d) −4.073 is a rational number because it is a terminating decimal.

PROBLEM 1-2 Determine whether each statement is true or false:

(a) $2 < 5$ (b) $-3 \le -3$ (c) $-5 \ge -8$ (d) $-4 > 1$

Solution: Refer to each number's coordinate on the real number line [see Example 1-2].

(a) $2 < 5$ is a true statement; 2 is to the left of 5 on the real number line.

(b) $-3 \le -3$ is a true statement; -3 equals -3.

(c) $-5 \ge -8$ is a true statement; -5 is to the right of -8 on the real number line.

(d) $-4 > 1$ is a false statement; -4 is to the left of 1 on the real number line, so it is less than 1.

PROBLEM 1-3 Find the absolute value of each of the following numbers:

(a) -4 (b) 0   (c) $\frac{1}{2}$ (d) -2.7

Solution: Recall that for any real number x, the absolute value of x is (a) x if x is positive, (b) if x is 0, and (c) $-x$ if x is negative [see Example 1-3].

(a) $|-4| = -(-4) = 4$ (b) $|0| = 0$ (c) $\left|\frac{1}{2}\right| = \frac{1}{2}$ (d) $|-2.7| = -(-2.7) = 2.7$

PROBLEM 1-4 Add the following signed numbers:

(a) -8 and -3 (b) -7 and 6   (c) $-\frac{3}{7}$ and $-\frac{2}{7}$ (d) $\frac{1}{2}$ and $\frac{1}{4}$

Solution [see Examples 1-4 and 1-5]:

(a) $-8 + (-3) = ?\,(|-8| + |-3|)$
$$= ?\,(8 + 3)$$
$$= ?\,11$$
$$= -11$$

To add two signed numbers with like signs, find the sum of the absolute values of the addends and write the same sign on the sum.

(b) $-7 + 6 = ?\,(|-7| - |6|)$
$$= ?\,(7 - 6)$$
$$= ?\,1$$
$$= -1$$

To add two signed numbers with opposite signs, find the difference of their absolute values and write the sign of the addend with the largest absolute value on the difference.

(c) $-\frac{3}{7} + -\frac{2}{7} = ?\,\left(\left|\frac{3}{7}\right| + \left|-\frac{2}{7}\right|\right)$
$$= ?\,\left(\frac{3}{7} + \frac{2}{7}\right)$$
$$= ?\,\frac{5}{7}$$
$$= -\frac{5}{7}$$

Adding two fractions with a common denominator produces a rational number whose numerator is the sum of the numerators, and whose denominator is the common denominator.

(d) $\frac{1}{2} + \frac{1}{4} = \frac{1}{2} \cdot \frac{2}{2} + \frac{1}{4}$
$$= \frac{2}{4} + \frac{1}{4}$$
$$= \frac{3}{4}$$

Before you can add fractions with different denominators, you must find a common denominator.

PROBLEM 1-5 Subtract the following signed numbers:

(a) -5 from 13   (b) -1 from -8 (c) $\frac{1}{5}$ from $\frac{4}{5}$ (d) $\frac{1}{2}$ from $-\frac{2}{3}$

Solution: Recall that to subtract signed numbers, you change the subtraction operation to one of addition, write the additive inverse of the second addend (formerly the subtrahend), and follow the rules for adding signed numbers [see Example 1-6].

(a) $13 - (-5) = 13 + (?)$ (b) $-8 - (-1) = -8 + (?)$
$$= 13 + 5 \qquad\qquad = -8 + 1$$
$$= 18 \qquad\qquad\quad = -7$$

(c) $\dfrac{4}{5} - \dfrac{1}{5} = \dfrac{4}{5} + (?)$

$= \dfrac{4}{5} + \left(-\dfrac{1}{5}\right)$

$= \dfrac{4 + (-1)}{5}$

$= \dfrac{3}{5}$

(d) $-\dfrac{2}{3} - \dfrac{1}{2} = -\dfrac{2}{3} + \left(-\dfrac{1}{2}\right)$

$= -\dfrac{2}{3}\cdot\dfrac{2}{2} + \left(-\dfrac{1}{2}\right)\cdot\dfrac{3}{3}$

$= \dfrac{-4 + (-3)}{6}$

$= \dfrac{-7}{6}$ or $-\dfrac{7}{6}$

PROBLEM 1-6 Multiply the following signed numbers:

(a) -4 times -7 (b) -3 times 8 (c) $-\dfrac{3}{4}$ times $\dfrac{1}{2}$ (d) $-\dfrac{2}{3}$ times $-\dfrac{7}{5}$

Solution: Recall that to multiply signed numbers, you find the product of their absolute values, and make the product positive if the original factors have like signs, and negative if the original factors have opposite signs [see Example 1-7].

(a) $-4 \cdot (-7) = ?\,|-4|\cdot|-7|$

$= ?\,(4\cdot 7)$

$= ?\,28$

$= 28$

(b) $-3\cdot 8 = ?\,|-3|\cdot|8|$

$= ?\,(3\cdot 8)$

$= ?\,24$

$= -24$

(c) $\left(-\dfrac{3}{4}\right)\cdot\dfrac{1}{2} = ?\,\left|-\dfrac{3}{4}\right|\cdot\left|\dfrac{1}{2}\right|$

$= ?\,\left(\dfrac{3}{4}\cdot\dfrac{1}{2}\right)$

$= ?\,\dfrac{3}{8}$

$= -\dfrac{3}{8}$

(d) $\left(-\dfrac{2}{3}\right)\left(-\dfrac{7}{5}\right) = ?\,\left|-\dfrac{2}{3}\right|\cdot\left|-\dfrac{7}{5}\right|$

$= ?\,\left(\dfrac{2}{3}\cdot\dfrac{7}{5}\right)$

$= ?\,\dfrac{14}{15}$

$= \dfrac{14}{15}$

PROBLEM 1-7 Divide the following numbers: (a) -27 by 9 (b) -51 by -3 (c) $-\dfrac{2}{5}$ by $\dfrac{7}{3}$

Solution: Recall that to divide signed numbers, you find the quotient of their absolute values, and make the quotient positive if the divisor and dividend have like signs, and negative if the divisor and dividend have opposite signs [see Example 1-8].

(a) $-27 \div 9 = ?\,(|-27| \div |9|)$

$= ?\,(27 \div 9)$

$= ?\,3$

$= -3$

(b) $-51 \div (-3) = ?\,(|-51| \div |-3|)$

$= ?\,(51 \div 3)$

$= ?\,17$

$= 17$

(c) $-\dfrac{2}{5}\div\dfrac{7}{3} = ?\,\left(\left|-\dfrac{2}{5}\right| \div \left|\dfrac{7}{3}\right|\right)$

$= ?\,\left(\dfrac{2}{5} \div \dfrac{7}{3}\right)$

$= ?\,\left(\dfrac{2}{5}\cdot\dfrac{3}{7}\right)$

$= ?\,\dfrac{6}{35}$

$= -\dfrac{6}{35}$

PROBLEM 1-8 Evaluate the following terms: (a) $(-1)^3$ (b) $(-5)^4$ (c) -5^4

Solution: Recall that to evaluate exponential notation, you can write the term as the product of repeated factors, then compute the product [see Example 1-10].

(a) $(-1)^3 = (-1)(-1)(-1)$

$= 1(-1)$

$= -1$

(b) $(-5)^4 = (-5)(-5)(-5)(-5)$

$= 25(-5)(-5)$

$= -125(-5)$

$= 625$

(c) $-5^4 = -(5\cdot 5\cdot 5\cdot 5)$

$= -(25\cdot 5\cdot 5)$

$= -(125\cdot 5)$

$= -625$

PROBLEM 1-9 Evaluate the following expressions:

(a) $(4 + 3\cdot 2) \div 2 + 3\cdot 5^2$ (b) $30 + [10 \div (-4 + 2)]^2$

Solution: Recall that to evaluate an expression that involves more than one operation, you follow the Order of Operations Agreement [see Examples 1-11 and 1-12].

$$\text{(a) } (4 + 3 \cdot 2) \div 2 + 3 \cdot 5^2 = (4 + 6) \div 2 + 3 \cdot 5^2$$
$$= 10 \div 2 + 3 \cdot 5^2$$
$$= 10 \div 2 + 3 \cdot 25$$
$$= 5 + 3 \cdot 25$$
$$= 5 + 75$$
$$= 80$$

$$\text{(b) } 30 + [10 \div (-4 + 2)]^2 = 30 + [10 \div (-2)]^2$$
$$= 30 + (-5)^2$$
$$= 30 + 25$$
$$= 55$$

PROBLEM 1-10 Evaluate $a[3b - 5(4b + 3) - 2] - a^2$, where $a = -3$ and $b = 5$.

Solution: Recall that to evaluate a variable expression, you substitute the given numerical values for the variables, then evaluate as before [see Example 1-13].

$$a[3b - 5(4b + 3) - 2] - a^2 = (-3)[3 \cdot 5 - 5(4 \cdot 5 + 3) - 2] - (-3)^2$$
$$= (-3)[3 \cdot 5 - 5(20 + 3) - 2] - (-3)^2$$
$$= (-3)[3 \cdot 5 - 5 \cdot 23 - 2] - (-3)^2$$
$$= (-3)[15 - 115 - 2] - (-3)^2$$
$$= (-3)(-102) - (-3)^2$$
$$= (-3)(-102) - 9$$
$$= 306 - 9$$
$$= 297$$

PROBLEM 1-11 Identify the property that is being illustrated in each of the following expressions:

(a) $(7 + 8) + 1 = 7 + (8 + 1)$ (b) $(4 + 3)2 = 2(4 + 3)$ (c) $5(7 + 3) = 5 \cdot 7 + 5 \cdot 3$

Solution [see Example 1-14]:

(a) The associative property of addition is being used here—the only change is the grouping of the symbols.
(b) The commutative property of multiplication is being used here—the only change is the order of the factors.
(c) The distributive property of multiplication over addition is being used here.

PROBLEM 1-12 Clear the parentheses in each of the following expressions:

(a) $-6(2x - 5)$ (b) $1 + (2y - 7)$ (c) $-5(5w - 3)$

Solution: Recall that you can clear parentheses by using the distributive property. To clear parentheses preceded by a plus sign, simply write the expression inside them. To clear parentheses preceded by a minus sign, write the opposite of each term inside them [see Example 1-15].

(a) $-6(2x - 5) = -12x + 30$ (b) $1 + (2y + 7) = 1 + 2y + 7$ (c) $-5(5w - 3) = -25w + 15$
$$= 8 + 2y$$

PROBLEM 1-13 Simplify the following expressions:

(a) $3x + 2xy + 7xy$ (b) $4p + 2pq - 7p + 11pq$

Solution: Recall that to simplify an expression with like terms, you combine the like terms [see Example 1-16].

(a) $3x + 2xy + 7xy = 3x + 9xy$

To add like terms, you add their numerical coefficients, then write the same literal part on the sum.

(b) $4p + 2pq - 7p + 11pq = 4p - 7p + 2pq + 11pq$

$$= (4p - 7p) + (2pq + 11pq)$$

$$= -3p + 13pq$$

First rearrange and regroup the terms if necessary, then combine like terms.

PROBLEM 1-14 Simplify $-5\{2x + 4[3(x - 2) + 5]\}$.

Solution: Recall that to simplify a variable expression, you use the distributive property to clear the grouping symbols, then combine like terms [see Example 1-17].

$$-5\{2x + 4[3(x - 2) + 5]\} = -5\{2x + 4[3x - 6 + 5]\}$$

$$= -5\{2x + 4[3x - 1]\}$$

$$= -5\{2x + 12x - 4\}$$

$$= -5\{14x - 4\}$$

$$= -70x + 20$$

PROBLEM 1-15 Determine whether each of the following statements is true or false:

(a) $3 \in Q$ (b) $0 \in N$ (c) $-3 \in W$ (d) $-7 \in Z$ (e) $3.14 \in I$

Solution: Recall the definitions of the various sets of real numbers [see Example 1-18].

(a) True. All the integers are rational numbers.
(b) False. Zero is not included in the natural numbers.
(c) False. The whole numbers do not include any negative numbers.
(d) True. The integers include the whole numbers and their additive inverses.
(e) False. Although it is used as an approximation of π, 3.14 is a terminating decimal, and is therefore a rational number.

PROBLEM 1-16 Graph each of the following sets of real numbers:

(a) $\{-2, 0, 3\}$ (b) $\{x \mid x \in N \text{ and } x \leq 3\}$ (c) $\{x \mid x \in W \text{ and } x < 5\}$
(d) $\{x \mid x \in R \text{ and } x \geq -2\}$

Solution: To graph a set of numbers, graph each number in the set [see Example 1-19].

(a)

Place a dot on the coordinates -2, 0, and 3.

(b)

The only natural numbers less than or equal to 3 are 1, 2, and 3.

(c)

The only whole numbers less than 5 are 0, 1, 2, 3, and 4.

(d)

Use a solid dot and shade the area to the right of -2 to indicate the set of all real number greater than or equal to -2.

PROBLEM 1-17 Find the intersection of each pair of sets:

(a) $\{1, 4\} \cap \{3, 4\}$ (b) $\{-1, 0, 1\} \cap \{0, 1, 2, 3\}$ (c) $\{a, b, c\} \cap \{a, d, e\}$ (d) $N \cap W$
(e) $Z \cap W$ (f) $N \cap Q$ (g) $\varnothing \cap 1, 7$

Solution: Recall that to find the intersection of two sets, you identify the numbers common to both sets [see Example 1-20].

(a) $\{1, 4\} \cap \{3, 4\} = \{4\}$ (b) $\{-1, 0, 1\} \cap \{0, 1, 2, 3\} = \{0, 1\}$
(c) $\{a, b, c\} \cap \{a, d, e\} = \{a\}$ (d) $N \cap W = N$ (e) $Z \cap W = W$
(f) $N \cap Q = N$ (g) $\varnothing \cap \{1, 7\} = \varnothing$

PROBLEM 1-18 Find the union of each pair of sets:

(a) $\{1, 3\} \cup \{5\}$ (b) $\{0, 1, 2\} \cup \{2, 3, 4\}$ (c) $\{a, b, c\} \cup \{c, d, e\}$ (d) $N \cup W$
(e) $Z \cup W$ (f) $Q \cup I$ (g) $\varnothing \cup \{3, 5\}$

Solution: Recall that to find the union of two sets, you identify the numbers which belong to the first set, to the second set, or to both sets [see Example 1-21].

(a) $\{1, 3\} \cup \{5\} = \{1, 3, 5\}$ (b) $\{0, 1, 2\} \cup \{2, 3, 4\} = \{0, 1, 2, 3, 4\}$
(c) $\{a, b, c\} \cup \{c, d, e\} = \{a, b, c, d, e\}$ (d) $N \cup W = W$ (e) $Z \cup W = Z$
(f) $Q \cup I = R$ (g) $\varnothing \cup \{3, 5\} = \{3, 5\}$

PROBLEM 1-19 Translate the following sentences into symbols using key words:

(a) The sum of 5 and 2.   (b) The difference between 5 and 2.
(c) The product of 5 and 2.   (d) The quotient of 5 and 2.
(e) Six more than the sum of 5 and 2.   (f) Six less than the difference of 5 and 2.
(g) Six times the product of 5 and 2.   (h) Six divided by the quotient of 5 and 2.
(i) 5 less 2   (j) 5 less than 2   (k) 5 subtracted from 2
(l) The product of 2 and a number, decreased by 5 equals 9.
(m) Five times the quantity, $x + 2$, is equal to 25.
(n) The product of 3 and a number less twice the same number is 20.
(o) One-half of a number, increased by 4, is the same as 8.

Solution: Recall that to help translate words to symbols, you can substitute the appropriate symbols for the corresponding key words [see Example 1-22].

(a) The sum of 5 and 2.

$5 + 2$ or $2 + 5$

(b) The difference between 5 and 2.

larger number $\longrightarrow$ $5 - 2$ $\longleftarrow$ smaller number

(c) The product of 5 and 2.

$5 \cdot 2$ or $2 \cdot 5$

(d) The quotient of 5 and 2.

first number given $\longrightarrow$ $5 \div 2$ $\longleftarrow$ second number given

(e) Six more than the sum of 5 and 2.

$6 + (5 + 2)$

(f) Six less than the difference between 5 and 2.

$(5 - 2) - 6$

(g) Six times the product of 5 and 2.

$6 \cdot (5 \cdot 2)$

(h) Six divided by the quotient of 5 and 2.

$6 \div (5 \div 2)$

(i) 5 less 2   (j) 5 less than 2   (k) 5 subtracted from 2

 $5 - 2$ $2 - 5$ $2 - 5$

(l) The product of 2 and a number, decreased by 5, equals 9.

$2 \cdot n$ $-$ 5 $=$ 9 or $2n - 5 = 9$

(m) Five times the quantity, $x + 2$, is equal to 25.

5 $\cdot$ $(x + 2)$ $=$ 25 or $5(x + 2) = 25$

(n) The product of 3 and a number less twice the same number is 20.

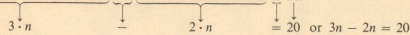

$$3 \cdot n \quad - \quad 2 \cdot n \quad = 20 \quad \text{or} \quad 3n - 2n = 20$$

(o) One-half of a number, increased by 4, is the same as 8.

$$\frac{1}{2} \quad \cdot \quad n \quad + \quad 4 \quad = \quad 8 \quad \text{or} \quad \frac{1}{2}n + 4 = 8 \quad \text{or} \quad \frac{n}{2} + 4 = 8$$

Supplementary Exercises

PROBLEM 1-20 Determine whether each of the following numbers is rational or irrational:

(a) $\frac{4}{5}$ **(b)** 2.561   **(c)** -6 **(d)** 8.434334333433334 . . .

PROBLEM 1-21 Determine whether each of the following statements is true or false:

(a) $1 < 7$ **(b)** $-5 \geq -5$ **(c)** $2 \geq -3$ **(d)** $150 > 200$ **(e)** $-4 \leq 5$ **(f)** $0 \geq -3$

PROBLEM 1-22 Find the absolute value of each of the following numbers:

(a) -14 **(b)** $-\frac{3}{4}$ **(c)** $6 + (-6)$ **(d)** 3.87   **(e)** 7

PROBLEM 1-23 Add each of the following pairs of signed numbers:

(a) $-7 + (-5)$ **(b)** $-8 + 3$ **(c)** $14 + (-10)$ **(d)** $-9 + 9$ **(e)** $-37 + (-12)$
(f) $-87 + 25$ **(g)** $132 + (-250)$ **(h)** $-145 + (-682)$ **(i)** $52 + (-52)$ **(j)** $-\frac{1}{5} + (-\frac{3}{5})$
(k) $-\frac{1}{5} + \frac{4}{5}$ **(l)** $\frac{3}{5} + \frac{1}{6}$ **(m)** $4.25 + 3.22$ **(n)** $8.7 + (-5.3)$ **(o)** $-6.4 + (-12.34)$

PROBLEM 1-24 Subtract each of the following pairs of signed numbers:

(a) $17 - (-12)$ **(b)** $-5 - (-7)$ **(c)** $16 - 10$ **(d)** $4 - 4$ **(e)** $-25 - (-34)$
(f) $-91 - 38$ **(g)** $155 - (-230)$ **(h)** $-322 - (-788)$ **(i)** $-34 - (-34)$ **(j)** $\frac{5}{7} - \frac{1}{7}$
(k) $-\frac{1}{7} - \frac{3}{7}$ **(l)** $\frac{3}{5} - \frac{1}{6}$ **(m)** $5.23 - 1.45$ **(n)** $9.1 - (-4.53)$ **(o)** $-5.3 - (-19.2)$

PROBLEM 1-25 Multiply each of the following pairs of signed numbers:

(a) $-3(-5)$ **(b)** $-7 \cdot 8$ **(c)** $-7 \cdot 0$ **(d)** $(-8)(-8)$ **(e)** $-22(-31)$ **(f)** $14(-34)$
(g) $-125(130)$ **(h)** $(100)(-432)$ **(i)** $-3245 \cdot 0$ **(j)** $(-\frac{5}{6})(\frac{1}{3})$ **(k)** $(-\frac{3}{4})(-\frac{3}{5})$ **(l)** $0 \cdot \frac{1}{4}$
(m) $(4.1) \cdot (-3.2)$ **(n)** $(-32.01)(-2.3)$ **(o)** $(-100.02)(-100)$

PROBLEM 1-26 Divide each of the following pairs of signed numbers:

(a) $(-36) \div 4$ **(b)** $(-36) \div (-4)$ **(c)** $-\frac{15}{5}$ **(d)** $\frac{-7}{-7}$ **(e)** $\frac{-160}{5}$ **(f)** $\frac{625}{-25}$

(g) $\frac{-800}{-200}$ **(h)** $\frac{1001}{-91}$ **(i)** $\frac{-151}{151}$ **(j)** $\left(-\frac{3}{4}\right) \div \left(-\frac{5}{7}\right)$ **(k)** $0 \div (-4)$ **(l)** $\frac{0}{17}$

(m) $5.1 \div (-3)$ **(n)** $-0.32 \div (-0.2)$ **(o)** $-31.7 \div (-31.7)$

PROBLEM 1-27 Evaluate the following expressions:

(a) $(-3)^3$ **(b)** 2^4 **(c)** $(-11)^2$ **(d)** -11^2 **(e)** $(-1)^5$ **(f)** $(-1)^6$

PROBLEM 1-28 Evaluate the following expressions:

(a) $(6 + 4 \div 2) + 6 \cdot 2^2 - 1$ (b) $-3^2 + \frac{8}{4} - [6 + 2(-5)]$

PROBLEM 1-29 Evaluate the following expression:

$m[2n - 3(4m - 5) - 6 \cdot n^2]$, where $m = 2$ and $n = -3$

PROBLEM 1-30 Identify the property that is being illustrated in each of the following equations:

(a) $3 \cdot (4 \cdot 5) = (3 \cdot 4) \cdot 5$ (b) $4 \cdot 3 + 7 = 3 \cdot 4 + 7$ (c) $6(8 + 1) = 6 \cdot 8 + 6 \cdot 1$
(d) $a = a \cdot \frac{3}{3}$

PROBLEM 1-31 Clear the parentheses in each of the following expressions:

(a) $-(5x + 3y - 7)$ (b) $4(5a + 2b - 7)$ (c) $+(5w - 8)$

PROBLEM 1-32 Simplify the following expressions:

(a) $4x + 3x - 5y$ (b) $3mn - 2m - 5mn + 8m$

PROBLEM 1-33 Simplify the following expressions:

(a) $-4\{3x + 2(x - 5) - 6\}$ (b) $-\{4w - [3(w - 5) - 2w] + 2\}$

PROBLEM 1-34 Determine whether each of the following statements is true or false:

(a) $\frac{1}{2} \in Q$ (b) $2 \in Z$ (c) $-8 \in W$ (d) $5 \in R$ (e) $-4 \in I$ (f) $5.4 \in Q$ (g) $1000 \in N$

PROBLEM 1-35 Graph each of the following sets of real numbers:

(a) $\{-4, -1, 3\}$ (b) $\{x \mid x \in Z \text{ and } x \geq -3\}$ (c) $\{x \mid x \in R \text{ and } x < 3\}$
(d) $\{x \mid x \in W \text{ and } x < 1\}$

PROBLEM 1-36 Find the intersection of each pair of sets:

(a) $\{-2, 7\} \cap \{-2, 0, 9\}$ (b) $\{-5, -3, -1, 1\} \cap \{-9, -3, -1, 1, 5\}$
(c) $\{x, y, z\} \cap \{p, q, x, z\}$ (d) $W \cap W$ (e) $\varnothing \cap N$
(f) $\{0, 2, 4, 6, 8, \ldots\} \cap \{1, 2, 3, 4, 5, \ldots\}$ (g) $\{\frac{1}{2}, \frac{3}{4}\} \cap \{\frac{2}{4}, \frac{1}{3}\}$

PROBLEM 1-37 Find the union of each pair of sets:

(a) $\{-2, 4\} \cup \{-3, 5\}$ (b) $\{-8, -7, -6\} \cup \{-6, -3\}$ (c) $\{x, y, z\} \cup \{p, q, x, z\}$

(d) $W \cup W$ (e) $W \cup \varnothing$ (f) $W \cup N$ (g) $\varnothing \cup \{4, 11\}$

PROBLEM 1-38 Translate each sentence into a mathematical statement:

(a) Twelve more than three times a number is thirty.
(b) A number added to twenty equals eleven.
(c) Eight less than a number is thirty-six.
(d) One-fourth of a number is eighteen.
(e) A number increased by sixteen is the same as nineteen.
(f) One-third of a number increased by twice the same number is thirty-five.
(g) Five times a number diminished by ten leaves seven.
(h) If six is subtracted from four times a number, the result is three.
(i) The product of a number and nine, less one, is forty-four.
(j) Half a number added to two times the same number is eighty.
(k) Thirteen more than an eighth of a number equals two and a half.
(l) The quotient of a number divided by four is ten.
(m) Six times the sum of two and some number gives one hundred sixty-eight.
(n) A number is multiplied by five and then increased by twenty to get forty-two.
(o) Three-fourths of a number is decreased by six to get twenty-one.

(p) Two-thirds of a number less thirteen is equal to two.

(q) A number is 8 more than twice the difference of the same number and 2.

(r) Three-eighths of a number is subtracted from one-half the same number to get negative five.

(s) A number increased by a third of itself is the same as eight.

(t) Half of a number equals a fourth of the same number plus six.

(u) Nine is added to a third of a number to leave nine.

(v) Three-fifths of a number joined with forty-seven will be eighty.

(w) Six subtracted from six times a number is seventy-two.

(x) Thirty subtracted from eight times a number is equal to five.

(y) One and a half is added to one-third of a number to get one-fourth.

(z) A number, when increased by four times its half, equals fifteen.

Answers to Supplementary Exercises

(1-20) (a) rational number   (b) rational number   (c) rational number   (d) irrational number

(1-21) (a) true   (b) true   (c) true   (d) false   (e) true   (f) true

(1-22) (a) 14   (b) $\frac{3}{4}$ (c) 0   (d) 3.87   (e) 7

(1-23) (a) -12 (b) -5 (c) 4   (d) 0   (e) -49 (f) -62 (g) -118 (h) -827
     (i) 0   (j) $-\frac{4}{5}$ (k) $\frac{3}{5}$ (l) $\frac{23}{30}$ (m) 7.47   (n) 3.4   (o) -18.74

(1-24) (a) 29   (b) 2   (c) 6   (d) 0   (e) 9   (f) -129 (g) 385   (h) 466   (i) 0
     (j) $\frac{4}{7}$ (k) $-\frac{4}{7}$ (l) $\frac{13}{30}$ (m) 3.78   (n) 13.63   (o) 13.9

(1-25) (a) 15   (b) -56 (c) 0   (d) 64   (e) 682   (f) -476 (g) $-16,250$
     (h) $-43,200$ (i) 0   (j) $-\frac{5}{18}$ (k) $\frac{9}{20}$ (l) 0   (m) -13.12 (n) 73.623
     (o) 10,002

(1-26) (a) -9 (b) 9   (c) -3 (d) 1   (e) -32 (f) -25 (g) 4   (h) -11
     (i) -1 (j) $\frac{21}{20}$ (k) 0   (l) 0   (m) -1.7 (n) 1.6   (o) 1

(1-27) (a) -27 (b) 16   (c) 121   (d) -121 (e) -1 (f) 1

(1-28) (a) 31   (b) -3

(1-29) -138

(1-30) (a) associative property of multiplication   (b) commutative property of multiplication
     (c) distributive property   (d) multiplication property of one

(1-31) (a) $-5x - 3y + 7$ (b) $20a + 8b - 28$ (c) $5w - 8$

(1-32) (a) $7x - 5y$ (b) $-2mn + 6m$

(1-33) (a) $-20x + 64$ (b) $-3w - 17$

(1-34) (a) true   (b) true   (c) false   (d) true   (e) false   (f) true   (g) true

(1-35) (a)   (b)

     (c)   (d)

(1-36) (a) $\{-2\}$ **(b)** $\{-3, -1, 1\}$ **(c)** $\{x, z\}$ **(d)** W **(e)** $\emptyset$ **(f)** $\{2, 4, 6, 8, \ldots\}$
(g) $\{\frac{1}{2}\}$

(1-37) (a) $\{-3, -2, 4, 5\}$ **(b)** $\{-8, -7, -6, -3\}$ **(c)** $\{p, q, x, y, z\}$ **(d)** W **(e)** W
(f) W **(g)** $\{4, 11\}$

(1-38) (a) $12 + 3n = 30$ **(b)** $n + 20 = 11$ **(c)** $n - 8 = 36$ **(d)** $\frac{1}{4}n = 18$
(e) $n + 16 = 19$ **(f)** $\frac{1}{3}n + 2n = 35$ **(g)** $5n - 10 = 7$ **(h)** $4n - 6 = 3$
(i) $9n - 1 = 44$ **(j)** $\frac{1}{2}n + 2n = 80$ **(k)** $13 + \frac{1}{8}n = 2\frac{1}{2}$ **(l)** $n/4 = 10$
(m) $6(2 + n) = 168$ **(n)** $5n + 20 = 42$ **(o)** $\frac{3}{4}n - 6 = 21$ **(p)** $\frac{2}{3}n - 13 = 2$
(q) $n = 8 + 2(n - 2)$ **(r)** $\frac{1}{2}n - \frac{3}{8}n = -5$ **(s)** $n + \frac{1}{3}n = 8$ **(t)** $\frac{1}{2}n = \frac{1}{4}n + 6$
(u) $\frac{1}{3}n + 9 = 9$ **(v)** $\frac{3}{5}n + 47 = 80$ **(w)** $6n - 6 = 72$ **(x)** $8n - 30 = 5$
(y) $\frac{1}{3}n + 1\frac{1}{2} = \frac{1}{4}$ **(z)** $n + [4(\frac{1}{2}n)] = 15$

2 EQUATIONS

THIS CHAPTER IS ABOUT

☑ Solving Equations Using the Principles
☑ Solving Equations Using the Distributive Property
☑ Solving Absolute Value Equations
☑ Solving Literal Equations
☑ Solving Formulas
☑ Solving Problems Using Equations

2-1. Solving Equations Using the Principles

The equation is the most important concept in mathematics. An **equation** is a mathematical statement that two *algebraic expressions* are equal. In the equation $3x - 5 = 4 + x$, the **left member** $3x - 5$ is equal to (=) the **right member** $4 + x$. An equation that can be written in the standard form $Ax + B = C$, where A, B, and C are real numbers, A is not equal to ($\neq$) 0, and x is any variable, is called a **first-degree equation in one variable** or a **linear equation in one variable**.

A number is **a solution of an equation** if the statement obtained by replacing the variable with that number is true. For example, the number 4 is a solution of the equation $3x - 7 = 13 - 2x$, because $(3 \cdot 4) - 7 = 13 - (2 \cdot 4)$. The solutions of an equation are also called its **roots**. The set of all solutions of an equation is called its **solution set. Solving an equation** means finding its solution set. A linear equation in one variable has exactly one solution.

A. Solving equations using the addition principle

To solve an equation of the form $x + B = C$, we isolate the variable x in one member, and the **constants** B and C in the other member. To do this, we use the **addition principle of equations.**

Note: In this form of the standard linear equation, A is understood to be 1.

Addition Principle of Equations

If $a = b$, then $a + c = b + c$ for all real numbers a, b, and c.

To solve an equation of the form $x + B = C$, we add the opposite of B ($-B$) to both members of the equation to isolate the variable x in one member.

EXAMPLE 2-1: Solve **(a)** $x - 5 = 3$ and **(b)** $y + 4 = 2$.

Solution: **(a)** $\qquad x - 5 = 3$

$\qquad x - 5 \mathbf{+ 5} = 3 \mathbf{+ 5}$ 　　Add the opposite of -5 (5) to both members to
$\qquad\qquad x + 0 = 8$ 　　　　　isolate the variable x in one member.

$\qquad\qquad\qquad x = 8 \longleftarrow$ proposed solution

$\quad Check: \quad x - 5 = 3 \longleftarrow$ original equation
$\qquad\qquad \overline{\quad 8 - 5 \quad 3\quad}$
$\qquad\qquad\qquad 3 \quad\ 3 \longleftarrow x = 8$ checks

Note: The ONLY solution of $x - 5 = 3$ is 8.

$$\text{(b)} \qquad y + 4 = 2$$

$$y + 4 + (-4) = 2 + (-4) \qquad \text{Add the opposite of } 4\,(-4)$$
$$y + 0 = -2 \qquad\qquad\qquad \text{to both members to isolate } y.$$

$$y = -2 \longleftarrow \text{proposed solution}$$

Check:
$$\begin{array}{c|c} y + 4 = 2 & \\ \hline -2 + 4 & 2 \\ 2 & 2 \end{array} \longleftarrow y = -2 \text{ checks}$$

Note: The ONLY solution of $y + 4 = 2$ is -2.

Notice that there is no subtraction principle of equations. We can apply the addition principle of equations to equations of this form either by adding the opposite of $B\,(-B)$ to both members, or by subtracting B from both members. Either operation will successfully isolate the variable.

B. Solving equations using the multiplication principle

To solve an equation of the form $Ax = C$, we again isolate the variable x in one member. To do this, we use the **multiplication principle of equations.**

Note: In this form of the standard linear equation, B is understood to be 0.

Multiplication Principle of Equations

If $a = b$, then $ca = cb$ for all real numbers a, b, and c.

To **solve an equation of the form $Ax = C$,** multiply both members by the **reciprocal** or **multiplicative inverse** of $A(1/A)$ to isolate the variable x in one member.

EXAMPLE 2-2: Solve (a) $-\frac{2}{3}x = 4$ and (b) $5y = 10$.

Solution: (a)
$$-\tfrac{2}{3}x = 4$$

$$-\tfrac{3}{2}\left(-\tfrac{2}{3}x\right) = -\tfrac{3}{2}(4) \qquad \text{Multiply both members by the reciprocal of}$$
$$1x = -6 \qquad\qquad\qquad -\tfrac{2}{3} - \tfrac{3}{2}.$$

$$x = -6 \longleftarrow \text{proposed solution}$$

Check:
$$\begin{array}{c|c} -\tfrac{2}{3}x = 4 & \longleftarrow \text{original equation} \\ \hline -\tfrac{2}{3}(-6) & 4 \\ 4 & 4 \end{array} \longleftarrow x = -6 \text{ checks}$$

The solution of $-\frac{2}{3}x = 4$ is -6.

$$\text{(b)} \qquad 5y = 10$$

$$\frac{1}{5} \cdot 5y = \frac{1}{5} \cdot 10 \text{ or } \frac{5y}{5} = \frac{10}{5} \qquad \text{Multiply both members by the reciprocal of 5, or just divide by 5.}$$

$$1y = 2$$

$$y = 2 \longleftarrow \text{proposed solution}$$

Check:
$$\begin{array}{c|c} 5y = 10 & \\ \hline 5(2) & 10 \\ 10 & 10 \end{array} \longleftarrow y = 2 \text{ checks}$$

The solution of $5y = 10$ is 2.

Notice that there is no division principle of equations. We can apply the multiplication principle of equations to equations of this form either by multiplying both members by the reciprocal of A, or by dividing both members by A. Either operation will isolate the variable.

C. Solving equations using the addition and multiplication principles together

To **solve an equation of the form** $Ax + B = C$, we must isolate the **variable term** Ax in one member using the addition principle of equations, then solve for the variable x using the multiplication principle of equations. To do this, you add the opposite of B to both members, then multiply by the reciprocal of A.

EXAMPLE 2-3: Solve (a) $3x - 5 = 7$ and (b) $3 = 5 - 4u$.

Solution: (a) $\qquad 3x - 5 = 7$

$3x - 5 + 5 = 7 + 5 \qquad\qquad$ Add the opposite of -5 (5) to both members to isolate $3x$ in one member.

$3x = 12$

$\dfrac{1}{3} \cdot 3x = \dfrac{1}{3} \cdot 12$ or $\dfrac{3x}{3} = \dfrac{12}{3} \qquad$ Multiply both members by the reciprocal of 3 (or just divide by 3) to isolate x in one member.

$x = 4$ ⟵ proposed solution $\qquad$ Check as before.

The solution of $3x - 5 = 7$ is 4.

(b) $\qquad 3 = 5 - 4u \qquad\qquad$ Add the opposite of 5 to both members to isolate $-4u$ in one member.

$3 - 5 = 5 - 5 - 4u$

$-2 = -4u$

$-\dfrac{1}{4}(-2) = -\dfrac{1}{4}(-4u)$ or $\dfrac{-2}{-4} = \dfrac{-4u}{-4} \qquad$ Multiply both members by the reciprocal of -4 or just divide by -4, to isolate u in one member.

$\dfrac{1}{2} = u$ ⟵ proposed solution $\qquad$ Check as before.

The solution of $3 = 5 - 4u$ is $\frac{1}{2}$.

D. Solving equations containing like terms

To **solve an equation containing like terms,** first collect and combine like terms, then use the addition and multiplication principles to solve as before.

EXAMPLE 2-4: Solve (a) $3x + 7 - 5x = -3$ and (b) $5t + 3 = 2 - 3t + 7t$.

Solution: (a) $\quad 3x + 7 - 5x = -3$

$-2x + 7 = -3 \qquad\qquad$ Combine like terms ($3x$ and $-5x$).

$-2x + 7 - 7 = -3 - 7 \qquad$ Isolate the variable term $-2x$ using the addition principle of equations.

$-2x = -10$

$-\frac{1}{2}(-2x) = -\frac{1}{2}(-10) \qquad$ Use the multiplication principle by multiplying by the reciprocal of -2.

$x = 5$ ⟵ proposed solution $\qquad$ Check as before.

(b) $\qquad 5t + 3 = 2 - 3t + 7t$

$5t + 3 = 2 + 4t \qquad\qquad$ Combine like terms ($-3t$ and $7t$).

$5t - 4t + 3 = 2 + 4t - 4t \qquad$ Use the addition principle by adding the opposite of $4t$ to both members.

$t + 3 = 2$

$t + 3 - 3 = 2 - 3 \qquad$ Use the addition principle again by adding the opposite of 3 to both sides.

$t = -1$ ⟵ proposed solution $\qquad$ Check as before.

Note: A equation that simplifies to a true number statement (such as 2 = 2) is called an **identity.** The **solution set of an identity** is the set of all real numbers. An equation that simplifies to a false statement (such as 3 = 7) has no solution. The **solution set of a false statement** is the empty set.

2-2. Solving Equations Using the Distributive Property

It is often necessary to simplify an equation before it can be solved. We often use the **distributive property** to simplify an equation to the standard form $Ax + B = C$. Recall that the distributive property states:

$$a(b + c) = ab + ac, \text{ for all real numbers } a, b, \text{ and } c$$

A. Solving equations containing parentheses

To solve an equation containing parentheses, we first use the distributive property to clear the parentheses. Then we can solve the equation using the addition and multiplication principles as before.

EXAMPLE 2-5: Solve the following equation using the distributive property: $4(2x - 3) = -2(3x - 5)$.

Solution:

$4(2x - 3) = -2(3x - 5)$	
$4(2x) + 4(-3) = (-2)(3x) + (-2)(-5)$	Clear parentheses using the distributive property.
$8x - 12 = -6x + 10$	
$14x - 12 = 10$	Combine like terms.
$14x = 22$	Add the opposite of -12 to both members.
$x = \frac{22}{14}$	Multiply by the reciprocal of 14.
$x = \frac{11}{7}$	Check as before.

B. Solving equations containing fractions

To **solve an equation containing fractions,** always clear the fractions in the equation first. To do this, first multiply both members by the least common denominator (LCD) of the terms, then use the distributive property to clear the parentheses. You can then solve the equation as before using the addition and multiplication principles.

EXAMPLE 2-6: Solve $\frac{3}{4} - \frac{2}{5}x = \frac{5x - 7}{10}$.

Solution: $\frac{3}{4} - \frac{2}{5}x = \frac{5x - 7}{10}$ Find the least common denominator (LCD).

The LCD of $\frac{3}{4}, \frac{2}{5}x$, and $\frac{5x - 7}{10}$ is 20.

$20\left(\frac{3}{4} - \frac{2}{5}x\right) = 20\left(\frac{5x - 7}{10}\right)$	Multiply both members by the LCD.
$20\left(\frac{3}{4}\right) - 20\left(\frac{2}{5}x\right) = 20\left(\frac{5x - 7}{10}\right)$	Clear fractions using the distributive property.
$15 - 8x = 2(5x - 7)$	
$15 - 8x = 10x - 14$	Clear parentheses.
$15 = 10x + 8x - 14$	Collect like terms.
$15 = 18x - 14$	Combine like terms.
$29 = 18x$	Add the opposite of -14 to both members.
$\frac{29}{18} = x$	Multiply by the reciprocal of 18. Check as before.

C. Solving equations containing decimals

To solve an equation containing decimals, we usually clear the decimals first. To do this, we multiply both members of the equation by the LCD of the terms. For decimals, the LCD is always a multiple of 10. After clearing the decimals we use the distributive property to clear the parentheses, and then the addition and multiplication principles to solve as before.

EXAMPLE 2-7: Solve the following equation: $2.7y + 0.47 = 8.57$.

Solution:

$2.7y + 0.47 = 8.57$	Find the LCD. (The LCD of decimals is always 100.)
$100(2.7y + 0.47) = 100(8.57)$	Multiply both members by the LCD.
$100(2.7y) + 100(0.47) = 100(8.57)$	Clear the decimals (and parentheses)
$270y + 47 = 857$	using the distributive property.
$270y = 810$	Add -47 to both members.
$y = \frac{810}{270}$	Multiply both members by $\frac{1}{270}$.
$y = 3$	Check as before.

D. Solving equations containing percents

To solve an equation containing percents, we rename the percents as either decimals or fractions, then solve as before. Recall that to rename a percent as a decimal, you divide by 100 or just move the decimal point two places to the left. For example, 153.2% is 1.532 in decimal form.

EXAMPLE 2-8: Solve the following equation by renaming percents first as decimals, and then as fractions: $20\%x + 30\%(100 - x) = 20$.

Solution: Renaming the percents as decimals, we get the following:

$20\%x + 30\%(100 - x) = 20$	
$0.20x + 0.30(100 - x) = 20$	Rename percents as decimals.
$20x + 30(100 - x) = 2000$	Clear decimals.
$20x + 3000 - 30x = 2000$	Clear parentheses.
$3000 - 10x = 2000$	Combine like terms.
$-10x = -1000$	Subtract 3000 from both members.
$x = 100$	Multiply by the reciprocal of -10. Check as before.

Renaming the percents as fractions, we get the following:

$20\%x + 30\%(100 - x) = 20$	
$\frac{20}{100}x + \frac{30}{100}(100 - x) = 20$	Rename percents as fractions.
$20x + 30(100 - x) = 2000$	Clear fractions.
$20x + 3000 - 30x = 2000$	Clear parentheses.
$3000 - 10x = 2000$	Combine like terms.
$-10x = -1000$	Subtract 3000 from both members.
$x = 100$	Multiply by the reciprocal of -10. Check as before.

Note: If a solution contains a fraction of a percent, you should always use a fractional form rather than a repeating or nonterminating decimal form. For example, write $66\frac{2}{3}\%$ as $66\frac{2}{3}/100$, or just $\frac{2}{3}$, rather than $0.\overline{6}$.

2-3. Solving Absolute Value Equations

You should recall that the absolute value of any real number x is written as $|x|$, and means

$$|x| = \begin{cases} x \text{ if } x \text{ is positive} \\ 0 \text{ if } x \text{ is zero} \\ -x \text{ if } x \text{ is negative} \end{cases}$$

From this definition of absolute value, we see that if $|x| = 4$, then either $x = 4$ or $x = -4$. To solve an absolute value equation of the form $|Ax + B| = C$, where $C \geq 0$, we let $Ax + B = C$ and $Ax + B = -C$, then solve both equations.

EXAMPLE 2-9: Solve the following equation: $|3x + 4| = 13$.

Solution:

$$|3x + 4| = 13$$

$$3x + 4 = 13 \text{ or } 3x + 4 = -13$$

$$3x = 9 \quad \text{or} \quad 3x = -17$$

$$x = 3 \quad \text{or} \quad x = -\tfrac{17}{3}$$

Check:

$|3x + 4| = 13$			$|3x + 4| = 13$	
$|3(3) + 4|$	13		$|3(-\tfrac{17}{3}) + 4|$	13
$|9 + 4|$	13		$|-17 + 4|$	13
$|13|$	13 ⟵ $x = 3$ checks		$|-13|$	13 ⟵ $x = -\tfrac{17}{3}$ checks

The two solutions of $|3x + 4| = 13$ are 3 and $-\tfrac{17}{3}$.

Sometimes it is necessary to isolate the absolute value expression in one member to solve an absolute value equation. You can do this by applying the addition and/or multiplication principles.

EXAMPLE 2-10: Solve the following equation: $3|5 - 2x| + 4 = 13$

Solution: $3|5 - 2x| + 4 = 13$

$$3|5 - 2x| = 9 \qquad\qquad \text{Subtract 4 from both members.}$$

$$|5 - 2x| = 3 \qquad\qquad \text{Divide both members by 3.}$$

$$5 - 2x = 3 \quad \text{or } 5 - 2x = -3$$

$$-2x = -2 \text{ or} \quad -2x = -8$$

$$x = 1 \quad \text{or} \qquad x = 4 \qquad\qquad \text{Check as before.}$$

Note 1: If $|Ax + B| = 0$, then $Ax + B = 0$, and $x = -B/A$ is the only solution.

Note 2: If $|Ax + B| = C$, then C cannot be negative.

2-4. Solving Literal Equations

An equation containing two or more variables is called a **literal equation.** In such equations the constants are usually (but not always) represented by letters from the beginning of the alphabet. For example, in the equation $Ax + By = C$, the letters A, B, and C represent the constants, while x and y are the variables.

Note: It is assumed in this chapter that the denominators of any of the literal equations do not equal zero.

A. Solving literal equations using the principles

To solve a literal equation for a given variable, we use the addition and multiplication principles and the distributive property as necessary to isolate the term containing that variable in one member of the equation. We then isolate that variable. The solution for one variable will always be in terms of the other variable(s). Check by substituting the solution in the original equation.

EXAMPLE 2-11: Solve the following literal equations for the indicated variables:

(a) $3x + 2y = 6$, for x (b) $ax - by = c$, for y

Solution: (a)
$$3x + 2y = 6$$

$$3x = -2y + 6 \qquad \text{Isolate the variable term } 3x \text{ in one member by adding } -2y \text{ to both members.}$$

$$x = -\frac{2}{3}y + 2 \qquad \text{Isolate the variable } x \text{ by multiplying by the reciprocal of 3.}$$

Check:
$$3x + 2y = 6 \longleftarrow \text{ original equation}$$

$$
\begin{array}{c|c}
3\left(-\frac{2}{3}y + 2\right) + 2y & 6 \\
-2y + 6 + 2y & 6 \\
6 & 6 \longleftarrow \; x = -\frac{2}{3}y + 2 \text{ checks}
\end{array}
$$

Substitute $-\frac{2}{3}y + 2$ for x.

Simplify.

(b) $ax - by = c$

$$-by = -ax + c \qquad \text{Isolate the variable term } -by \text{ in one member by adding } -ax \text{ to both members.}$$

$$y = \frac{-ax + c}{-b} \qquad \text{Isolate the variable } y \text{ by multiplying both the members by the reciprocal of } -b.$$

$$y = \frac{a}{b}x - \frac{c}{b} \qquad \text{Check as before.}$$

B. Solving literal equations containing like terms

To solve a literal equation containing like terms, we collect and combine in one member all the like terms in which the given variable appears, then isolate the variable using the distributive property.

EXAMPLE 2-12: Solve the following literal equations:

(a) $4p + 3q - 5 = 2q - p + 3$, for p (b) $ax - by - cx = dx + e$, for x

Solution: (a) $4p + p + 3q - 2q - 5 - 3 = 2q - 2q - p + p + 3 - 3$ Collect like terms.

$$5p + q - 8 = 0 \qquad \text{Combine like terms.}$$

$$5p = -q + 8 \qquad \text{Solve as before.}$$

$$p = \frac{-q + 8}{5}$$

$$p = -\frac{1}{5}q + \frac{8}{5} \qquad \text{Check as before.}$$

(b) $ax - cx - dx - by - e = dx - dx + e - e$ Collect like terms.

$$(a - c - d)x - by - e = 0 \qquad \text{Combine like terms.}$$

$$(a - c - d)x = by + e \qquad \text{Solve as before.}$$

$$x = \frac{by + e}{a - c - d} \qquad \text{Check as before.}$$

C. Solving literal equations containing parentheses

To **solve a literal equation containing parentheses,** we clear the parentheses and solve as before.

EXAMPLE 2-13: Solve the following equation for x: $a(bx + y) = b(y - ax)$

Solution:

$$a(bx + y) = b(y - ax)$$

$$abx + ay = by - abx \qquad \text{Clear parentheses.}$$

$$abx + abx + ay = -abx + abx \qquad \text{Solve as before.}$$

$$2abx + ay = by$$

$$2abx = by - ay$$

$$x = \frac{by - ay}{2ab} \text{ or } \frac{(b - a)y}{2ab} \qquad \text{Check as before.}$$

D. Solving literal equations containing fractions

To **solve a literal equation containing fractions,** we clear the fractions, then solve as before.

EXAMPLE 2-14: Solve $\dfrac{1}{a} = \dfrac{1}{x} - \dfrac{1}{y}$ for y.

Solution: The LCD of $\dfrac{1}{a}, \dfrac{1}{x}, \dfrac{1}{y}$ is axy.

$$axy\frac{1}{a} = axy\left(\frac{1}{x} - \frac{1}{y}\right) \qquad \text{Clear fractions by multiplying by the LCD.}$$

$$xy = ay - ax$$

$$xy - ay = -ax \qquad \text{Solve as before.}$$

$$(x - a)y = -ax$$

$$y = \frac{-ax}{x - a} \text{ or } -\frac{ax}{x - a} \text{ or } \frac{ax}{a - x} \qquad \text{Check as before.}$$

2-5. Solving Formulas

Recall: An equation containing more than one variable is called a literal equation.

Every **formula** is a literal equation because each contains two or more variables. To solve a formula for a given variable, proceed the same way as you would to solve a literal equation for a given variable.

EXAMPLE 2-15: Solve $P = 2(l + w)$ for w, where P is the perimeter, w is the width, and l is the length of a rectangle.

Solution:

$$P = 2(l + w) \longleftarrow \text{term containing } w$$

$$P = 2(l) + 2(w) \qquad \text{Clear parentheses.}$$

$$P = 2l + 2w \longleftarrow \text{parentheses cleared}$$

$$P - 2l = 2w \qquad \text{Isolate } w \text{ in one member.}$$

$$\frac{P - 2l}{2} = w \longleftarrow w \text{ is isolated}$$

$$w = \frac{P - 2l}{2} \text{ or } \frac{P}{2} - l$$

Note: The formula $w = \frac{1}{2}P - l$ means that the width (w) of a rectangle always equals one-half the perimeter (P) minus the length (l).

2-6. Solving Problems Using Equations

To **solve a problem using a linear equation in one variable,** we

1. *Read* the problem carefully several times.
2. *Draw a picture* (when appropriate) to help visualize the problem.
3. *Identify* the unknowns.
4. *Decide* how to represent the unknowns using one variable.
5. *Make a table* (when appropriate) to help represent the unknowns.
6. *Translate* the problem into a linear equation.
7. *Solve* the linear equation.
8. *Interpret* the solution of the linear equation with respect to each represented unknown to find the proposed solutions of the original problem.
9. *Check* to see if the proposed solutions satisfy all the conditions of the original problem.

A. Solving number problems using linear equations in one variable

Integers that differ by one are called **consecutive integers.** Even integers that differ by two are called **even consecutive integers,** and odd integers that differ by two are called **odd consecutive integers.** A special type of number problem is the **consecutive integer problem.**

EXAMPLE 2-16: Solve the following consecutive integer problem using a linear equation.

1. *Read:*   The sum of three consecutive even integers is 78. What are the integers?

2. *Identify:*   The unknown numbers are $\begin{cases} \text{the first consecutive even integer} \\ \text{the second consecutive even integer} \\ \text{the third consecutive even integer} \end{cases}$.

3. *Decide:*   Let n = the first consecutive even integer

   then $n + 2$ = the second consecutive even integer

   and $n + 4$ = the third consecutive even integer [see Note 1].

4. *Translate:*   The sum of three consecutive even integers is 78.

$$n + (n + 2) + (n + 4) = 78$$

$$n + n + 2 + n + 4 = 78 \quad \longleftarrow \text{ linear equation}$$

5. *Solve:*
$$3n + 6 = 78$$
$$3n = 72$$
$$n = 24$$

6. *Interpret:* $n = 24$ means the first consecutive even integer is 24.

   $n + 2 = 24 + 2 = 26$ means the second consecutive even integer is 26.

   $n + 4 = 24 + 4 = 28$ means the third consecutive even integer is 28.

7. *Check:*   Did you find three consecutive even integers? Yes: 24, 26, and 28.
   Do the three consecutive even integers have a sum of 78? Yes: $24 + 26 + 28 = 78$.

Solution: The three consecutive even integers are 24, 26, and 28.

Note 1: To represent three consecutive even (or odd) integers, you

   let n = the first consecutive even (or odd) integer

   then $n + 2$ = the second consecutive even (or odd) integer

   and $n + 4$ = the third consecutive even (or odd) integer

   because consecutive even (or odd) integers differ by two and $(n + 2) + 2 = n + 4$.

Note 2: To represent three consecutive integers, you

      let n = the first consecutive integer

      then $n + 1$ = the second consecutive integer

      and $n + 2$ = the third consecutive integer

because consecutive integers differ by one and $(n + 1) + 1 = n + 2$.

B. Solving geometry problems using linear equations in one variable

EXAMPLE 2-17: Solve the following geometry problem using a linear equation.

1. *Read:*   The perimeter of a rectangle is 64 feet. The length of the rectangle is 10 feet longer than the width. Find the area of the rectangle.

2. *Draw a picture:*

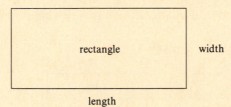

                rectangle   width

                    length

3. *Identify:*   The unknown measures are $\begin{cases} \text{the length of the rectangle} \\ \text{the width of the rectangle} \end{cases}$.

4. *Decide:*   Let l = the length of the rectangle

      and $l - 10$ = the width of the rectangle [see the following note].

5. *Translate:* $P = 2(l + w)$ ⟵ perimeter formula for a rectangle from Appendix Table 3

      $64 = 2[l + (l - 10)]$ ⟵ linear equation   **Replace w with $l - 10$.**

6. *Solve:* $64 = 2(l + l - 10)$

      $64 = 2(2l - 10)$

      $64 = 2(2l) - 2(10)$

      $64 = 4l - 20$

      $84 = 4l$

      $21 = l$

7. *Interpret:* $l = 21$ means the length of the rectangle is 21 feet [ft].
      $l - 10 = 21 - 10 = 11$ means the width of the rectangle is 11 feet.
      $l = 21$ ft and $w = 11$ ft means the area of the rectangle is:
        $A = lw = (21 \text{ ft})(11 \text{ ft}) = 231$ square feet ⟵ proposed solution

8. *Check:*   Did you find the length and the width? Yes: $l = 21$ feet and $w = 11$ feet
      Is the length of the rectangle 10 feet more than the width? Yes: $21 - 10 = 11$
      Is the perimeter of the rectangle 64 feet: Yes: $2(21 + 11) = 2(32) = 64$ [feet].

Solution: The area of the rectangle is 231 square feet.

Note: To represent the unknown length of a rectangle that is 10 ft longer than the width, you can

      let l = the length   let w = the width

                      or

    and $l - 10$ = the width   and $w + 10$ = the length

because l is 10 more than $l - 10$ and $w + 10$ is 10 more than w.

C. Solving currency value problems using linear equations in one variable

EXAMPLE 2-18: Solve the following currency value problem using a linear equation.

1. *Read:*   There are 37 dimes and quarters in all. The combined value of the dimes and quarters is $4.60. How many dimes are there? What is the value of the quarters?

2. *Identify:*   The unknown currency amounts are $\begin{Bmatrix} \text{the number of dimes} \\ \text{the number of quarters} \\ \text{the value of the dimes} \\ \text{the value of the quarters} \end{Bmatrix}$.

3. *Decide:*   Let d = the number of dimes

and $37 - d$ = the number of quarters [see the following note].

4. *Make a table:*

	denomination (d) [in cents]	number (n) [of coins]	value ($v = dn$) [in cents]
dimes	10	d	$10d$
quarters	25	$37 - d$	$25(37 - d)$
combined	—	37	460

5. *Translate:*   The value of the dimes combined with the value of the quarters is $4.60.

$$10d \qquad\qquad + \qquad\qquad 25(37 - d) \qquad = \qquad 460 \text{ [cents]}$$

$$10d + 25(37 - d) = 460 \longleftarrow \text{linear equation}$$

6. *Solve:*
$$10d + 25(37) - 25(d) = 460$$
$$10d + 925 - 25d = 460$$
$$925 - 15d = 460$$
$$-15d = -465$$
$$d = 31$$

7. *Interpret:* $d = 31$ means there are 31 dimes.
$37 - d = 37 - 31 = 6$ means there are 6 quarters.
$10d = 10(31) = 310$ means the value of the dimes is 310 cents or $3.10.
$25(37 - d) = 25(6) = 150$ means the value of the quarters is 150 cents or $1.50.

8. *Check:*   Are there 37 dimes and quarters in all? Yes: $31 + 6 = 37$.
Is the combined value of the dimes and quarters $4.60? Yes: $310 + 150 = 460$.

Solution: There are 31 dimes. The value of the quarters is $1.50.

Note: To represent 37 dimes and quarters, you can

let d = the number of dimes   let q = the number of quarters
                                    or
and $37 - d$ = the number of quarters   and $37 - q$ = the number of dimes

because $d + (37 - d) = 37$ and $q + (37 - q) = 37$.

D. Solving investment problems using linear equations in one variable

EXAMPLE 2-19: Solve this investment problem using a linear equation.

1. Read: Two amounts of principal, totaling $20,000, are invested. One amount is invested at 9% annual interest and the other at 12%. Together they earn $1920. How much is invested at 9%? How much does the 12% investment earn?

2. Identify: The unknown investment amounts are $\begin{cases} \text{the principal invested at 9\%} \\ \text{the principal invested at 12\%} \\ \text{the interest earned at 9\%} \\ \text{the interest earned at 12\%} \end{cases}$.

3. Decide: Let P = the principal invested at 9%

then $20,000 - P$ = the principal invested at 12% [see the following note].

4. Make a table:

	Principal (P) [in dollars]	rate (r) [as a percent]	time (t) [in years]	Interest ($I = Prt$) [in dollars]
principal at 9%	P	9%	1	$P(9\%)1$ or $9\%P$
principal at 12%	$20,000 - P$	12%	1	$(20,000 - P)(12\%)1$ or $12\%(20,000 - P)$
combined	20,000	—	1	1920

5. Translate: The principal at 9% plus the principal at 12% earns $1920.

$$9\%P + 12\%(20,000 - P) = 1920$$

$$9\%P + 12\%(20,000 - P) = 1920 \longleftarrow \text{linear equation}$$

6. Solve:
$$0.09P + 0.12(20,000 - P) = 1920$$
$$0.09P + 0.12(20,000) - 0.12P = 1920$$
$$0.09P + 2400 - 0.12P = 1920$$
$$2400 - 0.03P = 1920$$
$$-0.03P = -480$$
$$P = 16,000$$

7. Interpret: $P = 16,000$ means the principal invested at 9% is $16,000.
$20,000 - P = 20,000 - 16,000 = 4000$ means the principal invested at 12% is $4000.
$9\%P = (0.09)(16,000) = 1440$ means the interest earned at 9% is $1440.
$12\%(20,000 - P) = (0.12)(4000) = 480$ means the interest earned at 12% is $480.

8. Check: Do the two amounts of principal total $20,000? Yes: $16,000 + 4000 = 20,000$.
Do the two amounts of interest earned total $1920? Yes: $1440 + 480 = 1920$.

Solution: The principal invested at 9% is $16,000. The amount of interest earned by the 12% investment is $480.

Note: To represent two unknown amounts of principal totaling $20,000, you can

let P = the principal invested at 9%   let P = the principal invested at 12%
                or
and $20,000 - P$ = the principal invested at 12%   and $20,000 - P$ = the principal invested at 9%

because in both cases $P + (20,000 - P) = 20,000$.

E. Solving uniform motion problems using linear equations in one variable

EXAMPLE 2-20: Solve this uniform motion problem using a linear equation.

1. *Read:* Starting at the same place, two planes flew in opposite directions. The uniform rate of one plane was 100 miles per hour (mph) faster than the uniform rate of the other plane. In 6 hours the two planes were 4800 miles apart. Find the rate of the faster plane. How far did the slower plane travel?

2. *Draw a picture:*   distance for the slower plane   distance for the faster plane

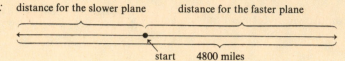

start   4800 miles

3. *Identify:*   The unknown rates and distances are $\left\{\begin{array}{l} \text{the rate of the slower plane} \\ \text{the rate of the faster plane} \\ \text{the distance for the slower plane} \\ \text{the distance for the faster plane} \end{array}\right\}$.

4. *Decide:*   Let r = the rate of the slower plane in mph

then $r + 100$ = the rate of the faster plane in mph [see the following note].

5. *Make a table:*

	rate (r) [in mph]	time (t) [in hours]	distance ($d = rt$) [in miles]
slower plane	r	6	$6r$
faster plane	$r + 100$	6	$6(r + 100)$
combined	—	6	4800

6. *Translate:*   Distance for slower plane plus distance for faster plane is 4800 miles.

$$6r \qquad + \qquad 6(r + 100) \qquad = \qquad 4800$$

$$6r + 6(r + 100) = 4800 \longleftarrow \text{linear equation}$$

7. *Solve:*
$$6r + 6(r) + 6(100) = 4800$$
$$12r + 600 = 4800$$
$$12r = 4200$$
$$r = 350$$

8. *Interpret:* $r = 350$ means the uniform rate of the slower plane is 350 mph.
$r + 100 = 350 + 100 = 450$ means the uniform rate of the faster plane is 450 mph.
$6r = 6(350) = 2100$ means the slower plane traveled 2100 miles.
$6(r + 100) = 6(450) = 2700$ means the faster plane traveled 2700 miles.

9. *Check:*   Is the rate of one plane 100 mph faster than the other plane?
      Yes: $450 - 100 = 350$.
Is the total distance traveled by the two planes 4800 miles?
      Yes: $2100 + 2700 = 4800$.

Solution: The rate of the faster plane is 450 mph.
      The distance traveled by the slower plane is 2100 miles.

Note: To represent an unknown rate that is 100 mph faster [or slower] than another unknown rate, you can

let r = the slower rate   let r = the faster rate
                                    or
and $r + 100$ = the faster rate   and $r - 100$ = the slower rate

because $r + 100$ is 100 more than r, and r is 100 more than $r - 100$.

RAISE YOUR GRADES

Can you ... ?

☑ solve equations using the addition principle
☑ solve equations using the multiplication principle
☑ solve equations using the addition and multiplication principles together
☑ solve equations containing like terms
☑ solve equations containing parentheses
☑ solve equations containing fractions
☑ solve equations containing decimals
☑ solve equations containing percents
☑ solve absolute value equations
☑ solve literal equations using the principles
☑ solve literal equations containing like terms
☑ solve literal equations containing parentheses
☑ solve literal equations containing fractions
☑ solve a formula for a given variable
☑ solve a number problem using a linear equation
☑ solve a geometry problem using a linear equation
☑ solve a currency value problem using a linear equation
☑ solve an investment problem using a linear equation
☑ solve a uniform motion problem using a linear equation

SOLVED PROBLEMS

PROBLEM 2-1 Solve the following equations:

(a) $z - 5 = 8$ (b) $y - 7 = 2$ (c) $x - 3 = 0$ (d) $-3 = w - 5$ (e) $-9 = v - 7$
(f) $u + 4 = 7$ (g) $t + 5 = 2$ (h) $-3 = r + 7$ (i) $-4 = p - 4$

Solution: Recall that to solve an equation of the form $x + B = C$, you add the opposite of B to both members to isolate the variable [see Example 2-1].

(a)
$$z - 5 = 8$$
$$z - 5 + 5 = 8 + 5$$
$$z + 0 = 13$$
$$z = 13$$

(b)
$$y - 7 = 2$$
$$y - 7 + 7 = 2 + 7$$
$$y + 0 = 9$$
$$y = 9$$

(c)
$$x - 3 = 0$$
$$x - 3 + 3 = 0 + 3$$
$$x + 0 = 3$$
$$x = 3$$

(d)
$$-3 = w - 5$$
$$-3 + 5 = w - 5 + 5$$
$$2 = w + 0$$
$$2 = w$$

(e)
$$-9 = v - 7$$
$$-9 + 7 = v - 7 + 7$$
$$-2 = v + 0$$
$$-2 = v$$

(f)
$$u + 4 = 7$$
$$u + 4 - 4 = 7 - 4$$
$$u + 0 = 3$$
$$u = 3$$

(g)
$$t + 5 = 2$$
$$t + 5 - 5 = 2 - 5$$
$$t + 0 = -3$$
$$t = -3$$

(h)
$$-3 = r + 7$$
$$-3 - 7 = r + 7 - 7$$
$$-10 = r + 0$$
$$-10 = r$$

(i)
$$-4 = p - 4$$
$$-4 + 4 = p - 4 + 4$$
$$0 = p + 0$$
$$0 = p$$

PROBLEM 2-2 Solve the following equations:

(a) $3z = 12$ (b) $4y = 7$ (c) $5x = 5$ (d) $14w = 7$ (e) $-4 = 2v$

(f) $-5 = 3u$ (g) $2 = -5t$ (h) $-9 = 9s$ (i) $-5 = -4r$ (j) $-\dfrac{2}{3} = \dfrac{4}{5}m$

(k) $\dfrac{j}{3} = -2$ (l) $-\dfrac{i}{4} = -\dfrac{5}{6}$

Solution: Recall that to solve an equation of the form $Ax = C$, we multiply both members of the equation by the reciprocal of A to isolate the variable [see Example 2-2].

(a) $3z = 12$

$$\frac{3z}{3} = \frac{12}{3}$$

$$z = 4$$

(b) $4y = 7$

$$\frac{4y}{4} = \frac{7}{4}$$

$$y = \frac{7}{4}$$

(c) $5x = 5$

$$\frac{5x}{5} = \frac{5}{5}$$

$$x = 1$$

(d) $14w = 7$

$$\frac{14w}{14} = \frac{7}{14}$$

$$w = \frac{1}{2}$$

(e) $-4 = 2v$

$$\frac{-4}{2} = \frac{2v}{2}$$

$$-2 = v$$

(f) $-5 = 3u$

$$\frac{-5}{3} = \frac{3u}{3}$$

$$-\frac{5}{3} = u$$

(g) $2 = -5t$

$$\frac{2}{-5} = \frac{-5t}{-5}$$

$$-\frac{2}{5} = t$$

(h) $-9 = 9s$

$$\frac{-9}{9} = \frac{9s}{9}$$

$$-1 = s$$

(i) $-5 = -4r$

$$\frac{-5}{-4} = \frac{-4r}{-4}$$

$$\frac{5}{4} = r$$

(j) $-\dfrac{2}{3} = \dfrac{4}{5}m$

$$\frac{5}{4}\left(-\frac{2}{3}\right) = \frac{5}{4}\cdot\frac{4}{5}m$$

$$-\frac{5}{6} = 1m$$

$$-\frac{5}{6} = m$$

(k) $\dfrac{j}{3} = -2$

$$3\cdot\frac{j}{3} = 3(-2)$$

$$1j = -6$$

$$j = -6$$

(l) $-\dfrac{i}{4} = -\dfrac{5}{6}$

$$-4\left(-\frac{i}{4}\right) = -4\left(-\frac{5}{6}\right)$$

$$1i = \frac{10}{3}$$

$$i = \frac{10}{3}$$

PROBLEM 2-3 Solve the following equations:

(a) $2z - 3 = 7$ (b) $3y - 4 = 6$ (c) $-1 = 4x - 3$ (d) $-6 = 5w - 2$
(e) $6 = 3u + 2$ (f) $-5 = 3 - 4s$ (g) $0 = 3 - 5r$ (h) $\frac{2}{3}p - 3 = 1$ (i) $4 - \frac{3}{5}n = 1$

Solution: Recall that to solve an equation of the form $Ax + B = C$, isolate the Ax term by adding the opposite of B to both members and isolate x by multiplying by the reciprocal of A or dividing by A [see Example 2-3].

(a) $2z - 3 = 7$

$$2z - 3 + 3 = 7 + 3$$

$$2z = 10$$

$$\frac{2z}{2} = \frac{10}{2}$$

$$z = 5$$

(b) $3y - 4 = 6$

$$3y - 4 + 4 = 6 + 4$$

$$3y = 10$$

$$\frac{3y}{3} = \frac{10}{3}$$

$$y = \frac{10}{3}$$

(c) $-1 = 4x - 3$

$$-1 + 3 = 4x - 3 + 3$$

$$2 = 4x$$

$$\frac{2}{4} = \frac{4x}{4}$$

$$\frac{1}{2} = x$$

(d) $-6 = 5w - 2$

$-6 + 2 = 5w - 2 + 2$

$-4 = 5w$

$\dfrac{-4}{5} = \dfrac{5w}{5}$

$-\dfrac{4}{5} = w$

(e) $6 = 3u + 2$

$6 - 2 = 3u + 2 - 2$

$4 = 3u$

$\dfrac{4}{3} = \dfrac{3u}{3}$

$\dfrac{4}{3} = u$

(f) $-5 = 3 - 4s$

$-5 - 3 = 3 - 3 - 4s$

$-8 = -4s$

$\dfrac{-8}{-4} = \dfrac{-4s}{-4}$

$2 = s$

(g) $0 = 3 - 5r$

$0 - 3 = 3 - 3 - 5r$

$-3 = -5r$

$\dfrac{-3}{-5} = \dfrac{-5r}{-5}$

$\dfrac{3}{5} = r$

(h) $\dfrac{2}{3}p - 3 = 1$

$\dfrac{2}{3}p - 3 + 3 = 1 + 3$

$\dfrac{2}{3}p = 4$

$\dfrac{3}{2}\cdot\dfrac{2}{3}p = \dfrac{3}{2}\cdot 4$

$p = 6$

(i) $4 - \dfrac{3}{5}n = 1$

$4 - 4 - \dfrac{3}{5}n = 1 - 4$

$-\dfrac{3}{5}n = -3$

$-\dfrac{5}{3}\left(-\dfrac{3}{5}n\right) = -\dfrac{5}{3}(-3)$

$n = 5$

PROBLEM 2-4 Solve the following equations:

(a) $3z + 5z = 4$ **(b)** $4y + 7y = 5$ **(c)** $5 = 4x - x$ **(d)** $5 = 5w - 2w$ **(e)** $6 = v - 4v$
(f) $7 = 3u - 5u$ **(g)** $5 - 3t = 4t$ **(h)** $4 - 5s = 2s$ **(i)** $3r - 5r = 9 - r$
(j) $5q - 9q = 16 - 2q$ **(k)** $3 - 4m + m - 5 = 2 + 4m - 1 - 7m$
(l) $5 - k - 7 + 3k = k - 5 + k + 3$

Solution: Recall that to solve an equation containing like terms, we collect and combine the like terms, then solve as before [see Example 2-4].

(a) $3z + 5z = 4$

$8z = 4$

$z = \tfrac{1}{2}$

(b) $4y + 7y = 5$

$11y = 5$

$y = \tfrac{5}{11}$

(c) $5 = 4x - x$

$5 = 3x$

$\tfrac{5}{3} = x$

(d) $5 = 5w - 2w$

$5 = 3w$

$\tfrac{5}{3} = w$

(e) $6 = v - 4v$

$6 = -3v$

$-2 = v$

(f) $7 = 3u - 5u$

$7 = -2u$

$\dfrac{7}{-2} = u$

$-\tfrac{7}{2} = u$

(g) $5 - 3t = 4t$

$5 - 3t + 3t = 4t + 3t$

$5 = 7t$

$\tfrac{5}{7} = t$

(h) $4 - 5s = 2s$

$4 - 5s + 5s = 2s + 5s$

$4 = 7s$

$\tfrac{4}{7} = s$

(i) $3r - 5r = 9 - r$

$-2r = 9 - r$

$-2r + r = 9 - r + r$

$-r = 9$

$r = \frac{9}{-1}$

$r = -9$

(j) $5q - 9q = 16 - 2q$

$-4q = 16 - 2q$

$-4q + 2q = 16 - 2q + 2q$

$-2q = 16$

$q = \frac{16}{-2}$

$q = -8$

(k) $3 - 4m + m - 5 = 2 + 4m - 1 - 7m$

$-3m - 2 = 1 - 3m$

$-2 = 1$

This equation has no solution.

(l) $5 - k - 7 + 3k = k - 5 + k + 3$

$-2 + 2k = 2k - 2$

$0 = 0$

The solution set is R (all real numbers).

PROBLEM 2-5 Solve the following equations:

(a) $3(z - 2) = 1$ (b) $2(y + 3) = 5$ (c) $3(2 - x) = 3x$ (d) $4w = 3(5 - w)$
(e) $3 - 2(5 - 2v) = 3v$ (f) $7 + 3(4 - u) = 5u$ (g) $3(s - 2) + 5 = 4(3 - s) - 2$
(h) $4(2 - 3p) + 5 = -3(4p - 2)$

Solution: Recall that to solve an equation containing parentheses, we clear parentheses using a distributive property, then solve as before [see Example 2-5].

(a) $3(z - 2) = 1$

$3z - 6 = 1$

$3z = 7$

$z = \frac{7}{3}$

(b) $2(y + 3) = 5$

$2y + 6 = 5$

$2y = -1$

$y = -\frac{1}{2}$

(c) $3(2 - x) = 3x$

$6 - 3x = 3x$

$6 = 6x$

$1 = x$

(d) $4w = 3(5 - w)$

$4w = 15 - 3w$

$7w = 15$

$w = \frac{15}{7}$

(e) $3 - 2(5 - 2v) = 3v$

$3 - 10 + 4v = 3v$

$-7 + 4v = 3v$

$-7 = -v$

$7 = v$

(f) $7 + 3(4 - u) = 5u$

$7 + 12 - 3u = 5u$

$19 - 3u = 5u$

$19 = 8u$

$\frac{19}{8} = u$

(g) $3(s - 2) + 5 = 4(3 - s) - 2$

$3s - 6 + 5 = 12 - 4s - 2$

$3s - 1 = 10 - 4s$

$7s = 11$

$s = \frac{11}{7}$

(h) $4(2 - 3p) + 5 = -3(4p - 2)$

$8 - 12p + 5 = -12p + 6$

$-12p + 13 = -12p + 6$

$13 = 6 \longleftarrow$ false statement

This equation has no solution.

PROBLEM 2-6 Solve the following equations:

(a) $\frac{1}{2}z - \frac{2}{3} = \frac{1}{6}$ (b) $\frac{3}{4}y - \frac{1}{5} = \frac{3}{10}$ (c) $2x - \frac{5}{6} = \frac{1}{3}x$ (d) $3w = \frac{2}{3} - \frac{5}{6}w$

(e) $\frac{4 - u}{3} = \frac{2u - 3}{4}$ (f) $\frac{2}{3}(t - 6) = \frac{3}{4}$

Solution: Recall that to solve an equation containing fractions, we eliminate the fractions by multiplying both members by the LCD of the terms [see Example 2-6].

(a) $\dfrac{1}{2}z - \dfrac{2}{3} = \dfrac{1}{6}$

The LCD is 6.

$$6\left(\dfrac{1}{2}z - \dfrac{2}{3}\right) = 6\left(\dfrac{1}{6}\right)$$

$$3z - 4 = 1$$

$$3z = 5$$

$$z = \dfrac{5}{3}$$

(b) $\dfrac{3}{4}y - \dfrac{1}{5} = \dfrac{3}{10}$

The LCD is 20.

$$20\left(\dfrac{3}{4}y - \dfrac{1}{5}\right) = 20\left(\dfrac{3}{10}\right)$$

$$15y - 4 = 6$$

$$15y = 10$$

$$y = \dfrac{2}{3}$$

(c) $2x - \dfrac{5}{6} = \dfrac{1}{3}x$

The LCD is 6.

$$6\left(2x - \dfrac{5}{6}\right) = 6\left(\dfrac{1}{3}x\right)$$

$$12x - 5 = 2x$$

$$-5 = -10x$$

$$\dfrac{1}{2} = x$$

(d) $3w = \dfrac{2}{3} - \dfrac{5}{6}w$

The LCD is 6.

$$6(3w) = 6\left(\dfrac{2}{3} - \dfrac{5}{6}w\right)$$

$$18w = 4 - 5w$$

$$23w = 4$$

$$w = \dfrac{4}{23}$$

(e) $\dfrac{4 - u}{3} = \dfrac{2u - 3}{4}$

The LCD is 12.

$$12\left(\dfrac{4 - u}{3}\right) = 12\left(\dfrac{2u - 3}{4}\right)$$

$$4(4 - u) = 3(2u - 3)$$

$$16 - 4u = 6u - 9$$

$$25 = 10u$$

$$\dfrac{5}{2} = u$$

(f) $\dfrac{2}{3}(t - 6) = \dfrac{3}{4}$

The LCD is 12.

$$12 \cdot \dfrac{2}{3}(t - 6) = 12 \cdot \dfrac{3}{4}$$

$$8(t - 6) = 9$$

$$8t - 48 = 9$$

$$8t = 57$$

$$t = \dfrac{57}{8}$$

PROBLEM 2-7 Solve the following equations:

(a) $2.8z = 1.4$ **(b)** $3.2y = 0.64$ **(c)** $4t = 3 - 2.7t$ **(d)** $0.4(2p - 2.3) = 0.01$

Solution: Recall that to solve an equation containing decimals, we usually eliminate the decimals by multiplying both members of the equation by the LCD of the terms, then solving as before [see Example 2-7].

(a) $2.8z = 1.4$

The LCD of 2.8 $(2\frac{8}{10})$ and 1.4 $(1\frac{4}{10})$ is 10

$$10(2.8z) = 10(1.4)$$

$$28z = 14$$

$$z = \tfrac{1}{2}$$

$$z = 0.5$$

(b) $3.2y = 0.64$

The LCD of 3.2 $(3\frac{2}{10})$ and 0.64 $(\frac{64}{100})$ is 100

$$100(3.2y) = 100(0.64)$$

$$320y = 64$$

$$y = \tfrac{1}{5}$$

$$y = 0.2$$

(c) $4t = 3 - 2.7t$

The LCD of 4, 3, and 2.7 is 10.

$$10(4t) = 10(3 - 2.7t)$$

$$40t = 30 - 27t$$

$$67t = 30$$

$$t = \tfrac{30}{67}$$

(d) $0.4(2p - 2.3) = 0.01$

$$0.8p - 0.92 = 0.01$$

The LCD of 0.8, 0.92, 0.01 is 100

$$100(0.8p - 0.92) = 100(0.01)$$

$$80p - 92 = 1$$

$$80p = 93$$

$$p = \tfrac{93}{80}$$

$$p = 1.1625$$

PROBLEM 2-8 Solve the following equations:

(a) $25\%z = 6$ (b) $10\%y = 4$ (c) $16\tfrac{2}{3}\%x + x = 7$ (d) $9 = 12\tfrac{1}{2}\%w + w$

Solution: Recall that to solve an equation containing percents, we write the percents as decimals or fractions, then solve as before [see Example 2-8].

(a) $25\%z = 6$

25% is 0.25 in decimal form.

$0.25z = 6$

The LCD is 100.

$25z = 600$

$z = 24$

(b) $10\%y = 4$

10% is 0.10 in decimal form.

$0.1y = 4$

The LCD is 10.

$1y = 40$

$y = 40$

(c) $16\tfrac{2}{3}\%x + x = 7$

$\tfrac{1}{6}$ is $16\tfrac{2}{3}\%$ in fraction form.

$\tfrac{1}{6}x + x = 7$

The LCD is 6.

$1x + 6x = 42$

$7x = 42$

$x = 6$

(d) $9 = 12\tfrac{1}{2}\%w + w$

$\tfrac{1}{8}$ is $12\tfrac{1}{2}\%$ in fraction form.

$9 = \tfrac{1}{8}w + w$

The LCD is 8.

$72 = 1w + 8w$

$72 = 9w$

$8 = w$

PROBLEM 2-9 Solve the following equations:

(a) $|z - 3| = 4$ (b) $|y + 4| = 3$ (c) $|2u + 3| = 5$ (d) $|2 - 3t| = 0$
(e) $|3r - 2| + 4 = 7$ (f) $3 - 4|2 - 3n| = 1$

Solution: Recall that to solve an absolute value equation of the form $|Ax + B| = C$, where $C \geq 0$, let $Ax + B = C$ or $Ax + B = -C$ and solve both equations [see Examples 2-9 and 2-10].

(a) $|z - 3| = 4$

$z - 3 = 4$ or $z - 3 = -4$

$z = 4 + 3$ or $z = -4 + 3$

$z = 7$ or $z = -1$

(b) $|y + 4| = 3$

$y + 4 = 3$ or $y + 4 = -3$

$y = 3 - 4$ or $y = -3 - 4$

$y = -1$ or $y = -7$

(c) $|2u + 3| = 5$

$2u + 3 = 5$ or $2u + 3 = -5$

$2u = 5 - 3$ or $2u = -5 - 3$

$2u = 2$ or $2u = -8$

$u = 1$ or $u = -4$

(d) $|2 - 3t| = 0$

$2 - 3t = 0$

$-3t = -2$

$t = \tfrac{2}{3}$

(e) $|3r - 2| + 4 = 7$

$|3r - 2| = 3$

$3r - 2 = 3$ or $3r - 2 = -3$

$3r = 5$ or $3r = -1$

$r = \tfrac{5}{3}$ or $r = -\tfrac{1}{3}$

(f) $3 - 4|2 - 3n| = 1$

$-4|2 - 3n| = -2$

$|2 - 3n| = \tfrac{1}{2}$

$2 - 3n = \tfrac{1}{2}$ or $2 - 3n = -\tfrac{1}{2}$

$-3n = -\tfrac{3}{2}$ or $-3n = -\tfrac{5}{2}$

$n = \tfrac{1}{2}$ or $n = \tfrac{5}{6}$

PROBLEM 2-10 Solve the following equations:

(a) $2x - 3y = 6$ for x (b) $2x - 3y = 6$ for y (c) $3x = 4y + 12$ for x
(d) $3x = 4y + 12$ for y (e) $3y = 2x - 4$ for x (f) $3y = 2x - 4$ for y
(g) $ax + by = c$ for x (h) $ax + by = c$ for y (i) $ax + by = a$ for x

Solution: Recall that to solve a literal equation, we isolate the given unknown term, then isolate the given unknown. The other member of the equation is the solution for the given unknown [see Example 2-11].

(a) Solve $2x - 3y = 6$ for x.

$$2x - 3y + 3y = 3y + 6$$
$$2x = 3y + 6$$
$$x = \frac{3y + 6}{2}$$
$$\text{or } x = \frac{3}{2}y + 3$$

(b) Solve $2x - 3y = 6$ for y.

$$2x - 2x - 3y = -2x + 6$$
$$-3y = -2x + 6$$
$$y = \frac{-2x + 6}{-3}$$
$$\text{or } y = \frac{2}{3}x - 2$$

(c) Solve $3x = 4y + 12$ for x.

$$x = \frac{4y + 12}{3}$$
$$\text{or } x = \frac{4}{3}y + 4$$

(d) Solve $3x = 4y + 12$ for y.

$$3x - 12 = 4y$$
$$\frac{3x - 12}{4} = y$$
$$\text{or } \frac{3}{4}x - 3 = y$$

(e) Solve $3y = 2x - 4$ for x.

$$3y + 4 = 2x$$
$$\frac{3y + 4}{2} = x$$
$$\text{or } \frac{3}{2}y + 2 = x$$

(f) Solve $3y = 2x - 4$ for y.

$$y = \frac{2x - 4}{3}$$
$$\text{or } y = \frac{2}{3}x - \frac{4}{3}$$

(g) Solve $ax + by = c$ for x.

$$ax = -by + c$$
$$x = \frac{-by + c}{a}$$
$$\text{or } x = -\frac{b}{a}y + \frac{c}{a}$$

(h) Solve $ax + by = c$ for y.

$$by = -ax + c$$
$$y = \frac{-ax + c}{b}$$
$$\text{or } y = -\frac{a}{b}x + \frac{c}{b}$$

(i) Solve $ax + by = a$ for x.

$$ax = -by + a$$
$$x = \frac{-by + a}{a}$$
$$\text{or } x = -\frac{b}{a}y + 1$$

PROBLEM 2-11 Solve the following equations:

(a) $3x + 2y - 4x + y = 3$ for x (b) $3x + 2y - 4x + y = 3$ for y
(c) $4r + s - 3r = 5 - 4s + 2r + s$ for r (d) $4r + s - 3r = 5 - 4s + 2r + s$ for s
(e) $ax + by = cy - dx$ for x (f) $ax + by = cy - dx$ for y (g) $ax + by = 3by - 2ax$ for x
(h) $ax + by = 3by - 2ax$ for y

Solution: Recall that to solve a literal equation, we collect and combine like terms, then solve as before [see Example 2-12].

(a) Solve $3x + 2y - 4x + y = 3$ for x.

$$-x + 3y = 3$$
$$-x = -3y + 3$$
$$x = \frac{-3y + 3}{-1}$$
$$\text{or } x = 3y - 3$$

(b) Solve $3x + 2y - 4x + y = 3$ for y.

$$-x + 3y = 3$$
$$3y = x + 3$$
$$y = \frac{x + 3}{3}$$
$$\text{or } y = \frac{1}{3}x + 1$$

(c) Solve $4r + s - 3r = 5 - 4s + 2r + s$ for r.

$$r + s = 5 - 3s + 2r$$

$$-r = -4s + 5$$

$$r = \frac{-4s + 5}{-1}$$

or $r = 4s - 5$.

(d) Solve $4r + s - 3r = 5 - 4s + 2r + s$ for s.

$$r + s = 5 - 3s + 2r$$

$$4s = r + 5$$

$$s = \frac{r + 5}{4}$$

or $s = \frac{1}{4}r + \frac{5}{4}$

(e) Solve $ax + by = cy - dx$ for x.

$$ax + dx = cy - by$$

$$(a + d)x = cy - by$$

$$x = \frac{cy - by}{a + d}$$

(f) Solve $ax + by = cy - dx$ for y.

$$ax + dx = cy - by$$

$$ax + dx = (c - b)y$$

$$\frac{ax + dx}{c - b} = y$$

(g) Solve $ax + by = 3by - 2ax$ for x.

$$ax + 2ax = 3by - by$$

$$3ax = 2by$$

$$x = \frac{2by}{3a}$$

(h) Solve $ax + by = 3by - 2ax$ for y.

$$ax + 2ax = 3by - by$$

$$3ax = 2by$$

$$\frac{3ax}{2b} = y$$

PROBLEM 2-12 Solve the following equations:

(a) $a(x - b) = c$ for x **(b)** $a(x - a) = b(a - x)$ for x **(c)** $a(x - b) = b(a - x)$ for x
(d) $a(x - b) = b(a - x)$ for a **(e)** $a(z - b) = cz$ for z **(f)** $ay = b(c - y)$ for y
(g) $a(m - n) = n(m - a)$ for n **(h)** $a(m - n) = n(m - a)$ for n

Solution: Recall that to solve a literal equation containing parentheses, we clear parentheses and solve as before [see Example 2-13].

(a) Solve $a(x - b) = c$ for x.

$$ax - ab = c$$

$$ax = ab + c$$

$$x = \frac{ab + c}{a}$$

(b) Solve $a(x - a) = b(a - x)$ for x.

$$ax - a^2 = ab - bx$$

$$ax + bx = a^2 + ab$$

$$(a + b)x = a^2 + ab$$

$$x = \frac{a^2 + ab}{a + b}$$

(c) Solve $a(x - b) = b(a - x)$ for x.

$$ax - ab = ab - bx$$

$$ax + bx = ab + ab$$

$$(a + b)x = 2ab$$

$$x = \frac{2ab}{a + b}$$

(d) Solve $a(x - b) = b(a - x)$ for a.

$$ax - ab = ab - bx$$

$$ax - ab - ab = -bx$$

$$ax - 2ab = -bx$$

$$a(x - 2b) = -bx$$

$$a = \frac{-bx}{x - 2b}$$

(e) Solve $a(z - b) = cz$ for z.

$$az - ab = cz$$

$$az - cz = ab$$

$$(a - c)z = ab$$

$$z = \frac{ab}{a - c}$$

(f) Solve $ay = b(c - y)$ for y.

$$ay = bc - by$$

$$ay + by = bc$$

$$(a + b)y = bc$$

$$y = \frac{bc}{a + b}$$

(g) Solve $a(m - n) = n(m - a)$ for n.   **(h)**   Solve $a(m - n) = n(m - a)$ for m.

$$am - an = mn - an$$

$$am - an + an = mn - an + an$$

$$am = mn$$

$$\frac{am}{m} = n$$

$$a = n$$

$$am - an = mn - an$$

$$am - an - mn + an = mn - an - mn + an$$

$$am - mn = 0$$

$$(a - n)m = 0$$

$$m = \frac{0}{a - n}$$

$$m = 0$$

PROBLEM 2-13 Solve the following equations:

(a) $\dfrac{1}{x} = \dfrac{1}{y}$ for x **(b)** $\dfrac{a}{x} - \dfrac{b}{y} = 0$ for y **(c)** $\dfrac{x + a}{y + b} = \dfrac{1}{a}$ for x

(d) $\dfrac{x + a}{y + b} = \dfrac{a}{b}$ for y **(e)** $\dfrac{1}{x} + \dfrac{1}{y} - \dfrac{1}{z} = 0$ for y **(f)** $\dfrac{1}{x} + \dfrac{1}{y} - \dfrac{1}{z} = 0$ for z

Solution: Recall that to solve a literal equation containing fractions, we clear the fractions by multiplying both members by the LCD of the terms [see Example 2-14].

(a) Solve $\dfrac{1}{x} = \dfrac{1}{y}$ for x.

The LCD is xy.

$$xy\frac{1}{x} = xy\frac{1}{y}$$

$$y = x$$

(b) Solve $\dfrac{a}{x} - \dfrac{b}{y} = 0$ for y.

The LCD is xy.

$$xy\frac{a}{x} - xy\frac{b}{y} = 0$$

$$ay - bx = 0$$

$$ay = bx$$

$$y = \frac{bx}{a}$$

(c) Solve $\dfrac{x + a}{y + b} = \dfrac{1}{a}$ for x.

The LCD is $a(y + b)$.

$$a(y + b)\frac{x + a}{y + b} = a(y + b)\frac{1}{a}$$

$$a(x + a) = y + b$$

$$ax + a^2 = y + b$$

$$ax = y + b - a^2$$

$$x = \frac{y + b - a^2}{a}$$

(d) Solve $\dfrac{x + a}{y + b} = \dfrac{a}{b}$ for y.

The LCD is $b(y + b)$.

$$b(y + b)\frac{x + a}{y + b} = b(y + b)\frac{a}{b}$$

$$b(x + a) = (y + b)a$$

$$bx + ab = ay + ab$$

$$bx + ab - ab = ay$$

$$\frac{bx + ab - ab}{a} = y$$

$$\frac{bx}{a} = y$$

(e) Solve $\dfrac{1}{x} + \dfrac{1}{y} - \dfrac{1}{z} = 0$ for y.

The LCD is xyz.

$$xyz\left(\frac{1}{x} + \frac{1}{y} - \frac{1}{z}\right) = xzy(0)$$

$$yz + xz - xy = 0$$

$$yz - xy = -xz$$

$$(z - x)y = -xz$$

$$y = \frac{-xz}{z - x}$$

or $y = \dfrac{xz}{x - z}$

(f) Solve $\dfrac{1}{x} + \dfrac{1}{y} + \dfrac{1}{z} = 0$ for z.

The LCD is xyz.

$$xyz\left(\frac{1}{x} + \frac{1}{y} - \frac{1}{z}\right) = xyz(0)$$

$$yz + xz - xy = 0$$

$$yz + xz = xy$$

$$(y + x)z = xy$$

$$z = \frac{xy}{x + y}$$

PROBLEM 2-14 Solve each formula for the indicated variable:

(a) *Distance Formula:* $d = rt$ for t **(b)** *Amount Formula:* $A = P(1 + rt)$ for t

Solution: Recall that to solve a formula for a given variable, proceed in the same way as you would to solve a literal equation in one variable [see Example 2-15].

(a) $d = rt$

$$\frac{d}{r} = \frac{rt}{r}$$

$$t = \frac{d}{r} \longleftarrow \text{solution}$$

(b) $A = P(1 + rt)$

$$A = P + Prt$$

$$A - P = Prt$$

$$\frac{A - P}{Pr} = \frac{Prt}{Pr}$$

$$t = \frac{A - P}{Pr} \text{ or } \frac{A}{Pr} - \frac{1}{r} \longleftarrow \text{solution}$$

PROBLEM 2-15 Solve each number problem using a linear equation:

(a) The sum of three numbers is 47. The second number is twice the first number. The third number is 2 more than the second number. Find the numbers.

(b) The sum of three consecutive odd integers is 135. Find the integers.

Solution: Recall that to solve a number problem using a linear equation, we:

1. *Read* the problem very carefully several times.
2. *Identify* the unknown numbers.
3. *Decide* how to represent the unknown numbers using one variable.
4. *Translate* the problem to a linear equation using key words.
5. *Solve* the linear equation.
6. *Interpret* the solution of the linear equation with respect to each represented unknown number to find the proposed solutions of the original problem.
7. *Check* to see if the proposed solutions satisfy all of the conditions of the original problem [see Example 2-16].

(a) *Identify:*   The unknown numbers are $\begin{cases} \text{the first number} \\ \text{the second number} \\ \text{the third number} \end{cases}$.

Decide: Let n = the first number

then $2n$ = the second number (because $2n$ is twice n)

and $2n + 2$ = the third number (because $2n + 2$ is 2 more than $2n$).

Translate: $\underbrace{\text{The sum of the three numbers}}$ is 47.

$$(n + 2n + (2n + 2)) = 47$$

$$n + 2n + 2n + 2 = 47 \longleftarrow \text{linear equation}$$

Solve:

$$5n + 2 = 47$$

$$5n = 45$$

Interpret:

$$\left.\begin{array}{r} n = 9 \\ 2n = 18 \\ 2n + 2 = 20 \end{array}\right\} \longleftarrow \text{solutions [check as before]}$$

(b) *Identify:*   The unknown numbers are $\begin{cases} \text{the first consecutive odd integer} \\ \text{the second consecutive odd integer} \\ \text{the third consecutive odd integer} \end{cases}$.

Decide: Let n = the first consecutive odd integer

then $n + 2$ = the second consecutive odd integer

and $n + 4$ = the third consecutive odd integer

because $(n + 2) + 2 = n + 4$.

Translate: The sum of three consecutive odd integers is 135.

$$n + (n + 2) + (n + 4) = 135$$

$$n + n + 2 + n + 4 = 135 \longleftarrow \text{linear equation}$$

Solve:
$$3n + 6 = 135$$
$$3n = 129$$

Interpret:
$$\left. \begin{array}{l} n = 43 \\ n + 2 = 45 \\ n + 4 = 47 \end{array} \right\} \longleftarrow \text{solutions [check as before]}$$

PROBLEM 2-16 Solve each geometry problem using a linear equation:

(a) The perimeter of a triangle is 41 meters. The longest side is 3 times the shortest side. The third side is 8 meters shorter than the longest side. How long is each side?

(b) The area of a rectangle is 192 in.^{2} [square inches]. Find the perimeter of the rectangle if its length is 16 inches.

Solution: Recall that to solve a geometry problem using a linear equation, you:

1. *Read* the problem very carefully several times.
2. *Draw a picture* if necessary to help visualize the problem.
3. *Identify* the unknown measures.
4. *Decide* how to represent the unknown measures using one variable.
5. *Translate* the problem to a linear equation using the correct geometry formula from **Appendix Table 5**.
6. *Solve* the linear equation.
7. *Interpret* the solution of the linear equation with respect to each represented unknown measure to find the proposed solutions of the original problem.
8. *Check* to see if the proposed solutions satisfy all the conditions of the original problem [see Example 2-17].

(a) *Draw a picture:*

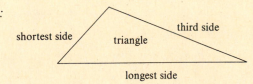

Identify: The unknown measures are $\left\{ \begin{array}{l} \text{the shortest side} \\ \text{the longest side} \\ \text{the third side} \end{array} \right\}$.

Decide: Let s = the shortest side

then $3s$ = the longest side (because $3s$ is 3 times s)

and $3s - 2$ = the third side (because $3s - 2$ is 2 less than $3s$).

Translate: $P = a + b + c \longleftarrow$ perimeter formula for a triangle from Appendix Table 5

$$47 = s + 3s + (3s - 2)$$

$$47 = s + 3s + 3s - 2 \longleftarrow \text{linear equation}$$

Solve:
$$47 = 7s - 2$$
$$49 = 7s$$

Interpret:

$$s = 7 \text{ [shortest side]}$$
$$3s = 21 \text{ [longest side]}$$
$$3s - 2 = 19 \text{ [third side]}$$

⟵ solutions [check as before]

(b) *Draw a picture:*

rectangle

area (192 in.²)   *width*

length (76 in.)

Identify: The unknown measures are $\begin{cases} \text{the width} \\ \text{the perimeter} \end{cases}$.

Decide: Use $P = 2(l + w)$ (perimeter formula for a rectangle)

and $A = lw$ (area formula for a rectangle)

from Appendix Table 5.

Translate: $A = lw$

$$192 = (16)w$$

$$16w = 192 \longleftarrow \text{linear equation}$$

Solve: $w = 12 \text{ [inches]}$

Interpret: $P = 2(l + w)$

$$= 2(16 + 12)$$

$$= 2(28)$$

$$= 56 \text{ [inches]} \longleftarrow \text{solution [check as before]}$$

PROBLEM 2-17 Solve each currency value problem using a linear equation:

(a) A person changed \$10 at a bank into dimes, nickels, and quarters. There were twice as many dimes as nickels and three times as many quarters as nickels. How many nickels and dimes were there? What was the value of the quarters?

(b) A theatre sold a total of 621 child and adult movie tickets for \$2725. A child ticket sold for \$3 and an adult ticket sold for \$5. How many child tickets were sold? What was the value of the adult tickets sold?

Solution: Recall that to solve a currency value problem using a linear equation, you:

1. *Read* the problem very carefully several times.
2. *Identify* the unknown currency amounts.
3. *Decide* how to represent the unknown number of each currency using one variable.
4. *Make a table* to help represent the unknown value of each currency using the following formula:

$$\begin{pmatrix} \text{denomination} \\ \text{of currency} \end{pmatrix} \begin{pmatrix} \text{amount of} \\ \text{currency} \end{pmatrix} = \begin{pmatrix} \text{value of} \\ \text{currency} \end{pmatrix}$$

5. *Translate* the problem to a linear equation.
6. *Solve* the linear equation.

7. *Interpret* the solution of the linear equation with respect to each represented unknown number and value of each currency to find the proposed solutions of the original problem.

8. *Check* to see if the proposed solutions satisfy all the conditions of the original problem [see Example 2-18].

(a) *Identify:* The unknown currency amounts are
$$\begin{cases} \text{the number of nickels} \\ \text{the number of dimes} \\ \text{the number of quarters} \\ \text{the value of the nickels} \\ \text{the value of the dimes} \\ \text{the value of the quarters} \end{cases}.$$

Decide: Let $n =$ the number of nickels

then $2n =$ the number of dimes

and $3n =$ the number of quarters.

Make a table:

	denomination (d) [in cents]	number (n) [of coins]	value ($v = dn$) [in cents]
nickels	5	n	$5n$
dimes	10	$2n$	$10(2n)$ or $20n$
quarters	25	$3n$	$25(3n)$ or $75n$
combined	—	—	1000

Translate: The combined value of the nickels, dimes, and quarters is $10.

$$(5n) + (20n) + (75n) = 1000 \text{ [cents]}$$

Solve:
$$5n + 20n + 75n = 1000 \longleftarrow \text{linear equation in one variable}$$
$$100n = 1000$$

number of nickels $\longrightarrow n = 10$ [coins]

number of dimes $\longrightarrow 2n = 20$ [coins] $\left. \right\} \longleftarrow$ solutions [check as before]

value of the quarters $\longrightarrow 75n = 750$ [cents]

(b) *Identify:* The unknown ticket amounts are
$$\begin{cases} \text{the number of child tickets} \\ \text{the number of adult tickets} \\ \text{the value of the child tickets} \\ \text{the value of the adult tickets} \end{cases}.$$

Decide: Let $n =$ the number of child tickets

and $621 - n =$ the number of adult tickets.

Make a table:

	denomination (d) [in dollars]	number (n) [of tickets]	value ($v = dn$) [in dollars]
child	3	n	$3n$
adult	5	$621 - n$	$5(621 - n)$
combined	—	621	2725

Translate: The value of the child tickets plus the value of the adult tickets is $2725.

$$3n \quad + \quad 5(621 - n) = 2725$$

$$3n + 5(621 - n) = 2725 \longleftarrow \text{linear equation}$$

Solve:

$$3n + 5(621) - 5(n) = 2725$$
$$3n + 3105 - 5n = 2725$$
$$3105 - 2n = 2725$$

Interpret:

$$-2n = -380$$

number of child tickets $\longrightarrow$ $n = 190$ [tickets]

value of adult tickets $\longrightarrow$ $5(621 - n) = 2155$ [dollars]

$\left.\begin{array}{l} \\ \\ \end{array}\right\}$ $\longleftarrow$ solutions [check as before]

PROBLEM 2-18 Solve each investment problem using a linear equation:

(a) Manuel invested in two different bonds each paying 8% interest. He invested $100 more in Bond A than in Bond B. The total annual interest from the bonds is $48. How much is invested in each type of bond?

(b) Part of a $6000 investment made a 12% profit. The other part made an 8% loss. The net gain was $400. How much was invested at 12%? How much was lost at 8%?

Solution: Recall that to solve an investment problem using a linear equation, we:

1. *Read* the problem very carefully several times.
2. *Identify* the unknown investment amounts.
3. *Decide* how to represent the unknown amounts of principal using one variable.
4. *Make a table* to help represent the unknown amounts of interest using the formula

$$\text{Interest} = (\text{Principal})(\text{rate})(\text{time}), \text{ or } I = Prt.$$

5. *Translate* the problem to a linear equation.
6. *Solve* the linear equation.
7. *Interpret* the solution of the linear equation with respect to each represented unknown amount to find the proposed solutions of the original problem.
8. *Check* to see if the proposed solutions satisfy all the conditions of the original problem [see Example 2-19].

(a) *Identify:* The unknown investment amounts are $\begin{cases} \text{the amount invested in Bond A} \\ \text{the amount invested in Bond B} \end{cases}$.

Decide: Let P = the amount invested in Bond A

and $P - 100$ = the amount invested in Bond B.

Make a table:

	Principal (P) [in dollars]	rate (r) [as a percent]	time (t) [in years]	Interest ($I = Prt$) [in dollars]
Bond A	P	8%	1	$8\%P$
Bond B	$P - 100$	8%	1	$8\%(P - 100)$
combined	100	8%	1	48

Translate: The amount of interest for Bond A plus amount of interest for Bond B equals $48.

$$8\%P \qquad + \qquad 8\%(P - 100) \qquad = \qquad 48$$

$$8\%P + 8\%(P - 100) = 48 \longleftarrow \text{linear equation}$$

Solve:

$$0.08P + 0.08(P - 100) = 48$$
$$0.08P + 0.08P - 8 = 48$$
$$0.16P - 8 = 48$$
$$0.16P = 56$$

Interpret: 　　amount invested in Bond A $\longrightarrow P = 350$ [dollars] $\Big\}\longleftarrow$ solutions [check as before]

　　　amount invested in Bond B $\longrightarrow P - 100 = 250$ [dollars]

(b) *Identify:* 　　The unknown investment amounts are $\left\{\begin{array}{l}\text{the principal that earned 12\% profit}\\\text{the principal that earned 8\% profit}\\\text{the amount of the 12\% profit}\\\text{the amount of the 8\% loss}\end{array}\right\}$.

Decide: 　　Let 　　$P =$ the principal that earned a 12% profit

and $6000 - P =$ the principal that earned an 8% loss.

Make a table:

	Principal (P) [in dollars]	rate (r) [as a percent]	time (t) [in years]	Interest ($I = Prt$) [in dollars]
12% profit	P	12%	1	$12\%P$
8% loss	$6000 - P$	8%	1	$8\%(6000 - P)$
combined	6000	—	1	400

Translate: 　　The amount of 12% profit minus the amount of 8% loss equals $400.

$$12\%P \quad - \quad 8\%(6000 - P) \quad = \quad 400$$

$$12\%P - 8\%(6000 - P) = 400 \longleftarrow \text{linear equation}$$

Solve:

$$0.12P - 0.08(6000 - P) = 400$$

$$0.12P - 0.08(6000) + 0.08P = 400$$

$$0.12P - 480 + 0.08P = 400$$

$$-480 + 0.2P = 400$$

Interpret: 　　principal that earned a 12% profit $\longrightarrow P = 4400$ [dollars] $\Big\}\longleftarrow$ solutions [check as before]

　　amount of the 8% loss $\longrightarrow 8\%(6000 - P) = 128$ [dollars]

PROBLEM 2-19 Solve each uniform motion problem using a linear equation:

(a) Two people are $3\frac{1}{2}$ miles apart and walking towards each other. Their uniform walking rates differ by 1 mph. They will meet in $\frac{1}{2}$ hour. What is the uniform rate of each walker? How much of the distance will each walker travel?

(b) A person drives from home to work at a constant rate in 45 minutes ($\frac{3}{4}$ hour). The same person drives from work to home at a constant rate that is 20 mph slower. The total commute time is 2 hours. How far is home from work? What is the constant rate on each part of the commute?

Solution: Recall that to solve a uniform motion problem using a linear equation, we:

1. *Read* the problem very carefully several times.
2. *Draw a picture* to help visualize the problem.
3. *Identify* the unknown distances, rates, and/or times.
4. *Decide* how to represent the unknown rates or times using one variable.
5. *Make a table* to help represent the unknown distances using the formula

$$\text{distance} = (\text{rate})(\text{time}), \text{ or } d = rt.$$

6. *Translate* the problem to a linear equation.
7. *Solve* the linear equation.
8. *Interpret* the solution of the linear equation with respect to each represented unknown distance, rate, and/or time to find the proposed solutions of the original problem.

9. *Check* to see if the proposed solutions satisfy all the conditions of the original problem [see Example 2-20].

(a) *Draw a picture:*

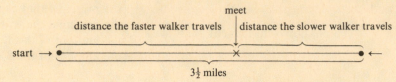

Identify: The unknown rates and distances are $\begin{cases} \text{the uniform rate of the faster walker} \\ \text{the uniform rate of the slower walker} \\ \text{the distance the faster walker travels} \\ \text{the distance the slower walker travels} \end{cases}$.

Decide: Let r = the uniform rate of the faster walker in mph

and $r - 1$ = the uniform rate of the slower walker in mph.

Make a table:

Make a table:

	rate (r) [in mph]	time (t) [in hours]	distance ($d = rt$) [in miles]
faster walker	r	$\frac{1}{2}$	$\frac{1}{2}r$
slower walker	$r - 1$	$\frac{1}{2}$	$\frac{1}{2}(r - 1)$
combined	—	$\frac{1}{2}$	$3\frac{1}{2}$

Translate: The faster walker's distance plus the slower walker's distance is $3\frac{1}{2}$ miles.

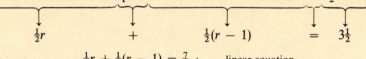

$$\tfrac{1}{2}r + \tfrac{1}{2}(r - 1) = \tfrac{7}{2} \longleftarrow \text{linear equation}$$

Solve:
$$2 \cdot \tfrac{1}{2}r + 2 \cdot \tfrac{1}{2}(r - 1) = 2 \cdot \tfrac{7}{2}$$
$$r + (r - 1) = 7$$
$$2r - 1 = 7$$
$$2r = 8$$

Interpret:

$$\left.\begin{array}{lr} \text{rate of the faster walker} & r = 4 \,[\text{mph}] \\ \text{rate of the slower walker} & r - 1 = 3 \,[\text{mph}] \\ \text{distance for faster walker} \longrightarrow \tfrac{1}{2}r = 2 \,[\text{miles}] \\ \text{distance for slower walker} \longrightarrow \tfrac{1}{2}(r - 1) = 1\tfrac{1}{2} \,[\text{miles}] \end{array}\right\} \leftarrow \begin{array}{l} \text{solutions [check} \\ \text{as before]} \end{array}$$

(b) *Draw a picture:*

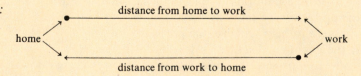

Identify: The unknown rates and distances are $\begin{cases} \text{the constant rate from home to work} \\ \text{the constant rate from work to home} \\ \text{the distance from work to home} \\ \text{the distance from home to work} \end{cases}$.

Decide: Let r = the constant rate from home to work in mph

then $r - 20$ = the constant rate from work to home in mph.

Make a table:

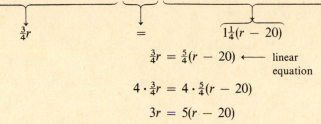

	rate (r) [in mph]	time (t) [in hours]	distance ($d = rt$) [in miles]
from home	r	$\frac{3}{4}$	$\frac{3}{4}r$
from work	$r - 20$	$1\frac{1}{4}$	$1\frac{1}{4}(r - 20)$

Translate: The distance from home to work equals the distance from work to home.

$$\frac{3}{4}r \qquad = \qquad 1\frac{1}{4}(r - 20)$$

$$\frac{3}{4}r = \frac{5}{4}(r - 20) \longleftarrow \text{linear equation}$$

Solve:

$$4 \cdot \frac{3}{4}r = 4 \cdot \frac{5}{4}(r - 20)$$

$$3r = 5(r - 20)$$

$$3r = 5(r) - 5(20)$$

$$3r = 5r - 100$$

$$-2r = -100$$

Interpret:

constant rate from home to work $\longrightarrow$ $r = 50$ [mph]

constant rate from work to home $\longrightarrow$ $r - 20 = 30$ [mph]

distance from home to work $\longrightarrow$ $\frac{3}{4}r = 37\frac{1}{2}$ [miles]

distance from work to home $\longrightarrow$ $1\frac{1}{4}(r - 20) = 37\frac{1}{2}$ [miles]

$\longleftarrow$ solutions [check as before]

Supplementary Exercises

PROBLEM 2-20 Solve the following equations:

(a) $a - 3 = 6$ (b) $b - 5 = 1$ (c) $-4 = c - 5$ (d) $-5 = d - 2$ (e) $e + 3 = 5$
(f) $f + 5 = 1$ (g) $-3 = 2 + g$ (h) $-1 = 4 + h$ (i) $-5 = -1 + i$ (j) $0 = 5 + j$
(k) $3 - k = 0$

PROBLEM 2-21 Solve the following equations:

(a) $4m = 8$ (b) $3n = 7$ (c) $2p = -6$ (d) $-9 = -3q$ (e) $12 = -5r$
(f) $-8 = -4s$ (g) $3 = -6t$ (h) $0 = -4u$ (i) $\frac{2}{3}v = \frac{4}{5}$ (j) $\frac{3}{4}w = \frac{1}{3}$
(k) $-\frac{1}{3}x = -2$ (l) $-\frac{3}{4}y = 3$

PROBLEM 2-22 Solve the following equations:

(a) $2z + 5 = 11$ (b) $5 + 6y = 29$ (c) $3 + 4x = 12$ (d) $3 = 4 + 2w$ (e) $3v - 4 = 5$
(f) $4u - 7 = 5$ (g) $5 = 2t - 6$ (h) $-3 = 4s - 6$ (i) $5 = 3 - 2r$ (j) $3 = 4 - 5q$
(k) $0 = 4 - 6p$ (l) $0 = 2 - 5n$ (m) $0 = 4 + 3m$ (n) $0 = 6 - 12k$
(o) $-3 - 4h = -1$

PROBLEM 2-23 Solve the following equations:

(a) $4z + 3z = 14$ (b) $5y - 2y = 6$ (c) $7x - x = 3$ (d) $5 = w - 6w$
(e) $-3 = v + 3v$ (f) $-3u = 4 - 2u$ (g) $4t = 3t - 7$ (h) $3 - 4s = 2 - 5s$
(i) $5 + 2r = 1 - 7r$ (j) $5q - 3 - 2q = 3q + 7 - 4q$ (k) $4p + 5 + 2p = 3p - 5 + 7p$
(l) $n - 3 + 4n - 3n = 8 - 2n + 5n$

PROBLEM 2-24 Solve the following equations:

(a) $2(3a - 1) = 4$ (b) $4(3 - 2y) = 5$ (c) $3x = 3(4x - 5)$ (d) $4w = 3(2w - 5)$
(e) $3(3 - 2v) = 2(3v - 5)$ (f) $4(2 - 3u) = 2(u + 4)$ (g) $2(3t - 1) = 3(1 + 2t)$
(h) $3(2s - 4) - 2(3s - 6) = 0$ (i) $4(2 - r) + 3(2r - 3) = 0$
(j) $3q + 2(3 - q) - 4(q + 3) = 0$

PROBLEM 2-25 Solve the following equations:

(a) $\dfrac{5}{7}a - 3 = 2$ (b) $\dfrac{3}{5}b - 1 = 3$ (c) $\dfrac{7}{4}c + 3 = 7$ (d) $\dfrac{2}{5}d + 3 = 7$ (e) $8 = 2 + \dfrac{3}{4}e$

(f) $\dfrac{1}{8} = \dfrac{1}{2} - \dfrac{7}{8}f$ (g) $\dfrac{1}{6} = \dfrac{1}{4}g - \dfrac{5}{6}$ (h) $\dfrac{1}{6} - \dfrac{1}{4}h = \dfrac{7}{12}$ (i) $\dfrac{2}{3}i + \dfrac{1}{2} = \dfrac{5}{6}i - \dfrac{2}{3}$

(j) $\dfrac{7}{6} - \dfrac{5}{2}j = \dfrac{2}{3}j + \dfrac{1}{2}$ (k) $\dfrac{4}{3}k + \dfrac{1}{6} = 2k + \dfrac{1}{2}$ (l) $\dfrac{3m + 2}{2} = \dfrac{1}{3} + \dfrac{m - 5}{6}$

(m) $\dfrac{5n - 2}{3} - \dfrac{3n - 2}{5} = \dfrac{2}{3}$ (n) $\dfrac{2}{5}(n - 2) - \dfrac{3}{4}(2n - 3) = \dfrac{1}{5}$ (o) $\dfrac{5}{4}(p - 3) = \dfrac{2}{3} + \dfrac{7}{6}(p - 2)$

PROBLEM 2-26 Solve the following equations:

(a) $1.7a = 0.34$ (b) $5.2b = 26$ (c) $c - 3.7 = 2.4$ (d) $-3.2 = 4.2 + d$
(e) $2.4e - 3.7 = 1.1$ (f) $4.1 - 1.7f = -1$ (g) $5.4(0.5g - 2) = 2.7$
(h) $3.2(1.4h - 2.3) = 1.6$ (i) $3.2i - 1.4 = 7.6 - 1.3i$ (j) $0.4j + 2.8 = 12.2 - 0.07j$

PROBLEM 2-27 Solve the following equations:

(a) $30\%a = 15$ (b) $30 = 15\%b$ (c) $12\frac{1}{2}\%c = 3$ (d) $16\frac{2}{3}\%d = 4$ (e) $e + 20\%e = 30$
(f) $f + 40\%f = 21$ (g) $8\%g - 3\%(210 - g) = 8$ (h) $25\%h - 10\%(12,000 - h) = 375$
(i) $6\%(60,000 - i) - 8\%i = 520$ (j) $3\%j + 2.4 = 8\%(j + 20)$
(k) $60\%k + 1.4 = 25\%(k + 14)$

PROBLEM 2-28 Solve the following equations:

(a) $|z - 1| = 3$ (b) $|y - 3| = 3$ (c) $|x + 2| = 2$ (d) $1 = |w + 3|$ (e) $5 = |2v - 3|$
(f) $8 = |3u + 2|$ (g) $|5 - 3t| = 2$ (h) $|1 - 5s| = 4$ (i) $2|i - 3| + 3 = 7$
(j) $3|j + 4| - 5 = 7$.

PROBLEM 2-29 Solve the following equations:

(a) $4x - 2y = 8$ for x (b) $4x - 2y = 8$ for y (c) $2x + 5y = 11$ for x
(d) $5x - 2y = 11$ for y (e) $ax + by = 3$ for x (f) $ax + by = 3$ for y
(g) $ax - by = c$ for x (h) $ax = by + c$ for y

PROBLEM 2-30 Solve the following equations:

(a) $ax + a = x$ for x (b) $ax + a = x$ for a (c) $ax + a = bx$ for x
(d) $4ay - 4a = y$ for y (e) $4pq + p = 2p + q$ for q (f) $4pq + p = 2p + q$ for p

(g) $p = 2m + 3n$ for m (h) $P = \dfrac{4rs}{3q}$ for s

PROBLEM 2-31 Solve the following equations:

(a) $a(x - 1) = x - 1$ for x (b) $a(x - 1) = 1 - ax$ for x (c) $a(x - 1) = a(1 - x)$ for x

(d) $a(x - b) = b(x - a)$ for x (e) $P = 2(m + n)$ for m (f) $mn - m(1 - n) = 1$ for n

(g) $mn - m(1 - 2n) = n$ for m (h) $g = h + (k - 1)d$ for k

PROBLEM 2-32 Solve the following equations:

(a) $\dfrac{1}{x} = \dfrac{1}{y} + \dfrac{1}{z}$ for y (b) $\dfrac{a}{x} = \dfrac{a}{y} - \dfrac{a}{z}$ for z (c) $\dfrac{a}{b} = \dfrac{c}{a} - \dfrac{b}{a}$ for c (d) $a = \dfrac{b - c}{b}$ for b

(e) $\dfrac{x - a}{b} = \dfrac{x + b}{a}$ for x (f) $\dfrac{y - a}{b} = \dfrac{b - y}{a}$ for y (g) $g = h\left(\dfrac{x + y}{x - z}\right)$ for y

(h) $g = h\left(\dfrac{x + y}{x - z}\right)$ for x (i) $g = h\left(\dfrac{x + y}{x - z}\right)$ for z

PROBLEM 2-33 Solve each formula for the indicated variable:

(a) *Percent formula:* $PB = A$ for B (b) *Volume formula:* $V = lwh$ for w

(c) *Interest formula:* $I = Prt$ for P (d) *Velocity formula:* $v = v_0 - 32t$ for t

(e) *Altitude/Temperature formula:* $a = 0.16(15 - t)$ for t

(f) *Area formula:* $A = \frac{1}{2}h(b_1 + b_2)$ for b_1

PROBLEM 2-34 Solve each number problem using a linear equation:

(a) The difference between two numbers is 14. The larger number is equal to 3 times the smaller number, less 4. Find the numbers.

(b) The sum of three numbers is 38. The second number is 2 more than three times the first number. The third number equals half the difference between the first two numbers. What are the numbers?

(c) The sum of three consecutive odd integers is 171. Find the integers.

(d) If the first of two consecutive integers is divided by 4 and the second consecutive integer is divided by 2, the sum of the quotients would be 5. What are the integers?

PROBLEM 2-35 Solve each geometry problem using a linear equation:

(a) The sum of three angles of a triangle is 180° [180 degrees]. One angle of a triangle is 4 times the second angle. The third angle of the triangle is 30° more than the second angle. Find the angles.

(b) Two angles of a triangle are equal. The third angle is equal to the sum of the other two angles. How many degrees are in each angle?

(c) The perimeter of a rectangle is 110 feet. The width is 5 feet shorter than twice the length. Find the area of the rectangle.

(d) The area of a rectangle is 240 m² [square meters]. The width of the rectangle is 15 m. Find the perimeter of the rectangle.

PROBLEM 2-36 Solve each currency value problem using a linear equation:

(a) There are $4.65 in dimes and quarters. There are 8 more dimes than quarters. How many dimes are there? What is the value of the quarters?

(b) A person changed $200 into five- and ten-dollar bills. The person then had 28 bills in all. How many ten-dollar bills were there? What was the value of the five-dollar bills?

(c) A stationery store sold 320 of two different types of pens for $190.55. One type of pen cost 25¢ and the other type cost 80¢. How many 80¢ pens were sold? What is the value of the 25¢ pens sold?

(d) The daily payroll for 100 skilled and unskilled laborers is $8320. Each unskilled laborer earns $64 per day. Each skilled laborer earns $112 per day. How many unskilled laborers are there? What is the value of the daily payroll for skilled laborers?

PROBLEM 2-37 Solve each investment problem using a linear equation:

(a) What amount of money invested at 12% will return $600 annually?

(b) After 1 year an 18% simple interest loan is paid off with $7080. What amount was originally borrowed?

(c) Part of an $8000 investment is at 5% per year, and the other part at 10%. The part invested at 10% earns $50 more than the part invested at 5%. What amount is invested at 10%? How much interest does the 5% investment earn per year?

(d) Part of a $6000 investment made a 12% profit. The other part made an 8% loss. The net loss was $200. How much was invested at 12%? How much was the 8% loss?

PROBLEM 2-38 Solve each uniform motion problem using a linear equation:

(a) Two cars pass each other traveling in opposite directions. One car averages 60 mph and the other averages 45 mph. How long will it take for the two cars to be 420 miles apart? How far will the faster car have traveled then?

(b) Two planes are 900 miles apart at noon and traveling towards each other. The constant rate of one plane is 166 mph and the other is 194 mph. What time will the planes pass each other? How far will the slower plane have traveled by then?

(c) Two planes leave the same airport 10 minutes apart traveling in the same direction. The uniform rate of the first plane is 258 km/h and the uniform rate of the other plane is 301 km/h. How long before the second plane overtakes the first plane? How far will they both have traveled by then?

(d) A car traveled at a constant rate for 2 hours. Road construction forced the car to travel at 30 km/h less for the next 30 minutes. The total distance driven was 185 km. What was the rate for each part of the trip? How far was each part of the trip?

(e) A person leaves home for another city at 9 am. The average rate going is 50 mph and returning home is 45 mph and 20 minutes longer over the same route. How far is the city driven to from home? How long did the return trip take?

(f) Two trains leave the same station 48 minutes apart traveling on parallel tracks in the same direction. Because the uniform rate of the second train is 10 mph faster than the first train, it takes 4 hours for the second train to catch up to the first train. Find the uniform rate of each train. How far did each train travel to the catch up point?

(g) Two people start from the same point and walk in opposite directions. One person walks $\frac{2}{3}$ mph faster than the other person. In 45 minutes they are 5 miles apart. How fast is each person walking? How much of the distance does each person cover?

(h) Two cars are 275 miles apart and traveling towards each other. Their average speeds differ by 10 mph. They will meet in $2\frac{1}{2}$ hours. What is the average speed for each car? How much of the distance will each car travel?

Answers to Supplementary Exercises

(2-20) **(a)** 9   **(b)** 6   **(c)** 1   **(d)** -3 **(e)** 2   **(f)** -4 **(g)** -5 **(h)** -5 **(i)** -4
      (j) -5 **(k)** 3

(2-21) **(a)** 2   **(b)** $\frac{7}{3}$ **(c)** -3 **(d)** 3   **(e)** $-\frac{12}{5}$ **(f)** 2   **(g)** $-\frac{1}{2}$ **(h)** 0   **(i)** $\frac{6}{5}$
      (j) $\frac{4}{9}$ **(k)** 6   **(l)** -4

(2-22) **(a)** 3   **(b)** 4   **(c)** $\frac{9}{4}$ **(d)** $-\frac{1}{2}$ **(e)** 3   **(f)** 3   **(g)** $\frac{11}{2}$ **(h)** $\frac{3}{4}$ **(i)** -1
      (j) $\frac{1}{5}$ **(k)** $\frac{2}{3}$ **(l)** $\frac{2}{5}$ **(m)** $-\frac{4}{3}$ **(n)** $\frac{1}{2}$ **(o)** $-\frac{1}{2}$

(2-23) **(a)** 2   **(b)** 2   **(c)** $\frac{1}{2}$ **(d)** -1 **(e)** $-\frac{3}{4}$ **(f)** -4 **(g)** -7 **(h)** -1 **(i)** $-\frac{4}{9}$
(j) $\frac{5}{2}$ **(k)** $\frac{5}{2}$ **(l)** -11

(2-24) **(a)** 1   **(b)** $\frac{7}{8}$ **(c)** $\frac{5}{3}$ **(d)** $\frac{15}{2}$ **(e)** $\frac{19}{12}$ **(f)** 0   **(g)** $\varnothing$ **(h)** $\{s \mid s \in R\}$
(i) $\frac{1}{2}$ **(j)** -2

(2-25) **(a)** 7   **(b)** $\frac{20}{3}$ **(c)** $\frac{16}{7}$ **(d)** 10   **(e)** 8   **(f)** $\frac{3}{7}$ **(g)** 4   **(h)** $-\frac{5}{3}$ **(i)** 7
(j) $\frac{4}{19}$ **(k)** $-\frac{1}{2}$ **(l)** $-\frac{9}{8}$ **(m)** $\frac{7}{8}$ **(n)** $\frac{25}{22}$ **(o)** 25

(2-26) **(a)** $\frac{1}{5}$ **(b)** 5   **(c)** 6.1   **(d)** -7.4 **(e)** 2   **(f)** 3   **(g)** 5   **(h)** 2   **(i)** 2
(j) 20

(2-27) **(a)** 50   **(b)** 200   **(c)** 24   **(d)** 24   **(e)** 25   **(f)** 15   **(g)** 130   **(h)** 4500
(i) 22,000   **(j)** 16   **(k)** 6

(2-28) **(a)** $-2, 4$ **(b)** $0, 6$ **(c)** $0, -4$ **(d)** $-2, -4$ **(e)** $-1, 4$ **(f)** $-\frac{10}{3}, 2$ **(g)** $1, \frac{7}{3}$
(h) $-\frac{3}{5}, 1$ **(i)** $1, 5$ **(j)** $0, -8$

(2-29) **(a)** $x = \frac{1}{2}y + 2$ **(b)** $y = 2x - 4$ **(c)** $x = -\frac{5}{2}y + \frac{11}{2}$ **(d)** $y = \frac{5}{2}x - \frac{11}{2}$

(e) $x = \dfrac{3 - by}{a}$ **(f)** $y = \dfrac{3 - ax}{b}$ **(g)** $x = \dfrac{by + c}{a}$ **(h)** $y = \dfrac{ax - c}{b}$

(2-30) **(a)** $x = \dfrac{-a}{a - 1}$ **(b)** $a = \dfrac{x}{x + 1}$ **(c)** $x = \dfrac{-a}{a - b}$ **(d)** $y = \dfrac{4a}{4a - 1}$

(e) $q = \dfrac{p}{4p - 1}$ **(f)** $p = \dfrac{q}{4q - 1}$ **(g)** $m = \dfrac{p - 3n}{2}$ **(h)** $s = \dfrac{3Pq}{4r}$

(2-31) **(a)** $x = 1$ **(b)** $x = \dfrac{a + 1}{2a}$ **(c)** $x = 1$ **(d)** $x = 0$ **(e)** $m = \dfrac{p - 2n}{2}$

(f) $n = \dfrac{m + 1}{2m}$ **(g)** $m = \dfrac{n}{3n - 1}$ **(h)** $k = \dfrac{g - h + d}{d}$

(2-32) **(a)** $y = \dfrac{xz}{z - x}$ **(b)** $z = \dfrac{-xy}{y - x}$ **(c)** $c = \dfrac{a^2 + b^2}{b}$ **(d)** $b = \dfrac{-c}{a - 1}$ **(e)** $x = \dfrac{a^2 + b^2}{a - b}$

(f) $y = \dfrac{a^2 + b^2}{a + b}$ **(g)** $y = \dfrac{gx - gz - hx}{h}$ **(h)** $x = \dfrac{gz + y}{g - 1}$ **(i)** $z = \dfrac{gx - hx - hy}{g}$

(2-33) **(a)** $B = \dfrac{A}{P}$ **(b)** $w = \dfrac{V}{lh}$ **(c)** $P = \dfrac{I}{rt}$ **(d)** $t = \dfrac{v - v_0}{-32}$ **(e)** $t = 15 - \dfrac{a}{0.16}$

(f) $b_1 = \dfrac{2A}{h} - b_2$

(2-34) **(a)** 23, 9   **(b)** 7, 23, 8   **(c)** 55, 57, 59   **(d)** 6, 7

(2-35) **(a)** $100°, 25°, 55°$ **(b)** $45°, 45°, 90°$ **(c)** $700 \, \text{ft}^2$ **(d)** 62 m

(2-36) **(a)** 19, $2.75   **(b)** 14($10 bills), $60   **(c)** 201, $29.75   **(d)** 60, $4880

(2-37) **(a)** $5000   **(b)** $6000   **(c)** $3000, $250   **(d)** $1400, $368

(2-38) **(a)** 4 h, 240 mi   **(b)** 2:30 PM, 415 mi   **(c)** 1 h, 301 mi
(d) 80 km/h, 50 km/h, 160 km, 25 km   **(e)** 150 mi, 3 h 20 min
(f) 50 mph, 60 mph, 240 mi   **(g)** 3 mph, $3\frac{2}{3}$ mph, $2\frac{1}{4}$ mi, $2\frac{3}{4}$ mi
(h) 50 mph, 60 mph, 125 mi, 150 mi

3 INEQUALITIES

THIS CHAPTER IS ABOUT

☑ Solving Inequalities Using the Principles
☑ Solving Inequalities Using the Distributive Property
☑ Solving Compound Inequalities
☑ Solving Absolute Value Inequalities

3-1. Solving Inequalities Using the Principles

In Section 1-1 we introduced the inequality symbols, and ordered relations between signed numbers. We can expand the idea of ordered relations to include algebraic expressions. When we establish a relationship of inequality between two algebraic expressions, we call the resulting mathematical statement an **inequality.** Each inequality consists of a **left member,** an **order relation** (inequality) **symbol,** and a **right member.** The following example shows the different relations of inequality.

EXAMPLE 3-1: Which of the following expressions are inequalities?

(a) $3x - 4 > 5 - 2x$ (b) $2y - 5 = 4 - 3y$
(c) $5z + 3 \geq 0$ (d) $3 - 2w \leq 4w + 3$
(e) $7 - 4v < 3v + 5$

Solution: (a) This is an inequality because the order relation is a "greater than" symbol.
           (b) This is not an inequality because the order relation is an equality symbol.
           (c) This is an inequality because the order relation is a "greater than or equal to" symbol.
           (d) This is an inequality because the order relation is a "less than or equal to" symbol.
           (e) This is an inequality because the order relation is a "less than" symbol.

An inequality in one variable is any expression that can be written in the form $Ax + B > C$, where A, B, and C are real numbers, and $A \neq 0$.

Note: Wherever the "greater than" symbol is used in a definition, it should be understood that the definition holds true for the other inequality symbols as well.

The process of solving inequalities is the same as that for equations—we isolate the variable in one member. A number is a solution of an inequality if the statement obtained by replacing the variable with that number is true. The set of all solutions of an inequality is called the **solution set** of that inequality. Two inequalities that have the same solution set are called **equivalent inequalities.**

EXAMPLE 3-2: (a) Is 2 a solution of $3x - 5 > 3 - 2x$?
               (b) Is -3 a solution of $4y + 1 \geq 2 - 3y$?

Solution: (a) $\underline{\quad 3x - 5 > 3 - 2x \quad}$ ⟵ original inequality

$$\begin{array}{c|c} 3(2) - 5 & 3 - 2(2) \\ 6 - 5 & 3 - 4 \\ 1 & -1 \end{array}$$ Replace x with 2.

              $x = 2$ checks

Since $1 > -1$, 2 is a solution of $3x - 5 > 3 - 2x$.

(b)

$$4y + 1 \geq 2 - 3y \longleftarrow \text{original inequality}$$

$4(-3) + 1$	$2 - 3(-3)$	Replace y with -3.
$-12 + 1$	$2 + 9$	
-11	$11 \longleftarrow$	$y = -3$ does not check

Since $-11 \not\geq 11$, -3 is not a solution of $4y + 1 \geq 2 - 3y$.

A. Solving inequalities using the addition principle

We can use the **addition principle** to **solve inequalities,** in a manner similar to the way we used it to solve equations. To solve an inequality in the form $x + B = C$, we isolate the variable x in one member and the constants B and C in the other, by adding the opposite of B to both members.

Addition Principle for Inequalities

If $a > b$, then $a + c > b + c$, or
if $a < b$, then $a + c < b + c$, for all real numbers a, b, and c.

EXAMPLE 3-3: Solve the following inequalities: **(a)** $x - 5 > 3$ **(b)** $4 < y + 3$ **(c)** $z - 3 \leq 1$

Solution: **(a)** $x - 5 > 3$

$\qquad x - 5 + 5 > 3 + 5$ Add the opposite of -5 to both members to isolate the variable x in one member.

$\qquad\qquad x + 0 > 8$

$\qquad\qquad\quad x > 8$

The solution set of $x - 5 > 3$ is $\{x \mid x > 8\}$.

Note 1: The solution set for the inequality in **(a)** is written in the form $\{x \mid x > 8\}$ and is read "The set of all x such that x is greater than 8." All solution sets for inequalities should be written in this form.

(b) $4 < y + 3$

$\qquad 4 + (-3) < y + 3 + (-3)$ Add the opposite of 3 to both members to isolate y.

$\qquad\qquad 4 - 3 < y + 3 - 3$

$\qquad\qquad\qquad 1 < y + 0$

$\qquad\qquad\qquad\quad 1 < y$

$\qquad\qquad 1 < y \text{ or } y > 1$

The solution set of $4 < y + 3$ is $\{y \mid y > 1\}$.

(c) $z - 3 \leq 1$

$\qquad z - 3 + 3 \leq 1 + 3$ Add the opposite of -3 to both members to isolate z.

$\qquad\qquad z + 0 \leq 4$

$\qquad\qquad\quad z \leq 4$

The solution set of $z - 3 \leq 1$ is $\{z \mid z \leq 4\}$.

Note 2: We can apply the addition principle for inequalities by adding the opposite of the numerical addend to both members, or by simply subtracting the numerical addend from both members.

B. Solving inequalities using the multiplication principle

To solve an inequality in the form $Ax > C$, where $A > 0$, we use the **multiplication principle for inequalities** to multiply both members of the inequality by the reciprocal of A, to get $x > C/A$. If $A < 0$ (i.e., if A is negative) we still multiply both members by the reciprocal of A, but at the same time we must reverse the inequality symbol, so that the solution is $x < C/A$. This reversal of the inequality sign is a crucial difference between use of the multiplication principle for equations and for inequalities.

Caution: Whenever we multiply or divide an equality by a negative number, we MUST REVERSE THE INEQUALITY SYMBOL.

Multiplication Principle for Inequalities

For all real numbers a, b, and c:

If $a > b$ and $c > 0$, then $ac > bc$.   If $a > b$ and $c < 0$, then $ac < bc$.

If $a < b$ and $c > 0$, then $ac < bc$.   If $a < b$ and $c < 0$, then $ac > bc$.

EXAMPLE 3-4: Multiply the inequality $3 > -4$ by (a) 5 and (b) -2.

Solution: (a) $3 > -4$ is true and $5(3) > 5(-4)$ is true because $15 > -20$.

(b) $3 > -4$ is true and $-2(3) < -2(-4)$ is true because $-6 < 8$.

Note: $3 > -4$ is true but, $-2(3) > -2(-4)$ is false because $-6 \not> 8$.

EXAMPLE 3-5: Solve (a) $\frac{3}{2}x > 9$, (b) $-4 < 8y$, and (c) $-6z < -2$.

Solution: (a) $\frac{3}{2}x > 9$

$\frac{2}{3} \cdot \frac{3}{2}x > \frac{2}{3} \cdot 9$ Multiply both members of the inequality by the reciprocal of $\frac{3}{2}$ to isolate the variable x in one member.

$1x > 6$

$x > 6$

The solution of $\frac{3}{2}x > 9$ is $\{x \mid x > 6\}$.

(b) $-4 \geq 8y$

$\frac{1}{8}(-4) \geq \frac{1}{8}(8y)$ Multiply both members of the inequality by the reciprocal of 8 to isolate y.

$\frac{-4}{8} \geq \frac{8y}{8}$

$-\frac{1}{2} \geq 1y$

$-\frac{1}{2} \geq y$

The solution set of $-4 \geq 8y$ is $\{y \mid y \leq -\frac{1}{2}\}$.

(c) $-6z < -2$

$\frac{-6z}{-6} > \frac{-2}{-6}$ Divide both members of the inequality by -6 (or multiply by the reciprocal of -6) and reverse the inequality symbol to isolate z.

$z > \frac{1}{3}$

The solution set of $-6z < -2$ is $\{z \mid z > \frac{1}{3}\}$.

Note: We can apply the multiplication principle for inequalities by multiplying both members of the inequality by the reciprocal of the coefficient of the variable, or by simply dividing both members by the coefficient of the variable.

C. Solving inequalities using the addition and multiplication principles together

To **solve an inequality of the form** $Ax + B > C$, we isolate the variable term Ax in one member using the addition principle for inequalities, then isolate the variable x using the multiplication principle for inequalities. To do this, we add the opposite of B to both members, then multiply by the reciprocal of A. (The process for solving inequalities is the same as that for equations, with the exception of reversing the inequality symbol when multiplying or dividing both members by a negative number.)

EXAMPLE 3-6: Solve (a) $2x - 5 > 3$ and (b) $4 - 5y \leq -11$.

Solution: (a)

$$2x - 5 > 3$$

$$2x - 5 + 5 > 3 + 5 \qquad \text{Add the opposite of } -5 \text{ to both members to}$$
$$\qquad\qquad\qquad\qquad\qquad \text{isolate the variable term } 2x \text{ in one member.}$$
$$2x > 8$$

$$\frac{2x}{2} > \frac{8}{2} \qquad \text{Multiply both members by the reciprocal of 2}$$
$$\qquad\qquad\qquad \text{(or just divide both members by 2) to isolate } x.$$

$$x > 4$$

The solution set of $2x - 5 > 3$ is $\{x \mid x > 4\}$.

(b)

$$4 - 5y \leq -11$$

$$4 - 4 - 5y \leq -11 - 4 \qquad \text{Add the opposite of 4 to both members}$$
$$\qquad\qquad\qquad\qquad\qquad \text{to isolate } -5y \text{ in one member.}$$
$$-5y \leq -15$$

$$\frac{-5y}{-5} \geq \frac{-15}{-5} \qquad \text{Divide both members by } -5 \text{ and reverse}$$
$$\qquad\qquad\qquad\quad \text{the inequality symbol to isolate } y.$$

$$y \geq 3$$

The solution set of $4 - 5y \leq -11$ is $\{y \mid y \geq 3\}$.

D. Solving inequalities containing like terms

To **solve an inequality containing like terms**, we collect and combine like terms, then solve as before.

EXAMPLE 3-7: Solve the following inequalities:

(a) $5x - 4 + x < 2$ (b) $4y - 7 \geq 2 - 5y$ (c) $4 - 3z \geq 2z - 7 - 5z$

Solution: (a) $5x - 4 + x < 2$

$$6x - 4 < 2 \qquad \text{Combine like terms.}$$

$$6x - 4 + 4 < 2 + 4 \qquad \text{Isolate the variable term } 4x \text{ using the}$$
$$\qquad\qquad\qquad\qquad \text{addition principle for inequalities.}$$
$$6x < 6$$

$$\frac{6x}{6} < \frac{6}{6} \qquad \text{Isolate the variable } x \text{ using the multiplication}$$
$$\qquad\qquad\qquad \text{principle for inequalities.}$$

$$x < 1$$

The solution set of $5x - 4 + x < 2$ is $\{x \mid x < 1\}$.

(b) $4y - 7 \geq 2 - 5y$

$4y - 4y - 7 \geq 2 - 5y - 4y$ Collect like terms using the addition principle for inequalities.

$-7 \geq 2 - 9y$ Combine like terms

$-7 - 2 \geq 2 - 2 - 9y$ Isolate the variable term $9y$ using the addition principle for inequalities.

$-9 \geq -9y$

$\dfrac{-9}{-9} \leq \dfrac{-9y}{-9}$ Divide by -9, reversing the inequality symbol.

$1 \leq y$

The solution set of $4y - 7 \geq 2 - 5y$ is $\{y|y \geq 1\}$.

(c) $4 - 3z \geq 2z - 7 - 5z$

$4 - 3z \geq -3z - 7$

$4 \geq -7$ ⟵ true statement

The solution set of $4 - 3z \geq 2z - 7 - 5z$ is $\{z|z \in R\}$.

Note: A true statement (like $4 \geq -7$) means that all real numbers are solutions of the original inequality, and the solution set is therefore the set of all real numbers. If the derived equivalent inequality was a false statement (like $3 < -1$), then there would be no solution, and the solution set of the original inequality would be the empty set.

3-2. Solving Inequalities Using the Distributive Property

Just as was necessary with equations, it is often necessary to simplify inequalities before solving them. In many cases you can use the **distributive property** to simplify an inequality.

A. Solving inequalities containing parentheses

To solve an inequality containing parentheses, we clear the parentheses using the distributive property and solve as before.

EXAMPLE 3-8: Solve the following inequalities using the distributive property:

(a) $3(2x - 5) > 2(3 - 2x)$ **(b)** $-(3 - 4y) \leq 2(2y - 3)$

Solution: **(a)** $3(2x - 5) > 2(3 - 2x)$

$6x - 15 > 6 - 4x$ Clear parentheses using the distributive property.

$6x + 4x > 6 + 15$ Collect like terms.

$10x > 21$ Combine like terms.

$x > \dfrac{21}{10}$ Isolate the variable by dividing by 10.

The solution set of $3(2x - 5) > 2(3 - 2x)$ is $\{x|x > \frac{21}{10}\}$.

(b) $-(3 - 4y) \leq 2(2y - 3)$

$-3 + 4y \leq 4y - 6$ Clear parentheses using the distributive property.

$-3 + 4y - 4y \leq 4y - 4y - 6$ Collect like terms.

$-3 \leq -6$ ⟵ false statement

Note: The solution set of $-(3 - 4y) \leq 2(2y - 3)$ is $\emptyset$ (the empty set) because $-3 \leq -6$ is a false statement. So the original inequality has no solution.

Recall: If the original inequality simplifies to a true statement, its solution set is the set of all real numbers.

B. Solving inequalities containing fractions

To **solve an inequality containing fractions,** we clear the fractions first. To clear fractions in an inequality, we multiply both members by the least common denominator (LCD) of the terms, then solve as before.

EXAMPLE 3-9: Solve **(a)** $\frac{1}{2}x - \frac{2}{3} \leq \frac{5}{6}$ and **(b)** $\frac{2}{3}(2y - 5) > \frac{5}{6}(3 - 4y)$.

Solution: **(a)** $\frac{1}{2}x - \frac{2}{3} \leq \frac{5}{6}$

The LCD is 6.	Find the least common denominator (LCD).
$6(\frac{1}{2}x - \frac{2}{3}) \leq 6 \cdot \frac{5}{6}$	Multiply both members by the LCD.
$3x - 4 \leq 5$	Simplify using the distributive property.
$3x \leq 9$	Add 4 to both members.
$x \leq 3$	Divide both members by 3.

The solution set of $\frac{1}{2}x - \frac{2}{3} \leq \frac{5}{6}$ is $\{x \mid x \leq 3\}$.

(b) $\frac{2}{3}(2y - 5) > \frac{5}{6}(3 - 4y)$

The LCD is 6.	Find the LCD of $\frac{2}{3}$ and $\frac{5}{6}$.
$6 \cdot \frac{2}{3}(2y - 5) > 6 \cdot \frac{5}{6}(3 - 4y)$	Multiply both members by 6.
$4(2y - 5) > 5(3 - 4y)$	
$8y - 20 > 15 - 20y$	Clear parentheses using the distributive property.
$-35 > -28y$	Collect and combine like terms.
$\frac{5}{4} < y$	Divide by -28 and reverse the inequality symbol.

The solution set of $\frac{2}{3}(2y - 5) > \frac{5}{6}(3 - 4y)$ is $\{y \mid y > \frac{5}{4}\}$.

3-3. Solving Compound Inequalities

A **compound inequality** is composed of two or more inequalities. We **solve compound inequalities** by using the addition and multiplication principles for inequalities to find a solution set for each simple inequality. We then use the set operations of union and intersection to find the solution set for the compound inequality.

A. Graphing compound inequalities

To **graph a compound inequality,** we graph each simple inequality and form the union or the intersection of the solution sets of the simple inequalities. Whenever *or* is used in a compound inequality, we form the union. Whenever *and* is used in a compound inequality, we form the intersection.

Note: An inequality in the form of $A < x < B$ may also be written in the form $A < x$ *and* $x < B$.

EXAMPLE 3-10: Graph the following compound inequalities:

(a) $x > 2$ *or* $x < -2$ **(b)** $y \geq 2$ *and* $y < 4$ **(c)** $-2 < z \leq 1$

Solution: (a) $x > 2$ *or* $x < -2$

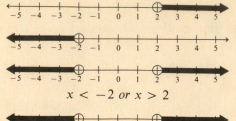

Graph $x > 2$ with a hollow circle on 2.

Graph $x < -2$ with a hollow circle on -2.

Graph the union of the two solution sets.

$x < -2$ *or* $x > 2$

The graph of $x > 2$ *or* $x < -2$.

(b) $y \geq 2$ *and* $y < 4$

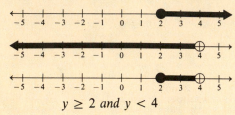

Graph $y \geq 2$ with a solid circle on 2.

Graph $y < 4$ with a hollow circle on 4.

Graph the intersection of the two solution sets.

$y \geq 2$ *and* $y < 4$

The graph of $y \geq 2$ *and* $y < 4$.

(c) $-2 < z \leq 1$

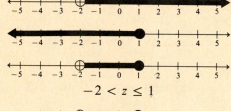

Graph $-2 < z$.

Graph $z \leq 1$.

Graph the intersection of the two solution sets.

$-2 < z \leq 1$

The graph of $-2 < z \leq 1$.

B. Solving compound-*or* inequalities

A **compound-*or* inequality** is a statement in which either the left *or* the right inequality is true, or both inequalities are true. You **solve a compound-*or* inequality** by solving both simple inequalities and forming the union of their solution sets.

EXAMPLE 3-11: Solve (a) $3x - 5 < 0$ *or* $3x - 5 \geq 3$, (b) $2y + 3 < 4$ *or* $2y + 3 > 0$.

Solution: (a) $3x - 5 < 0$ *or* $3x - 5 \geq 3$

$\qquad 3x < 5$ *or* $\qquad 3x \geq 8$ $\qquad$ Isolate the variable term $3x$.

$\qquad x < \frac{5}{3}$ *or* $\qquad x \geq \frac{8}{3}$ $\qquad$ Isolate the variable x.

$\{x \mid x < \frac{5}{3}\} \cup \{x \mid x \geq \frac{8}{3}\}$ $\qquad$ Form the union of the solution sets.

The solution set of $3x - 5 < 0$ or $3x - 5 \geq 3$ is $\{x \mid x < \frac{5}{3}$ or $x \geq \frac{8}{3}\}$.

$\qquad$ (b) $2y + 3 < 4$ *or* $2y + 3 > 0$

$\qquad 2y < 1$ *or* $\qquad 2y > -3$ $\qquad$ Isolate the variable term $2y$.

$\qquad y < \frac{1}{2}$ *or* $\qquad y > \frac{-3}{2}$ $\qquad$ Isolate the variable y.

$\{y \mid y < \frac{1}{2}\} \cup \{y \mid y > -\frac{3}{2}\}$ $\qquad$ Form the union of the solution sets.

The solution set of $3y - 5 < 0$ or $3y - 5 > 3$ is $\{y \mid y \in R\}$.

C. Solving compound-*and* inequalities

A **compound-*and* inequality** is a statement in which both the left inequality and the right inequality are true. We **solve a compound-*and* inequality** by solving both simple inequalities, then forming the intersection of their solution sets.

EXAMPLE 3-12: Solve (a) $3x - 2 < 4$ *and* $3x - 2 > -1$, (b) $-3 \leq 2 - 5y < 12$.

Solution: (a) $3x - 2 < 4$ *and* $3x - 2 > -1$

$\qquad\qquad 3x < 6$ *and* $\quad 3x > 1$ $\qquad\qquad$ Isolate the variable term $3x$.

$\qquad\qquad x < 2$ *and* $\quad x > \frac{1}{3}$ $\qquad\qquad$ Isolate the variable x.

$\qquad\qquad \{x \mid x < 2\} \cap \{x \mid x > \frac{1}{3}\}$ $\qquad\qquad$ Form the intersection of the solution sets.

The solution set of $3x - 2 < 4$ *and* $3x - 2 > -1$ is $\{x \mid \frac{1}{3} < x < 2\}$.

$\qquad$ (b) $-3 \leq 2 - 5y < 12$

$\qquad\qquad -5 \leq -5y < 10$ $\qquad\qquad$ Isolate the variable term by adding -2 to all three members.

$\qquad\qquad 1 \geq y > -2$ $\qquad\qquad$ Isolate the variable by dividing by -5 and reversing the inequality symbols.

The solution set of $-3 \leq 2 - 5y < 12$ is $\{y \mid -2 < y \leq 1\}$.

3-4. Solving Absolute Value Inequalities

You have already learned that $|x| = 3$ has two solutions: -3 and 3. On the number line, $|x| = 3$ represents the distance that x or $-x$ is from zero: 3 units.

The absolute value inequalities $|x| > 3$ and $|x| < 3$ also represent distances on the number line. $|x| > 3$ represents all numbers whose distance from zero is *greater* than 3 units. $|x| < 3$ represents all numbers whose distance from zero is *less* than 3 units.

A. Solving absolute value inequalities in the form $|E| > r$

E represents an expression such as $Ax + B$. To solve an absolute value inequality in the form $|E| > r$, where $r > 0$, you can write it in the form of a compound-*or* inequality:

$$E < -r \text{ or } E > r$$

Now that the absolute value inequality is in the form of a compound-*or* inequality, it can be solved as before.

Caution: You can only use the compound-*or* inequality as an equivalent expression when the absolute value in the inequality is **greater than,** or **greater than or equal to,** the number to which it is being compared.

EXAMPLE 3-13: Solve the following absolute value inequalities: (a) $|2x - 3| > 5$ $\qquad$ (b) $3 \leq |2 - 3y|$

Solution: (a) $|2x - 3| > 5$

$\qquad\qquad 2x - 3 > 5$ *or* $2x - 3 < -5$ $\qquad\qquad$ Write as a compound-*or* inequality.

$\qquad\qquad 2x > 8$ *or* $\quad 2x < -2$ $\qquad\qquad$ Solve each inequality.

$\qquad\qquad x > 4$ *or* $\quad x < -1$ $\qquad\qquad$ Simplify.

The solution set of $|2x - 3| > 5$ is $\{x \mid x < -1 \text{ or } x > 4\}$.

(b) $3 \le |2 - 3y|$

$2 - 3y \ge 3$ *or* $2 - 3y \le -3$	Write as a compound-*or* inequality.
$-3y \ge 1$ *or* $-3y \le -5$	Solve each inequality.
$y \le -\frac{1}{3}$ *or* $y \ge \frac{5}{3}$	Isolate y in each inequality.

The solution set of $3 \le |2 - 3y|$ is $\{y \mid y \le -\frac{1}{3}$ or $y \ge \frac{5}{3}\}$.

B. Solving absolute value inequalities in the form $|E| < r$

To solve an absolute value inequality in the form $|E| < r$, where $r > 0$, you can write it in the form of a compound-*and* inequality:

$$-r < E < r$$

Now that the absolute value inequality is in the form of a compound-*and* inequality, it can be solved as before.

Caution: You can only use the compound-*and* inequality as an equivalent expression when the absolute value expression in the inequality is **less than,** or **less than or equal to,** the number to which it is being compared.

EXAMPLE 3-14: Solve the following absolute value inequalities: **(a)** $|2x + 5| < 3$ **(b)** $|3 - 2y| \le 1$

Solution: **(a)** $|2x + 5| < 3$

$-3 < 2x + 5 < 3$	Write as a compound-*and* inequality.
$-8 < 2x < -2$	Add -5 to all three members.
$-4 < x < -1$	Divide all three members by 2.

The solution set of $|2x + 5| < 3$ is $\{x \mid -4 < x < -1\}$.

(b) $|3 - 2y| \le 1$

$-1 \le 3 - 2y \le 1$	Write as a compound-*and* inequality.
$-4 \le -2y \le -2$	Add -3 to all three members.
$2 \ge y \ge 1$	Divide all three members by -2 and reverse the inequality symbols.

The solution set of $|3 - 2y| \le 1$ is $\{y \mid 1 \le y \le 2\}$.

RAISE YOUR GRADES

Can you . . . ?

☑ determine if a given number is a solution of an inequality
☑ solve inequalities using the addition principle
☑ solve inequalities using the multiplication principle
☑ solve inequalities using the addition and multiplication principles together
☑ solve inequalities containing like terms
☑ solve inequalities containing parentheses
☑ solve inequalities containing fractions
☑ graph compound inequalities
☑ solve compound-*or* inequalities
☑ solve compound-*and* inequalities
☑ solve absolute value inequalities of the form $|E| > r$ and $|E| \ge r$, where $r > 0$
☑ solve absolute value inequalities of the form $|E| < r$ and $|E| \le r$, where $r > 0$

SOLVED PROBLEMS

PROBLEM 3-1 For which of the following is 3 a solution?

(a) $3x + 4 > 7$ **(b)** $2y - 3 > 12$ **(c)** $0 < 4 - 3z$ **(d)** $4 \geq 2 - 3w$ **(e)** $-2 \leq 4v - 3$
(f) $-3 > 3t - 12$

Solution: Recall that to determine if a number is a solution of an inequality, you replace the variable with that number. If the statement obtained is true, the number is a solution. If the statement obtained is false, the number is not a solution [see Example 3-2].

(a)

$3x + 4 > 7$	
$3 \cdot 3 + 4$	7
$9 + 4$	7
13	7

$13 > 7 \longleftarrow$ true

3 is a solution.

(b)

$2y - 3 > 12$	
$2 \cdot 3 - 3$	12
$6 - 3$	12
3	12

$3 > 12 \longleftarrow$ false

3 is not a solution.

(c)

$0 < 4 - 3z$	
0	$4 - 3 \cdot 3$
0	$4 - 9$
0	-5

$0 < -5 \longleftarrow$ false

3 is not a solution.

(d)

$4 \geq 2 - 3w$	
4	$2 - 3 \cdot 3$
4	$2 - 9$
4	-7

$4 \geq -7 \longleftarrow$ true

3 is a solution.

(e)

$-2 \leq 4v - 3$	
-2	$4 \cdot 3 - 3$
-2	$12 - 3$
-2	9

$-2 \leq 9 \longleftarrow$ true

3 is a solution.

(f)

$-3 > 3t - 12$	
-3	$3 \cdot 3 - 12$
-3	$9 - 12$
-3	-3

$-3 > -3 \longleftarrow$ false

3 is not a solution.

PROBLEM 3-2 Solve the following inequalities:

(a) $x - 6 > 3$ **(b)** $4 < y - 2$ **(c)** $6 < z + 4$ **(d)** $-4 > w - 3$ **(e)** $v + 4 < 1$
(f) $u - 3 \geq 2$ **(g)** $1 \leq t + 3$ **(h)** $s - 5 \leq -3$ **(i)** $-2 \geq r + 7$

Solution: Recall that to solve an inequality in the form $x + B > C$, you add the opposite of B to both members using the addition principle for inequalities [see Example 3-3].

(a)
$$x - 6 > 3$$
$$x - 6 + 6 > 3 + 6$$
$$x + 0 > 9$$
$$x > 9$$
$$\{x \mid x > 9\}$$

(b)
$$4 < y - 2$$
$$4 + 2 < y - 2 + 2$$
$$6 < y + 0$$
$$6 < y$$
$$\{y \mid y > 6\}$$

(c)
$$6 < z + 4$$
$$6 - 4 < z + 4 - 4$$
$$2 < z + 0$$
$$2 < z$$
$$\{z \mid z > 2\}$$

(d)
$$-4 > w - 3$$
$$-4 + 3 > w - 3 + 3$$
$$-1 > w + 0$$
$$-1 > w$$
$$\{w \mid w < -1\}$$

(e)
$$v + 4 < 1$$
$$v + 4 - 4 < 1 - 4$$
$$v + 0 < -3$$
$$v < -3$$
$$\{v \mid v < -3\}$$

(f)
$$u - 3 \geq 2$$
$$u - 3 + 3 \geq 2 + 3$$
$$u + 0 \geq 5$$
$$u \geq 5$$
$$\{u \mid u \geq 5\}$$

(g)
$$1 \leq t + 3$$
$$1 - 3 \leq t + 3 - 3$$
$$-2 \leq t + 0$$
$$-2 \leq t$$
$$\{t \mid t \geq -2\}$$

(h)
$$s - 5 \leq -3$$
$$s - 5 + 5 \leq -3 + 5$$
$$s + 0 \leq 2$$
$$s \leq 2$$
$$\{s \mid s \leq 2\}$$

(i)
$$-2 \geq r + 7$$
$$-2 - 7 \geq r + 7 - 7$$
$$-9 \geq r + 0$$
$$-9 \geq r$$
$$\{r \mid r \leq -9\}$$

PROBLEM 3-3 Solve the following inequalities:

(a) $\frac{2}{3}z > 6$ (b) $\frac{4}{5}y < -4$ (c) $-\frac{3}{5}x \le 6$ (d) $-\frac{4}{3}w \ge -6$ (e) $-w > -3$

(f) $5u < 5$ (g) $-6 \le -15s$ (h) $\frac{-8q}{3} > 16$ (i) $-\frac{6}{5} \ge -\frac{3}{5}p$

Solution: Recall that to solve an inequality of the form $Ax > C$, where $A > 0$, you multiply both members of the inequality by the reciprocal of A. To solve an inequality of the form $Ax > C$, where $A < 0$, you multiply both members of the inequality by the reciprocal of A and reverse the inequality symbol [see Example 3-5].

(a) $\frac{2}{3}z > 6$

$$\frac{3}{2} \cdot \frac{2}{3}z > \frac{3}{2} \cdot 6$$

$$1z > 9$$

$$z > 9$$

$$\{z \mid z > 9\}$$

(b) $\frac{4}{5}y < -4$

$$\frac{5}{4} \cdot \frac{4}{5}y < \frac{5}{4}(-4)$$

$$1y < \frac{-20}{4}$$

$$y < -5$$

$$\{y \mid y < -5\}$$

(c) $-\frac{3}{5}x \le 6$

$$-\frac{5}{3}\left(-\frac{3}{5}x\right) \ge -\frac{5}{3} \cdot 6$$

$$1x \ge -10$$

$$x \ge -10$$

$$\{x \mid x \ge -10\}$$

(d) $-\frac{4}{3}w \ge -6$

$$-\frac{3}{4}\left(-\frac{4}{3}w\right) \le -\frac{3}{4}(-6)$$

$$1w \le \frac{9}{2}$$

$$w \le \frac{9}{2}$$

$$\left\{w \mid w \le \frac{9}{2}\right\}$$

(e) $-w > -3$

$$\frac{-w}{-1} < \frac{-3}{-1}$$

$$1w < 3$$

$$w < 3$$

$$\{w \mid w < 3\}$$

(f) $5u < 5$

$$\frac{5u}{5} < \frac{5}{5}$$

$$1u < 1$$

$$u < 1$$

$$\{u \mid u < 1\}$$

(g) $-6 \le -15s$

$$\frac{-6}{-15} \ge \frac{-15s}{-15}$$

$$\frac{2}{5} \ge s$$

$$\left\{s \mid s \le \frac{2}{5}\right\}$$

(h) $\frac{-8q}{3} > 16$

$$-\frac{3}{8}\left(-\frac{8}{3}q\right) < -\frac{3}{8} \cdot 6$$

$$q < -6$$

$$\{q \mid q < -6\}$$

(i) $-\frac{6}{5} \ge -\frac{3}{5}p$

$$-\frac{5}{3}\left(-\frac{6}{5}\right) \le -\frac{5}{3}\left(-\frac{3}{5}p\right)$$

$$2 \le p$$

$$\{p \mid p \ge 2\}$$

PROBLEM 3-4 Solve the following inequalities:

(a) $3z - 2 > 7$ (b) $2y - 4 \le 2$ (c) $-3 \le 4x + 3$ (d) $-5 > 3w - 7$ (e) $3 \le 4 - 2v$
(f) $-3 \le 5 - 4u$ (g) $-2 \ge 3r + 4$ (h) $-4 \ge 5 - 3q$ (i) $4 > 4 - 3p$

Solution: Recall that to solve an inequality of the form $Ax + B > C$, you add the opposite of B to both members, then multiply both members by the reciprocal of A [see Example 3-6].

(a) $\quad 3z - 2 > 7$

$3z - 2 + 2 > 7 + 2$

$\quad\quad 3z > 9$

$\quad\quad \dfrac{3z}{3} > \dfrac{9}{3}$

$\quad\quad z > 3$

$\{z \mid z > 3\}$

(b) $\quad 2y - 4 \le 2$

$2y - 4 + 4 \le 2 + 4$

$\quad\quad 2y \le 6$

$\quad\quad \dfrac{2y}{2} \le \dfrac{6}{2}$

$\quad\quad y \le 3$

$\{y \mid y \le 3\}$

(c) $\quad -3 \le 4x + 3$

$-3 - 3 \le 4x + 3 - 3$

$\quad\quad -6 \le 4x$

$\quad\quad \dfrac{-6}{4} \le \dfrac{4x}{4}$

$\quad\quad -\dfrac{3}{2} \le x$

$\left\{x \mid x \ge -\dfrac{3}{2}\right\}$

(d) $\quad -5 > 3w - 7$

$-5 + 7 > 3w - 7 + 7$

$\quad\quad 2 > 3w$

$\quad\quad \dfrac{2}{3} > \dfrac{3w}{3}$

$\quad\quad \dfrac{2}{3} > w$

$\left\{w \mid w < \dfrac{2}{3}\right\}$

(e) $\quad 3 \le 4 - 2v$

$3 - 4 \le 4 - 4 - 2v$

$\quad\quad -1 \le -2v$

$\quad\quad \dfrac{-1}{-2} \ge \dfrac{-2v}{-2}$

$\quad\quad \dfrac{1}{2} \ge v$

$\left\{v \mid v \le \dfrac{1}{2}\right\}$

(f) $\quad -3 \le 5 - 4u$

$-3 - 5 \le 5 - 5 - 4u$

$\quad\quad -8 \le -4u$

$\quad\quad \dfrac{-8}{-4} \ge \dfrac{-4u}{-4}$

$\quad\quad 2 \ge u$

$\{u \mid u \le 2\}$

(g) $\quad -2 \ge 3r + 4$

$\quad -6 \ge 3r$

$\quad -2 \ge r$

$\{r \mid r \le -2\}$

(h) $\quad -4 \ge 5 - 3q$

$\quad -9 \ge -3q$

$\quad 3 \le q$

$\{q \mid q \ge 3\}$

(i) $\quad 4 > 4 - 3p$

$\quad 0 > -3p$

$\quad 0 < p$

$\{p \mid p > 0\}$

PROBLEM 3-5 Solve the following inequalities:

(a) $3z + 5z > 16$ (b) $4y - y \le 12$ (c) $2x - 5x < 6$ (d) $2 \le 3w - 7w$
(e) $3v + 5 - v \ge 2v + 6$ (f) $6u - 3 + 2u \le 2 + 7u$ (g) $4t - 10 < 3t - 8 + 2t$
(h) $2s + 6 - 6s \ge 18$ (i) $5p - 10 < 9p - 1 - 4p$

Solution: Recall that to solve an inequality which contains like terms, you collect and combine like terms, then solve as before [see Example 3-7].

(a) $3z + 5z > 16$

$\quad 8z > 16$

$\quad z > 2$

$\{z \mid z > 2\}$

(b) $4y - y \le 12$

$\quad 3y \le 12$

$\quad y \le 4$

$\{y \mid y \le 4\}$

(c) $2x - 5x < 6$

$\quad -3x < 6$

$\quad x > -2$

$\{x \mid x > -2\}$

(d) $\quad 2 \le 3w - 7w$

$\quad 2 \le -4w$

$\quad -\tfrac{1}{2} \ge w$

$\{w \mid w \le -\tfrac{1}{2}\}$

(e) $\quad 3v + 5 - v \ge 2v + 6$

$\quad 2v + 5 \ge 2v + 6$

$\quad 2v - 2v + 5 \ge 2v - 2v + 6$

$\quad 5 \ge 6$

No solution, or $\varnothing$.

(f) $6u - 3 + 2u \le 2 + 7u$

$\quad 8u - 3 \le 2 + 7u$

$\quad u - 3 \le 2$

$\quad u \le 5$

$\{u \mid u \le 5\}$

(g) $4t - 10 < 3t - 8 + 2t$

$\quad 4t - 10 < 5t - 8$

$\quad\quad -10 < t - 8$

$\quad\quad\quad -2 < t$

$\quad \{t \mid t > -2\}$

(h) $2s + 6 - 6s \geq 18$

$\quad 6 - 4s \geq 18$

$\quad\quad -4s \geq 12$

$\quad\quad s \leq -3$

$\quad \{s \mid s \leq -3\}$

(i) $5p - 10 < 9p - 1 - 4p$

$\quad 5p - 10 < 5p - 1$

$\quad\quad -10 < -1$

All real numbers, or
$\{p \mid p \in R\}.$

PROBLEM 3-6 Solve the following inequalities:

(a) $3(2 - z) > 4$ **(b)** $y > 2(y - 1)$ **(c)** $2(x - 1) \geq 3(2 - 3x)$ **(d)** $3(2w - 5) \leq 4(5w + 3)$
(e) $3(5 - 2t) \geq 3 - 6t$ **(f)** $4(3s - 2) \geq 6(2s - 1)$

Solution: Recall that to solve an inequality containing parentheses, you clear the parentheses using a distributative property, then solve as before [see Example 3-8].

(a) $3(2 - z) > 4$

$\quad 6 - 3z > 4$

$\quad -3z > -2$

$\quad z < \frac{2}{3}$

$\{z \mid z < \frac{2}{3}\}$

(b) $y > 2(y - 1)$

$\quad y > 2y - 2$

$\quad -y > -2$

$\quad y < 2$

$\{y \mid y < 2\}$

(c) $2(x - 1) \geq 3(2 - 3x)$

$\quad 2x - 2 \geq 6 - 9x$

$\quad 11x \geq 8$

$\quad x \geq \frac{8}{11}$

$\{x \mid x \geq \frac{8}{11}\}$

(d) $3(2w - 5) \leq 4(5w + 3)$

$\quad 6w - 15 \leq 20w + 12$

$\quad -27 \leq 14w$

$\quad -\frac{27}{14} \leq w$

$\{w \mid w \geq -\frac{27}{14}\}$

(e) $3(5 - 2t) \geq 3 - 6t$

$\quad 15 - 6t \geq 3 - 6t$

$\quad 15 \geq 3$

$\{t \mid t \in R\}$

(f) $4(3s - 2) \geq 6(2s - 1)$

$\quad 12s - 8 \geq 12s - 6$

$\quad -8 \geq -6$

$\quad \varnothing$

PROBLEM 3-7 Solve the following inequalities:

(a) $\frac{1}{2}z + \frac{2}{3} > \frac{1}{6}$ **(b)** $\frac{3}{4}y + \frac{1}{6} \leq \frac{1}{2}$ **(c)** $\frac{2}{5}x - \frac{1}{4} \leq \frac{3}{4}x$ **(d)** $\frac{1}{2}v - \frac{2}{3} \geq \frac{3}{4} - \frac{5}{6}v$

(e) $\frac{2u - 3}{2} > \frac{3u + 2}{3}$ **(f)** $\frac{3 + 4t}{4} < \frac{3t + 4}{3}$

Solution: Recall that to solve an inequality containing fractions, you clear the fractions by multiplying each member by the LCD of the terms, then solve as before [see Example 3-9].

(a) $\frac{1}{2}z + \frac{2}{3} > \frac{1}{6}$

The LCD is 6.

$6\left(\frac{1}{2}z + \frac{2}{3}\right) > 6 \cdot \frac{1}{6}$

$6 \cdot \frac{1}{2}z + 6 \cdot \frac{2}{3} > 6 \cdot \frac{1}{6}$

$3z + 4 > 1$

$3z > -3$

$z > -1$

$\{z \mid z > -1\}$

(b) $\frac{3}{4}y + \frac{1}{6} \leq \frac{1}{2}$

The LCD is 12.

$12\left(\frac{3}{4}y + \frac{1}{6}\right) \leq 12 \cdot \frac{1}{2}$

$12 \cdot \frac{3}{4}y + 12 \cdot \frac{1}{6} \leq 12 \cdot \frac{1}{2}$

$9y + 2 \leq 6$

$9y \leq 4$

$y \leq \frac{4}{9}$

$\left\{y \mid y \leq \frac{4}{9}\right\}$

(c) $\frac{2}{5}x - \frac{1}{4} \leq \frac{3}{4}x$

The LCD is 20.

$20\left(\frac{2}{5}x - \frac{1}{4}\right) \leq 20\left(\frac{3}{4}x\right)$

$8x - 5 \leq 15x$

$-5 \leq 7x$

$-\frac{5}{7} \leq x$

$\left\{x \mid x \geq -\frac{5}{7}\right\}$

(d) $\dfrac{1}{2}v - \dfrac{2}{3} \ge \dfrac{3}{4} - \dfrac{5}{6}v$

$\quad$ The LCD is 12.

$$12\left(\dfrac{1}{2}v - \dfrac{2}{3}\right) \ge 12\left(\dfrac{3}{4} - \dfrac{5}{6}v\right)$$

$$6v - 8 \ge 9 - 10v$$

$$16v \ge 17$$

$$v \ge \dfrac{17}{16}$$

$$\left\{v \,\middle|\, v \ge \dfrac{17}{16}\right\}$$

(e) $\dfrac{2u - 3}{2} > \dfrac{3u + 2}{3}$

$\quad$ The LCD is 6.

$$6\left(\dfrac{2u - 3}{2}\right) > 6\left(\dfrac{3u + 2}{3}\right)$$

$$3(2u - 3) > 2(3u + 2)$$

$$6u - 9 > 6u + 4$$

$$-9 > 4$$

$$\varnothing$$

(f) $\dfrac{3 + 4t}{4} < \dfrac{3t + 4}{3}$

$\quad$ The LCD is 12.

$$12\left(\dfrac{3 + 4t}{4}\right) < 12\left(\dfrac{3t + 4}{3}\right)$$

$$3(3 + 4t) < 4(3t + 4)$$

$$9 + 12t < 12t + 16$$

$$9 < 16$$

$$\{t \,|\, t \in R\}$$

PROBLEM 3-8 Graph the following compound inequalities:

(a) $x > 3$ *or* $x < 1$ $\qquad$ **(b)** $y > 0$ *or* $y \le -2$ $\qquad$ **(c)** $x > 0$ *and* $x \le 3$ $\qquad$ **(d)** $-2 < w \le 1$

(e) $v > 1$ *or* $v < 1$ $\qquad$ **(f)** $u \ge 3$ *or* $u \le 2$ $\qquad$ **(g)** $t > 0$ *or* $t \le 1$ $\qquad$ **(h)** $s \le 2$ *and* $s > 0$

(i) $r \ge 3$ *and* $r \le -1$ $\qquad$ **(j)** $0 \le q < 3$

Solution: Recall that to graph a compound inequality, you graph each simple inequality, then form the union or intersection of the solution sets of the simple inequalities. Whenever *or* is used, you form the union, and whenever *and* is used, you form the intersection. Remember that a hollow circle means that the point is not included in the solution set [see Example 3-10].

(a) $x > 3$ *or* $x < 1$

$x > 3$
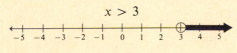

$x < 1$
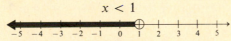

$x < 1$ *or* $x > 3$
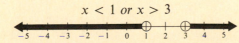

(b) $y > 0$ *or* $y \le -2$

$y > 0$
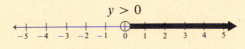

$y \le -2$
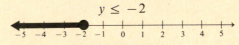

$y \le -2$ *or* $y > 0$
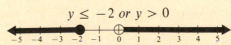

(c) $x > 0$ *and* $x \le 3$

$x > 0$
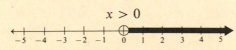

$x \le 3$
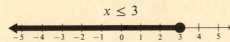

$x > 0$ *and* $x \le 3$
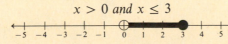

(d) $-2 < w \le 1$

$-2 < w$
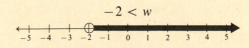

$w \le 1$

$-2 < w \le 1$
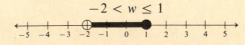

(e) $v > 1$ *or* $v < 1$

$v > 1$
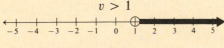

$v < 1$
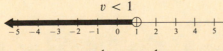

$v > 1$ *or* $v < 1$
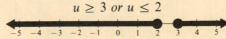

(f) $u \ge 3$ *or* $u \le 2$

$u \ge 3$
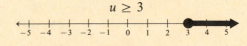

$u \le 2$
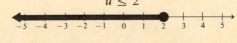

$u \ge 3$ *or* $u \le 2$

(g) $t > 0$ *or* $t \le 1$ **(h)** $s \le 2$ *and* $s > 0$

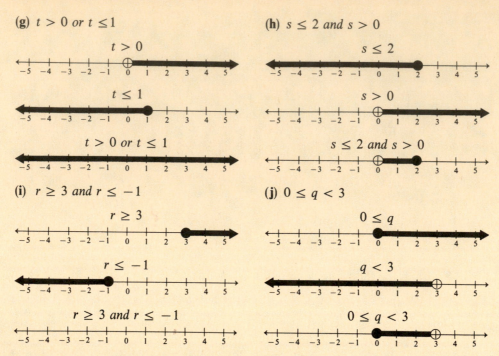

$t > 0$

$t \le 1$

$t > 0$ *or* $t \le 1$

$s \le 2$

$s > 0$

$s \le 2$ *and* $s > 0$

(i) $r \ge 3$ *and* $r \le -1$ **(j)** $0 \le q < 3$

$r \ge 3$

$r \le -1$

$r \ge 3$ *and* $r \le -1$

$0 \le q$

$q < 3$

$0 \le q < 3$

PROBLEM 3-9 Solve the following compound-*or* inequalities:

(a) $3z - 2 < 1$ *or* $3z - 2 > 7$ **(b)** $2y + 1 < -3$ *or* $2y + 1 \ge 3$
(c) $2x - 1 \le -5$ *or* $2x - 1 > 2$ **(d)** $6w - 2 \le -3$ *or* $6w - 2 > 1$
(e) $3 - 4u \le -1$ *or* $3 - 4u > 2$ **(f)** $3s - 2 < 7$ *or* $s \ge 3$ **(g)** $r < 2$ *or* $4 - r \ge 2r - 2$
(h) $3 - q \ge 3 + 3q$ *or* $q < 0$ **(i)** $2 + 3p < 5$ *or* $3 - 2p > 4 - 2p$

Solution: Recall that to solve a compound-*or* inequality, we solve the simple inequalities and form the union of the solution sets [see Example 3-11].

(a) $3z - 2 < 1$ *or* $3z - 2 > 7$ **(b)** $2y + 1 < -3$ *or* $2y + 1 \ge 3$ **(c)** $2x - 1 \le -5$ *or* $2x - 1 > 2$

$\quad 3z < 3$ *or* $\quad 3z > 9$ $2y < -4$ *or* $\quad 2y \ge 2$ $2x \le -4$ *or* $\quad 2x > 3$

$\quad z < 1$ *or* $\quad\quad z > 3$ $y < -2$ *or* $\quad y \ge 1$ $x \le -2$ *or* $\quad x > \frac{3}{2}$

$\{z \mid z < 1\} \cup \{z \mid z > 3\}$ $\{y \mid y < -2\} \cup \{y \mid y \ge 1\}$ $\{x \mid x \le -2\} \cup \{x \mid x > \frac{3}{2}\}$

$\{z \mid z < 1 \text{ or } z > 3\}$ $\{y \mid y < -2 \text{ or } y \ge 1\}$ $\{x \mid x \le -2 \text{ or } x > \frac{3}{2}\}$

(d) $6w - 2 \le -3$ *or* $6w - 2 > 1$ **(e)** $3 - 4u \le -1$ *or* $3 - 4u > 2$ **(f)** $3s - 2 < 7$ *or* $s \ge 3$

$\quad 6w \le -1$ *or* $\quad 6w > 3$ $-4u \le -4$ *or* $-4u > -1$ $3s < 9$ *or* $s \ge 3$

$\quad w \le -\frac{1}{6}$ *or* $\quad w > \frac{1}{2}$ $u \ge 1$ *or* $\quad u < \frac{1}{4}$ $s < 3$ *or* $s \ge 3$

$\{w \mid w \le -\frac{1}{6}\} \cup \{w \mid w > \frac{1}{2}\}$ $\{u \mid u \ge 1\} \cup \{u \mid u < \frac{1}{4}\}$ $\{s \mid s < 3\} \cup \{s \mid s \ge 3\}$

$\{w \mid w \le -\frac{1}{6} \text{ or } w > \frac{1}{2}\}$ $\{u \mid u \ge 1 \text{ or } u < \frac{1}{4}\}$ $\{s \mid s \in R\}$

(g) $r < 2$ *or* $4 - r \ge 2r - 2$ **(h)** $3 - q \ge 3 + 3q$ *or* $q < 0$ **(i)** $2 + 3p < 5$ *or* $3 - 2p > 4 - 2$

$\quad r < 2$ *or* $\quad 6 \ge 3r$ $0 \ge 4q$ *or* $q < 0$ $3p < 3$ *or* $\quad 3 > 4$

$\quad r < 2$ *or* $\quad 2 \ge r$ $0 \ge q$ *or* $q < 0$ $p < 1$ *or* $\quad 3 > 4$

$\{r \mid r < 2\} \cup \{r \mid r \le 2\}$ $\{q \mid q \le 0\} \cup \{q \mid q < 0\}$ $\{p \mid p < 1\} \cup \varnothing$

$\{r \mid r \le 2\}$ $\{q \mid q \le 0\}$ $\{p \mid p < 1\}$

PROBLEM 3-10 Solve the following compound-*and* inequalities:

(a) $z - 4 > -2$ *and* $z - 4 < 1$ **(b)** $2y - 5 < 4$ *and* $2y - 5 \ge 3$ **(c)** $-3 < 2x - 1 \le 4$
(d) $-1 \le 3 - 4w < 8$ **(e)** $1 - 2v < 3$ *and* $1 - 2v \ge 6$ **(f)** $2 - 3u < 3$ *and* $3u - 2 > 0$
(g) $2 \le 2 - 7r < 4$ **(h)** $0 < 4 - 3q < 6$

Solution: Recall that to solve a compound-*and* inequality, you solve each of the simple inequalities, then form the intersection of their solution sets [see Example 3-12].

(a) $z - 4 > -2$ and $z - 4 < 1$

$\qquad z > 2 \quad$ and $\quad z < 5$

$\qquad \{z|z > 2\} \cap \{z|z < 5\}$

$\qquad \{z|2 < z < 5\}$

$\qquad$ or $\{z|z > 2 \text{ and } z < 5\}$

(b) $2y - 5 < 4$ and $2y - 5 \geq 3$

$\qquad 2y < 9 \quad$ and $\qquad 2y \geq 8$

$\qquad y < \frac{9}{2} \quad$ and $\qquad y \geq 4$

$\qquad \{y|y < \frac{9}{2}\} \cap \{y|y \geq 4\}$

$\qquad \{y|4 \leq y < \frac{9}{2}\}$

$\qquad$ or $\{y|y \geq 4 \text{ and } y < \frac{9}{2}\}$

(c) $-3 < 2x - 1 \leq 4$

$\qquad -2 < 2x \leq 5$

$\qquad -1 < x \leq \frac{5}{2}$

$\qquad \{x|-1 < x \leq \frac{5}{2}\}$

$\qquad$ or $\{x|x > -1 \text{ and } x \leq \frac{5}{2}\}$

(d) $-1 \leq 3 - 4w < 8$

$\qquad -4 \leq -4w < 5$

$\qquad 1 \geq w > -\frac{5}{4}$

$\qquad \{w|-\frac{5}{4} < w \leq 1\}$

$\qquad$ or $\{w|w > -\frac{5}{4} \text{ and } w \leq 1\}$

(e) $1 - 2v < 3 \quad$ and $1 - 2v \geq 6$

$\qquad -2v < 2 \quad$ and $\quad -2v \geq 5$

$\qquad v > -1 \text{ and} \qquad v \leq -\frac{5}{2}$

$\qquad \{v|v > -1\} \cap \{v|v \leq -\frac{5}{2}\}$

$\qquad \emptyset$

(f) $2 - 3u < 3 \quad$ and $3u - 2 > 0$

$\qquad -3u < 1 \quad$ and $\qquad 3u > 2$

$\qquad u > -\frac{1}{3} \text{ and} \qquad u > \frac{2}{3}$

$\qquad \{u|u > -\frac{1}{3}\} \cap \{u|u > \frac{2}{3}\}$

$\qquad \{u|u > \frac{2}{3}\}$

(g) $2 \leq 2 - 7r < 4$

$\qquad 0 \leq -7r < 2$

$\qquad 0 \geq r > -\frac{2}{7}$

$\qquad \{r|-\frac{2}{7} < r \leq 0\}$

(h) $0 < 4 - 3q < 6$

$\qquad -4 < -3q < 2$

$\qquad \frac{4}{3} > q > -\frac{2}{3}$

$\qquad \{q|-\frac{2}{3} < q < \frac{4}{3}\}$

PROBLEM 3-11 Solve the following absolute value inequalities:

(a) $|3z| > 6$ $\quad$ **(b)** $|x - 4| \geq 3$ $\quad$ **(c)** $4 < |w + 4|$ $\quad$ **(d)** $3 \leq |2v - 3|$ $\quad$ **(e)** $0 < |3 - 2s|$
(f) $|4 - 3s| - 5 \geq 3$

Solution: Recall that to solve an absolute value inequality of the form $|E| > r$, you write $E < -r$ or $E > r$, then solve the compound-*or* inequality as before [see Example 3-13].

(a) $|3z| > 6$

$\quad 3z < -6$ or $3z > 6$

$\quad z < -2$ or $\ z > 2$

$\quad \{z|z < -2 \text{ or } z > 2\}$

(b) $|x - 4| \geq 3$

$\quad x - 4 \leq -3$ or $x - 4 \geq 3$

$\quad x \leq 1 \quad$ or $\quad x \geq 7$

$\quad \{x|x \leq 1 \text{ or } x \geq 7\}$

(c) $4 < |w + 4|$

$\quad w + 4 < -4$ or $w + 4 > 4$

$\quad w < -8$ or $\quad w > 0$

$\quad \{w|w < -8 \text{ or } w > 0\}$

(d) $3 \leq |2v - 3|$

$\quad 2v - 3 \leq -3$ or $2v - 3 \geq 3$

$\quad 2v \leq 0 \quad$ or $\quad 2v \geq 6$

$\quad v \leq 0 \quad$ or $\quad v \geq 3$

$\quad \{v|v \leq 0 \text{ or } v \geq 3\}$

(e) $0 < |3 - 2s|$

$\quad 3 - 2s < 0 \quad$ or $3 - 2s > 0$

$\quad -2s < -3$ or $\ -2s > -3$

$\quad s > \frac{3}{2} \quad$ or $\quad s < \frac{3}{2}$

$\quad \{s|s \neq \frac{3}{2}\}$

(f) $|4 - 3s| - 5 \geq 3$

$\quad |4 - 3s| \geq 8$

$\quad 4 - 3s \leq -8$ or $4 - 3s \geq 8$

$\quad -3s \leq -12$ or $\ -3s \geq 4$

$\quad s \geq 4 \qquad$ or $\qquad s \leq -\frac{4}{3}$

$\quad \{s|s \geq 4 \text{ or } s \leq -\frac{4}{3}\}$

PROBLEM 3-12 Solve the following absolute value inequalities:

(a) $|2z| < 4$ (b) $|5y| \le 6$ (c) $|x - 4| < 1$ (d) $|w + 3| \le 3$ (e) $3 > |5 - 2t|$
(f) $|4 - 3s| + 5 \le 5$

Solution: Recall that to solve an absolute value inequality of the form $|E| < r$, you write $-r < E < r$, then solve the compound-*and* inequality as before [see Example 3-14].

(a) $|2z| < 4$

$-4 < 2z < 4$

$-2 < z < 2$

$\{z | -2 < z < 2\}$

(b) $|5y| \le 6$

$-6 \le 5y \le 6$

$-\frac{6}{5} \le y \le \frac{6}{5}$

$\{y | -\frac{6}{5} \le y \le \frac{6}{5}\}$

(c) $|x - 4| < 1$

$-1 < x - 4 < 1$

$3 < x < 5$

$\{x | 3 < x < 5\}$

(d) $|w + 3| \le 3$

$-3 \le w + 3 \le 3$

$-6 \le w \le 0$

$\{w | -6 \le w \le 0\}$

(e) $3 > |5 - 2t|$

$-3 < 5 - 2t < 3$

$-8 < -2t < -2$

$4 > t > 1$

$\{t | 1 < t < 4\}$

(f) $|4 - 3s| + 5 \le 5$

$|4 - 3s| \le 0$

$-0 \le 4 - 3s \le 0$

$-4 \le -3s \le -4$

$\frac{4}{3} \ge s \ge \frac{4}{3}$

$\{s | s = \frac{4}{3}\}$

Supplementary Exercises

PROBLEM 3-13 Of which of the following is -2 a solution?

(a) $2z + 3 > -2$ (b) $3y + 5 \ge 0$ (c) $4 \le 5x - 3$ (d) $10 \ge 4w - 5$ (e) $4 > 2v - 3$
(f) $4u + 3 \le 5$ (g) $5t - 3 \le 6 + 2t$ (h) $3 - 2s > 4s + 8 - 2s$ (i) $8r - 9 \ge 4 - 7r$
(j) $3q - 6 > q + 7 - 4q$

PROBLEM 3-14 Solve the following inequalities:

(a) $z - 5 < 2$ (b) $y - 6 \le 4$ (c) $x - 4 > -3$ (d) $-4 \le w - 3$ (e) $0 < v + 4$
(f) $0 \le u + 3$ (g) $3 + t \ge 8$ (h) $3 \le 7 + s$ (i) $-5 > 6 + r$ (j) $0 \ge 7 + q$

PROBLEM 3-15 Solve the following inequalities:

(a) $\frac{5}{3}z > 15$ (b) $\frac{4}{7}y \ge 2$ (c) $\frac{7}{9}x \le 7$ (d) $-\frac{2}{5}w < 3$ (e) $-\frac{4}{7}v \ge 6$ (f) $-\frac{5}{6} > -\frac{4}{5}u$
(g) $-\frac{4}{3} \le -\frac{5}{6}t$ (h) $6s < 3$ (i) $-3 < 4r$ (j) $-8 \le -12q$ (k) $-6p \le 4$

PROBLEM 3-16 Solve the following inequalities:

(a) $3z - 2 > 7$ (b) $4y - 3 \ge 5$ (c) $2x + 3 < 9$ (d) $6 \le 4w + 10$ (e) $5 - 3v > 2$
(f) $2 - 3u \le 7$ (g) $5 > 2t + 4$ (h) $-4 \ge 5 - 3s$ (i) $-3 \le 5 - 2r$ (j) $4 < 3 - 5q$
(k) $3 \le 5 - 7p$

PROBLEM 3-17 Solve the following inequalities:

(a) $3z + 5z > 16$ (b) $4y - 2y \ge 12$ (c) $3x - 5x \le 8$ (d) $5w - 9w < 6$
(e) $5v - 3 + 2v \ge 4v$ (f) $7u + 4 - 2u > 3u$ (g) $5t - 7 - 9t \le 3 - 2t$
(h) $2s - 5 - 7s < 3 + 4s$ (i) $4 - 3r - 7 \ge r + 5 - 5r$ (j) $3q - 3 + 4q > 9q + 3 - 2q$
(k) $3 - 4p + p < 4p - 7p + 4$

PROBLEM 3-18 Solve the following inequalities:

(a) $4(z + 3) > 6$ (b) $2y > 3(y - 2)$ (c) $3(x - 3) \geq 4(2 - x)$ (d) $4(5 - 2w) < 3(3w + 2)$
(e) $5(3v + 1) \geq 3 - 2(3 - 6v)$ (f) $3(1 - 2u) + 5 \leq 4(u - 5)$ (g) $4(4 - 3t) \geq 6(3 - 2t)$
(h) $3(5 - 2r) \leq 3(5 - 2r)$

PROBLEM 3-19 Solve the following inequalities:

(a) $z + \dfrac{3}{4} > \dfrac{3}{2}$ (b) $\dfrac{4}{5}y - \dfrac{3}{4} \geq \dfrac{1}{2}$ (c) $\dfrac{3}{4}x - \dfrac{5}{6} < \dfrac{2}{3}x$ (d) $\dfrac{5}{6}w \leq \dfrac{2}{3} - \dfrac{3}{4}w$

(e) $\dfrac{2}{3}v - \dfrac{3}{5} \leq \dfrac{3}{5}v + \dfrac{1}{6}$ (f) $\dfrac{3 - 2u}{4} < \dfrac{3u + 4}{3}$ (g) $\dfrac{3 - 4t}{3} \geq \dfrac{2}{3}t - \dfrac{1}{4}$ (h) $\dfrac{2}{3}s - \dfrac{4}{5} > \dfrac{3s - 2}{2}$

PROBLEM 3-20 Graph the following compound inequalities:

(a) $z < 0$ or $z > 2$ (b) $y \leq -2$ or $y > 0$ (c) $x < -4$ or $x \geq -1$ (d) $w > -1$ or $w \leq 1$
(e) $v < 2$ and $v \geq -2$ (f) $u < 0$ and $u \geq -1$ (g) $t \geq -4$ and $t < -2$
(h) $-4 < s < -1$ (i) $-3 < r \leq 2$ (j) $4 \leq q < 5$

PROBLEM 3-21 Solve the following compound-*or* inequalities:

(a) $2z + 5 < 3$ or $2z + 5 > 5$ (b) $4y + 5 < 0$ or $4y + 5 \geq 3$
(c) $5x - 3 < -2$ or $5x - 3 > 3$ (d) $3w + 5 \geq 0$ or $3w + 5 < -4$
(e) $5v - 3 \geq 2$ or $5v - 3 \leq 0$ (f) $4u - 2 \geq 4$ or $3u - 2 < 1$
(g) $3 - 2t > -3$ or $t \leq -3$ (h) $4 - 3s > 1$ or $s \leq 1$ (i) $3 - 2r > 5$ or $3r - 2 \leq 1$
(j) $4 - 3q > 4$ or $3 - 4q \leq 3$ (k) $p \geq 4$ or $2 - 3p \leq 5$

PROBLEM 3-22 Solve the following compound-*and* inequalities:

(a) $3z + 4 < 1$ and $3z + 4 > -2$ (b) $2y + 3 < 5$ and $2y + 3 > 1$
(c) $2x - 3 > 1$ and $2x - 3 < 5$ (d) $4w - 5 \geq 0$ and $4w - 5 \leq 3$
(e) $3v - 2 \leq -5$ and $3v - 2 > -8$ (f) $-3 < 4 - 2u \leq 2$ (g) $0 \leq 3t - 2 < 4$
(h) $4 < 5 - 2s < 7$ (i) $-8 \leq 2 - 5r \leq -3$ (j) $3 - 2q \geq 4$ and $3 - 2q \leq 4$
(k) $3 - 4p \geq 3$ and $3 - 2p < 1$

PROBLEM 3-23 Solve the following absolute value inequalities:

(a) $|5z| > 15$ (b) $|4y| \geq 3$ (c) $|x - 5| > 2$ (d) $2 \leq |w + 7|$ (e) $5 \leq |3v - 4|$
(f) $3 < |4 - 3u|$ (g) $0 < |5 - 3t|$ (h) $5 \leq 3 + |3s - 4|$

PROBLEM 3-24 Solve the following absolute value inequalities:

(a) $|3z| < 15$ (b) $|5y| \leq 8$ (c) $|x - 3| < 2$ (d) $2 \geq |w + 1|$ (e) $3 \geq |3v - 4|$
(f) $|3u - 1| \leq 4$ (g) $3 > |3 - 4t|$ (h) $|5 - 2s| < 4$ (i) $|3 - 2r| + 3 \leq 5$

Answers to Supplementary Exercises

(3-13) (a) yes   (b) no   (c) no   (d) yes   (e) yes   (f) yes   (g) yes   (h) yes
(i) no   (j) no

(3-14) (a) $\{z \mid z < 7\}$ (b) $\{y \mid y \leq 10\}$ (c) $\{x \mid x > 1\}$ (d) $\{w \mid w \geq -1\}$
(e) $\{v \mid v > -4\}$ (f) $\{u \mid u \geq -3\}$ (g) $\{t \mid t \geq 5\}$ (h) $\{s \mid s \geq -4\}$
(i) $\{r \mid r < -11\}$ (j) $\{q \mid q \leq -7\}$

(3-15) (a) $\{z \mid z > 9\}$ (b) $\{y \mid y \geq \frac{7}{2}\}$ (c) $\{x \mid x \leq 9\}$ (d) $\{w \mid w > -\frac{15}{2}\}$
(e) $\{v \mid v \leq -\frac{21}{2}\}$ (f) $\{u \mid u > \frac{25}{24}\}$ (g) $\{t \mid t \leq \frac{8}{3}\}$ (h) $\{s \mid s < \frac{1}{2}\}$
(i) $\{r \mid r > -\frac{3}{4}\}$ (j) $\{q \mid q \leq \frac{2}{3}\}$ (k) $\{p \mid p \geq -\frac{2}{3}\}$

(3-16) **(a)** $\{z \mid z > 3\}$　　**(b)** $\{y \mid y \geq 2\}$　　**(c)** $\{x \mid x < 3\}$　　**(d)** $\{w \mid w \geq -1\}$
　　　(e) $\{v \mid v < 1\}$　　**(f)** $\{u \mid u \geq -\frac{5}{3}\}$　　**(g)** $\{t \mid t < \frac{1}{2}\}$　　**(h)** $\{s \mid s \geq 3\}$
　　　(i) $\{r \mid r \leq 4\}$　　**(j)** $\{q \mid q < -\frac{1}{3}\}$　　**(k)** $\{p \mid p \leq \frac{2}{7}\}$

(3-17) **(a)** $\{z \mid z > 2\}$　　**(b)** $\{y \mid y \geq 6\}$　　**(c)** $\{x \mid x \geq -4\}$　　**(d)** $\{w \mid w > -\frac{3}{2}\}$
　　　(e) $\{v \mid v \geq 1\}$　　**(f)** $\{u \mid u > -2\}$　　**(g)** $\{t \mid t \geq -5\}$　　**(h)** $\{s \mid s > -\frac{8}{9}\}$
　　　(i) $\{r \mid r \geq 8\}$　　**(j)** $\varnothing$　　**(k)** $\{p \mid p \in R\}$

(3-18) **(a)** $\{z \mid z > -\frac{3}{2}\}$　　**(b)** $\{y \mid y < 6\}$　　**(c)** $\{x \mid x \geq \frac{17}{7}\}$
　　　(d) $\{w \mid w > \frac{14}{17}\}$　　**(e)** $\{v \mid v \geq -\frac{8}{3}\}$　　**(f)** $\{u \mid u \geq \frac{14}{5}\}$
　　　(g) $\varnothing$　　　　**(h)** $\{r \mid r \in R\}$

(3-19) **(a)** $\{z \mid z > \frac{3}{4}\}$　　**(b)** $\{y \mid y \geq \frac{25}{16}\}$　　**(c)** $\{x \mid x < 10\}$
　　　(d) $\{w \mid w \leq \frac{8}{19}\}$　　**(e)** $\{v \mid v \leq \frac{23}{2}\}$　　**(f)** $\{u \mid u > -\frac{7}{18}\}$
　　　(g) $\{t \mid t \leq \frac{5}{8}\}$　　**(h)** $\{s \mid s < \frac{6}{25}\}$

(3-20)
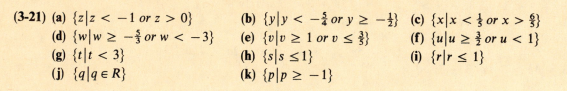

(3-21) **(a)** $\{z \mid z < -1 \text{ or } z > 0\}$　　**(b)** $\{y \mid y < -\frac{5}{4} \text{ or } y \geq -\frac{1}{2}\}$　**(c)** $\{x \mid x < \frac{1}{5} \text{ or } x > \frac{6}{5}\}$
　　　(d) $\{w \mid w \geq -\frac{5}{3} \text{ or } w < -3\}$　　**(e)** $\{v \mid v \geq 1 \text{ or } v \leq \frac{3}{5}\}$　　**(f)** $\{u \mid u \geq \frac{3}{2} \text{ or } u < 1\}$
　　　(g) $\{t \mid t < 3\}$　　　　**(h)** $\{s \mid s \leq 1\}$　　　　**(i)** $\{r \mid r \leq 1\}$
　　　(j) $\{q \mid q \in R\}$　　　**(k)** $\{p \mid p \geq -1\}$

(3-22) **(a)** $\{z \mid -2 < z < -1\}$　**(b)** $\{y \mid -1 < y < 1\}$　　**(c)** $\{x \mid 2 < x < 4\}$
　　　(d) $\{w \mid \frac{5}{4} \leq w \leq 2\}$　**(e)** $\{v \mid -2 < v \leq -1\}$　　**(f)** $\{u \mid 1 \leq u < \frac{7}{2}\}$
　　　(g) $\{t \mid \frac{2}{3} \leq t < 2\}$　　**(h)** $\{s \mid -1 < s < \frac{1}{2}\}$　　**(i)** $\{r \mid 1 \leq r \leq 2\}$
　　　(j) $\{q \mid q = -\frac{1}{2}\}$　　**(k)** $\varnothing$

(3-23) **(a)** $\{z \mid z < -3 \text{ or } z > 3\}$　　**(b)** $\{y \mid y \leq -\frac{3}{4} \text{ or } y \geq \frac{3}{4}\}$　　**(c)** $\{x \mid x < 3 \text{ or } x > 7\}$
　　　(d) $\{w \mid w \leq -9 \text{ or } w \geq -5\}$　　**(e)** $\{v \mid v \leq -\frac{1}{3} \text{ or } v \geq 3\}$　　**(f)** $\{u \mid u > \frac{7}{3} \text{ or } u < \frac{1}{3}\}$
　　　(g) $\{t \mid t \neq \frac{5}{3}\}$　　　　**(h)** $\{s \mid s \leq \frac{2}{3} \text{ or } s \geq 2\}$

(3-24) **(a)** $\{z \mid -5 < z < 5\}$　**(b)** $\{y \mid -\frac{8}{5} \leq y \leq \frac{8}{5}\}$　**(c)** $\{x \mid 1 < x < 5\}$
　　　(d) $\{w \mid -3 \leq w \leq 1\}$　**(e)** $\{v \mid \frac{1}{3} \leq v \leq \frac{7}{3}\}$　　**(f)** $\{u \mid -1 \leq u \leq \frac{5}{3}\}$
　　　(g) $\{t \mid 0 < t < \frac{3}{2}\}$　　**(h)** $\{s \mid \frac{1}{2} < s < \frac{9}{2}\}$　　**(i)** $\{r \mid \frac{1}{2} \leq r \leq \frac{5}{2}\}$

EQUATIONS AND GRAPHS

4-1. The Rectangular Coordinate System

When we construct two perpendicular number lines, called axes, that intersect on a plane, we have a **rectangular** or **Cartesian coordinate system,** which was introduced by Rene Descartes (1596–1650) to establish a relationship between equations and the graphs of equations. With each point on the plane we associate an **ordered pair** of real numbers (x, y) called the **coordinates of the point.** The point where the number lines intersect is called the **origin,** which has coordinates $(0, 0)$. The **abscissa** (first coordinate or x-coordinate) of a point describes the *horizontal distance* from the vertical or y-axis, and the **ordinate** (second coordinate or y-coordinate) describes the *vertical distance* from the horizontal or x-axis.

A. Graphing ordered pairs

To **graph an ordered pair,** we plot a dot at the point on the coordinate system associated with that pair.

Caution: The order of the coordinates is important; the first coordinate *always* describes the horizontal, and the second the vertical, distance from the origin. This is why the coordinates are called an *ordered* pair.

EXAMPLE 4-1: Graph the following ordered pairs: **(a)** $(2, 4)$ and $(-3, -1)$ **(b)** $(-2, 0)$ and $(0, 3)$

Solution: **(a)** $(2, 4)$ and $(-3, -1)$

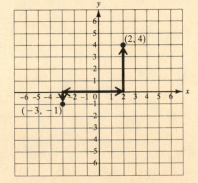

Plot the graph of $(2, 4)$ 2 units to the right of the y-axis and 4 units above the x-axis.

Plot the graph of $(-3, -1)$ 3 units to the left of the y-axis and 1 unit below the x-axis.

(b) $(-2, 0)$ and $(0, 3)$

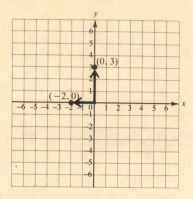

Plot the graph of $(-2, 0)$ 2 units to the left of the y-axis on the x-axis.

Plot the graph of $(0, 3)$ on the y-axis 3 units above the x-axis.

B. Finding the coordinates of points

To **find the coordinates of a point** in a rectangular coordinate system, we project lines from the point that are perpendicular to each axis, and read the resulting coordinates.

EXAMPLE 4-2: Find the coordinates of points A, B, and C on the following graph.

Solution:

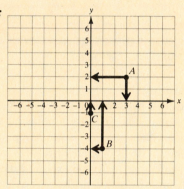

Project A vertically onto the x-axis at 3, and horizontally onto the y-axis at 2. So A is the graph of the ordered pair $(3, 2)$.

Project B vertically onto the x-axis at 1, and horizontally onto the y-axis at -4. So B is the graph of the ordered pair $(1, -4)$.

Project C vertically onto the x-axis at 0, and horizontally onto the y-axis at -1. So C is the graph of the ordered pair $(0, -1)$.

4-2. Solving Equations with Two Variables

A. Determining whether an ordered pair is a solution of an equation in two variables

An ordered pair (a, b) is **a solution of an equation in two variables x and y** if the statement obtained by replacing the variable x with a and the variable y with b is true.

EXAMPLE 4-3: Determine whether **(a)** $(2, -3)$ and **(b)** $(0, 6)$ are solutions of $3x - 2y = 12$.

Solution: **(a)**

$$3x - 2y = 12 \longleftarrow \text{original equation}$$

$$
\begin{array}{c|c}
3(2) - 2(-3) & 12 \\
6 + 6 & 12 \\
12 & 12 \longleftarrow (2, 3) \text{ is a solution}
\end{array}
$$

Replace x with 2 and y with -3.

(b)

$$3x - 2y = 12 \longleftarrow \text{original equation}$$

$$
\begin{array}{c|c}
3(0) - 2(6) & 12 \\
0 - 12 & 12 \\
-12 & 12 \longleftarrow (0, 6) \text{ is not a solution}
\end{array}
$$

Replace x with 0 and y with 6.

Note: An equation in two variables may have more than one solution.

B. Finding indicated solutions for given equations

To find the **indicated solutions for an equation in two variables,** we substitute the given numerical values for one variable and solve the resulting equations for the other variable.

EXAMPLE 4-4: Find the solutions of $3x - 2y = 12$ for $x = 4, 0$, and -2.

Solution:
$$3x - 2y = 12$$
$$-2y = -3x + 12 \qquad \text{Solve for } y.$$
$$y = \tfrac{3}{2}x - 6$$

x	$\tfrac{3}{2}x - 6 = y$		Make a table.
4	$\tfrac{3}{2} \cdot 4 - 6 = 0$	⟵ Replace x with 4.	
0	$\tfrac{3}{2} \cdot 0 - 6 = -6$	⟵ Replace x with 0.	
-2	$\tfrac{3}{2}(-2) - 6 = -9$	⟵ Replace x with -2.	

So $(4, 0)$, $(0, -6)$, and $(-2, -9)$ are solutions of $3x - 2y = 12$.

C. Finding the intercepts of the graphs of equations

The point where the graph of an equation intersects the x-axis is called **the x-intercept,** and the point where the graph intersects the y-axis is called **the y-intercept.** The x-intercept has coordinates of the form $(a, 0)$ and the y-intercept has coordinates of the form $(0, b)$. To **find an intercept** for one variable, we replace the other variable with 0.

Caution: The graphs of some equations do not have both intercepts.

EXAMPLE 4-5: Find **(a)** the x-intercept and **(b)** the y-intercept of the graph of $3x - 4y = 6$.

Solution: **(a)** $3x - 4y = 6$ ⟵ original equation
$$3x - 4(0) = 6 \qquad \text{Replace } y \text{ with 0.}$$
$$3x - 0 = 6$$
$$3x = 6$$
$$x = 2$$

So $(2, 0)$ is the x-intercept of the graph of $3x - 4y = 6$.

(b) $3x - 4y = 6$ ⟵ original equation
$$3(0) - 4y = 6 \qquad \text{Replace } x \text{ with 0.}$$
$$0 - 4y = 6$$
$$-4y = 6$$
$$y = -\tfrac{3}{2}$$

So $(0, -\tfrac{3}{2})$ is the y-intercept of the graph of $3x - 4y = 6$.

4-3. Graphing Linear Equations

A **linear equation** is an equation that can be written in the form $Ax + By = C$, where A, B, and C are real numbers and A and B are not both 0.

A. Graphing linear equations in two variables by graphing two solutions

The graph of the set of ordered pairs that is the solution set of a linear equation is a straight line. Since any two points determine a line, we can **graph a linear equation** by plotting any two solutions (points) of the equation and drawing a straight line through those two points. The straight line is **the graph of the linear equation.** We can check the graph by plotting a third solution (point) of the equation, which should also lie on the line.

EXAMPLE 4-6: Graph the following linear equations: **(a)** $2x + 3y = 6$ **(b)** $4x - 3y = 6$

Solution: **(a)**

$$2x + 3y = 6$$

$$3y = -2x + 6 \qquad \text{Solve for } y.$$

$$y = -\tfrac{2}{3}x + 2$$

x	$-\tfrac{2}{3}x + 2 = y$
3	$-\tfrac{2}{3} \cdot 3 + 2 = 0$
0	$-\tfrac{2}{3} \cdot 0 + 2 = 2$
-3	$-\tfrac{2}{3}(-3) + 2 = 4$

Find three solutions.

$(3, 0), (0, 2), (-3, 4)$ ⟵ solutions of $2x + 3y = 6$

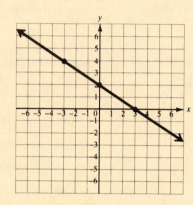

Plot the three solutions.
Draw a straight line through the points.

(b)

$$4x - 3y = 6$$

$$-3y = -4x + 6 \qquad \text{Solve for } y.$$

$$y = \tfrac{4}{3}x - 2$$

x	$\tfrac{4}{3}x - 2 = y$
3	$\tfrac{4}{3} \cdot 3 - 2 = 2$
0	$\tfrac{4}{3} \cdot 0 - 2 = -2$
-3	$\tfrac{4}{3}(-3) - 2 = -6$

Find three solutions.

$(3, 2), (0, -2), (-3, -6)$ ⟵ solutions of $4x - 3y = 6$

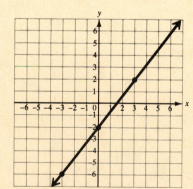

Plot the three solutions.
Draw a straight line through the points.

B. Graphing linear equations using the intercept method

To **graph a linear equation using the intercept method,** we plot both intercepts and draw a straight line through these two points.

EXAMPLE 4-7: Graph the following linear equations using the intercept method:
(a) $2x - y = 4$ **(b)** $x + 4y = 4$

Solution: **(a)** $2x - y = 4$

$$2(0) - y = 4 \qquad \text{Replace } x \text{ with 0 to find the } y\text{-intercept.}$$

$$0 - y = 4$$

$$y = -4 \longleftarrow (0, -4) \text{ is the } y\text{-intercept}$$

$$2x - y = 4$$

$$2x - 0 = 4 \qquad \text{Replace } y \text{ with 0 to find the } x\text{-intercept.}$$

$$2x = 4$$

$$x = 2 \longleftarrow (2, 0) \text{ is the } x\text{-intercept}$$

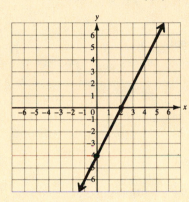

Plot the intercepts.

Draw a straight line through the two points.

(b) $x + 4y = 4$

$$0 + 4y = 4 \qquad \text{Replace } x \text{ with 0 to find the } y\text{-intercept.}$$

$$4y = 4$$

$$y = 1 \longleftarrow (0, 1) \text{ is the } y\text{-intercept}$$

$$x + 4y = 4$$

$$x + 4(0) = 4 \qquad \text{Replace } y \text{ with 0 to find the } x\text{-intercept.}$$

$$x + 0 = 4$$

$$x = 4 \longleftarrow (4, 0) \text{ is the } x\text{-intercept}$$

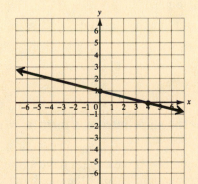

Plot the intercepts.

Draw a straight line through the two points.

C. Graphing linear equations having graphs parallel to an axis

An equation in the form $Ax + By = C$, where $B = 0$, can be written simply as $x = C/A$. An equation of the form $Ax + By = C$, where $A = 0$, can be written simply as $y = C/B$. The graphs of such equations are straight lines parallel to an axis.

EXAMPLE 4-8: Graph the following linear equations: **(a)** $x = 3$ **(b)** $y + 2 = 0$

Solution: **(a)**

$$x = 3$$

$$x + 0y = 3 \qquad \text{Write the equation in standard form.}$$

$$x = 0y + 3 \qquad \text{Solve for } x.$$

y	$0y + 3 = x$	Find two solutions.
2	$0(2) + 3 = 3$	
-2	$0(-2) + 3 = 3$	← (3, 2) and (3, −2) are solutions of $x = 3$

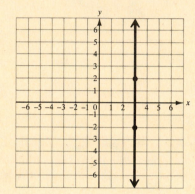

Plot the two solutions.

Draw a straight line through the points.

(b)

$$y + 2 = 0$$

$$0x + y = -2 \qquad \text{Write the equation in standard form.}$$

$$y = 0x - 2 \qquad \text{Solve for } y.$$

x	$0x - 2 = y$	Find two solutions.
2	$0 \cdot 2 - 2 = -2$	
-2	$0(-2) - 2 = -2$	← (2, −2) and (−2, −2) are solutions of $y + 2 = 0$

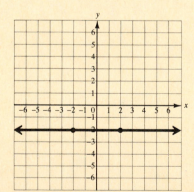

Plot the two points.

Draw a straight line through the points.

D. Finding the slopes and *y*-intercepts of equations in the slope-intercept form $y = mx + b$

Any linear equation in the standard form $Ax + By = C$ may also be written in the form $y = mx + b$ by solving for *y*. This is the **slope–intercept form,** where the number *m* is the **slope** and the number *b* is the **y-intercept.** To **find the slope and y-intercept** of the graph of an equation, we solve the standard form of the equation for *y*.

EXAMPLE 4-9: Find the slope and *y*-intercept of the following equations:

(a) $3x + 2y = 6$ **(b)** $2x + 5y = 0$ **(c)** $y + 5 = 0$ **(d)** $x - 2 = 0$

Solution:

(a) $3x + 2y = 6$

$\qquad 2y = -3x + 6 \qquad$ Solve for y.

$\qquad y = \dfrac{-3x + 6}{2}$

$\qquad y = -\frac{3}{2}x + 3$

$\qquad\qquad b = 3 \longleftarrow$ 3 is the y-intercept

$\qquad m = -\frac{3}{2} \longleftarrow -\frac{3}{2}$ is the slope

(b) $2x + 5y = 0$

$\qquad 5y = -2x + 0 \qquad$ Solve for y.

$\qquad y = -\frac{2}{5}x + 0$

$\qquad\qquad b = 0 \longleftarrow$ 0 is the y-intercept

$\qquad m = -\frac{2}{5} \longleftarrow -\frac{2}{5}$ is the slope

(c) $y + 5 = 0$

$\qquad y = 0x - 5 \qquad$ Solve for y.

$\qquad\qquad b = -5 \longleftarrow -5$ is the y-intercept

$\qquad m = 0 \longleftarrow$ 0 is the slope

(d) $x - 2 = 0$

$\qquad x = 2$

The graph is a vertical line.
The line has *no* slope.

E. Graphing linear equations using the *y*-intercepts and slopes

To **graph a linear equation using the *y*-intercept and slope,** we first plot the *y*-intercept. We then use the denominator of the slope to determine the horizontal distance of a second point on the line from the *y*-axis, and the numerator to determine the vertical distance of that point from the *y*-intercept.

EXAMPLE 4-10: Graph the following linear equations using the *y*-intercept and slope:

(a) $y = \frac{3}{2}x - 2$ **(b)** $2x + 3y = 6$

Solution: **(a)** $y = \frac{3}{2}x - 2$

$\qquad\qquad (0, -2)$ is the *y*-intercept.

$\qquad\qquad m = \frac{3}{2}$ is the slope.

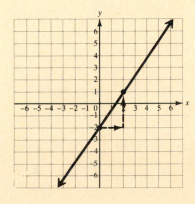

Plot the *y*-intercept.

Plot a second point 2 units to the right of and 3 units above the *y*-intercept, that is, $(0 + 2, -2 + 3) = (2, 1)$.

Draw a straight line through the two points.

(b) $2x + 3y = 6$

$\qquad\qquad 3y = -2x + 6 \qquad$ Solve for y.

$\qquad\qquad y = -\frac{2}{3}x + 2$

$\qquad (0, 2)$ is the *y*-intercept.

$\qquad m = -\frac{2}{3} = \frac{-2}{3}$ is the slope.

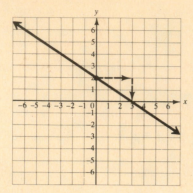

Plot the *y*-intercept.

Plot a second point 3 units to the right of and 2 units below the *y*-intercept.

Draw a straight line through the two points.

Note: In this example, $m = -\frac{2}{3}$ may also be written as $m = \frac{2}{-3}$. Then, if we plot the *y*-intercept and a second point 3 units to the *left* of and 2 units *above* the *y*-intercept, the straight line graph will be the same.

4-4. Finding the Slopes of Lines

The **slope of a line** is determined by its "rise" (or change in vertical distance) and its "run" (or change in horizontal distance). If (x_1, y_1) and (x_2, y_2) are two points on a line, the slope of the line is

$$m = \frac{\text{change in } y}{\text{change in } x} = \frac{y_2 - y_1}{x_2 - x_1}$$

A. Finding the slope of the line given the coordinates of two of its points

EXAMPLE 4-11: Find the slope of the line that contains the following points:

(a) $(2, 3)$ and $(4, 7)$ **(b)** $(-3, 4)$ and $(4, -2)$

Solution:

(a) $m = \dfrac{y_2 - y_1}{x_2 - x_1}$

$= \dfrac{7 - 3}{4 - 2}$ Replace (x_2, y_2) with $(4, 7)$ and (x_1, y_1) with $(2, 3)$.

$= \dfrac{4}{2}$

$= 2$ ⟵ 2 is the slope of the line containing $(2, 3)$ and $(4, 7)$

(b) $m = \dfrac{y_2 - y_1}{x_2 - x_1}$

$= \dfrac{-2 - 4}{4 - (-3)}$ Replace (x_2, y_2) with $(4, -2)$ and (x_1, y_1) with $(-3, 4)$.

$= \dfrac{-6}{7}$ ⟵ $-\dfrac{6}{7}$ is the slope of the line containing $(-3, 4)$ and $(4, -2)$

Note: It does not matter which ordered pair replaces (x_2, y_2) and which one replaces, (x_1, y_1).

B. Finding the slope of a line parallel to an axis

Any two points (x_1, y_1) and (x_2, y_2) on a *horizontal line* have the same *y*-coordinate $(y_1 = y_2)$, so $y_2 - y_1$ will equal 0. Therefore, the slope of any horizontal line is 0, because

$$m = \frac{y_2 - y_1}{x_2 - x_1} = \frac{0}{x_2 - x_1} = 0$$

Any two points (x_1, y_1) and (x_2, y_2) on a *vertical line* have the same *x*-coordinate $(x_2 = x_1)$, so $x_2 - x_1$ will equal 0. Therefore, any vertical line is said to have no slope, because

$$m = \frac{y_2 - y_1}{x_2 - x_1} = \frac{y_2 - y_1}{0}$$

and division by 0 is undefined.

Caution: "A slope of 0" is NOT THE SAME AS "no slope."

EXAMPLE 4-12: Find the slope of the line containing (a) (2, 4) and (−3, 4) and (b) (3, 3) and (3, −1).

Solution:

(a) $m = \dfrac{y_2 - y_1}{x_2 - x_1}$

$= \dfrac{4 - 4}{2 - (-3)}$ Replace (x_2, y_2) with (2, 4) and (x_1, y_1) with (−3, 4).

$= \dfrac{0}{5}$

$= 0$

So this is a line parallel to the *x*-axis (i.e., a *horizontal* line), which has a slope of 0.

(b) $m = \dfrac{y_2 - y_1}{x_2 - x_1}$

$= \dfrac{-1 - 3}{3 - 3}$ Replace (x_2, y_2) with (3, −1) and (x_1, y_1) with (3, 3).

$= \dfrac{-4}{0}$ ⟵ undefined

So this is a line parallel to the *y*-axis (i.e., a *vertical* line), which has no slope.

4-5. Finding Linear Equations

If (x, y) is any point on a line and (x_1, y_1) is any other point on the same line, then the slope of the line is

$$m = \frac{y - y_1}{x - x_1}$$

We can rewrite this equation as

$$y - y_1 = m(x - x_1)$$

which is the **point–slope form** of the equation of a line.

A. Finding an equation of a line given its slope and the coordinates of one point on the line

EXAMPLE 4-13: Find the standard form of an equation of a line with slope (a) $\frac{2}{3}$ and containing (2, 3), (b) −4 and containing (−3, 4), (c) 0 and containing (2, −1), (d) no slope and containing (−4, −1).

Solution: (a) $y - y_1 = m(x - x_1)$

$y - 3 = m(x - 2)$ Replace (x_1, y_1) with (2, 3).

$y - 3 = \frac{2}{3}(x - 2)$ Replace m with $\frac{2}{3}$.

$y - 3 = \frac{2}{3}x - \frac{4}{3}$ ⟵ point–slope form

$y = \frac{2}{3}x + \frac{5}{3}$ ⟵ slope–intercept form

$3y = 2x + 5$

$\left.\begin{array}{l} -2x + 3y = 5 \text{ or} \\ 2x - 3y = -5 \end{array}\right\}$ ⟵ standard form

(b) $y - y_1 = m(x - x_1)$

$y - 4 = m(x - (-3))$ Replace (x_1, y_1) with (−3, 4).

$y - 4 = -4(x + 3)$ Replace m with −4.

$y - 4 = -4x - 12$ ⟵ point–slope form

$y = -4x - 8$ ⟵ slope–intercept form

$4x + y = -8$ ⟵ standard form

(c) $y - y_1 = m(x - x_1)$

$y - (-1) = 0(x - 2)$ Replace (x_1, y_1) with $(2, -1)$ and m with 0.

$y + 1 = 0$ ←——— point–slope form

$y = -1$ ←——— slope–intercept and standard form

(d) Since a line with no slope is a vertical line, and has an equation of the form $x = C$, standard form of the equation of the vertical line containing $(-4, -1)$ is $x = -4$.

B. Finding the equation of a line given two points

To find **the equation of a line containing two points,** we find the slope, then use the point–slope form to find the standard form of the equation.

EXAMPLE 4-14: Find an equation of the line containing the following points:
(a) $(2, 3)$ and $(4, 6)$ (b) $(-3, 2)$ and $(4, 1)$ (c) $(-3, 2)$ and $(1, 2)$ (d) $(2, 1)$ and $(2, 3)$

Solution: (a) $m = \dfrac{y_2 - y_1}{x_2 - x_1}$ ←——— slope formula

$= \dfrac{6 - 3}{4 - 2}$ Replace (x_2, y_2) with $(6, 4)$ and (x_1, y_1) with $(2, 3)$.

$= \dfrac{3}{2}$

$y - y_1 = m(x - x_1)$ ←——— point–slope form

$y - 3 = \dfrac{3}{2}(x - 2)$ Replace (x_1, y_1) with $(2, 3)$ and m with $\dfrac{3}{2}$.

$y - 3 = \dfrac{3}{2}x - 3$

$y = \dfrac{3}{2}x$ or $2y = 3x$

$3x - 2y = 0$

(b) $m = \dfrac{y_2 - y_1}{x_2 - x_1}$

$= \dfrac{1 - 2}{4 - (-3)}$ Replace (x_2, y_2) with $(4, 1)$ and (x_1, y_1) with $(-3, 2)$.

$= -\dfrac{1}{7}$

$y - y_1 = m(x - x_1)$

$y - 1 = -\dfrac{1}{7}(x - 4)$ Replace (x_1, y_1) with $(4, 1)$ and m with $-\dfrac{1}{7}$.

$y - 1 = -\dfrac{1}{7}x + \dfrac{4}{7}$

$y = -\dfrac{1}{7}x + \dfrac{11}{7}$ or $7y = -x + 11$

$x + 7y = 11$

(c)
$$m = \frac{y_2 - y_1}{x_2 - x_1}$$

$$= \frac{2 - 2}{-3 - 1} \qquad \text{Replace } (x_1, y_1) \text{ with } (-3, 2) \text{ and } (x_1, y_1) \text{ with } (1, 2).$$

$$= \frac{0}{-4}$$

$$= 0$$

$$y - y_1 = m(x - x_1)$$

$$y - 2 = 0(x - 2)$$

$$y - 2 = 0$$

$$y = 2$$

(d) $m = \dfrac{y_2 - y_1}{x_2 - x_1}$

$$= \frac{3 - 1}{2 - 2}$$

$$= \frac{2}{0} \longleftarrow m \text{ is undefined}$$

If the slope is undefined, the line is vertical. Therefore the equation of this line is $x = 2$.

C. Finding equations of lines parallel to a given line

Two lines are parallel if they have the same slope but different y-intercepts. Lines with the same slope and y-intercept are identical.

EXAMPLE 4-15: Determine whether the graphs of (a) $3x - 2y = 4$ and $6x - 4y = 0$ and (b) $3x + 2y = 4$ and $x + 2y = 1$ are parallel.

Solution: (a) $3x - 2y = 4$ Find the slope.

$$-2y = -3x + 4$$

$$y = \tfrac{3}{2}x - 2 \longleftarrow m = \tfrac{3}{2} \text{ and } y\text{-intercept } -2$$

$$6x - 4y = 0$$

$$-4y = -6x + 0$$

$$y = \tfrac{-6}{-4}x + 0 \longleftarrow m = \tfrac{3}{2} \text{ and } y\text{-intercept } 0$$

Since both graphs have slopes of $\tfrac{3}{2}$ and different y-intercepts, the lines are parallel.

(b) $3x + 2y = 4$ Find the slope.

$$2y = -3x + 4$$

$$y = -\tfrac{3}{2}x + 2 \longleftarrow m = -\tfrac{3}{2} \text{ and } y\text{-intercept } 2$$

$$x + 2y = 1$$

$$2y = -x + 1$$

$$y = -\tfrac{1}{2}x + \tfrac{1}{2} \longleftarrow m = -\tfrac{1}{2} \text{ and } y\text{-intercept } \tfrac{1}{2}$$

Since both slopes are different, the lines are not parallel.

To **find an equation of a line parallel to a given line,** we find the slope of the given line and use it as the slope of all the lines parallel to the given line. To do this, we use $y - y_1 = m(x - x_1)$ and substitute the slope of the given line for m and the coordinates of the given point for x_1 and y_1, then simplify.

EXAMPLE 4-16: Find an equation of a line parallel to the graph of (**a**) $3x - 2y = 4$ and containing $(0, 3)$ and (**b**) $y = 4$ and containing $(0, 0)$.

Solution: (**a**) $\qquad 3x - 2y = 4 \qquad\qquad\qquad$ Find the slope.

$$-2y = -3x + 4$$

$$y = \tfrac{3}{2}x - 2 \longleftarrow m = \tfrac{3}{2}$$

$$y - y_1 = m(x - x_1) \longleftarrow \text{point-slope form}$$

$$y - 3 = \tfrac{3}{2}(x - 0) \qquad\qquad \text{Replace } (x_1, y_1) \text{ with } (0, 3) \text{ and } m \text{ with } \tfrac{3}{2}.$$

$$y - 3 = \tfrac{3}{2}x$$

$$y = \tfrac{3}{2}x + 3 \text{ or } 3x - 2y = -6 \qquad \text{Forms of the equation of the line parallel to the graph of } 3x - 2y = 4 \text{ and containing } (0, 3).$$

(**b**) $\qquad\quad y = 4 \qquad\qquad\qquad$ Find the slope.

$$y = 0x + 4 \longleftarrow m = 0$$

$$y - y_1 = m(x - x_1) \longleftarrow \text{point–slope form}$$

$$y - 0 = 0(x - 0) \qquad\qquad \text{Replace } (x_1, y_1) \text{ with } (0, 0) \text{ and } m \text{ with } 0.$$

$$y = 0 \qquad\qquad\qquad \text{The equation of the line parallel to the graph of } y = 4 \text{ and containing } (0, 0).$$

D. Finding equations of lines perpendicular to a given line

Two lines are perpendicular if the product of their slopes is -1, or, in other words, if the slope of one line is the negative reciprocal of the slope of the other line. If a line has slope m, the lines perpendicular to it have slope $-(1/m)$.

EXAMPLE 4-17: Determine whether the graphs of (**a**) $2x - 3y = 6$ and $3x - 2y = 6$, (**b**) $3x - y = 4$ and $x + 3y = 0$ and (**c**) $x = 4$ and $y = -2$ are perpendicular.

Solution: (**a**) $2x - 3y = 6 \qquad\qquad\qquad\qquad\qquad\qquad 3x - 2y = 6$

$$-3y = -2x + 6 \qquad \text{Find the slope.} \qquad -2y = -3x + 6$$

$$y = \tfrac{2}{3}x - 2 \longleftarrow m_1 = \tfrac{2}{3} \qquad\qquad m_2 = \tfrac{3}{2} \longrightarrow y = \tfrac{3}{2}x - 3$$

$$m_1 \cdot m_2 = \tfrac{2}{3} \cdot \tfrac{3}{2} = 1 \qquad \text{The product of the two slopes is not equal to } -1;$$
$$\text{therefore the lines are not perpendicular.}$$

(**b**) $3x - y = 4 \qquad\qquad\qquad\qquad\qquad\qquad x + 3y = 0$

$$-y = -3x + 4 \qquad \text{Find the slope.} \qquad 3y = -x$$

$$y = 3x - 4 \longleftarrow m_1 = 3 \qquad\qquad m_2 = -\tfrac{1}{3} \longrightarrow y = -\tfrac{1}{3}x$$

$$m_1 \cdot m_2 = 3(-\tfrac{1}{3}) = -1 \qquad \text{The product of the two slopes is } -1; \text{ therefore the lines are perpendicular.}$$

(**c**) The graph of $x = 4$ has no slope and is a vertical line. The graph of $y = -2$ has a slope of 0 and is a horizontal line. Since all horizontal lines are perpendicular to vertical lines, these lines are perpendicular

To **find an equation of a line perpendicular to a given line,** we find the slope of the given line, then use the negative reciprocal of that slope as the slope of any line perpendicular to the given line. To do this, we use $y - y_1 = m(x - x_1)$ and substitute the negative reciprocal of the slope of the given line for m, and the coordinates of the given point for x_1 and y_1, then simplify.

EXAMPLE 4-18: Find an equation of the line containing $(1, 3)$ and perpendicular to the graph of the following equations: **(a)** $3x - 2y = 4$ **(b)** $x + 4y = 2$ **(c)** $x + 2 = 0$

Solution: **(a)**

$$3x - 2y = 4 \qquad \text{Find the slope.}$$

$$-2y = -3x + 4$$

$$y = \tfrac{3}{2}x - 2 \longleftarrow m_1 = \tfrac{3}{2}$$

$$m_2 = -\tfrac{2}{3} \qquad \text{Find the negative reciprocal of } \tfrac{3}{2}.$$

$$y - y_1 = m(x - x_1) \longleftarrow \text{point–slope formula}$$

$$y - 3 = -\tfrac{2}{3}(x - 1) \qquad \text{Replace } (x_1, y_1) \text{ with } (1, 3) \text{ and } m \text{ with } -\tfrac{2}{3}.$$

$$y - 3 = -\tfrac{2}{3}x + \tfrac{2}{3}$$

$$y = -\tfrac{2}{3}x + \tfrac{11}{3} \quad \text{or} \qquad 3y = -2x + 11$$

$$2x + 3y = 11$$

(b)

$$x + 4y = 2 \qquad \text{Find the slope.}$$

$$4y = -x + 2$$

$$y = -\tfrac{1}{4}x + \tfrac{1}{2} \qquad m = -\tfrac{1}{4}$$

$$m_2 = 4 \qquad \text{Find the negative reciprocal of } -\tfrac{1}{4}.$$

$$y - y_1 = m(x - x_1) \longleftarrow \text{point–slope formula}$$

$$y - 3 = 4(x - 1) \qquad \text{Replace } (x_1, y_1) \text{ with } (1, 3) \text{ and } m \text{ with } 4.$$

$$y - 3 = 4x - 4$$

$$y = 4x - 1 \quad \text{or} \quad 4x - y = 1$$

(c)

$$x + 2 = 0 \qquad \text{Find the slope.}$$

$$x = -2 \qquad \text{The graph has no slope. The graph is a vertical line.}$$

$$m_2 = 0 \qquad \text{The slope of a horizontal line is 0.}$$

$$y - y_1 = m(x - x_1) \longleftarrow \text{point–slope formula}$$

$$y - 3 = 0(x - 1) \qquad \text{Replace } (x_1, y_1) \text{ with } (1, 3) \text{ and } m \text{ with } 0.$$

$$y = 3$$

4-6. Solving Problems Using Linear Relationships

A relationship between two unlike measures that can be described by a linear equation in two variables is called a **linear relationship.**

To solve a problem using a linear relationship, we:

1. *Read* the problem carefully, several times.
2. *Identify* the two unlike measures.
3. *Decide* how to represent the two unlike measures using two variables.
4. *Find the slope* using the two unlike measures to form a rate.
5. *Translate* the problem to a linear equation in two variables using the point–slope formula.
6. *Substitute* the given measure that is associated with the unknown measure in the linear equation in two variables to form a linear equation in one variable.
7. *Solve* the linear equation in one variable.
8. *Interpret* the solution of the linear equation in one variable with respect to the unknown measure to find the proposed solution of the original problem.
9. *Check* to see if the proposed solution satisfies all the conditions of the original problem.

EXAMPLE 4-19: Solve the following problem using a linear relationship.

1. *Read:*   A certain company makes a profit of $700 when 500 units are produced, and a profit of $550 when 400 units are produced. Assuming the profit and the number of units produced is a linear relationship, what is the break-even point [the point at which the profit, or loss, is $0]?

2. *Identify:*   The two unlike measures are $\begin{cases} \text{the number of units produced} \\ \text{the amount of profit in dollars} \end{cases}$.

3. *Decide:*   Let $x =$ the number of units produced

and $y =$ the amount of profit in dollars [see Note 1].

4. *Find the slope:* $m = \dfrac{y_1 - y_2}{x_1 - x_2}$ ⟵ slope formula

$$= \frac{700 - 550}{500 - 400} \qquad \text{Substitute: } (x_1, y_1) = (500, 700) \text{ and } (x_2, y_2) = (400, 550).$$

$$= \frac{150}{100}$$

$$= \frac{3}{2} \quad \text{⟵ slope of the linear equation in two variables [see Note 2]}$$

5. *Translate:* $y - y_1 = m(x - y_1)$ ⟵ point–slope formula

$y - 700 = \frac{3}{2}(x - 500)$ Substitute $m = \frac{3}{2}$ and $(x_1, y_1) = (500, 700)$.

$y - 700 = \frac{3}{2}x - 750$ Write slope–intercept form.

$y = \frac{3}{2}x - 50$ ⟵ linear equation in two variables that describes the linear relationship between profit (y) and units (x) [see Note 3]

6. *Substitute:* $(0) = \frac{3}{2}x - 50$ Find the break-even point [profit (y) = $0].

$0 = \frac{3}{2}x - 50$ ⟵ linear equation in one variable

7. *Solve:* $50 = \frac{3}{2}x$

$33\frac{1}{3} = x$ ⟵ proposed break-even point [see Note 4]

8. *Interpret:* $x = 33\frac{1}{3}$ means the break-even point will result when $33\frac{1}{3}$ units are produced.

9. *Check:*   Does producing $33\frac{1}{3}$ units (x) result in $0 profit ($y$) using $y = \frac{3}{2}x - 50$?
Yes: $\frac{3}{2}x - 50 = \frac{3}{2}(33\frac{1}{3}) - 50 = \frac{3}{2} \cdot \frac{100}{3} - 50 = 50 - 50 = 0$.

Does producing 500 units (x) result in $700 profit ($y$) using $y = \frac{3}{2}x - 50$?
Yes: $\frac{3}{2}x - 50 = \frac{3}{2}(500) - 50 = 750 - 50 = 700$.

Does producing 400 units (x) result in $550 profit ($y$) using $y = \frac{3}{2}x - 50$?
Yes: $\frac{3}{2}x - 50 = \frac{3}{2}(400) - 50 = 600 - 50 = 550$.

Note 1: If you let $x =$ the amount of profit in dollars and $y =$ the number of units produced in Step 3 of Example 4-19, you will get the same solution [$33\frac{1}{3}$ units] through use of a different linear equation in two variables: $y = \frac{2}{3}x + \frac{100}{3}$.

Note 2: The positive slope $\frac{3}{2}$ means the profit increases by $3 for each increase in production of 2 units.

Note 3: For $y = \frac{3}{2}x - 50$, the negative y-intercept [$x = 0$] shows that there will be a loss of $50 when no [zero] units are produced.

Note 4: For $y = \frac{3}{2}x - 50$, the x-intercept [$y = 0$] represents the break-even point.

4-7. Graphing Linear Inequalities

If we graph a linear equation of the form $Ax + By = C$, the coordinate plane will consist of the graph of the equation and two half-planes. One half-plane is the graph of $Ax + By > C$ and the other is the graph of $Ax + By < C$. We can determine which half-plane is the solution set of $Ax + By > C$ by determining if any one point in that half-plane is a solution of the inequality.

A. Determining whether ordered pairs are solutions of linear inequalities

An ordered pair is **a solution of a linear inequality** if the statement obtained by replacing the variables with the coordinates is true.

EXAMPLE 4-20: Determine whether $(-1, 2)$ is a solution of **(a)** $3x - 2y < 6$ **(b)** $2x + 3y \geq 12$.

Solution: **(a)**

$$3x - 2y < 6 \longleftarrow \text{original inequality}$$

$$\begin{array}{c|c} 3(-1) - 2(2) & 6 \\ -3 - 4 & 6 \\ -7 & 6 \end{array}$$ Replace x with -1 and y with 2.

$$-7 < 6 \longleftarrow (-1, 2) \text{ is a solution of } 3x - 2y < 6$$

(b)

$$2x + 3y \geq 12 \longleftarrow \text{original inequality}$$

$$\begin{array}{c|c} 2(-1) + 3(2) & 12 \\ -2 + 6 & 12 \\ 4 & 12 \end{array}$$ Replace x with -1 and y with 2.

$$4 \geq 12 \longleftarrow \text{false}$$

So $(-1, 2)$ is not a solution of $2x + 3y \geq 12$.

B. Graphing linear inequalities of the form $Ax + By \geq C$ or $Ax + By \leq C$ where A and B are not zero

To **graph a linear inequality,** we graph the equation form of the inequality and use a test point to determine which half-plane is the graph of the linear inequality. We then shade that half-plane to indicate that it is the solution. $Ax + By \leq C$ means $Ax + By = C$ or $Ax + By < C$.

EXAMPLE 4-21: Graph **(a)** $3x + 2y \leq 6$ **(b)** $x - 4y \geq 4$.

Solution: **(a)**

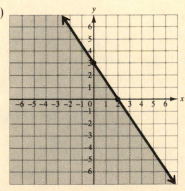

Graph $3x + 2y = 6$ using a solid line, since it is part of the solution.

Use $(0, 0)$ as a test point and find that $3(0) + 2(0) \leq 6$ means $(0, 0)$ is a solution.

Shade the half-plane that contains $(0, 0)$.

(b)

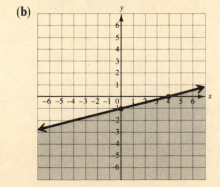

Graph $x - 4y = 4$ using a solid line since it is part of the solution.

Use $(0, 0)$ as a test point and find that $0 - 4(0) \geq 4$ means $(0, 0)$ is not a solution.

Shade the half-plane that does not contain $(0, 0)$.

C. Graphing linear inequalities of the form $Ax + By < C$ or $Ax + By > C$

We indicate that $Ax + By = C$ is not part of the solution set of $Ax + By < C$ by plotting its graph with a dashed line.

EXAMPLE 4-22: Graph $3x + 4y > 12$.

Solution:

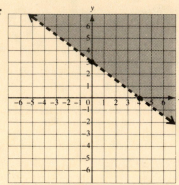

Graph $3x + 4y = 12$ using a dashed line, since it is not part of the solution.

Use $(0, 0)$ as a test point and find that $3(0) + 4(0) > 12$, meaning that $(0, 0)$ is not a solution.

Shade the half-plane that does not contain $(0, 0)$.

Note: If the graph of an ordered pair is on the line, we are not able to use it as a test point. We must choose a point that is in one of the half-planes.

D. Graphing linear equations of the form $x \geq C$ or $y < C$

Earlier we graphed simple inequalities such as $x > C$ on a number line. Now we graph similar inequalities on rectangular coordinates by writing the inequality in the form $x + 0y > C$.

EXAMPLE 4-23: Graph **(a)** $x \geq -2$ and **(b)** $y < 2$.

Solution: **(a)**

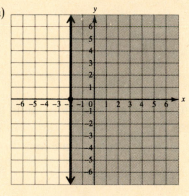

Graph $x = -2$ using a solid line because it is part of the solution.

Use $(0, 0)$ as a test point and find that $0 \geq -2$, meaning that $(0, 0)$ is a solution.

Shade the half-plane that contains $(0, 0)$.

(b)

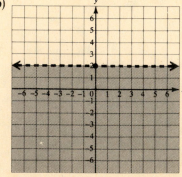

Graph $y = 2$ using a dashed line because it is not part of the solution.

Use $(0, 0)$ as a test point and find that $0 < 2$, meaning that $(0, 0)$ is a solution.

Shade the half-plane that contains $(0, 0)$.

4-8. Graphing Absolute Value Inequalities

To graph an absolute value inequality, we write it as a compound inequality.

A. Graphing compound-*or* inequalities

To graph a compound-*or* inequality, we graph the two simple inequalities. The graph of the compound-*or* inequality is the union of the graphs of the simple inequalities.

EXAMPLE 4-24: Graph the following inequalities: (a) $x < 2$ or $y \geq -1$
(b) $x + y \leq 3$ or $3x - 2y > 6$

Solution: (a)

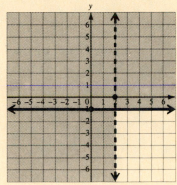

Graph $x < 2$.
Graph $y \geq -1$.
Shade the union of the two graphs.

(b)

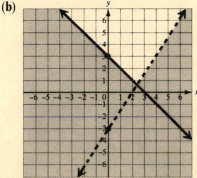

Graph $x + y \leq 3$.
Graph $3x - 2y > 6$.
Shade the union of the two graphs.

B. Graphing compound-*and* inequalities

To graph a compound-*and* inequality, we graph the two simple inequalities. The graph of the compound-*and* inequality is the intersection of the graphs of the simple inequalities.

EXAMPLE 4-25: Graph (a) $2x - 3y < 6$ and $2x - 3y \geq 0$ and (b) $-3 \leq x < 2$.

Solution: (a)

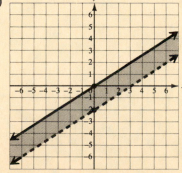

Graph $2x - 3y < 6$.
Graph $2x - 3y \geq 0$.
Shade the intersection of the two graphs.

(b)

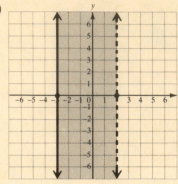

Graph $-3 \le x$.

Graph $x < 2$.

Shade the intersection of the two graphs.

C. Graphing absolute value inequalities of the form $|Ax + By| \ge C$

To graph an absolute value inequality of the form $|Ax + By| \ge C$, where $C \ge 0$, we graph $Ax + By \ge C$ and $Ax + By \le -C$. The union of the two graphs is the graph of the absolute value inequality.

EXAMPLE 4-26: Graph the following inequalities: **(a)** $|y + 2| \ge 1$ **(b)** $|2x + 3y| > 6$

Solution: **(a)** $y + 2 \ge 1$ or $y + 2 \le -1$

$\qquad\qquad y \ge -1$ or $\qquad y \le -3$

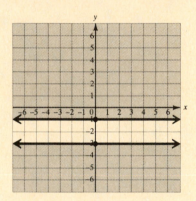

Graph $y \ge -1$.

Graph $y \le -3$.

Shade the union of the two graphs.

(b) $2x + 3y > 6$ or $2x + 3y < -6$

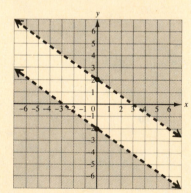

Graph $2x + 3y > 6$.

Graph $2x + 3y < -6$.

Shade the union of the two graphs.

D. Graphing absolute value inequalities of the form $|Ax + By| \le C$

To graph an absolute value inequality of the form $|Ax + By| \le C$, we graph $-C \le Ax + By \le C$. The intersection of the two graphs is the graph of the absolute value inequality.

EXAMPLE 4-27: Graph the following inequalities: **(a)** $|x - 2| \le 2$ **(b)** $|3x + 4y| < 12$

Solution: **(a)** $-2 \le x - 2 \le 2$

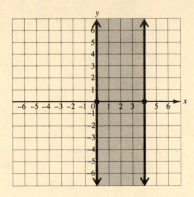

Graph $0 \le x$.

Graph $x \le 4$.

Shade the intersection of the two graphs.

(b) $-12 < 3x + 4y < 12$

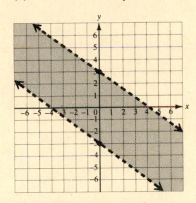

Graph $-12 < 3x + 4y$.

Graph $3x + 4y < 12$.

Shade the intersection of the two graphs.

RAISE YOUR GRADES

Can you . . . ?

☑ graph ordered pairs
☑ find the coordinates of points
☑ determine whether an ordered pair is a solution of an equation
☑ find indicated solutions for given equations
☑ find the intercepts of the graphs of equations
☑ graph linear equations by graphing two solutions
☑ graph linear equations using the intercept method
☑ graph linear equations having graphs parallel to an axis
☑ find the slopes and y-intercepts of equations of the form $y = mx + b$
☑ graph linear equations using the y-intercept and slopes
☑ find the slopes of lines
☑ find the slopes of lines parallel to an axis
☑ find an equation of a line given its slope and the coordinates of a point on the line
☑ find an equation of a line containing two points
☑ determine whether two lines are parallel
☑ find an equation of a line parallel to a given line and containing a given point
☑ determine whether two lines are perpendicular
☑ find an equation of a line perpendicular to a given line and containing a given point
☑ solve a problem using a linear relationship
☑ determine whether ordered pairs are solutions of linear inequalities
☑ graph linear inequalities
☑ graph compound inequalities
☑ graph absolute value inequalities

SOLVED PROBLEMS

PROBLEM 4-1 Graph the following ordered pairs:

(a) $A(3, 2)$, $B(-2, 3)$, $C(-3, -4)$, $D(1, -2)$
(b) $E(1, 0)$, $F(0, 2)$, $G(-3, 0)$, $H(0, -4)$
(c) $I(-2, 1)$, $J(3, -4)$, $K(-1, -3)$, $L(2, -4)$

Solution: Recall that to graph an ordered pair, we plot a dot at the point on the coordinate system associated with that pair, using the abscissa as a horizontal measure and the ordinate as a vertical measure from the origin [see Example 4-1].

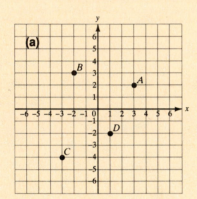 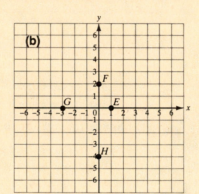 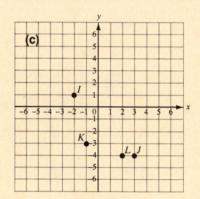

PROBLEM 4-2 Find the coordinates of the following points:

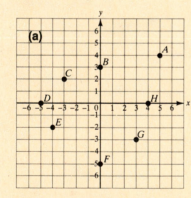 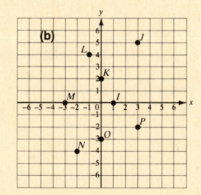 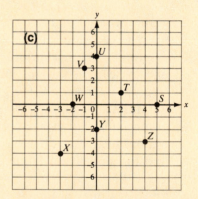

Solution: Recall that to find the coordinates of a point, we project lines from the point that are perpendicular to each axis, and read the resulting coordinates [see Example 4-2].

(a) $A(5, 4)$, $B(0, 3)$, $C(-3, 2)$,
$D(-5, 0)$, $E(-4, -2)$,
$F(0, -5)$, $G(3, -3)$,
$H(4, 0)$

(b) $I(1, 0)$, $J(3, 5)$, $K(0, 2)$,
$L(-1, 4)$, $M(-3, 0)$,
$N(-2, -4)$, $O(0, -3)$,
$P(3, -2)$

(c) $S(5, 0)$, $T(2, 1)$, $U(0, 4)$,
$V(-1, 3)$, $W(-2, 0)$,
$X(-3, -4)$, $Y(-2, 0)$,
$Z(4, -3)$

PROBLEM 4-3 Determine whether the given ordered pair is a solution of the given equation:

(a) $(3, -2)$, $4x + 3y = 6$
(b) $(3, 4)$, $4x - 3y = 0$
(c) $(-3, 4)$, $4x - 3y = 0$
(d) $(3, \frac{1}{3})$, $3y = 2x - 5$
(e) $(-1, -\frac{1}{2})$, $2y = 4x + 3$
(f) $(0, \frac{2}{3})$, $3y = 5x + 2$

Solution: Recall that an ordered pair is a solution of a given equation if the statement obtained by replacing the variables with the coordinates is true [see Example 4-3].

(a)

$$4x + 3y = 6$$

$4(3) + 3(-2)$	6
$12 - 6$	6
6	6

$$6 = 6$$

$(3, -2)$ is a solution.

(b)

$$4x - 3y = 0$$

$4(3) - 3(4)$	0
$12 - 12$	0
0	0

$$0 = 0$$

$(3, 4)$ is a solution.

(c)

$$4x - 3y = 0$$

$4(-3) - 3(4)$	0
$-12 - 12$	0
-24	0

$$-24 \neq 0$$

$(-3, 4)$ is not a solution.

(d)

$$3y = 2x - 5$$

$3(\frac{1}{3})$	$2(3) - 5$
1	$6 - 5$
1	1

$$1 = 1$$

$(3, \frac{1}{3})$ is a solution.

(e)

$$2y = 4x + 3$$

$2(-\frac{1}{2})$	$4(-1) + 3$
-1	$-4 + 3$
-1	-1

$$-1 = -1$$

$(-1, -\frac{1}{2})$ is a solution.

(f)

$$3y = 5x + 2$$

$3(\frac{2}{3})$	$5(0) + 2$
2	$0 + 2$
2	2

$$2 = 2$$

$(0, \frac{2}{3})$ is a solution.

PROBLEM 4-4 Find the indicated solution for each equation:

(a) $4x + 3y = 6$ for $x = 3, 0, -3$

(b) $3x - 4y = 12$ for $y = 3, 0, -3$

(c) $x + 3y = 4$ for $x = 4, 1, -2$

(d) $5x + 2y = 0$ for $x = 2, 0, -1$

Solution: Recall that to find the indicated solutions for an equation in two variables, we substitute the numerical values for one variable and solve the resulting equation for the other variable [see Example 4-4].

(a) $4x + 3y = 6$

$$3y = -4x + 6$$
$$y = -\tfrac{4}{3}x + 2$$

x	$-\tfrac{4}{3}x + 2 = y$
3	$-\tfrac{4}{3}(3) + 2 = -2$
0	$-\tfrac{4}{3}(0) + 2 = 2$
-3	$-\tfrac{4}{3}(-3) + 2 = 6$

$(3, -2), (0, 2), (-3, 6)$ are solutions.

(b) $3x - 4y = 12$

$$3x = 4y + 12$$
$$x = \tfrac{4}{3}y + 4$$

y	$\tfrac{4}{3}y + 4 = x$
3	$\tfrac{4}{3}(3) + 4 = 8$
0	$\tfrac{4}{3}(0) + 4 = 4$
-3	$\tfrac{4}{3}(-3) + 4 = 0$

$(8, 3), (4, 0), (0, -3)$ are solutions.

(c) $x + 3y = 4$

$$3y = -x + 4$$
$$y = -\tfrac{1}{3}x + \tfrac{4}{3}$$

x	$-\tfrac{1}{3}x + \tfrac{4}{3} = y$
4	$-\tfrac{1}{3}(4) + \tfrac{4}{3} = 0$
1	$-\tfrac{1}{3}(1) + \tfrac{4}{3} = 1$
-2	$-\tfrac{1}{3}(-2) + \tfrac{4}{3} = 2$

$(4, 0), (1, 1), (-2, 2)$ are solutions.

(d) $5x + 2y = 0$

$$2y = -5x$$
$$y = -\tfrac{5}{2}x$$

x	$-\tfrac{5}{2}x = y$
2	$-\tfrac{5}{2}(2) = -5$
0	$-\tfrac{5}{2}(0) = 0$
-1	$-\tfrac{5}{2}(-1) = \tfrac{5}{2}$

$(2, -5), (0, 0), (-1, \tfrac{5}{2})$ are solutions.

PROBLEM 4-5 Find the x-intercept and the y-intercept of the graph of:

(a) $3x + 2y = 6$ **(b)** $3x - y = 6$ **(c)** $y = \tfrac{2}{3}x + 4$

(d) $y = \tfrac{3}{4}x - 3$ **(e)** $\tfrac{2}{3}x + \tfrac{4}{5}y = 4$ **(f)** $\tfrac{4}{5}x - \tfrac{2}{3}y = 4$

Solution: Recall that to find an intercept for one variable, we replace the other variable with 0 [see Example 4-5].

(a) $3x + 2y = 6$

$3x + 2(0) = 6$

$3x + 0 = 6$

$3x = 6$

$x = 2$

$3x + 2y = 6$

$3(0) + 2y = 6$

$0 + 2y = 6$

$2y = 6$

$y = 3$

$(2, 0)$ is the *x*-intercept and $(0, 3)$ is the *y*-intercept.

(b) $3x - y = 6$

$3x - 0 = 6$

$3x = 6$

$x = 2$

$3x - y = 6$

$3(0) - y = 6$

$-y = 6$

$y = -6$

$(2, 0)$ is the *x*-intercept and $(0, -6)$ is the *y*-intercept.

(c) $y = \frac{2}{3}x + 4$

$0 = \frac{2}{3}x + 4$

$-4 = \frac{2}{3}x$

$-6 = x$

$y = \frac{2}{3}x + 4$

$y = \frac{2}{3}(0) + 4$

$y = 4$

$(-6, 0)$ is the *x*-intercept and $(0, 4)$ is the *y*-intercept.

(d) $y = \frac{3}{4}x - 3$

$0 = \frac{3}{4}x - 3$

$3 = \frac{3}{4}x$

$4 = x$

$y = \frac{3}{4}x - 3$

$y = \frac{3}{4}(0) - 3$

$y = -3$

$(4, 0)$ is the *x*-intercept and $(0, -3)$ is the *y*-intercept.

(e) $\frac{2}{3}x + \frac{4}{5}y = 4$

$\frac{2}{3}x + \frac{4}{5}(0) = 4$

$\frac{2}{3}x = 4$

$x = 6$

$\frac{2}{3}x + \frac{4}{5}y = 4$

$\frac{2}{3}(0) + \frac{4}{5}y = 4$

$\frac{4}{5}y = 4$

$y = 5$

$(6, 0)$ is the *x*-intercept and $(0, 5)$ is the *y*-intercept.

(f) $\frac{4}{5}x - \frac{2}{3}y = 4$

$\frac{4}{5}x - \frac{2}{3}(0) = 4$

$\frac{4}{5}x = 4$

$x = 5$

$\frac{4}{5}x - \frac{2}{3}y = 4$

$\frac{4}{5}(0) - \frac{2}{3}y = 4$

$-\frac{2}{3}y = 4$

$y = -6$

$(5, 0)$ is the *x*-intercept and $(0, -6)$ is the *y*-intercept.

PROBLEM 4-6 Graph the following linear equations:

(a) $3x + 2y = 6$ **(b)** $3x - 2y = 6$ **(c)** $2x + y = 4$

(d) $y = \frac{1}{2}x + 2$ **(e)** $y = \frac{2}{3}x + 2$ **(f)** $\frac{1}{3}x + \frac{1}{2}y = 1$

Solution: Recall that we graph a linear equation by plotting any two solutions of the equation and drawing a straight line through the two points. A third point can be used as a check [see Example 4-6].

(a) $3x + 2y = 6$

$2y = -3x + 6$

$y = -\frac{3}{2}x + 3$

x	$-\frac{3}{2}x + 3 = y$
2	$-\frac{3}{2}(2) + 3 = 0$
0	$-\frac{3}{2}(0) + 3 = 3$
4	$-\frac{3}{2}(4) + 3 = -3$

$(2, 0), (0, 3), (4, -3)$ are solutions.

(b) $3x - 2y = 6$

$-2y = -3x + 6$

$y = \frac{3}{2}x - 3$

x	$\frac{3}{2}x - 3 = y$
2	$\frac{3}{2}(2) - 3 = 0$
0	$\frac{3}{2}(0) - 3 = -3$
4	$\frac{3}{2}(4) - 3 = 3$

$(2, 0), (0, -3), (4, 3)$ are solutions.

(c) $2x + y = 4$

$y = -2x + 4$

x	$-2x + 4$
0	$-2(0) + 4 = 4$
2	$-2(2) + 4 = 0$
4	$-2(4) + 4 = -4$

$(0, 4), (2, 0), (4, -4)$, are solutions.

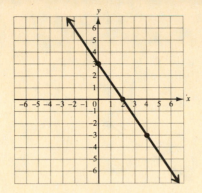 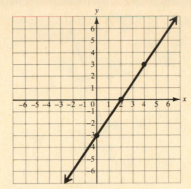 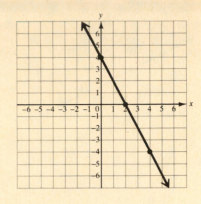

(d) $y = \frac{1}{2}x + 2$

x	$\frac{1}{2}x + 2 = y$
0	$\frac{1}{2}(0) + 2 = 2$
2	$\frac{1}{2}(2) + 2 = 3$
4	$\frac{1}{2}(4) + 2 = 4$

$(0, 2), (2, 3), (4, 4)$ are solutions.

(e) $y = \frac{2}{3}x + 2$

x	$\frac{2}{3}x + 2 = y$
0	$\frac{2}{3}(0) + 2 = 2$
3	$\frac{2}{3}(3) + 2 = 4$
-3	$\frac{2}{3}(-3) + 2 = 0$

$(0, 2), (3, 4), (-3, 0)$ are solutions.

(f) $\frac{1}{3}x + \frac{1}{2}y = 1$

$$\frac{1}{2}y = -\frac{1}{3}x + 1$$
$$y = -\frac{2}{3}x + 2$$

x	$-\frac{2}{3}x + 2 = y$
0	$-\frac{2}{3}(0) + 2 = 2$
3	$-\frac{2}{3}(3) + 2 = 0$
-3	$-\frac{2}{3}(-3) + 2 = 4$

$(0, 2), (3, 0), (-3, 4)$ are solutions.

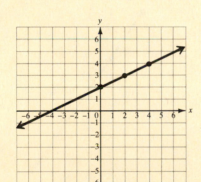 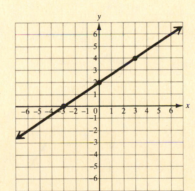 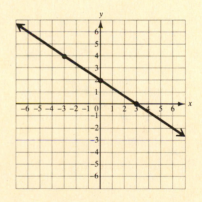

PROBLEM 4-7 Graph the following linear equations using the intercept method:

(a) $x + 2y = 4$ **(b)** $3x - y = 3$ **(c)** $2x - 3y = 6$
(d) $y = \frac{2}{3}x - 2$ **(e)** $y = -\frac{3}{2}x + 3$ **(f)** $\frac{1}{2}x + \frac{1}{3}y = 1$.

Solution: Recall that to graph a linear equation using the intercept method, we plot both intercepts and draw a straight line through these two points [see Example 4-7].

(a) $x + 2y = 4$
$x + 2(0) = 4$
$x = 4$
$0 + 2y = 4$
$y = 2$
$(4, 0)$ and $(0, 2)$ are the intercepts.

(b) $3x - y = 3$
$3x - 0 = 3$
$x = 1$
$3(0) - y = 3$
$y = -3$
$(1, 0)$ and $(0, -3)$ are the intercepts.

(c) $2x - 3y = 6$
$2x - 3(0) = 6$
$x = 3$
$2(0) - 3y = 6$
$y = -2$
$(3, 0)$ and $(0, -2)$ are the intercepts.

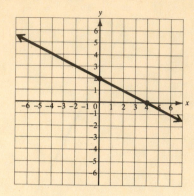

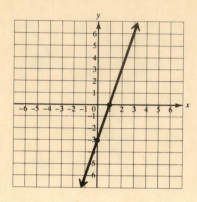

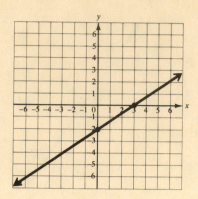

(d) $y = \frac{2}{3}x - 2$

$0 = \frac{2}{3}x - 2$

$3 = x$

$y = \frac{2}{3}(0) - 2$

$y = -2$

$(3, 0)$ and $(0, -2)$ are the intercepts.

(e) $y = -\frac{3}{2}x + 3$

$0 = -\frac{3}{2}x + 3$

$2 = x$

$y = -\frac{3}{2}(0) + 3$

$y = 3$

$(2, 0)$ and $(0, 3)$ are the intercepts.

(f) $\frac{1}{2}x + \frac{1}{3}y = 1$

$\frac{1}{2}x + \frac{1}{3}(0) = 1$

$x = 2$

$\frac{1}{2}(0) + \frac{1}{3}y = 1$

$y = 3$

$(2, 0)$ and $(0, 3)$ are the intercepts.

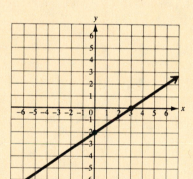

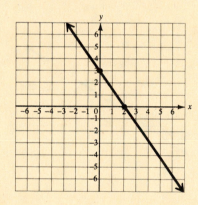

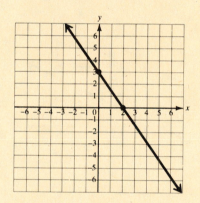

PROBLEM 4-8 Graph the following linear equations:

(a) $x = 4$ **(b)** $y = 2$ **(c)** $x = -3$
(d) $y = -1$ **(e)** $y + 4 = 0$ **(f)** $x = 0$

Solution: Recall that to graph any linear equation we graph two of its solutions and draw a straight line through the two points [see Example 4-8].

(a) $x = 4$

$x = 0y + 4$

y	$0y + 4 = x$
2	$0(2) + 4 = 4$
0	$0(0) + 4 = 4$

$(4, 2)$ and $(4, 0)$ are two solutions.

(b) $y = 2$

$y = 0x + 2$

x	$0x + 2 = y$
2	$0(2) + 2 = 2$
-2	$0(-2) + 2 = 2$

$(2, 2)$ and $(-2, 2)$ are two solutions.

(c) $x = -3$

$x = 0y - 3$

y	$0y - 3 = x$
4	$0(4) - 3 = -3$
-4	$0(-4) - 3 = -3$

$(-3, 4)$ and $(-3, -4)$ are two solutions.

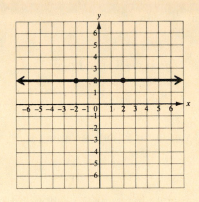

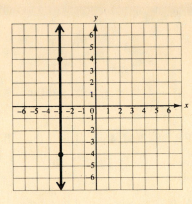

(d) $y = -1$

$y = 0x - 1$

x	$0x - 1 = y$
4	$0(4) - 1 = -1$
-4	$0(-4) - 1 = -1$

$(4, -1)$ and $(-4, -1)$ are two solutions.

(e) $y + 4 = 0$

$y = 0x - 4$

x	$0x - 4 = y$
4	$0(4) - 4 = -4$
-3	$0(-3) - 4 = -4$

$(4, -4)$ and $(-3, -4)$ are two solutions.

(f) $x = 0$

$x = 0y + 0$

y	$0y + 0 = x$
4	$0(4) + 0 = 0$
-3	$0(-3) + 0 = 0$

$(0, 4)$ and $(0, -3)$ are two solutions.

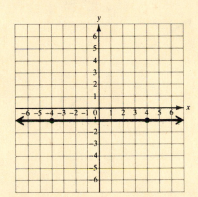

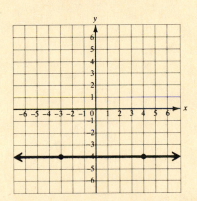

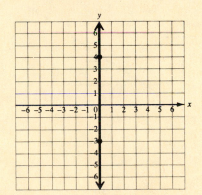

PROBLEM 4-9 Find the slope and y-intercept of the graph of the following linear equations:

(a) $3x + 2y = 12$ **(b)** $3x - 2y = 12$ **(c)** $2x - 5y = 10$
(d) $5x + 2y = 10$ **(e)** $3x - 4y = 6$ **(f)** $3x - 4y = 0$
(g) $2x + 5y = 0$ **(h)** $y = 3$ **(i)** $x = -3$

Solution: Recall that to find the slope and y-intercept of the graph of an equation in standard form, we solve for y and read the coefficient of x as the slope and the constant as the y-intercept. [see Example 4-9].

(a) $3x + 2y = 12$

$2y = -3x + 12$

$y = -\frac{3}{2}x + 6$

The slope is $-\frac{3}{2}$ and the y-intercept is 6.

(b) $3x - 2y = 12$

$-2y = -3x + 12$

$y = \frac{3}{2}x - 6$

The slope is $\frac{3}{2}$ and the y-intercept is -6.

(c) $2x - 5y = 10$

$-5y = -2x + 10$

$y = \frac{2}{5}x - 2$

The slope is $\frac{2}{5}$ and the y-intercept is -2.

(d) $5x + 2y = 10$

$2y = -5x + 10$

$y = -\frac{5}{2}x + 5$

The slope is $-\frac{5}{2}$ and the y-intercept is 5.

(e) $3x - 4y = 6$

$-4y = -3x + 6$

$y = \frac{3}{4}x - \frac{3}{2}$

The slope is $\frac{3}{4}$ and the y-intercept is $-\frac{3}{2}$.

(f) $3x - 4y = 0$

$-4y = -3x + 0$

$y = \frac{3}{4}x + 0$

The slope is $\frac{3}{4}$ and the y-intercept is 0.

(g) $2x + 5y = 0$

$$5y = -2x + 0$$

$$y = -\tfrac{2}{5}x + 0$$

The slope is $-\tfrac{2}{5}x$ and the y-intercept is 0.

(h) $y = 3$

$$y = 0x + 3$$

The slope is 0 and the y-intercept is 3.

(i) $x = -3$

The graph of this equation is a vertical line and therefore, has no slope and no y-intercept.

PROBLEM 4-10 Graph the following linear equations, using the y-intercept and slope method:

(a) $y = \tfrac{2}{3}x + 2$ **(b)** $y = -\tfrac{4}{5}x - 1$ **(c)** $y = -\tfrac{3}{4}x + 1$
(d) $4x + 3y = 12$ **(e)** $3x - y = 2$ **(f)** $3x + 4y = 2$

Solution: Recall that to graph a linear equation using the y-intercept and slope, we plot the y-intercept, then use the denominator of the slope as the abscissa to determine the horizontal distance of a second point from the y-axis, and the numerator as the ordinate to determine the vertical distance from the y-intercept. We then draw a straight line through the two points. [see Example 4-10].

(a) $y = \tfrac{2}{3}x + 2$

The slope is $\tfrac{2}{3}$ and the y-intercept is 2.

(b) $y = -\tfrac{4}{5}x - 1$

The slope is $\tfrac{-4}{5}$ and the y-intercept is -1.

(c) $y = -\tfrac{3}{4}x + 1$

The slope is $\tfrac{-3}{4}$ and the y-intercept is 1.

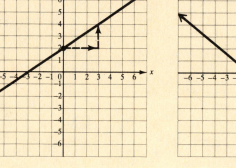
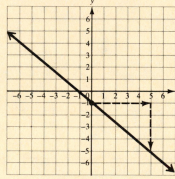
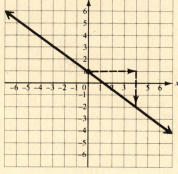

(d) $4x + 3y = 12$

$$3y = -4x + 12$$

$$y = -\tfrac{4}{3}x + 4$$

The slope is $-\tfrac{4}{3}$ and the y-intercept is 4.

(e) $3x - y = 2$

$$-y = -3x + 2$$

$$y = 3x - 2$$

The slope is $\tfrac{3}{1}$ and the y-intercept is -2.

(f) $3x + 4y = 2$

$$4y = -3x + 2$$

$$y = -\tfrac{3}{4}x + \tfrac{1}{2}$$

The slope is $-\tfrac{3}{4}$ and the y-intercept is $\tfrac{1}{2}$.

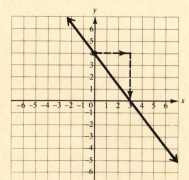
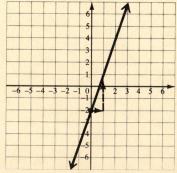
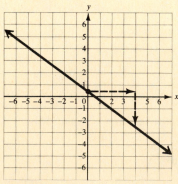

PROBLEM 4-11 Find the slope of the line containing the following points:

(a) (2, 3) and (5, 4)   **(b)** (3, -2) and (-2, 4)   **(c)** (4, 2) and (-2, -3)
(d) (4, -3) and (3, -4)   **(e)** (0, -4) and (2, 0)   **(f)** (4, 3) and (0, 2)

Solution: Recall that to find the slope of the line containing two given points, we use the following formula

$$m = \frac{y_2 - y_1}{x_2 - x_1}$$

and substitute the given coordinates for (x_1, y_1) and (x_2, y_2) [see Example 4-11].

(a) $m = \dfrac{y_2 - y_1}{x_2 - x_1}$

$= \dfrac{4 - 3}{5 - 2}$

$= \dfrac{1}{3}$

The slope is $\frac{1}{3}$.

(b) $m = \dfrac{y_2 - y_1}{x_2 - x_1}$

$= \dfrac{4 - (-2)}{-2 - 3}$

$= \dfrac{6}{-5}$

The slope is $-\frac{6}{5}$.

(c) $m = \dfrac{y_2 - y_1}{x_2 - x_1}$

$= \dfrac{-3 - 2}{-2 - 4}$

$= \dfrac{-5}{-6}$

The slope is $\frac{5}{6}$.

(d) $m = \dfrac{y_2 - y_1}{x_2 - x_1}$

$= \dfrac{-4 - (-3)}{3 - 4}$

$= \dfrac{-1}{-1}$

The slope is 1.

(e) $m = \dfrac{y_2 - y_1}{x_2 - x_1}$

$= \dfrac{0 - (-4)}{2 - 0}$

$= \dfrac{4}{2}$

The slope is 2.

(f) $m = \dfrac{y_2 - y_1}{x_2 - x_1}$

$= \dfrac{2 - 3}{0 - 4}$

$= \dfrac{-1}{-4}$

The slope is $\frac{1}{4}$.

PROBLEM 4-12 Find the slope of the line parallel to an axis and containing the following points:

(a) (3, 3) and (3, 0)   **(b)** (3, 3) and (0, 3)   **(c)** (2, 3) and (2, −3)
(d) (−2, 0) and (2, 0)   **(e)** (−2, −2) and (−2, 3)   **(f)** (−1, −1) and (−1, 0)

Solution: Recall that we find the slope of a line parallel to an axis and containing two given points by using the formula

$$m = \frac{y_2 - y_1}{x_2 - x_1}$$

and substituting the given coordinates for (x_2, y_2) and (x_1, y_1) [see Example 4-12].

(a) $m = \dfrac{y_2 - y_1}{x_2 - x_1}$

$= \dfrac{3 - 0}{3 - 3}$

$= \dfrac{3}{0}$

The line has no slope.

(b) $m = \dfrac{y_2 - y_1}{x_2 - x_1}$

$= \dfrac{3 - 3}{3 - 0}$

$= \dfrac{0}{3}$

The slope is 0.

(c) $m = \dfrac{y_2 - y_1}{x_2 - x_1}$

$= \dfrac{-3 - 3}{2 - 2}$

$= \dfrac{-6}{0}$

The line has no slope.

(d) $m = \dfrac{y_2 - y_1}{x_2 - x_1}$

$= \dfrac{0 - 0}{-2 - 2}$

$= \dfrac{0}{-4}$

The slope is 0.

(e) $m = \dfrac{y_2 - y_1}{x_2 - x_1}$

$= \dfrac{-2 - 3}{-2 - (-2)}$

$= \dfrac{-5}{0}$

The line has no slope.

(f) $m = \dfrac{y_2 - y_1}{x_2 - x_1}$

$= \dfrac{-1 - 0}{-1 - (-1)}$

$= \dfrac{-1}{0}$

The line has no slope.

PROBLEM 4-13 Find an equation of a line containing the given point and having the given slope:

(a) $(1, 2)$ and $m = 3$ (b) $(2, 3)$ and $m = -4$ (c) $(-2, 3)$ and $m = \frac{2}{3}$
(d) $(-4, 1)$ and $m = \frac{3}{4}$ (e) $(2, 0)$ and $m = -\frac{2}{5}$ (f) $(0, 0)$ and $m = \frac{5}{3}$
(g) $(-3, 3)$ and $m = 0$ (h) $(0, -1)$ and $m = 0$ (i) $(2, 0)$ and having no slope

Solution: Recall that to find the equation of a line containing a given point and having a given slope, we use the point–slope form of the equation and substitute for (x_1, y_1) and m [see Example 4-13].

(a) $y - y_1 = m(x - x_1)$

$y - 2 = 3(x - 1)$

$y - 2 = 3x - 3$

$y = 3x - 1$

$3x - y = 1$

(b) $y - y_1 = m(x - x_1)$

$y - 3 = -4(x - 2)$

$y - 3 = -4x + 8$

$y = -4x + 11$

$4x + y = 11$

(c) $y - y_1 = m(x - x_1)$

$y - 3 = \frac{2}{3}[x - (-2)]$

$y - 3 = \frac{2}{3}x + \frac{4}{3}$

$y = \frac{2}{3}x + \frac{13}{3}$

or $2x - 3y = -13$

(d) $y - y_1 = m(x - x_1)$

$y - 1 = \frac{3}{4}[x - (-4)]$

$y - 1 = \frac{3}{4}x + 3$

$y = \frac{3}{4}x + 4$

or $3x - 4y = -16$

(e) $y - y_1 = m(x - x_1)$

$y - 0 = -\frac{2}{5}(x - 2)$

$y - 0 = -\frac{2}{5}x + \frac{4}{5}$

$y = -\frac{2}{5}x + \frac{4}{5}$

or $2x + 5y = 4$

(f) $y - y_1 = m(x - x_1)$

$y - 0 = \frac{5}{3}(x - 0)$

$y - 0 = \frac{5}{3}x - 0$

$y = \frac{5}{3}x$

or $5x - 3y = 0$

(g) $y - y_1 = m(x - x_1)$

$y - 3 = 0[x - (-3)]$

$y - 3 = 0$

$y = 3$

(h) $y - y_1 = m(x - x_1)$

$y - (-1) = 0(x - 0)$

$y + 1 = 0$

$y = -1$

(i) Since the line has no slope, it is a vertical line and has an equation of $x = x_1$.

$x = 2$

PROBLEM 4-14 Find an equation of the line containing the following points:

(a) $(2, 3)$ and $(5, 4)$ (b) $(3, -2)$ and $(-2, 4)$ (c) $(4, 2)$ and $(-2, -3)$
(d) $(4, 3)$ and $(0, 3)$ (e) $(-3, -1)$ and $(3, -1)$ (f) $(3, 3)$ and $(3, 0)$

Solution: Recall that to find an equation of a line, we find the slope, then use the point–slope form to find the standard form of the equation [see Example 4-14].

(a) $m = \dfrac{4 - 3}{5 - 2}$

$= \dfrac{1}{3}$

$y - y_1 = m(x - x_1)$

$y - 4 = \dfrac{1}{3}(x - 5)$

$y - 4 = \dfrac{1}{3}x - \dfrac{5}{3}$

$y = \dfrac{1}{3}x + \dfrac{7}{3}$

$x - 3y = -7$

(b) $m = \dfrac{-2 - 4}{3 - (-2)}$

$= \dfrac{-6}{5}$

$y - y_1 = m(x - x_1)$

$y - 4 = -\dfrac{6}{5}(x - (-2))$

$y - 4 = -\dfrac{6}{5}x - \dfrac{12}{5}$

$y = -\dfrac{6}{5}x + \dfrac{8}{5}$

$6x + 5y = 8$

(c) $m = \dfrac{2 - (-3)}{4 - (-2)}$

$= \dfrac{5}{6}$

$y - 2 = \dfrac{5}{6}(x - 4)$

$y - 2 = \dfrac{5}{6}x - \dfrac{10}{3}$

$y = \dfrac{5}{6}x - \dfrac{4}{3}$

$5x - 6y = 8$

(d) $m = \dfrac{3-3}{4-0}$

$= 0$

$y - 3 = 0(x - 0)$

$y - 3 = 0$

$y = 3$

(e) $m = \dfrac{-1-(-1)}{-3-3}$

$= 0$

$y - (-1) = 0(x - 3)$

$y + 1 = 0$

$y = -1$

(f) $m = \dfrac{3-0}{3-3}$

$= \dfrac{3}{0}$

The line with no slope is vertical and has an equation of the form $x = x_1$. The equation is $x = 3$.

PROBLEM 4-15 Determine whether the graphs of the following linear equations are parallel:

(a) $3x - 2y = 4$ and $6x - 4y = 4$

(b) $2x + 3y = 4$ and $3x + 2y = 4$

(c) $4x - 2y = 1$ and $y = 2x - \frac{1}{2}$

(d) $y = \frac{2}{3}x - 2$ and $3x - 2y = 2$

Solution: Recall that to determine whether the graphs of two equations are parallel, we determine whether they have the same slope but different y-intercepts [see Example 4-15].

(a) $3x - 2y = 4$ and $6x - 4y = 4$

$-2y = -3x + 4$ $-4y = -6x + 4$

$y = \frac{3}{2}x - 2$ $y = \frac{3}{2}x - 1$

With slopes of $\frac{3}{2}$ and different y-intercepts, the lines are parallel.

(b) $2x + 3y = 4$ and $3x + 2y = 4$

$3y = -2x + 4$ $2y = -3x + 4$

$y = -\frac{2}{3}x + \frac{4}{3}$ $y = -\frac{3}{2}x + 2$

With different slopes, the lines are not parallel.

(c) $4x - 2y = 1$ and $y = 2x - \frac{1}{2}$

$-2y = -4x + 1$

$y = 2x - \frac{1}{2}$

With slopes of 2 and y-intercepts of $-\frac{1}{2}$, the lines are not parallel because they are identical.

(d) $y = \frac{2}{3}x - 2$ and $3x - 2y = 2$

$-2y = -3x + 2$

$y = \frac{3}{2}x - 1$

With different slopes, the lines are not parallel.

PROBLEM 4-16 Find an equation of a line that contains the given point and is parallel to the given line:

(a) $(2, 3)$, $y = 4x - 3$

(b) $(-2, 3)$, $2x - 3y = 4$

(c) $(-2, -3)$, $3x + 4y = 12$

(d) $(-1, 0)$, $3x = y - 2$

(e) $(2, 2)$, $y = 3$

(f) $(0, 3)$, $x = 3$

Solution: Recall that to find an equation of a line that contains a given point and is parallel to a given line, we find the slope of the given line, then use the point–slope form to find an equation of the parallel line [see Example 4-16].

(a) $y = 4x - 3$

$m = 4$

$y - y_1 = m(x - x_1)$

$y - 3 = 4(x - 2)$

$y - 3 = 4x - 8$

$y = 4x - 5$

(b) $-3y = -2x + 4$

$y = \frac{2}{3}x - \frac{4}{3}$

$y - y_1 = m(x - x_1)$

$y - 3 = \frac{2}{3}[x - (-2)]$

$y - 3 = \frac{2}{3}x + \frac{4}{3}$

$y = \frac{2}{3}x + \frac{13}{3}$

(c) $3x + 4y = 12$

$4y = -3x + 12$

$y = -\frac{3}{4}x + 3$

$y - y_1 = m(x - x_1)$

$y - (-3) = -\frac{3}{4}(x - (-2))$

$y + 3 = -\frac{3}{4}x - \frac{3}{2}$

$y = -\frac{3}{4}x - \frac{9}{2}$

(d) $3x + 2 = y$

$y - y_1 = m(x - x_1)$

$y - 0 = 3(x - (-1))$

$y = 3x + 3y - 2 = 0$

(e) $y = 0x + 3$

$y - y_1 = m(x - x_1)$

$y - 2 = 0(x - 2)$

$y - 2 = 0$

$y = 2$

(f) $x = 3$

The line has no slope.
The equation of a vertical
line has the form $x = x_1$.

$x = 0$

PROBLEM 4-17 Determine whether the graphs of the following linear equations are perpendicular:

(a) $3x - 2y = 4$ and $2x + 3y = 4$

(b) $3x - 2y = 4$ and $2x - 3y = 4$

(c) $3x - 2y = 2$ and $x = 3$

(d) $y = 3$ and $x = 3$

Solution: Recall that to determine whether the graphs are perpendicular, we determine whether the product of their slopes is -1 [see Example 4-17].

(a) $3x - 2y = 4$

$-2y = -3x + 4$

$y = \frac{3}{2}x - 2$

$2x + 3y = 4$

$3y = -2x + 4$

$y = -\frac{2}{3}x + \frac{4}{3}$

$m_1 \cdot m_2 = \frac{3}{2}(-\frac{2}{3})$

$= -1$

The lines are perpendicular.

(b) $3x - 2y = 4$

$-2y = -3x + 4$

$y = \frac{3}{2}x - 2$

$2x - 3y = 4$

$-3y = -2x + 4$

$y = \frac{2}{3}x - \frac{4}{3}$

$m_1 \cdot m_2 = \frac{3}{2} \cdot \frac{2}{3}$

$= 1$

The lines are not perpendicular.

(c) $3x - 2y = 2$

$-2y = -3x + 2$

$y = \frac{3}{2}x - 1$

$x = 3$

no slope

$\frac{3}{2}$ is the slope of a line perpendicular to a line of slope $-\frac{2}{3}$. Therefore, it is not perpendicular to a line with no slope.

(d) $y = 3$

0 slope

$x = 3$

no slope

The graph of $y = 3$ is a horizontal line. The graph of $x = 3$ is a vertical line. Therefore, the lines are perpendicular.

PROBLEM 4-18 Find an equation of a line that contains a given point and is perpendicular to a given line:

(a) $(1, 3), 3x + 2y = 6$

(b) $(2, -3), 3x + 2y = 6$

(c) $(-2, 3), x + 4y = 4$

(d) $(-2, 1), y = 3$

(e) $(2, -1), x = 3$

(f) $(-1, 2), x + 2 = 0$

Solution: Recall that to find an equation of a line that contains a given point and is perpendicular to a given line, we find the slope of the given line. We then use the negative reciprocal of that slope as the slope of the perpendicular line, and thus find the equation using the point–slope form [see Example 4-18].

(a) $3x + 2y = 6$

$2y = -3x + 6$

$y = -\frac{3}{2}x + 3$

$m_2 = \frac{2}{3}$

$y - y_1 = m(x - x_1)$

$y - 3 = \frac{2}{3}(x - 1)$

$y - 3 = \frac{2}{3}x - \frac{2}{3}$

$y = \frac{2}{3}x + \frac{7}{3}$

(b) $3x + 2y = 6$

$2y = -3x + 6$

$y = -\frac{3}{2}x + 3$

$m_2 = \frac{2}{3}$

$y - y_1 = m(x - x_1)$

$y - (-3) = \frac{2}{3}(x - 2)$

$y + 3 = \frac{2}{3}x - \frac{4}{3}$

$y = \frac{2}{3}x - \frac{13}{3}$

(c) $x + 4y = 4$

$4y = -x + 4$

$y = -\frac{1}{4}x + 1$

$m_2 = \frac{4}{1}$

$y - 3 = 4[x - (-2)]$

$y - 3 = 4x + 8$

$y = 4x + 11$

(d) $y = 3$

The line is horizontal. The line perpendicular to it is vertical, with an equation in the form $x = x_1$.

$x = -2$

(e) $x = 3$

The line is vertical. The line perpendicular to it is horizontal with an equation of the form $y = y_1$.

$y = -1$

(f) $x + 2 = 0$

$x = -2$

The line is vertical. The line perpendicular to it is horizontal with an equation of the form $y = y_1$.

$y = 2$

PROBLEM 4-19 Solve each problem using a linear relationship.

(a) A company truck is purchased for $30,000. It depreciates at a constant rate of $6000 per year (**straight-line depreciation** for tax purposes). Find the linear equation in two variables that describes this linear relationship. What is the **book value** (the value of the truck after depreciation) after 3 years?

(b) A certain taxi charges $5 plus an additional $1.50 for each $\frac{1}{4}$ miles driven. Find the linear equation that describes the charge for taking the taxi a given number of miles. How far can a person take this taxi for $20?

Solution: To solve a problem using a linear relationship, you use the steps given on **page 89**.

(a) *Identity:* The two unlike measures are $\left\{ \begin{array}{l} \text{the amount of time after purchase} \\ \text{the book value of the truck} \end{array} \right\}$.

Decide: Let $x =$ the amount of time after purchase

and $y =$ the book value of the truck.

Find the slope: $m = \dfrac{y}{x}$ ⟵ the slope in **rate form**

$= \dfrac{-6000}{1}$ ⟵ the book value (y) decreases over time (x) at the rate of $6000 per 1 year

$= -6000$ ⟵ slope

Translate: $y - y_1 = m(x - x_1)$ ⟵ point–slope formula

$y - 30,000 = -6000(x - 0)$ ⟵ the book value (y) is $30,000 at the time of purchase

$y - 30,000 = -6000x$ Write slope-intercept form.

$y = -6000x + 30,000$ ⟵ linear equation in two variables that describes the linear relationship between the book value (y) and the amount of time (x) after purchase

Substitute: $y = -6000(3) + 30,000$ Find the book value (y) after a time (x) of 3 years.

Solve: $y = -18,000 + 30,000$
$y = 12,000 \, [\text{dollars}]$ ⟵ book value after 3 years

(b) *Identify* The two unlike measures are $\left\{ \begin{array}{l} \text{the number of miles driven} \\ \text{the taxi fare} \end{array} \right\}$.

Decide: Let $x =$ the number of miles driven

and $y =$ the taxi fare.

$$\text{Find the slope:} \quad m = \frac{y}{x} \longleftarrow \text{slope in rate form}$$

$$= \frac{1.50}{\frac{1}{4}} \longleftarrow \text{the taxi rate } (y) \text{ is \$1.50 for each } \frac{1}{4} \text{ miles driven } (x)$$

$$= \frac{1.50}{0.25}$$

$$= 6 \longleftarrow \text{slope}$$

$$\text{Translate:} \quad y - y_1 = m(x - x_1) \longleftarrow \text{point–slope formula}$$

$$y - 5 = 6(x - 0) \longleftarrow \begin{array}{l}\text{the taxi fare } (y) \text{ is \$5 when the number of miles} \\ \text{driven } (x) \text{ is } 0\end{array}$$

$$y - 5 = 6x \qquad \text{Write slope–intercept form.}$$

$$y = 6x + 5 \longleftarrow \begin{array}{l}\text{linear equation in two variables that} \\ \text{describes the linear relationship between} \\ \text{the taxi fare } (y) \text{ and the number of miles} \\ \text{driven } (x)\end{array}$$

$$\text{Substitute:} \quad 20 = 6x + 5 \qquad \begin{array}{l}\text{Find the number of miles driven} \\ (x) \text{ when the taxi fare } (y) \text{ is \$20.}\end{array}$$

$$\text{Solve:} \quad 15 = 6x$$

$$\frac{15}{6} = x$$

$$x = 2\tfrac{1}{2} \,[\text{miles}] \longleftarrow \text{solution}$$

PROBLEM 4-20 Determine whether a given point is a solution of the given linear inequality:

(a) $(0, 0)$, $2x - 3y < 6$ (b) $(1, 0)$, $3x + 2y \geq 6$ (c) $(2, 3)$, $x - 4y > 4$

(d) $(2, 0)$, $3x \geq 5$ (e) $(3, -2)$, $2y > 3$ (f) $(3, 0)$, $2x \geq 5y$

Solution: Recall that an ordered pair is a solution of a linear inequality if the statement obtained by replacing the variables with the coordinates is true [see Example 4-20].

(a)
$$\begin{array}{c|c} 2x - 3y < 6 & \\ \hline 2(0) - 3(0) & 6 \\ 0 - 0 & 6 \\ 0 & 6 \end{array}$$

$$0 < 6 \longleftarrow \text{true}$$

$(0, 0)$ is a solution.

(b)
$$\begin{array}{c|c} 3x + 2y \geq 6 & \\ \hline 3(1) + 2(0) & 6 \\ 3 + 0 & 6 \\ 3 & 6 \end{array}$$

$$3 \geq 6 \longleftarrow \text{false}$$

$(1, 0)$ is not a solution.

(c)
$$\begin{array}{c|c} x - 4y > 4 & \\ \hline 2 - 4(3) & 4 \\ 2 - 12 & 4 \\ -10 & 4 \end{array}$$

$$-10 > 4 \longleftarrow \text{false}$$

$(2, 3)$ is not a solution.

(d)
$$\begin{array}{c|c} 3x \geq 5 & \\ \hline 3(2) & 5 \\ 6 & 5 \end{array}$$

$$6 \geq 5 \longleftarrow \text{true}$$

$(2, 0)$ is a solution.

(e)
$$\begin{array}{c|c} 2y > 3 & \\ \hline 2(-2) & 3 \\ -4 & 3 \end{array}$$

$$-4 > 3 \longleftarrow \text{false}$$

$(3, -2)$ is not a solution.

(f)
$$\begin{array}{c|c} 2x \geq 5y & \\ \hline 2(3) & 5(0) \\ 6 & 0 \end{array}$$

$$6 \geq 0 \longleftarrow \text{true}$$

$(3, 0)$ is a solution.

PROBLEM 4-21 Graph the following linear inequalities:

(a) $3x - 2y \leq 6$ (b) $2x + 3y \geq 6$ (c) $4y \leq 3x + 8$

(d) $x \leq 3y + 3$ (e) $3x \geq 4y$ (f) $y \geq 4x$

Solution: Recall that to graph a linear inequality, we graph the equation form of the inequality and use a test point to determine which half-plane is the solution of the linear inequality. The equation is graphed with a solid line because the line is part of the solution [see Example 4-21].

(a) Graph $3x - 2y = 6$ using a solid line. Use $(0, 0)$ as a test point and find that $3(0) - 2(0) \leq 6$ means $(0, 0)$ is a solution. Shade the half-plane that contains $(0, 0)$.

(b) Graph $2x + 3y = 6$ using a solid line. Use $(0, 0)$ as a test point and find that $2 \cdot 0 + 3 \cdot 0 \not\geq 6$ means $(0, 0)$ is not a solution. Shade the half-plane that does not contain $(0, 0)$.

(c) Graph $4y = 3x + 8$ using a solid line. Use $(0, 0)$ as a test point and find that $4(0) \leq 3(0) + 8$ means $(0, 0)$ is a solution. Shade the half-plane that contains $(0, 0)$.

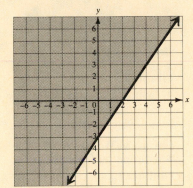 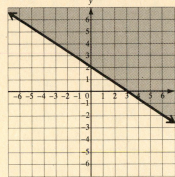 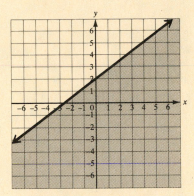

(d) Graph $x = 3y + 3$ using a solid line. Use $(0, 0)$ as a test point and find that $0 \leq 3(0) + 3$ means $(0, 0)$ is a solution. Shade the half-plane that contains $(0, 0)$.

(e) Graph $3x = 4y$ using a solid line. Use $(3, 0)$ as a test point and find that $3(3) \geq 4(0)$ means $(3, 0)$ is a solution. Shade the half-plane that contains $(3, 0)$.

(f) Graph $y = 4x$ using a solid line. Use $(3, 0)$ as a test point and find that $0 \not\geq 4(3)$ means $(3, 0)$ is a solution. Shade the half-plane that does not contain $(3, 0)$.

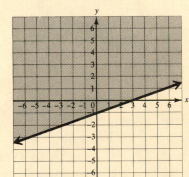 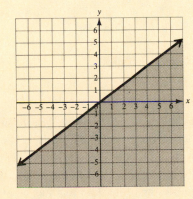 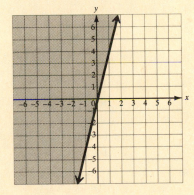

PROBLEM 4-22 Graph the following inequalities:

(a) $2x - 3y < 6$ **(b)** $3x + 2y > 12$ **(c)** $3x - 4y > 12$
(d) $3y > 4x + 9$ **(e)** $3x > 4y - 8$ **(f)** $4x - 3y > 0$

Solution: Recall that to graph a linear inequality of this form, we graph the equation form of the inequality, using a dashed line because the equation is not part of the solution [see Example 4-22].

(a) Graph $2x - 3y = 6$ using a dashed line. Use $(0, 0)$ as a test point and find that $2(0) - 3(0) < 6$, meaning that $(0, 0)$ is a solution. Shade the half-plane that contains $(0, 0)$.

(b) Graph $3x + 2y = 12$ using a dashed line. Use $(0, 0)$ as a test point and find that $3(0) + 2(0) \not> 12$, meaning that $(0, 0)$ is not a solution. Shade the half-plane that does not contain $(0, 0)$.

(c) Graph $3x - 4y = 12$ using a dashed line. Use $(0, 0)$ as a test point and find that $3(0) - 4(0) \not> 12$, meaning that $(0, 0)$ is not a solution. Shade the half-plane that does not contain $(0, 0)$.

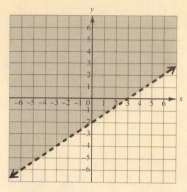
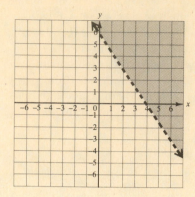
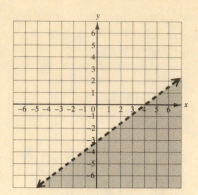

(d) Graph $3y = 4x + 9$ using a dashed line. Use $(0, 0)$ as a test point and find that $3(0) \not> 4(0) + 9$, meaning that $(0, 0)$ is not a solution. Shade the half-plane that does not contain $(0, 0)$.

(e) Graph $3x = 4y - 8$ using a dashed line. Use $(0, 0)$ as a test point and find that $3(0) > 4(0) - 8$, meaning that $(0, 0)$ is a solution. Shade the half-plane that contains $(0, 0)$.

(f) Graph $4x - 3y = 0$ using a dashed line. Use $(3, 0)$ as a test point and find that $4(3) - 3(0) > 0$, meaning that $(3, 0)$ is a solution. Shade the half-plane that contains $(3, 0)$.

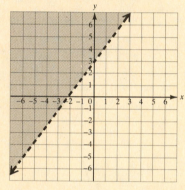
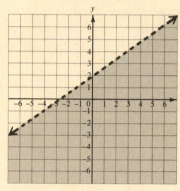
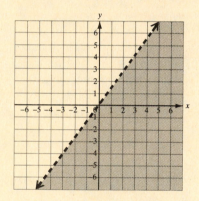

PROBLEM 4-23 Graph the following inequalities:

(a) $x \geq 3$ **(b)** $y > -3$ **(c)** $-4 > x$
(d) $1 \geq y$ **(e)** $2 \leq x$ **(f)** $0 < y$

Solution: Recall that to graph an inequality of this form, we graph the equation form of the inequality and use a test point to determine which half-plane is the graph of the linear inequality. [see Example 4-23].

(a) Graph $x = 3$ using a solid line. Use $(0, 0)$ as a test point and find that $0 \geq 3$, meaning that $(0, 0)$ is not a solution. Shade the half-plane that does not contain $(0, 0)$.

(b) Graph $y = -3$ using a dashed line. Use $(0, 0)$ as a test point and find that $0 > -3$, meaning that $(0, 0)$ is a solution. Shade the half-plane that contains $(0, 0)$.

(c) Graph $-4 = x$ using a dashed line. Use $(0, 0)$ as a test point and find that $-4 > 0$, meaning that $(0, 0)$ is not a solution. Shade the half-plane that does not contain $(0, 0)$.

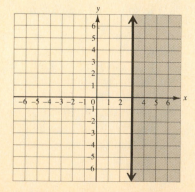
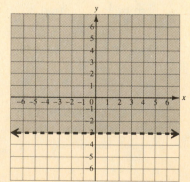
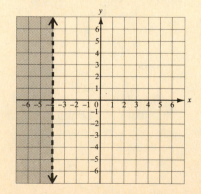

(d) Graph $1 = y$ using a solid line. Use $(0, 0)$ as a test point and find that $1 \geq 0$, meaning that $(0, 0)$ is a solution. Shade the half-plane that contains $(0, 0)$.

(e) Graph $2 = x$ using a solid line. Use $(0, 0)$ as a test point and find that $2 \leq 0$, meaning that $(0, 0)$ is not a solution. Shade the half-plane that does not contain $(0, 0)$.

(f) Graph $0 = y$ using a dashed line. Use $(0, 2)$ as a test point and find that $0 < 2$, meaning that $(0, 2)$ is a solution. Shade the half-plane that contains $(0, 2)$.

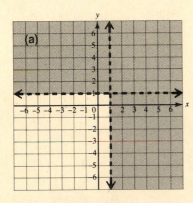

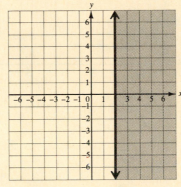

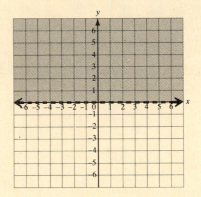

PROBLEM 4-24 Graph the following inequalities:

(a) $x > 1$ or $y > 1$

(b) $x \leq -2$ or $y \leq -3$

(d) $x + 2y \geq 4$ or $2x + y < 4$

(c) $x + y \leq 0$ or $x - y \leq 0$

Solution: Recall that to graph a compound *or* inequality, we graph the simple inequalities first. The graph of the compound *or* inequality is then the union of the graphs of the simple inequalities [see Example 4-24].

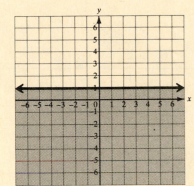

Graph $x > 1$.
Graph $y > 1$.
Shade the union.

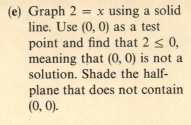

Graph $x \leq -2$.
Graph $y \leq -3$.
Shade the union.

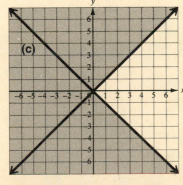

Graph $x + y \leq 0$.
Graph $x - y \leq 0$.
Shade the union.

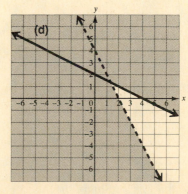

Graph $x + 2y \geq 4$.
Graph $2x + y < 4$.
Shade the union.

PROBLEM 4-25 Graph the following inequalities:

(a) $x \leq 1$ and $y \geq 1$

(b) $x > -2$ and $y < -1$

(c) $-6 < 2x + 3y < 0$

(d) $-12 < 4x - 3y \leq 12$

Solution: Recall that to graph a compound *and* inequality, we graph the two simple inequalities first. The graph of the compound *and* inequality is then the intersection of the graphs of the simple inequalities [see Example 4-25].

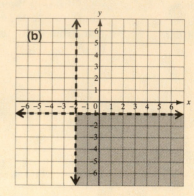

(a) Graph $x \leq 1$.
Graph $y \geq 1$.
Shade the intersection.

(b) Graph $x > -2$.
Graph $y < -1$.
Shade the intersection.

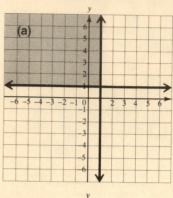

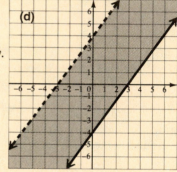

(c) Graph $-6 < 2x + 3y$.
Graph $2x + 3y < 0$.
Shade the intersection.

(d) Graph $-12 < 4x - 3y$.
Graph $4x - 3y \leq 12$.
Shade the intersection.

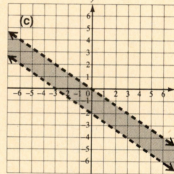

PROBLEM 4-26 Graph the following inequalities:

(a) $|x| > 2$ **(b)** $|y| \geq 3$ **(c)** $|x - 3| \geq 2$ **(d)** $|2x + 3y| > 6$

Solution: Recall that to graph an absolute value inequality of the form $|Ax + By| > C$, we graph $Ax + By > C$ and $Ax + By < -C$. The union of the two graphs is the graph of the absolute value inequality [see Example 4-26].

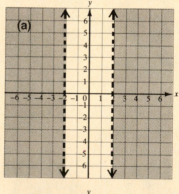

(a) Graph $x > 2$.
Graph $x < -2$.
Shade the union.

(b) Graph $y \geq 3$.
Graph $y \leq -3$.
Shade the union.

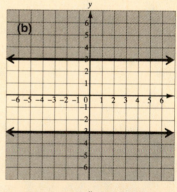

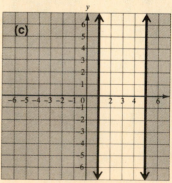

(c) Graph $x - 3 \geq 2$.
Graph $x - 3 \leq -2$.
Shade the union.

(d) Graph $2x + 3y > 6$.
Graph $2x + 3y < -6$.
Shade the union.

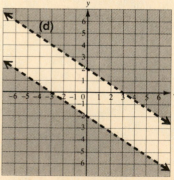

PROBLEM 4-27 Graph the following inequalities:

(a) $|x| \leq 4$ (b) $|y| < 3$ (c) $|y - 2| \leq 0$ (d) $|x - y| < 2$

Solution: Recall that to graph an absolute value inequality of the form $|Ax + By| < C$, we graph $-C < Ax + By < C$ [see Example 4-27].

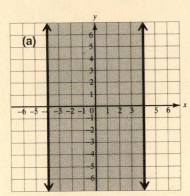

Graph $x \leq 4$

Graph $-4 \leq x$.

Shade the intersection.

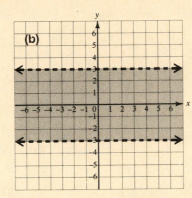

Graph $y < 3$.

Graph $-3 < y$.

Shade the intersection.

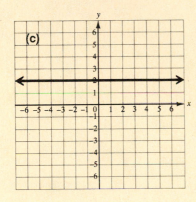

Graph $y - 2 \leq 0$.

Graph $0 \leq y - 2$.

Shade the intersection.

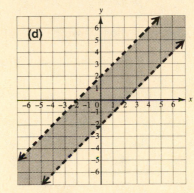

Graph $x - y < 2$.

Graph $-2 < x - y$.

Shade the intersection.

Supplementary Exercises

PROBLEM 4-28 Graph the following ordered pairs:

(a) $A(4, 4)$, $B(-3, 2)$, $C(-1, -4)$, $D(4, -2)$ (b) $E(-1, 0)$, $F(0, 1)$, $G(0, -3)$, $H(0, 3)$
(c) $M(1, 4)$, $N(-2, 4)$, $P(-3, -2)$, $Q(3, -2)$

PROBLEM 4-29 Find the coordinates of the following points:

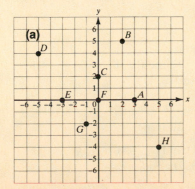

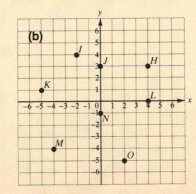

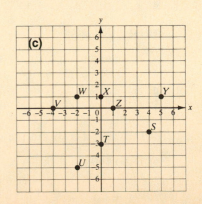

PROBLEM 4-30 Determine whether the given ordered pair is a solution of the given equation:

(a) $(-2, 4)$, $2x + 3y = 6$ (b) $(3, 0)$, $2x + 3y = 6$ (c) $(-1, -2)$, $3x - y = 1$
(d) $(1, 2)$, $3x - y = 1$ (e) $(2, 4)$, $3x - y = 1$ (f) $(\frac{1}{2}, 0)$, $y = 4x - 2$
(g) $(0, \frac{2}{3})$, $2y = x + 3$ (h) $(\frac{1}{2}, \frac{2}{3})$, $3y = 2x + 1$

PROBLEM 4-31 Find the indicated solution for each equation:

(a) $3x - 2y = 6$ for $x = 2, 0, -2$ (b) $3x - 2y = 6$ for $y = 3, 0, -3$
(c) $2x + 3y = 12$ for $x = 6, 3, 0$ (d) $2x + 3y = 12$ for $y = 4, 2, 0$
(e) $2x + 3y = 12$ for $x = 0, -3, -6$ (f) $x + 3y = 0$ for $y = 1, 0, -1$
(g) $x + 3y = 0$ for $x = 3, 0, -3$ (h) $2x = 3y - 3$ for $y = 1, 0, -1$
(i) $2x = 3y - 3$ for $x = 0, -1, -2$ (j) $y = \frac{2}{3}x + 3$ for $x = 3, 0, -\frac{3}{2}$
(k) $y = \frac{3}{2}x + \frac{1}{3}$ for $x = 2, 0, -\frac{1}{3}$

PROBLEM 4-32 Find the x-intercept and y-intercept of the graph of the following equations:

(a) $4x + 3y = 12$ (b) $4x - 3y = 12$ (c) $x - 4y = 12$ (d) $x + 4y = 12$
(e) $3x - y = 12$ (f) $y = 2x - 4$ (g) $y = 5x + 3$ (h) $y = \frac{1}{2}x + 2$
(i) $y = -\frac{2}{3}x - 2$ (j) $\frac{1}{3}x + \frac{1}{5}y = 2$ (k) $\frac{2}{3}x - \frac{2}{5}y = 1$

PROBLEM 4-33 Graph the following equations:

(a) $2x - 3y = 6$ (b) $2x - y = 4$ (c) $x + 2y = 4$
(d) $3x + 4y = 12$ (e) $x - 3y = 6$ (f) $3x + y = 6$
(g) $y = \frac{1}{3}x + 1$ (h) $y = \frac{2}{3}x - 2$ (i) $\frac{1}{3}x - \frac{1}{2}y = 1$

PROBLEM 4-34 Graph the following linear equations using the intercept method:

(a) $x - 2y = 4$ (b) $x + 3y = 3$ (c) $3x - 2y = 6$ (d) $4x + 3y = 12$
(e) $3x - 4y = 12$ (f) $4x + 5y = 20$ (g) $4y = 5x + 20$ (h) $y = \frac{1}{2}x - 2$
(i) $y = \frac{2}{3}x + 2$

PROBLEM 4-35 Graph the following equations:

(a) $x = 1$ (b) $y = 3$ (c) $x = -2$
(d) $y = -4$ (e) $x + 5 = 0$ (f) $y + 1 = 0$
(g) $x - 5 = 0$ (h) $y = 0$ (i) $x = 0$

PROBLEM 4-36 Find the slope and y-intercept of:

(a) $4x + 5y = 20$ (b) $3x - 2y = 6$ (c) $5x - 2y = 10$ (d) $2x + 5y = 10$
(e) $5x + 3y = 3$ (f) $4x - 3y = 6$ (g) $3x + 4y = 3$ (h) $4x - 3y = 0$
(i) $5x + 2y = 0$ (j) $y = 4$ (k) $x = -2$

PROBLEM 4-37 Graph the following linear equations using the y-intercept and the slope:

(a) $y = \frac{2}{3}x + 3$ (b) $y = \frac{4}{5}x - 2$ (c) $y = -\frac{3}{4}x + 2$
(d) $y = \frac{3}{4}x$ (e) $y = -\frac{5}{2}x$ (f) $y = 3x - 1$
(g) $y = -4x + 2$ (h) $3x - 2y = 4$ (i) $3x + 4y = 12$

PROBLEM 4-38 Find the slope of the line containing:

(a) $(1, 3)$ and $(4, 4)$ (b) $(2, -1)$ and $(-2, 4)$ (c) $(3, 2)$ and $(-3, -3)$
(d) $(-2, 2)$ and $(5, 1)$ (e) $(1, -3)$ and $(2, 0)$ (f) $(-3, -1)$ and $(0, 3)$
(g) $(1, -3)$ and $(0, -4)$ (h) $(0, -1)$ and $(1, 0)$ (i) $(2, 3)$ and $(0, 1)$
(j) $(-1, 1)$ and $(2, -2)$ (k) $(2, 2)$ and $(3, -1)$

PROBLEM 4-39 Find the slope of the line parallel to an axis containing:

(a) $(1, 1)$ and $(1, -2)$ (b) $(1, 1)$ and $(-2, 1)$ (c) $(4, 1)$ and $(4, -1)$
(d) $(-1, 0)$ and $(3, 0)$ (e) $(0, 2)$ and $(0, -5)$ (f) $(-3, 2)$ and $(-3, -2)$

(g) $(-1, -4)$ and $(2, -4)$ **(h)** $(-3, 2)$ and $(-3, 0)$ **(i)** $(-1, -1)$ and $(3, -1)$
(j) $(0, 0)$ and $(-3, 0)$ **(k)** $(2, 2)$ and $(2, -3)$

PROBLEM 4-40 Find an equation of a line containing the given point and having the given slope:

(a) $(3, 2)$ and $m = 2$ **(b)** $(1, 4)$ and $m = -2$ **(c)** $(-1, 2)$ and $m = \frac{3}{2}$
(d) $(-3, 2)$ and $m = \frac{2}{5}$ **(e)** $(-3, -1)$ and $m = \frac{2}{3}$ **(f)** $(0, 4)$ and $m = -\frac{1}{4}$
(g) $(3, 0)$ and $m = -\frac{3}{4}$ **(h)** $(0, 0)$ and $m = -\frac{4}{3}$ **(i)** $(-1, 1)$ and $m = 0$
(j) $(0, -3)$ and $m = 0$ **(k)** $(-2, 0)$ and having no slope

PROBLEM 4-41 Find an equation of the line containing:

(a) $(1, 2)$ and $(4, 3)$ **(b)** $(1, -2)$ and $(-4, 4)$ **(c)** $(3, 1)$ and $(-3, -4)$
(d) $(-2, 1)$ and $(3, 0)$ **(e)** $(1, -2)$ and $(2, 0)$ **(f)** $(-3, 0)$ and $(0, 3)$
(g) $(3, -2)$ and $(2, -3)$ **(h)** $(1, 0)$ and $(0, -3)$ **(i)** $(2, 0)$ and $(1, 0)$
(j) $(-1, 2)$ and $(1, 2)$ **(k)** $(2, 2)$ and $(2, 0)$

PROBLEM 4-42 Determine whether the graphs of the following pairs of equations are parallel:

(a) $3x - 4y = 8$ and $6x - 8y = 8$ **(b)** $4x + 3y = 6$ and $3x + 4y = 6$
(c) $2x + 3y = 4$ and $4x + 6y = 6$ **(d)** $4x + 3y = 6$ and $3x - 4y = 6$
(e) $4x - 3y = 6$ and $4x - 3y = 12$ **(f)** $2x - 3y = 6$ and $y = \frac{2}{3}x - 2$
(g) $2x - y = 1$ and $y = 2x + 1$ **(h)** $y = \frac{2}{3}x + 4$ and $2x - 3y = 4$
(i) $3x - y = 2$ and $x + 3y = 2$ **(j)** $2x + y = 0$ and $y = 2x + 3$

PROBLEM 4-43 Find an equation of a line containing a given point and parallel to a given line:

(a) $(1, 2)$, $y = 3x + 4$ **(b)** $(-1, 2)$, $2x + 3y = 6$ **(c)** $(-2, -1)$, $2x - 3y = 6$
(d) $(0, 0)$, $x + y = 1$ **(e)** $(2, 1)$, $x + 3y = 0$ **(f)** $(3, 2)$, $x = 2y - 3$
(g) $(0, -3)$, $2x - 5y = 1$ **(h)** $(-3, 0)$, $y = 3x + 2$ **(i)** $(2, -2)$, $0 = 3y - 2x$
(j) $(-2, 3)$, $y = 4$ **(k)** $(3, -1)$, $x = -3$

PROBLEM 4-44 Determine whether the graphs of the following pairs of equations are perpendicular:

(a) $4x + 3y = 12$ and $3x - 4y = 12$ **(b)** $3x + 4y = 12$ and $4x + 3y = 12$
(c) $x + 3y = 0$ and $3x - y = 0$ **(d)** $4x - y = 0$ and $x + 4y = 0$
(e) $2x - 3y = 0$ and $3x + 2y = 4$ **(f)** $x = 3y + 4$ and $y = 4 - 3x$
(g) $y = \frac{2}{3}x - 3$ and $y = 3 - \frac{3}{2}x$ **(h)** $3x = 2y - 1$ and $2x = 3 - 2y$
(i) $x + y = 0$ and $y = 3$ **(j)** $x = 4$ and $y = 3$

PROBLEM 4-45 Find an equation of a line containing a given point and perpendicular to a given line:

(a) $(3, 1)$, $2x - 3y = 6$ **(b)** $(2, -1)$, $2x + 3y = 6$ **(c)** $(-1, 0)$, $3x - 4y = 2$
(d) $(2, 0)$, $x + y = 1$ **(e)** $(0, 1)$, $x - y = 4$ **(f)** $(0, 0)$, $2x + 3y = 12$
(g) $(-2, 3)$, $y = 3x + 1$ **(h)** $(2, -1)$, $x = 3y - 1$ **(i)** $(1, 1)$, $3x - 5y = 0$
(j) $(-1, -2)$, $y = 2$ **(k)** $(5, 0)$, $x = 2$

PROBLEM 4-46 Solve each problem using a linear relationship:

(a) The cost to produce 100 units is $500. The cost to produce 1000 to $4000. Assuming there is a linear relationship between the cost and the number of units produced, find a linear equation in two variables that describes this linear relationship. Find the cost to produce 5000 units, to the nearest cent. How many units can be produced for $39,000?

(b) A computer is purchased for $1000 and depreciated, using straight line depreciation, at $120 per year. Find a linear equation in two variables that describes this linear relationship. What is the book value after $4\frac{1}{2}$ years? How long did it take the computer to reach a $100 **scrap value** (the value at which the computer is sold for scrap)?

(c) A certain car can be rented for 20¢ per mile plus a $5 drop-off fee. Find a linear equation in two variables that describes the linear rela-

(d) A certain long distance call costs $3 for the first 5 minutes and $5 for the first 10 minutes. Find the linear equation in two variables that

tionship between the rental car charge and the number of miles driven. What is the charge to rent the car for a trip of 200 miles? How far was the car driven if the rental charge is $100?

describes the linear relationship between the cost and the number of minutes called. What does it cost to call for 15 minutes? How long can a person call for $10?

PROBLEM 4-47 Determine whether a given ordered pair is a solution of the given linear inequality:

(a) $(0, 0)$, $3x - 4y \leq 3$ (b) $(1, 3)$, $3x - 4y \geq 3$ (c) $(0, 2)$, $4x + 3y > 4$
(d) $(3, 0)$, $4x + 3y < 2$ (e) $(3, 1)$, $x + 3y \leq 3$ (f) $(2, 2)$, $3x + y \geq 2$
(g) $(0, 5)$, $y < 3x + 3$ (h) $(-3, -2)$, $3y \leq 4x$ (i) $(2, -2)$, $y > 3x - 2$
(j) $(-1, 3)$, $y \leq \frac{3}{2}x + 2$ (k) $(-4, -1)$, $y \leq \frac{4}{5} - \frac{1}{5}x$

PROBLEM 4-48 Graph the following inequalities:

(a) $2x + 3y \leq 6$ (b) $3x - 4y \geq 12$ (c) $4y \leq 3x + 4$ (d) $3x \geq 2y + 4$
(e) $3x \leq 2y$ (f) $x \geq 3y$ (g) $3x - 2 \leq 4y$ (h) $x + 4 \leq 2y$
(i) $x - y \leq 0$

PROBLEM 4-49 Graph the following inequalities:

(a) $4x - 3y < 12$ (b) $4x + 3y > 12$ (c) $2x - y > 4$ (d) $x + 3y < 3$
(e) $x < 3y$ (f) $4x < -5y$ (g) $3x > 2y + 4$ (h) $3y > 2x + 6$
(i) $x + y > 0$

PROBLEM 4-50 Graph the following inequalities:

(a) $x \leq 5$ (b) $y \leq -1$ (c) $x \geq -3$ (d) $y > 4$
(e) $0 > x$ (f) $1 < y$ (g) $-3 < x$ (h) $-3 \geq y$
(i) $x > 2$

PROBLEM 4-51 Graph the following inequalities:

(a) $x \geq 0$ or $y \geq 3$ (b) $x > -1$ or $y < -2$ (c) $x - y < 0$ or $x + y < 0$
(d) $3x + 2y > 6$ or $x - y \leq -3$ (e) $2x - 3y < 6$ or $3x + 2y > 6$
(f) $3x + y \leq 3$ or $x - 3y > 0$ (g) $2x + y \leq 4$ or $x < 3$
(h) $y < -2$ or $x - 2y < 4$ (i) $x + 2y \geq 0$ or $x < 1$

PROBLEM 4-52 Graph the following inequalities:

(a) $x \leq 3$ and $y > -1$ (b) $x > -2$ and $y \geq 3$ (c) $x + y \leq 3$ and $x - y < -3$
(d) $2x + y \leq 4$ and $x - 2y \geq 0$ (e) $x + 3y \geq 3$ and $3x - y > 3$ (f) $0 < 3x - 4y < 12$
(g) $-12 \leq 4x + 3y \leq 0$ (h) $-6 < 3x + 2y \leq 6$ (i) $4 \leq 2y - x < 6$

PROBLEM 4-53 Graph the following inequalities:

(a) $|x| \geq 1$ (b) $|y| > 1$ (c) $|x + 1| > 3$ (d) $|y - 1| \geq 2$
(e) $|3y - 2x| > 6$ (f) $|3x + 4y| \geq 12$ (g) $|4x - 6y| \geq 12$ (h) $|x + 3y| > 6$
(i) $|x - y| > 0$

PROBLEM 4-54 Graph the following inequalities:

(a) $|x| \leq 2$ (b) $|y| < 4$ (c) $|x - 3| < 2$ (d) $|y + 3| \leq 3$
(e) $|2x + 3y| < 6$ (f) $|4x - 3y| \leq 12$ (g) $|4x - y| \leq 4$ (h) $|x - 3y| < 6$
(i) $|x + y| \leq 0$

Answers to Supplementary Exercises

(4-28)

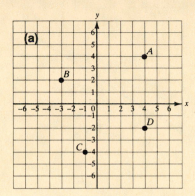
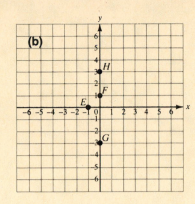
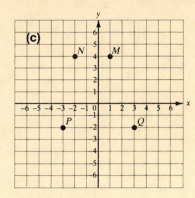

(4-29) **(a)** $A(3, 0)$, $B(2, 5)$, $C(0, 2)$, $D(-5, 4)$, $E(-3, 0)$, $F(0, 0)$, $G(-1, -2)$, $H(5, -4)$

      (b) $H(4, 3)$, $I(-2, 4)$, $J(0, 3)$, $K(-5, 1)$, $L(4, 0)$, $M(-4, -4)$, $N(0, -1)$, $O(2, -5)$

      (c) $S(4, -2)$, $T(0, -3)$, $U(-2, -5)$, $V(-4, 0)$, $W(-2, 1)$, $X(0, 1)$, $Y(5, 1)$, $Z(1, 0)$

(4-30) **(a)** No   **(b)** Yes   **(c)** No   **(d)** Yes   **(e)** No   **(f)** Yes   **(g)** No   **(h)** Yes

(4-31) **(a)** $(2, 0)$, $(0, -3)$, $(-2, -6)$ **(b)** $(4, 3)$, $(2, 0)$, $(0, -3)$ **(c)** $(6, 0)$, $(3, 2)$, $(0, 4)$

      (d) $(0, 4)$, $(3, 2)$, $(6, 0)$ **(e)** $(0, 4)$, $(-3, 6)$, $(-6, 8)$ **(f)** $(-3, 1)$, $(0, 0)$, $(3, -1)$

      (g) $(3, -1)$, $(0, 0)$, $(-3, 1)$ **(h)** $(0, 1)$, $(-\frac{3}{2}, 0)$, $(-3, -1)$ **(i)** $(0, 1)$, $(-1, \frac{1}{3})$, $(-2, -\frac{1}{3})$

      (j) $(3, 5)$, $(0, 3)$, $(-\frac{3}{2}, 2)$ **(k)** $(2, \frac{10}{3})$, $(0, \frac{1}{3})$, $(-\frac{1}{3}, -\frac{1}{6})$

(4-32) **(a)** $(3, 0)$, $(0, 4)$ **(b)** $(3, 0)$, $(0, -4)$ **(c)** $(12, 0)$, $(0, -3)$ **(d)** $(12, 0)$, $(0, 3)$

      (e) $(4, 0)$, $(0, -12)$ **(f)** $(2, 0)$, $(0, -4)$ **(g)** $(-\frac{3}{5}, 0)$, $(0, 3)$ **(h)** $(-4, 0)$, $(0, 2)$

      (i) $(-3, 0)$, $(0, -2)$ **(j)** $(6, 0)$, $(0, 10)$ **(k)** $(\frac{3}{2}, 0)$, $(0, -\frac{5}{2})$

(4-33)

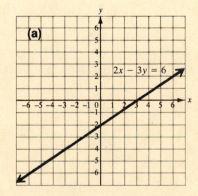
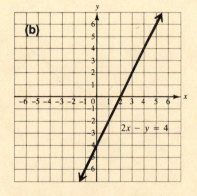
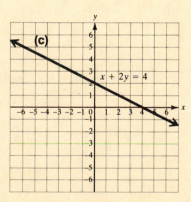

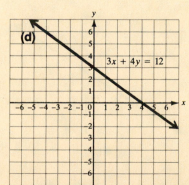
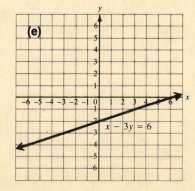
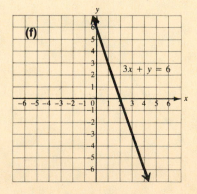

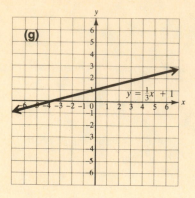

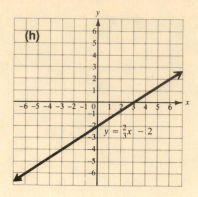

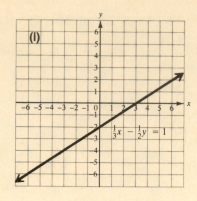

(4-34)

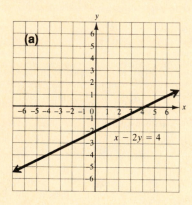

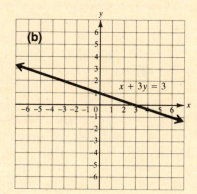

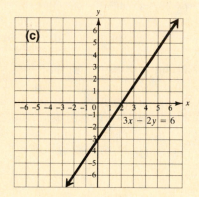

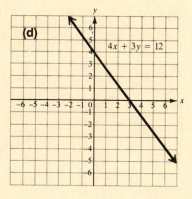

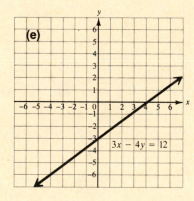

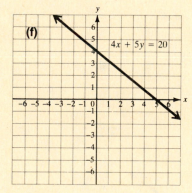

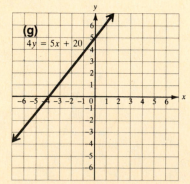

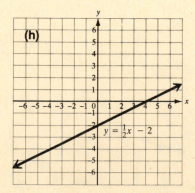

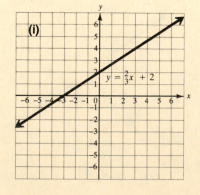

(4-35)

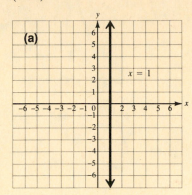

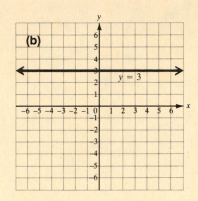

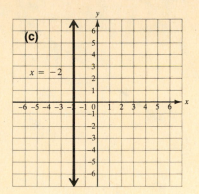

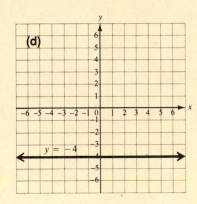

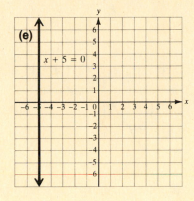

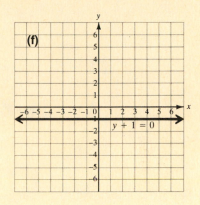

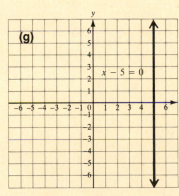

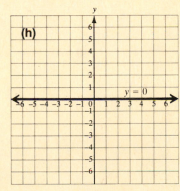

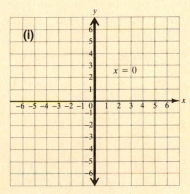

(4-36) (a) $-\frac{4}{5}$, $(0, 4)$ (b) $\frac{3}{2}$, $(0, -3)$ (c) $\frac{5}{2}$, $(0, -5)$ (d) $-\frac{2}{5}$, $(0, 2)$ (e) $-\frac{5}{3}$, $(0, 1)$

(f) $\frac{4}{3}$, $(0, -2)$ (g) $-\frac{3}{4}$, $(0, \frac{3}{4})$ (h) $\frac{4}{3}$, $(0, 0)$ (i) $-\frac{5}{2}$, $(0, 0)$ (j) 0, $(0, 4)$

(k) no slope, no y-intercept.

(4-37)

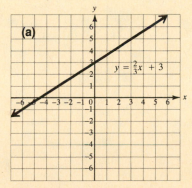

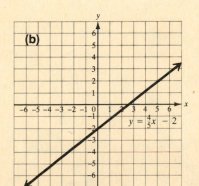

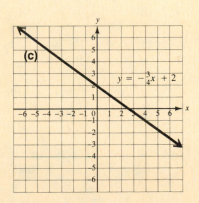

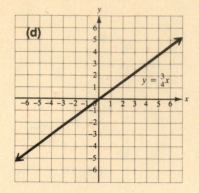
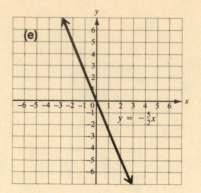
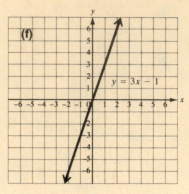

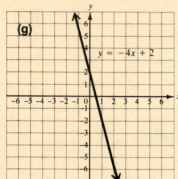
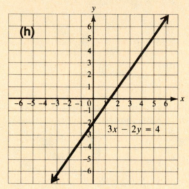
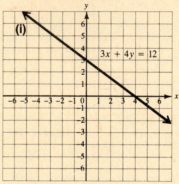

(4-38) **(a)** $\frac{1}{3}$ **(b)** $-\frac{5}{4}$ **(c)** $\frac{5}{6}$ **(d)** $-\frac{1}{7}$ **(e)** 3 **(f)** $\frac{4}{3}$ **(g)** 1 **(h)** 1 **(i)** 1
(j) -1 **(k)** -3

(4-39) **(a)** No slope **(b)** 0 **(c)** No slope **(d)** 0 **(e)** No slope **(f)** No slope
(g) 0 **(h)** No slope **(i)** 0 **(j)** 0 **(k)** No slope

(4-40) **(a)** $y = 2x - 4$ **(b)** $y = -2x + 6$ **(c)** $y = \frac{3}{2}x + \frac{7}{2}$ **(d)** $y = \frac{2}{5}x + \frac{16}{5}$
(e) $y = \frac{2}{3}x + 1$ **(f)** $y = -\frac{1}{4}x + 4$ **(g)** $y = -\frac{3}{4}x + \frac{9}{4}$ **(h)** $y = -\frac{4}{3}x$
(i) $y = 1$ **(j)** $y = -3$ **(k)** $x = -2$

(4-41) **(a)** $y = \frac{1}{3}x + \frac{5}{3}$ **(b)** $y = -\frac{6}{5}x - \frac{4}{5}$ **(c)** $y = \frac{5}{6}x - \frac{3}{2}$ **(d)** $y = -\frac{1}{5}x + \frac{3}{5}$
(e) $y = 2x - 4$ **(f)** $y = x + 3$ **(g)** $y = x - 5$ **(h)** $y = 3x - 3$ **(i)** $y = 0$
(j) $y = 2$ **(k)** $x = 2$

(4-42) **(a)** Yes **(b)** No **(c)** Yes **(d)** No **(e)** Yes **(f)** No, identical **(g)** Yes
(h) Yes **(i)** No **(j)** No

(4-43) **(a)** $y = 3x - 1$ **(b)** $y = -\frac{2}{3}x + \frac{4}{3}$ **(c)** $y = \frac{2}{3}x + \frac{1}{3}$ **(d)** $y = -x$
(e) $y = -\frac{1}{3}x + \frac{5}{3}$ **(f)** $y = \frac{1}{2}x + \frac{1}{2}$ **(g)** $y = \frac{2}{5}x - 3$ **(h)** $y = 3x + 9$
(i) $y = \frac{2}{3}x - \frac{10}{3}$ **(j)** $y = 3$ **(k)** $x = 3$

(4-44) **(a)** Yes **(b)** No **(c)** Yes **(d)** Yes **(e)** Yes **(f)** Yes **(g)** Yes **(h)** No
(i) No **(j)** Yes

(4-45) **(a)** $3x + 2y = 11$ **(b)** $3x - 2y = 8$ **(c)** $4x + 3y = -4$ **(d)** $x - y = 2$
(e) $x + y = 1$ **(f)** $3x - 2y = 0$ **(g)** $x + 3y = 7$ **(h)** $3x + y = 5$
(i) $5x + 3y = 8$ **(j)** $x = -1$ **(k)** $y = 0$

(4-46) **(a)** $y = \frac{35}{9}x + \frac{1000}{9}$ or $y = \frac{9}{35}x - \frac{200}{7}$, \$19,555.56, 10,000 units
(b) $y = -120x + 1000$ or $y = -\frac{1}{120}x + \frac{25}{3}$, \$460, $7\frac{1}{2}$ years.
(c) $y = 0.2x + 5$ or $y = 5x - 25$, \$45, 475 miles.
(d) $y = \frac{2}{5}x + 1$ or $y = 10x - 5$, \$7, $22\frac{1}{2}$ minutes.

(4-47) (a) Yes (b) No (c) Yes (d) No (e) No (f) Yes (g) No (h) No
      (i) No (j) No (k) Yes

(4-48)

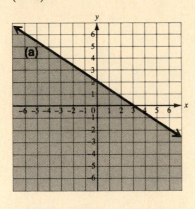
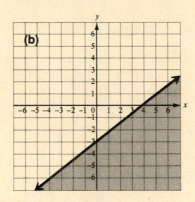
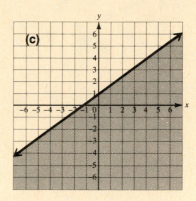

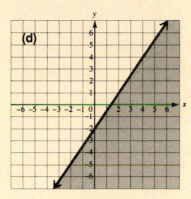
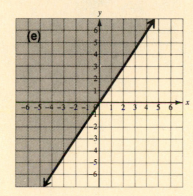
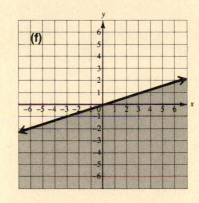

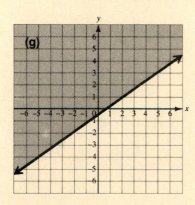
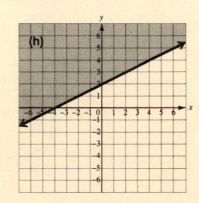
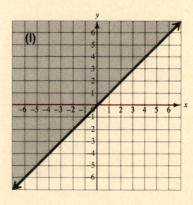

(4-49)

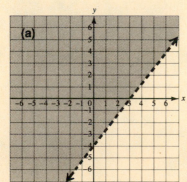
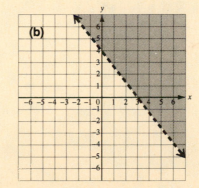
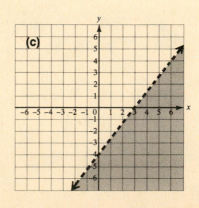

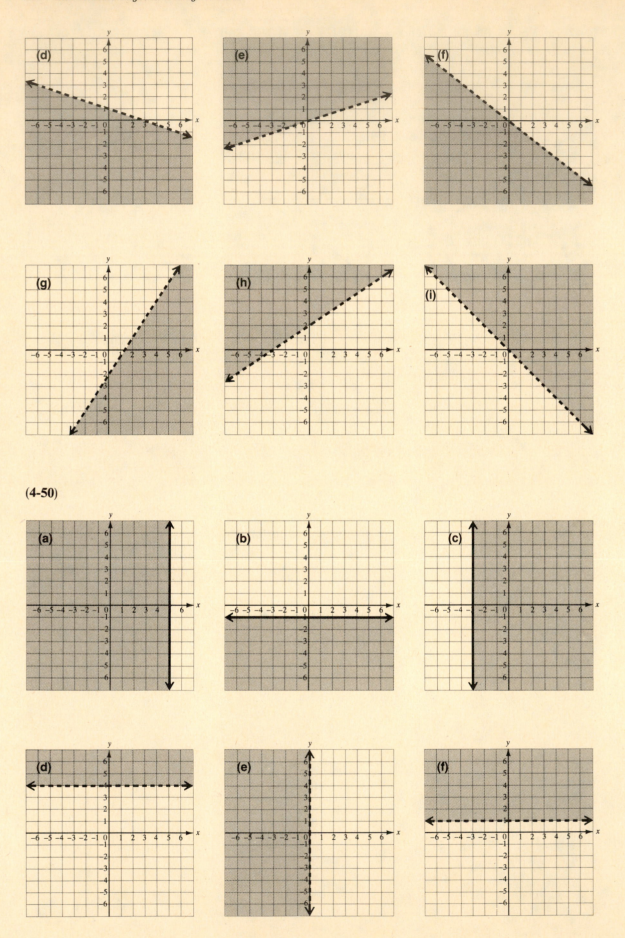

(4-50)

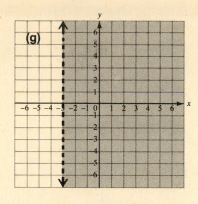

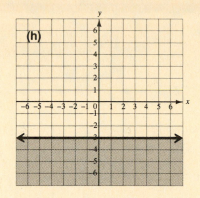

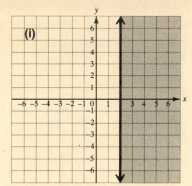

(4-51)

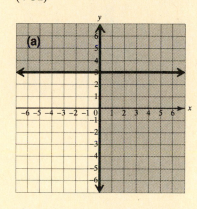

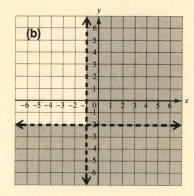

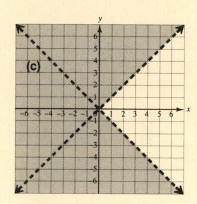

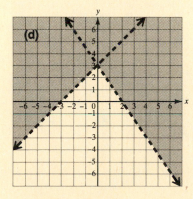

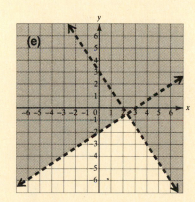

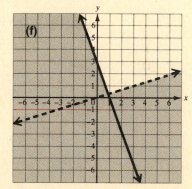

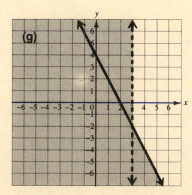

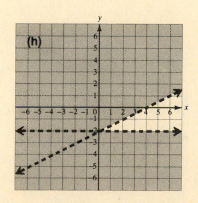

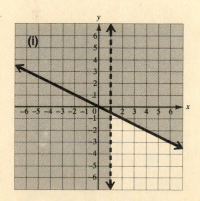

(4-52)

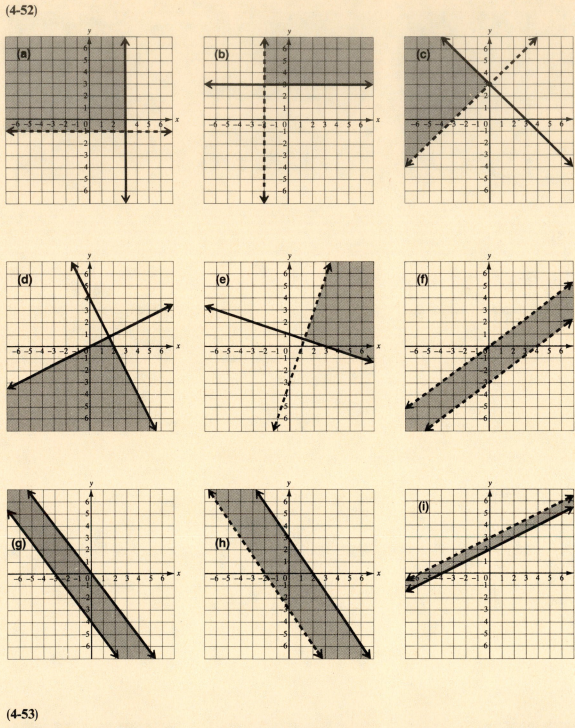

(4-53)

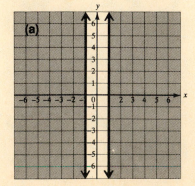
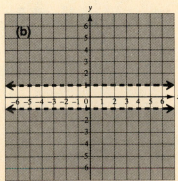
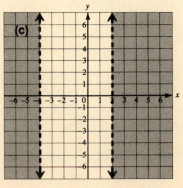

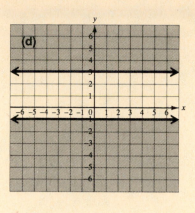

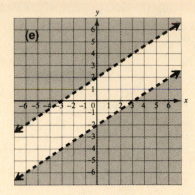

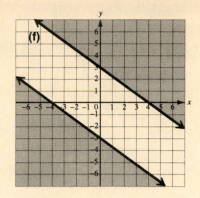

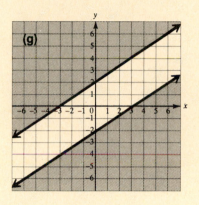

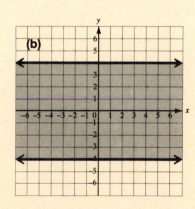

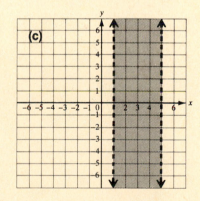

(4-54)

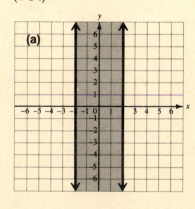

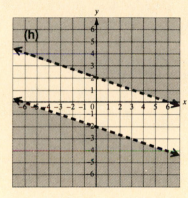

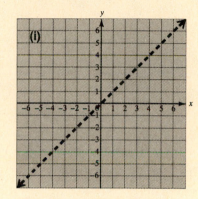

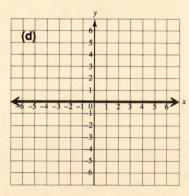

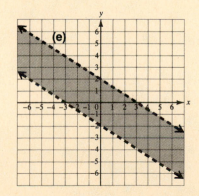

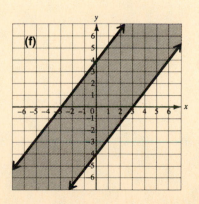

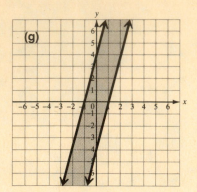

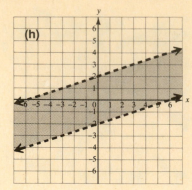

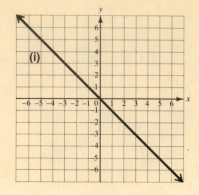

5 SYSTEMS OF EQUATIONS

THIS CHAPTER IS ABOUT

☑ Solving Systems of Equations by Graphing
☑ Solving Systems Using Algebraic Methods
☑ Solving Third-Order Systems
☑ Determinants
☑ Solving Systems Using Cramer's Rule
☑ Solving Problems Using Systems

5-1. Solving Systems of Equations by Graphing

A **system of equations** is two or more equations considered simultaneously. A **second-order system** is a system of two linear equations in the same two variables.

A. Determining whether ordered pairs are solutions of second-order systems

An ordered pair is **a solution of a second-order system** if it is a solution of each equation. To determine whether an ordered pair is a solution of a second-order system, you must determine if it is a solution of each equation of the system.

EXAMPLE 5-1: Determine if $(2, 3)$ is a solution of the following second-order systems:

(a) $\begin{cases} 3x + 2y = 12 \\ 2x + 3y = 13 \end{cases}$ (b) $\begin{cases} x + 2y = 8 \\ x - y = 1 \end{cases}$

Solution: (a)

$$\begin{array}{r|l} 3x + 2y = 12 \\ \hline 3(2) + 2(3) & 12 \\ 6 + 6 & 12 \\ \hline 12 = 12 \end{array}$$ ⟵ original equations

Substitute $(2, 3)$.

⟵ $(2, 3)$ checks

$$\begin{array}{r|l} 2x + 3y = 13 \\ \hline 2(2) + 3(3) & 13 \\ 4 + 9 & 13 \\ \hline 13 = 13 \end{array}$$

⟶ $13 = 13$

Since $(2, 3)$ is a solution of each equation, it is a solution of the system.

(b)

$$\begin{array}{r|l} x + 2y = 8 \\ \hline 2 + 2(3) & 8 \\ \hline 8 = 8 \end{array}$$ ⟵ original equations ⟶

Substitute $(2, 3)$.

⟵ $(2, 3)$ checks $(2,3)$ does not check ⟶

$$\begin{array}{r|l} x - y = 1 \\ \hline 2 - 3 & 1 \\ \hline -1 \neq 1 \end{array}$$

Since $(2, 3)$ is not a solution of $x - y = 1$, it is not a solution of the system.

B. Solving systems by graphing

To **solve a system of linear equations by graphing,** we graph each equation, then consider the intersection of the graphs. Each point of intersection is a solution to the system. If there is no point of intersection (if the lines are parallel), the system has no solution and is called an **inconsistent system.** A **consistent system** has one or more solutions. If it has exactly one, it is called an **independent system.** If it has more than one solution it is called a **dependent system.**

127

EXAMPLE 5-2: Solve the following systems by graphing them:

(a) $\left\{\begin{array}{l} x - y = 4 \\ x + y = 4 \end{array}\right\}$ (b) $\left\{\begin{array}{l} 2x - y = 4 \\ 4x - 2y = 6 \end{array}\right\}$ (c) $\left\{\begin{array}{l} x + 2y = 4 \\ 2x + 4y = 8 \end{array}\right\}$

Solution: (a)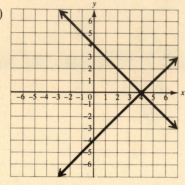

Graph $x - y = 4$.

Graph $x + y = 4$.

Determine that the graphs intersect at $(4, 0)$.

$(4, 0)$ is the solution of $\left\{\begin{array}{l} x - y = 4 \\ x + y = 4 \end{array}\right\}$.

(b)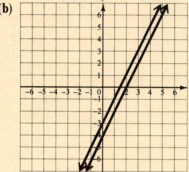

Graph $2x - y = 4$.

Graph $4x - 2y = 6$.

Determine that the lines do not intersect.

$\left\{\begin{array}{l} 2x - y = 4 \\ 4x - 2y = 6 \end{array}\right\}$ has no solution and is an inconsistent system.

(c)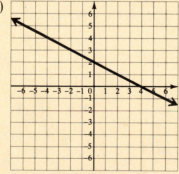

Graph $x + 2y = 4$.

Graph $2x + 4y = 8$.

Determine that the lines are identical.

$\left\{\begin{array}{l} x + 2y = 4 \\ 2x + 4y = 8 \end{array}\right\}$ has many solutions and is a dependent system.

5-2. Solving Systems Using Algebraic Methods

The method of solving systems by graphing them is of little use unless the coordinates of the point of intersection are easy to read. Therefore, it is better to use algebraic methods, such as the substitution method or the addition (elimination) method.

A. Solving second-order systems by the substitution method

To **solve a second-order system by the substitution method,** we:

1. Solve one of the equations for one of the variables in terms of the other variable.
2. Substitute the solution obtained in step 1 for that variable in the second equation, in order to obtain an equation in one variable.
3. Solve the equation obtained in step 2.
4. Substitute the numerical value obtained in step 3 in the first equation.
5. Solve the equation obtained in step 4.
6. Check the proposed solution in the system.

EXAMPLE 5-3: Solve $\begin{cases} 2x + 3y = 12 \\ 3x - y = 7 \end{cases}$ by the substitution method.

Solution:

$$2x + 3y = 12 \qquad \text{Solve the first equation for } y.$$
$$3y = -2x + 12$$
$$y = -\tfrac{2}{3}x + 4$$

$$3x - y = 7 \qquad \text{Substitute } y \text{ in second equation.}$$
$$3x - (-\tfrac{2}{3}x + 4) = 7 \qquad \text{Solve for } x.$$
$$3x + \tfrac{2}{3}x - 4 = 7$$
$$3x + \tfrac{2}{3}x = 11$$
$$\tfrac{11}{3}x = 11$$
$$x = 11 \cdot \tfrac{3}{11}$$
$$x = 3$$

$$2x + 3y = 12 \qquad \text{Substitute for } x \text{ in first equation.}$$
$$2(3) + 3y = 12$$
$$3y = 6$$
$$y = 2$$

$$(3, 2) \longleftarrow \text{proposed solution}$$

$$
\begin{array}{c|c}
2x + 3y = 12 \longleftarrow \text{original equations} \longrightarrow & 3x - y = 7 \\
\hline
2 \cdot 3 + 3 \cdot 2 \quad | \quad 12 & 3 \cdot 3 - 2 \quad | \quad 7 \\
6 + 6 \quad | \quad 12 & 9 - 2 \quad | \quad 7 \\
\end{array}
$$

$$12 = 12 \longleftarrow (3, 2) \text{ checks} \longrightarrow 7 = 7$$

$(3, 2)$ is the solution of $\begin{cases} 2x + 3y = 12 \\ 3x - y = 7 \end{cases}$.

B. Solving second-order systems by the addition method

This method is also referred to as the **elimination method** or the **linear combination method.** To **solve a second-order system by the addition method,** we:

1. Multiply both equations by appropriate numbers so that the coefficients of one of the variables are opposite numbers.
2. Add the new equations to get an equation in one variable.
3. Solve the equation for that variable.
4. Substitute the solution obtained in one of the original equations and solve for the other variable.
5. Check the proposed solution in both of the given equations.

EXAMPLE 5-4: Solve the following systems by the addition method:

(a) $\begin{cases} 2x + 3y = 12 \\ 3x + 2y = 13 \end{cases}$ **(b)** $\begin{cases} 3x - 2y = 4 \\ 6x - 4y = 6 \end{cases}$ **(c)** $\begin{cases} 2x - 3y = 4 \\ 4x - 6y = 8 \end{cases}$

Solution: **(a)** $\begin{cases} 2x + 3y = 12 \\ 3x + 2y = 13 \end{cases}$

$$4x + 6y = 24$$
$$-9x - 6y = -39$$

Multiply $2x + 3y = 12$ by 2 and $3x + 2y = 13$ by -3 to get $6y$ and $-6y$.

$$\begin{array}{r} 4x + 6y = 24 \\ -9x - 6y = -39 \\ \hline -5x = -15 \end{array}$$

Add the equations to get an equation in one variable.

$$x = 3$$ Solve for x.

$$3(x) + 2y = 13$$ Substitute 3 for x in one of the original equations.

$$3(3) + 2y = 13$$

$$9 + 2y = 13$$ Solve for y.

$$2y = 4$$

$$y = 2$$

$$\begin{array}{c|c} 2x + 3y = 12 \\ \hline 2(3) + 3(2) & 12 \\ 6 + 6 & 12 \\ 12 & 12 \end{array}$$ ⟵ original equations ⟶ $\begin{array}{c|c} 3x + 2y = 13 \\ \hline 3(3) + 2(2) & 13 \\ 9 + 4 & 13 \\ 13 & 13 \end{array}$

⟵ (3, 2) checks ⟶

$(3, 2)$ is the solution of $\begin{cases} 2x + 3y = 12 \\ 3x + 2y = 13 \end{cases}$.

(b) $\begin{cases} 3x - 2y = 4 \\ 6x - 4y = 6 \end{cases}$

$$-6x + 4y = -8$$

Multiply $3x - 2y = 4$ by -2 to get $4y$ and $-4y$.

$$6x - 4y = 6$$

$$\begin{array}{r} -6x + 4y = -8 \\ 6x - 4y = 6 \\ \hline 0 = -2 \end{array}$$

Add the equations to eliminate one variable.

$$0 = -2$$ ⟵ false statement

Note: A second-order system that reduces to a false statement is an inconsistent system, and has no solution.

(c) $\begin{cases} 2x - 3y = 4 \\ 4x - 6y = 8 \end{cases}$

$$-4x + 6y = -8$$

Multiply $2x - 3y = 4$ by -2 to get $6y$ and $-6y$.

$$4x - 6y = 8$$

$$\begin{array}{r} -4x + 6y = -8 \\ 4x - 6y = 8 \\ \hline 0 = 0 \end{array}$$

Add the equations to eliminate one variable.

$$0 = 0$$ ⟵ true statement

Note: A second-order system that reduces to a true statement is a consistent dependent system, and has an infinite number of solutions.

5-3. Solving Third-Order Systems

A **third-order system** is a system of three linear equations in the same three variables. The graph of a linear equation in three variables is a plane in three-dimensional space.

A. Determine whether ordered triples are solutions of third-order systems

An ordered triple, which has the form (x, y, z), is **a solution of a third-order system** if it is a solution of all three equations of the system.

EXAMPLE 5-5: Determine whether $(2, 0, -1)$ is a solution of the following systems:

(a) $\begin{cases} 3x - 2y + z = 5 \\ x + 4y - 3z = 5 \\ 2x - y - 2z = 6 \end{cases}$ (b) $\begin{cases} x + 3y - 2z = 4 \\ 2x - 3y + 3z = 1 \\ 3x + y - z = 3 \end{cases}$

Solution: Substitute the given ordered triple in the original equations.

(a)

$$\begin{array}{c|c} 3x - 2y + z = 5 & \\ \hline 3 \cdot 2 - 2 \cdot 0 + (-1) & 5 \\ 6 - 0 - 1 & 5 \\ \hline 5 = 5 & \end{array} \qquad \begin{array}{c|c} x + 4y - 3z = 5 & \\ \hline 2 + 4 \cdot 0 - 3(-1) & 5 \\ 2 + 0 + 3 & 5 \\ \hline 5 = 5 & \end{array} \qquad \begin{array}{c|c} 2x - y - 2z = 6 & \\ \hline 2 \cdot 2 - 0 - 2(-1) & 6 \\ 4 - 0 + 2 & 6 \\ \hline 6 = 6 & \end{array}$$

$(2, 0, -1)$ checks. Since $(2, 0, -1)$ is a solution of each equation, it is a solution of the system.

(b)

$$\begin{array}{c|c} x + 3y - 2z = 4 & \\ \hline 2 + 3 \cdot 0 - 2(-1) & 4 \\ 2 + 0 + 2 & 4 \\ \hline 4 = 4 & \end{array} \qquad \begin{array}{c|c} 2x - 3y + 3z = 1 & \\ \hline 2 \cdot 2 - 3 \cdot 0 + 3(-1) & 1 \\ 4 + 0 - 3 & 1 \\ \hline 1 = 1 & \end{array} \qquad \begin{array}{c|c} 3x + y - z = 3 & \\ \hline 3 \cdot 2 + 0 - (-1) & 3 \\ 6 + 0 + 1 & 3 \\ \hline 7 \neq 3 & \end{array}$$

Since $(2, 0, -1)$ is not a solution of $3x + y - z = 3$, it is not a solution of the system.

B. Solve third-order systems by the addition method

It may be possible to solve a third-order system by graphing, but this requires a great deal of accuracy. Generally, the substitution method is a more complicated method of solving third-order systems; the addition method is more efficient when extended to solve them. To **solve a third-order system by the addition method,** we:

1. Eliminate a variable from any two equations, in the manner used in a second-order system, to get an equation in two variables.
2. Eliminate the same variable from the remaining equation and either of the first two equations to get another equation in two variables.
3. Use the addition method to solve the new second-order system obtained from steps 1 and 2.
4. Substitute the solutions obtained in step 3 in one of the original equations and solve it for the third variable.
5. Check the proposed solution in each of the three equations.

EXAMPLE 5-6: Solve the following third-order systems by the addition method:

(a) $\begin{cases} 2x + 3y + z = 11 \\ 3x - 2y + 2z = 5 \\ 4x + 3y - 3z = 1 \end{cases}$ (b) $\begin{cases} 2x - 3y + z = 4 \\ 4x - 6y + 2z = 8 \\ 2x + 2y - z = 4 \end{cases}$

Solution: (a) $\begin{cases} 2x + 3y + z = 11 \\ 4x + 3y - 3z = 1 \\ 3x - 2y + 2z = 5 \end{cases}$

$$\begin{array}{ll} -4x - 6y - 2z = -22 & \text{Multiply } 2x + 3y + z = 11 \text{ by } -2. \\ \underline{3x - 2y + 2z = 5} & \\ -x - 8y = -17 & \text{Add the equations to eliminate } z. \end{array}$$

$$6x + 9y + 3z = 33$$
$$4x + 3y - 3z = 1$$
$$\overline{10x + 12y \quad\quad = 34}$$

Multiply $2x + 3y + z = 11$ by 3.

Add the equations to eliminate z.

$$\begin{cases} -x - 8y = -17 \\ 10x + 12y = 34 \end{cases}$$

Solve the new second-order system.

$$-10x - 80y = -170$$
$$10x + 12y = 34$$
$$\overline{-68y = -136}$$

Multiply $-x - 8y = -17$ by 10.

$$y = 2$$
$$-x - 8y = -17$$
$$-x - 8(2) = -17$$
$$-x - 16 = -17$$
$$-x = -1$$
$$x = 1$$

$$2x + 3y + z = 11$$
$$2 \cdot 1 + 3 \cdot 2 + z = 11$$
$$2 + 6 + z = 11$$
$$z = 3$$

Substitute 1 for x and 2 for y in one of the original equations.

$$2x + 3y + z = 11$$
$$\begin{array}{c|c} 2 \cdot 1 + 3 \cdot 2 + 3 & 11 \\ 2 + 6 + 3 & 11 \end{array}$$
$$11 = 11 \longleftarrow (1, 2, 3) \text{ checks}$$

Check $(1, 2, 3)$ in each equation.

$$3x - 2y + 2z = 5$$
$$\begin{array}{c|c} 3 \cdot 1 - 2 \cdot 2 + 2 \cdot 3 & 5 \\ 3 - 4 + 6 & 5 \end{array}$$
$$5 = 5 \longleftarrow (1, 2, 3) \text{ checks}$$

$$4x + 3y - 3z = 1$$
$$\begin{array}{c|c} 4 \cdot 1 + 3 \cdot 2 - 3 \cdot 3 & 1 \\ 4 + 6 - 9 & 1 \end{array}$$
$$1 = 1 \longleftarrow (1, 2, 3) \text{ checks}$$

$(1, 2, 3)$ is the solution of the system.

(b) $\begin{cases} 2x - 3y + z = 4 \\ 4x - 6y + 2z = 8 \\ 2x + 2y - z = 4 \end{cases}$

$$-4x + 6y - 2z = -8$$
$$4x - 6y + 2z = 8$$
$$\overline{0 = 0}$$

Multiply $2x - 3y + z = 4$ by -2.

Add to eliminate a variable.

$$0 = 0 \longleftarrow \text{a true statement may indicate a dependent system, but you must check the other pairs of equations}$$

$$2x - 3y + z = 4$$
$$2x + 2y - z = 4$$
$$\overline{4x - y \quad = 8}$$

Add to eliminate a variable.

$$4x - 6y + 2z = 8$$
$$4x + 4y - 2z = 8$$
$$\overline{8x - 2y \qquad = 0}$$

Multiply $2x + 2y - z = 4$ by 2.

Add to eliminate z.

$$4x - y = 8$$

Solve the second-order system.

$$8x - 2y = 0$$

$$-8x + 2y = -16$$
$$\underline{8x - 2y = \quad 0}$$
$$0 = -16 \longleftarrow \text{a false statement indicates the system is an inconsistent system}$$

The system has no solution.

Note: With a third-order system, one true statement does not necessarily imply a dependent system, but one false statement *does* imply an inconsistent system.

5-4. Determinants

A **determinant** is a square array of numbers written inside vertical lines.

A. Evaluating second-order determinants

If a standard second-order system is given as

$$\begin{cases} a_1x + b_1y = c_1 \\ a_2x + b_2y = c_2 \end{cases}$$

then a second-order determinant is written in the form

$$\begin{vmatrix} a_1 & b_1 \\ a_2 & b_2 \end{vmatrix}$$

To **evaluate the determinant,** we find the number $a_1b_2 - a_2b_1$.

EXAMPLE 5-7: Evaluate the following second-order determinants:

(a) $\begin{vmatrix} 3 & 4 \\ 2 & 5 \end{vmatrix}$ (b) $\begin{vmatrix} 2 & -3 \\ 0 & -1 \end{vmatrix}$ (c) $\begin{vmatrix} -2 & 3 \\ -1 & -4 \end{vmatrix}$

Solution: (a) $\begin{vmatrix} 3 & 4 \\ 2 & 5 \end{vmatrix}$

$$= 3(5) - 2(4)$$
$$= 7$$

Multiply the diagonal elements and indicate the difference of their products.

(b) $\begin{vmatrix} 2 & -3 \\ 0 & -1 \end{vmatrix}$

$$= 2(-1) - 0(-3)$$
$$= -2$$

Multiply the diagonal elements and indicate the difference of their products.

(c) $\begin{vmatrix} -2 & 3 \\ -1 & -4 \end{vmatrix}$

$$= -2(-4) - (-1)(3)$$
$$= 11$$

Multiply the diagonal elements and indicate the difference of their products.

B. Expanding third-order determinants

A **third-order determinant** has the form

$$\begin{vmatrix} a_1 & b_1 & c_1 \\ a_2 & b_2 & c_2 \\ a_3 & b_3 & c_3 \end{vmatrix}$$

Each element of the third-order determinant has a corresponding **minor,** which is defined as the remaining four elements when the row and column to which the original element belongs are deleted. For example, the minor that corresponds to a_2 is

$$\begin{vmatrix} b_1 & c_1 \\ b_3 & c_3 \end{vmatrix}$$

The most convenient way to evaluate a third-order determinant is to **expand** it about any convenient row or column. To do this, we first form the product of each member of that row or column with its corresponding minor, then add or subtract those products according to the value given to them by the **sign array** for a third-order determinant, which is

$$\begin{vmatrix} + & - & + \\ - & + & - \\ + & - & + \end{vmatrix}$$

EXAMPLE 5-8: Expand the following third-order determinants about the given row or column:

(a) $\begin{vmatrix} 2 & -1 & 3 \\ 1 & 0 & 2 \\ 3 & -2 & 1 \end{vmatrix}$ about row 1, and (b) $\begin{vmatrix} -1 & 3 & 0 \\ 2 & 1 & 3 \\ -3 & -2 & -1 \end{vmatrix}$ about column 2.

Solution: (a) $\quad 2\begin{vmatrix} 0 & 2 \\ -2 & 1 \end{vmatrix} \quad -1\begin{vmatrix} -1 & 3 \\ -2 & 1 \end{vmatrix} \quad 3\begin{vmatrix} -1 & 3 \\ 0 & 2 \end{vmatrix}$ Form the products of elements and their minors.

$\quad = +2\begin{vmatrix} 0 & 2 \\ -2 & 1 \end{vmatrix} - (-1)\begin{vmatrix} -1 & 3 \\ -2 & 1 \end{vmatrix} + 3\begin{vmatrix} -1 & 3 \\ 0 & 2 \end{vmatrix}$ Insert the appropriate signs according to the sign array.

(b) $\quad 3\begin{vmatrix} 2 & 3 \\ -3 & -1 \end{vmatrix} \quad 1\begin{vmatrix} -1 & 0 \\ -3 & -1 \end{vmatrix} \quad (-2)\begin{vmatrix} -1 & 0 \\ 2 & 3 \end{vmatrix}$ Form the products of elements and their minors.

$\quad = -3\begin{vmatrix} 2 & 3 \\ -3 & -1 \end{vmatrix} + 1\begin{vmatrix} -1 & 0 \\ -3 & -1 \end{vmatrix} - (-2)\begin{vmatrix} -1 & 0 \\ 2 & 3 \end{vmatrix}$ Insert the appropriate signs according to the sign array.

C. Evaluate third-order determinants

To **evaluate a third-order determinant,** we expand it, then evaluate the resulting expression.

EXAMPLE 5-9: Evaluate the following third-order determinants:

(a) $\begin{vmatrix} 2 & 3 & 1 \\ -1 & 1 & 0 \\ -2 & 3 & 2 \end{vmatrix}$ (b) $\begin{vmatrix} -1 & 0 & 3 \\ 2 & -2 & 0 \\ -3 & 0 & 1 \end{vmatrix}$

Solution: (a) $\begin{vmatrix} 2 & 3 & 1 \\ -1 & 1 & 0 \\ -2 & 3 & 2 \end{vmatrix}$

$\quad = 2\begin{vmatrix} 1 & 0 \\ 3 & 2 \end{vmatrix} - 3\begin{vmatrix} -1 & 0 \\ -2 & 2 \end{vmatrix} + 1\begin{vmatrix} -1 & 1 \\ -2 & 3 \end{vmatrix}$ Expand about row 1.

$\quad = 2(2 - 0) - 3(-2 - 0) + 1(-3 + 2)$ Evaluate the second-order determinants.

$\quad = 9$ Evaluate the expression.

(b) $\begin{vmatrix} -1 & 0 & 3 \\ 2 & -2 & 0 \\ -3 & 0 & 1 \end{vmatrix}$

$= -0 \begin{vmatrix} 2 & 0 \\ -3 & 1 \end{vmatrix} + (-2) \begin{vmatrix} -1 & 3 \\ -3 & 1 \end{vmatrix} - (0) \begin{vmatrix} -1 & 3 \\ 2 & 0 \end{vmatrix}$ — Expand about column 2 (two zeros in column 2 mean it will be easier to evaluate).

$= -0(2 - 0) - 2(-1 + 9) - 0(0 - 6)$ — Evaluate the second-order determinants.

$= -16$ — Evaluate the expression.

5-5. Solving Systems Using Cramer's Rule

We can use special determinants to solve systems of equations. If you recall that a standard second-order system is given as

$$\begin{cases} a_1 x + b_1 y = c_1 \\ a_2 x + b_2 y = c_2 \end{cases}$$

we can then define certain standard determinants. Determinate D consists of the coefficients of the variables of the equations:

$$\begin{vmatrix} a_1 & b_1 \\ a_2 & b_2 \end{vmatrix}$$

We get determinant D_x by replacing the x-coefficients with the constants,

$$\begin{vmatrix} c_1 & b_1 \\ c_2 & b_2 \end{vmatrix}$$

and determinant D_y by replacing the y-coefficients with the constants

$$\begin{vmatrix} a_1 & c_1 \\ a_2 & c_2 \end{vmatrix}$$

For a third-order system, determinants D, D_x, D_y, and D_z are defined in a similar manner.

A. Solving second-order systems using Cramer's rule

Cramer's rule states that to find the solution of a second-order system we set

$$x = \frac{D_x}{D} \quad \text{and} \quad y = \frac{D_y}{D}$$

and simplify the expressions.

EXAMPLE 5-10: Solve the following second-order systems:

(a) $\begin{cases} 3x + 2y = 12 \\ 2x + 3y = 13 \end{cases}$ **(b)** $\begin{cases} 3x - 2y = 6 \\ 6x - 4y = 12 \end{cases}$ **(c)** $\begin{cases} x - 2y = 4 \\ 2x - 4y = 4 \end{cases}$

Solution: **(a)** $\begin{cases} 3x + 2y = 12 \\ 2x + 3y = 13 \end{cases}$

$D = \begin{vmatrix} 3 & 2 \\ 2 & 3 \end{vmatrix} = 9 - 4 = 5$ — Write and evaluate the coefficient determinant D.

$D_x = \begin{vmatrix} 12 & 2 \\ 13 & 3 \end{vmatrix} = 36 - 26 = 10$ — Write and evaluate the determinant D_x.

$D_y = \begin{vmatrix} 3 & 12 \\ 2 & 13 \end{vmatrix} = 39 - 24 = 15$ — Write and evaluate the determinant D_y.

$$x = \frac{D_x}{D} = \frac{10}{5} = 2 \qquad\qquad \text{Solve for } x.$$

$$y = \frac{D_y}{D} = \frac{15}{5} = 3 \qquad\qquad \text{Solve for } y.$$

The solution of the system is (2, 3). Check as before.

(b) $\begin{cases} 3x - 2y = 6 \\ 6x - 4y = 12 \end{cases}$

$$D = \begin{vmatrix} 3 & -2 \\ 6 & -4 \end{vmatrix} = -12 + 12 = 0 \qquad\qquad \begin{array}{l} \text{Write and evaluate the coefficient} \\ \text{determinant } D. \end{array}$$

$$D_x = \begin{vmatrix} 6 & -2 \\ 12 & -4 \end{vmatrix} = -24 + 24 = 0 \qquad\qquad \text{Write and evaluate the determinant } D_x.$$

$$D_y = \begin{vmatrix} 3 & 6 \\ 6 & 12 \end{vmatrix} = 36 - 36 = 0 \qquad\qquad \text{Write and evaluate the determinant } D_y.$$

Because all the determinants are 0, the system is dependent, and has an infinite number of solutions.

(c) $\begin{cases} x - 2y = 4 \\ 2x - 4y = 4 \end{cases}$

$$D = \begin{vmatrix} 1 & -2 \\ 2 & -4 \end{vmatrix} = -4 + 4 = 0$$

$$D_x = \begin{vmatrix} 4 & -2 \\ 4 & -4 \end{vmatrix} = -16 + 8 = -8$$

$$D_y = \begin{vmatrix} 1 & 4 \\ 2 & 4 \end{vmatrix} = 4 - 8 = -4$$

Because $D = 0$ and at least one of the other determinants is not 0, the system is inconsistent and has no solution.

B. Solving third-order systems using Cramer's rule

We can also use Cramer's rule to solve third-order systems. If a standard third-order system is given as

$$\begin{cases} a_1 x + b_1 y + c_1 z = d_1 \\ a_2 x + b_2 y + c_2 z = d_2 \\ a_3 x + b_3 y + c_3 z = d_3 \end{cases}$$

then the standard determinants for a third-order system are:

$$D = \begin{vmatrix} a_1 & b_1 & c_1 \\ a_2 & b_2 & c_2 \\ a_3 & b_3 & c_3 \end{vmatrix} \qquad D_x = \begin{vmatrix} d_1 & b_1 & c_1 \\ d_2 & b_2 & c_2 \\ d_3 & b_3 & c_3 \end{vmatrix}$$

$$D_y = \begin{vmatrix} a_1 & d_1 & c_1 \\ a_2 & d_2 & c_2 \\ a_3 & d_3 & c_3 \end{vmatrix} \qquad D_z = \begin{vmatrix} a_1 & b_1 & d_1 \\ a_2 & b_2 & d_2 \\ a_3 & b_3 & d_3 \end{vmatrix}$$

As with a second-order system, we evaluate each determinant, then follow Cramer's rule for third-order systems to arrive at a solution:

$$x = \frac{D_x}{D} \qquad y = \frac{D_y}{D} \qquad z = \frac{D_z}{D}$$

EXAMPLE 5-11: Solve the following third-order system using Cramer's rule:

$$\begin{cases} 2x - 3y + z = 6 \\ 3x + 2y - z = 0 \\ x - y + 2z = 4 \end{cases}$$

Solution: $D = \begin{vmatrix} 2 & -3 & 1 \\ 3 & 2 & -1 \\ 1 & -1 & 2 \end{vmatrix}$ Write D.

$= 2\begin{vmatrix} 2 & -1 \\ -1 & 2 \end{vmatrix} - 3\begin{vmatrix} -3 & 1 \\ -1 & 2 \end{vmatrix} + 1\begin{vmatrix} -3 & 1 \\ 2 & -1 \end{vmatrix}$ Evaluate D by expanding about the first column.

$= 2(3) - 3(-5) + 1(1)$

$= 22$

$D_x = \begin{vmatrix} 6 & -3 & 1 \\ 0 & 2 & -1 \\ 4 & -1 & 2 \end{vmatrix}$ Write D_x.

$= 6\begin{vmatrix} 2 & -1 \\ -1 & 2 \end{vmatrix} - 0\begin{vmatrix} -3 & 1 \\ -1 & 2 \end{vmatrix} + 4\begin{vmatrix} -3 & 1 \\ 2 & -1 \end{vmatrix}$ Evaluate D_x by expanding about the first column.

$= 6(3) - 0(-5) + 4(1)$

$= 22$

$D_y = \begin{vmatrix} 2 & 6 & 1 \\ 3 & 0 & -1 \\ 1 & 4 & 2 \end{vmatrix}$ Write D_y.

$= 2\begin{vmatrix} 0 & -1 \\ 4 & 2 \end{vmatrix} - 3\begin{vmatrix} 6 & 1 \\ 4 & 2 \end{vmatrix} + 1\begin{vmatrix} 6 & 1 \\ 0 & -1 \end{vmatrix}$ Evaluate D_y by expanding about the first column.

$= 2(4) - 3(8) + 1(-6)$

$= 8 - 24 - 6$

$= -22$

$D_z = \begin{vmatrix} 2 & -3 & 6 \\ 3 & 2 & 0 \\ 1 & -1 & 4 \end{vmatrix}$ Write D_z.

$= 2\begin{vmatrix} 2 & 0 \\ -1 & 4 \end{vmatrix} - 3\begin{vmatrix} -3 & 6 \\ -1 & 4 \end{vmatrix} + 1\begin{vmatrix} -3 & 6 \\ 2 & 0 \end{vmatrix}$ Evaluate D_z by expanding about the first column.

$= 2(8) - 3(-6) + 1(-12)$

$= 16 + 18 - 12$

$= 22$

$x = \dfrac{D_x}{D} = \dfrac{22}{22} = 1$ Solve for x.

$y = \dfrac{D_y}{D} = \dfrac{-22}{22} = -1$ Solve for y.

$z = \dfrac{D_z}{D} = \dfrac{22}{22} = 1$ Solve for z.

The solution of the system is $(1, -1, 1)$. Check as before.

5-6. Solving Problems Using Systems

To **solve a problem using a system,** we:

1. *Read* the problem carefully, several times.
2. *Identify* the unknowns.
3. *Decide* how to represent the unknowns using two or more variables.
4. *Make a table* when appropriate to help represent the unknowns.
5. *Translate* the problem to a system.
6. *Solve* the system.
7. *Interpret* the solutions of the system with respect to each represented unknown to find the proposed solutions of the original problem.
8. *Check* to see if the proposed solutions satisfy all the conditions of the original problem.

A. Solving age problems using systems

EXAMPLE 5-12: Solve the following age problem using a system:

1. *Read:* Four years from now, Bob will be three times the age that Chris is now. Three years ago, the sum of their ages was 10. How old was Bob three years ago? How old will Chris be four years from now?

2. *Identify:* The unknown ages are $\begin{cases} \text{the current age of Bob} \\ \text{the current age of Chris} \\ \text{their ages four years from now} \\ \text{their ages three years ago} \end{cases}$.

3. *Decide:* Let b = the current age of Bob

and c = the current age of Chris [see Note 1].

4. *Make a table:*

	now	four years from now	three years ago	
Bob	b	$b + 4$	$b - 3$	[See Note 2.]
Chris	c	$c + 4$	$c - 3$	

5. *Translate:*

Four years from now, Bob will be three times the age that Chris is now.

$b + 4 \qquad = \qquad 3 \quad \cdot \qquad c$

Three years ago, the sum of their ages was 10.

$(b - 3) + (c - 3) \qquad = 10$

6. *Solve:*
$$\begin{cases} b + 4 = 3c \\ (b - c) + (c - 3) = 10 \end{cases} \text{write equations in standard form} \longrightarrow \begin{cases} b - 3c = -4 \\ b + c = 16 \end{cases}$$

$b + c = 16 \longleftarrow$ one of the system equations

$b = 16 - c \longleftarrow$ solved equation

$b - 3c = -4 \longleftarrow$ the other system equation

$(16 - c) - 3c = -4$ Substitute to eliminate a variable.

$16 - c - 3c = -4$ Solve for c.

$16 - 4c = -4$

$-4c = -20$

$c = 5$

$$b = 16 - c \longleftarrow \text{solved equation}$$

$$b = 16 - 5 \qquad \text{Substitute for } c.$$

$$b = 11 \qquad \text{Solve for } b.$$

7. *Interpret:* $b = 11$ means that Bob is now 11 years old.
$c = 5$ means that Chris is now 5 years old.
$b - 3 = 11 - 3 = 8$ means that three years ago Bob was 8 years old.
$c + 4 = 5 + 4 = 9$ means that four years from now Chris will be 9 years old.

8. *Check:*   Is the age of Bob four years from now three times the age of Chris now?
Yes: $b + 4 = 11 + 4 = 15$ and $3c = 3(5) = 15$.
Was the sum of their ages 10 three years ago?
Yes: $(b - 3) + (c - 3) = (11 - 3) + (5 - 3) = 8 + 2 = 10$.

Note 1: To solve an age problem using a system, you represent each different unknown current age with a different variable.

Note 2: To represent any unknown ages that are in the future, you add to the current-age variables. For those in the past, you subtract from the current-age variables.

B. Solving mixture problem using systems

EXAMPLE 5-13: Solve the following mixture problem using a system:

1. *Read:*   How much pure antifreeze must be added to 12 gallons of a 20%-antifreeze solution (20% antifreeze and 80% water) to obtain a 60%-antifreeze solution?

2. *Identify:*   The unknown base amounts are $\begin{cases} \text{the amount of antifreeze to be added} \\ \text{the amount of 60\%-antifreeze solution} \end{cases}$.

3. *Decide:*   Let $x =$ the amount of antifreeze to be added

and $y =$ the amount of 60%-antifreeze solution [see Note 1].

4. *Make a table:*

	percent [of antifreeze]	base amount [in gallons]	amount of antifreeze [in gallons]	
20%-antifreeze solution	20%	12	20%(12) or 2.4	[See Note 2.]
added antifreeze	100%	x	100%x or 1x	
60%-antifreeze solution	60%	y	60%y or 0.6 y	

5. *Translate:*   Total amount of 20% solution plus added antifreeze equals 60% solution.

$$12 \qquad + \qquad x \qquad = \qquad y$$

Amount of antifreeze in 20% solution plus added antifreeze equals 60% solution.

$$2.4 \qquad + \qquad 1x \qquad = \qquad 0.6y$$

6. *Solve:*
$$12 + x = y \longrightarrow 12 + x = y$$

$$2.4 + 1x = 0.6y \underset{\text{clear decimals}}{\longrightarrow} 24 + 10x = 6y \qquad \text{Multiply by 10.}$$

$$y = x + 12 \longleftarrow \text{solved equation}$$

$$24 + 10x = 6y \longleftarrow \text{other system equation}$$

$$24 + 10x = 6(12 + x) \qquad \text{Substitute to eliminate a variable.}$$

$$24 + 10x = 72 + 6x \qquad \text{Solve for } x.$$

$$10x = 48 + 6x$$

$$4x = 48$$

$$x = 12$$

$$y = x + 12 \longleftarrow \text{solved equation}$$

$$y = (12) + 12 \qquad \text{Substitute for } x.$$

$$y = 24 \qquad \text{Solve for } y.$$

7. *Interpret:* $x = 12$ means the amount of antifreeze to be added is 12 gallons.
          $y = 24$ means the amount of 60%-antifreeze solution is 24 gallons.

8. *Check:*   Does the total amount of 20% solution plus added antifreeze equal the total amount
          of 60% solution?
          Yes: Total amount of $\underset{12}{\underline{20\% \text{ solution}}}$ + $\underset{12}{\underline{\text{added antifreeze}}}$ = $\underset{24}{\underline{60\% \text{ solution}}}$

          Does the amount of antifreeze in the 20% solution, and added antifreeze, equal
          the amount of antifreeze in the 60% solution?
          Yes: Amount of antifreeze in $\underset{\underbrace{20\%(12)}}{\underline{20\% \text{ solution}}}$ + $\underset{\underbrace{100\%(12)}}{\underline{\text{added antifreeze}}}$ = $\underset{\underbrace{60\%(24)}}{\underline{60\% \text{ solution}}}$

$$2.4 \quad + \quad 12 \quad = \quad 14.4$$

Note 1: To solve a mixture problem using a system, you represent each different unknown base
          amount with a different variable.

Note 2: To represent an unknown amount of ingredient in an unknown base amount, you multiply
          the known percent of ingredient times the corresponding base-amount variable.

C. Solving uniform motion problems involving opposite rates using systems

An object that moves at a constant rate is said to move with **uniform motion.** The formula used with
uniform motion problems is $d = rt$, where d is the distance traveled, r is the rate at which a vehicle
travels, and t is the time needed to travel that distance. The rate at which a plane or boat travels
may be altered by forces such as wind or water currents. For example, if the **airspeed** of a plane (the
rate of the plane in still air) is represented by r, and the **wind velocity** (the constant speed and direction
of the blowing wind) is represented by w, then the **groundspeed** of the plane (its **true speed** in relation
to the ground) is represented as $r + w$ when the plane is flying with the wind, and as $r - w$ when
the plane is flying against the wind; "with the wind" and "against the wind" are **opposite wind rates.**

EXAMPLE 5-14: Solve the following uniform motion problem involving opposite wind rates using
a system.

1. *Read:*   Flying at a constant airspeed, a pilot takes 2 hours to fly 660 miles with the wind,
          and 3 hours to make the return trip against the same wind. Assuming the wind
          velocity to be constant during the entire trip, what was the constant airspeed of

the plane, and the wind velocity? What was the groundspeed (true speed) of the plane with the wind and against the wind?

2. *Identify:*   The unknown rates are $\begin{cases} \text{the constant airspeed of the plane} \\ \text{the constant wind velocity} \\ \text{the groundspeed with the wind} \\ \text{the groundspeed against the wind} \end{cases}$.

3. *Decide:*   Let a = the constant airspeed of the plane

and w = the constant wind velocity [see the following note].

4. *Make a table:*

	distance (d) [in miles]	rate (r) [groundspeed in mph]	time (t) [in hours]
with the wind	660	$a + w$	2
against the wind	660	$a - w$	3

5. *Translate:*   With the wind, the distance equals the rate times the time.

$$660 = (a + w) \cdot 2$$

Against the wind, the distance equals the rate times the time.

$$660 = (a - w) \cdot 3$$

6. *Solve:* $\begin{cases} 660 = (a + w)2 \\ 660 = (a - w)3 \end{cases}$ write equations in standard form $\longrightarrow$

$a + w = 330$
$\underline{a - w = 220}$
$2a + 0 = 550$ Use addition method to eliminate a variable.

$2a - 550$ Solve for a.
$a = 275$

$a + w = 330$ $\longleftarrow$ one of the system equations

$(275) + w = 330$ Substitute for a.

$w = 55$ Solve for w.

7. *Interpret:* $a = 275$ means the airspeed of the plane is 275 mph.
$w = 55$ means the velocity of the wind is 55 mph.
$a + w = 275 + 55 = 330$ means the groundspeed with the wind is 330 mph.
$a - w = 275 - 55 = 220$ means the groundspeed against the wind is 220 mph.

8. *Check:*   Does it take 2 hours to fly 660 miles with the wind if the groundspeed is 330 mph?
Yes: $d = rt$ or $660 = (330)(2)$.
Does it take 3 hours to fly 660 miles against the wind if the groundspeed is 220 mph?
Yes: $d = rt$ or $660 = (220)(3)$.

Note: To solve a uniform motion problem involving opposite wind rates using a system, you represent the unknown constant airspeed and wind velocity with different variables.

If the **waterspeed** of a boat (the speed of the boat in still water) is represented by r, and the **current rate** (the constant speed of the flowing river water) is represented by c, then the **landspeed** of the boat (the true speed of the boat with respect to land) is represented as $r + c$ when the boat is traveling downstream, and as $r - c$ when the boat is traveling upstream; "downstream" and "upstream" are **opposite current rates.**

EXAMPLE 5-15: Solve the following uniform motion problem involving opposite current rates:

1. *Read:*   Rowing at a constant rate, it takes Carol 3 hours to row 27 miles downstream, and 9 hours to make the return trip. What is the waterspeed of the boat, as well as the current rate? What is the landspeed of the boat going downstream and upstream?

2. *Identify:*   The unknown rates are $\begin{cases} \text{the constant waterspeed of the boat} \\ \text{the current rate} \\ \text{the landspeed going downstream} \\ \text{the landspeed going upstream} \end{cases}$.

3. *Decide:*   Let $x =$ the constant waterspeed of the boat

and $y =$ the constant current rate [see the following note].

4. *Make a table:*

	distance (d) [in miles]	rate (r) [landspeed in mph]	time (t) [in hours]
downstream	27	$x + y$	3
upstream	27	$x - y$	9

5. *Translate:*   Rowing downstream, the distance equals, the rate times the time.

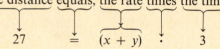

$$27 = (x + y) \cdot 3$$

Rowing upstream, the distance equals the rate times the time.

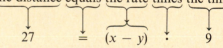

$$27 = (x - y) \cdot 9$$

6. *Solve:* $\begin{cases} 27 = (x + y)3 \\ 27 = (x - y)9 \end{cases}$ $\xrightarrow[\text{divide by 9}]{\text{divide by 3}}$ $\begin{array}{l} x + y = 9 \\ x - y = 3 \\ \hline 2x + 0 = 12 \end{array}$ ⟵ standard form

                                          Use the addition method to eliminate a variable.

$$2x = 12$$ Solve for x.

$$x = 6$$

$x + y = 9$ ⟵ one of the system equations

$6 + y = 9$ Substitute for x.

$y = 3$ Solve for y.

7. *Interpret:* $x = 6$ means the constant waterspeed of the boat is 6 mph.
$y = 3$ means the constant current rate is 3 mph.
$x + y = 6 + 3 = 9$ means the landspeed [true speed] downstream is 9 mph.
$x - y = 6 - 3 = 3$ means the landspeed [true speed] upstream is 3 mph.

8. *Check:*   Does it take 3 hours to row 27 miles downstream if the landspeed is 9 mph?
Yes: $d = rt$ or $27 = (9)(3)$
Does it take 9 hours to row 27 miles upstream if the landspeed is 3 mph?
Yes: $d = rt$ or $27 = (3)(9)$

Note: To solve a uniform motion problem involving opposite current rates using a system, you represent the unknown constant waterspeed of the boat and the current rate with different variables.

RAISE YOUR GRADES

Can you . . . ?

☑ determine whether ordered pairs are solutions of second-order systems
☑ solve second-order systems by graphing
☑ solve second-order systems by the substitution method
☑ solve second-order systems by the addition (elimination) method
☑ determine whether ordered triples are solutions of third-order systems
☑ solve third-order systems by the addition method
☑ evaluate second-order determinants
☑ expand third-order determinants
☑ evaluate third-order determinants
☑ solve second-order systems using Cramer's rule
☑ solve third-order systems using Cramer's rule
☑ solve a number problem using a system
☑ solve an age problem using a system
☑ write expanded notation
☑ reverse the digits of a number
☑ solve a digit problem using a system
☑ solve a mixture problem using a system
☑ solve a uniform motion problem involving opposite rates using a system

---

SOLVED PROBLEMS

PROBLEM 5-1 Determine whether the given ordered pair is a solution of the given second-order system:

(a) $(3, 1)$, $\begin{cases} x - 2y = 1 \\ 2x - 3y = 3 \end{cases}$ **(b)** $(-2, 3)$, $\begin{cases} 3x + 2y = 0 \\ x + y = 1 \end{cases}$ **(c)** $(2, 3)$, $\begin{cases} 3x - 4y = 12 \\ 2x + 3y = 4 \end{cases}$

Solution: Recall that to solve a system of linear equations by graphing, you graph each equation, then consider their intersection. Each point of intersection is a solution to the system [see Example 5-2].

(a) $\begin{cases} x - 2y = 1 \\ 2x - 3y = 3 \end{cases}$ **(b)** $\begin{cases} 3x + 2y = 0 \\ x + y = 1 \end{cases}$ **(c)** $\begin{cases} 3x - 4y = 12 \\ 2x + 3y = 4 \end{cases}$

$$\begin{array}{c|c} x - 2y = 1 & \\ \hline 3 - 2(1) & 1 \\ 3 - 2 & 1 \end{array}$$

$$\begin{array}{c|c} 3x + 2y = 0 & \\ \hline 3(-2) + 2(3) & 0 \\ -6 + 6 & 0 \end{array}$$

$$\begin{array}{c|c} 3x - 4y = 12 & \\ \hline 3(2) - 4(3) & 12 \\ 6 - 12 & 12 \end{array}$$

$1 = 1$ (3, 1) checks.

$0 = 0$ $(-2, 3)$ checks.

$-6 \neq 12$ (2, 3) does not check.

(2, 3) is not a solution.

$$\frac{2x - 3y = 3}{\begin{array}{c|c} 2(3) - 3(1) & 3 \\ 6 - 3 & 3 \end{array}}$$

$$\frac{x + y = 1}{\begin{array}{c|c} -2 + 3 & 1 \end{array}}$$

$3 = 3$ (3, 1) checks. $1 = 1$ (−2, 3) checks.

(3, 1) is a solution.   (−2, 3) is a solution.

PROBLEM 5-2 Solve the following systems by graphing:

(a) $\begin{cases} x + y = 1 \\ 3x + 2y = 0 \end{cases}$ (b) $\begin{cases} 2x + 3y = 5 \\ 3x - 2y = -12 \end{cases}$ (c) $\begin{cases} 3x + 2y = 8 \\ 2x - y = 3 \end{cases}$

Solution: Recall that to determine whether an ordered pair is a solution of a second-order system, you substitute the ordered pair in each equation and determine if the resulting statements are true [see Example 5-1].

(a) $\begin{cases} x + y = 1 \\ 3x + 2y = 0 \end{cases}$ (b) $\begin{cases} 2x + 3y = 5 \\ 3x - 2y = -12 \end{cases}$ (c) $\begin{cases} 3x + 2y = 8 \\ 2x - y = 3 \end{cases}$

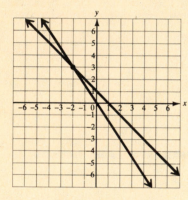   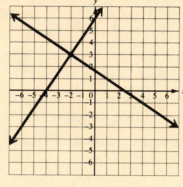   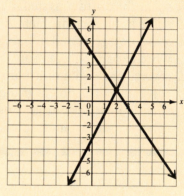

(−2, 3) is the solution.   (−2, 3) is the solution.   (2, 1) is the solution.

PROBLEM 5-3 Solve the following second-order systems by the substitution method:

(a) $\begin{cases} 3x + 2y = 0 \\ x + y = -1 \end{cases}$ (b) $\begin{cases} 4x - y = 1 \\ x + y = 4 \end{cases}$ (c) $\begin{cases} 3x - 2y = 5 \\ 3x + 4y = 8 \end{cases}$

(d) $\begin{cases} x - y = 0 \\ x + y = 4 \end{cases}$ (e) $\begin{cases} 3x + 2y = 12 \\ 2x + 3y = 13 \end{cases}$ (f) $\begin{cases} x - 4y = 0 \\ 2x - 8y = 0 \end{cases}$

Solution: Recall that to solve a second-order system by the substitution method, you:

1. Solve one of the equations for one of the variables in terms of the other variable.
2. Substitute the solution obtained in step 1 for that variable in the second equation, in order to obtain an equation in one variable.
3. Solve the equation obtained in step 2.
4. Substitute the numerical value obtained in step 3 in the first equation.
5. Solve the equation obtained in step 4.
6. Check the proposed solution in the system [see Example 5-3].

(a) $x + y = -1$ (b) $x + y = 4$ (c) $3x - 2y = 5$

$\quad y = -x - 1$ $\quad y = -x + 4$ $\quad 3x = 2y + 5$

$3x + 2(-x - 1) = 0$ $4x - (-x + 4) = 1$ $(2y + 5) + 4y = 8$

$3x - 2x - 2 = 0$ $4x + x - 4 = 1$ $2y + 5 + 4y = 8$

$$x = 2$$
$$x + y = -1$$
$$2 + y = -1$$
$$y = -3$$

$$\frac{3x + 2y = 0}{3(2) + 2(-3) \quad 0}$$
$$6 - 6 \quad 0$$
$$0 = 0$$

$$\frac{x + y = -1}{2 + (-3) \mid -1}$$
$$-1 = -1$$

$(2, -3)$ is the solution.

$$5x = 5$$
$$x = 1$$
$$x + y = 4$$
$$1 + y = 4$$
$$y = 3$$

$$\frac{4x - y = 1}{4(1) - 3 \mid 1}$$
$$4 - 3 \mid 1$$
$$1 = 1$$

$$\frac{x + y = 4}{1 + 3 \mid 4}$$
$$4 = 4$$

$(1, 3)$ is the solution.

$$6y = 3$$
$$y = \tfrac{1}{2}$$
$$3x - 2y = 5$$
$$3x - 2(\tfrac{1}{2}) = 5$$
$$3x = 6$$
$$x = 2$$

$$\frac{3x - 2y = 5}{3(2) - 2(\tfrac{1}{2}) \mid 5}$$
$$6 - 1 \mid 5$$
$$5 = 5$$

$$\frac{3x + 4y = 8}{3(2) + 4(\tfrac{1}{2}) \mid 8}$$
$$6 + 2 \mid 8$$
$$8 = 8$$

$(2, \tfrac{1}{2})$ is the solution.

(d) $x - y = 0$
$$x = y$$
$$x + y = 4$$
$$y + y = 4$$
$$2y = 4$$
$$y = 2$$
$$x - y = 0$$
$$x - 2 = 0$$
$$x = 2$$

$$\frac{x - y = 0}{2 - 2 \quad 0}$$
$$0 = 0$$

$$\frac{x + y = 4}{2 + 2 \quad 4}$$
$$4 = 4$$

$(2, 2)$ is the solution.

(e) $3x + 2y = 12$
$$y = -\tfrac{3}{2}x + 6$$
$$2x + 3y = 13$$
$$2x + 3(-\tfrac{3}{2}x + 6) = 13$$
$$2x - \tfrac{9}{2}x + 18 = 13$$
$$-\tfrac{5}{2}x = -5$$
$$x = 2$$
$$3x + 2y = 12$$
$$3(2) + 2y = 12$$
$$2y = 6$$
$$y = 3$$

$$\frac{3x + 2y = 12}{3(2) + 2(3) \mid 12}$$
$$6 + 6 \mid 12$$
$$12 = 12$$

$$\frac{2x + 3y = 13}{2(2) + 3(3) \mid 13}$$
$$4 + 9 \mid 13$$
$$13 = 13$$

$(2, 3)$ is the solution.

(f) $x - 4y = 0$
$$x = 4y$$
$$2x - 8y = 0$$
$$2(4y) - 8y = 0$$
$$8y - 8y = 0$$
$$0 = 0$$

Since the system reduces to a true statement, it is a dependent system and has an infinite number of solutions.

PROBLEM 5-4 Solve the following systems by the addition method:

(a) $\begin{cases} x + y = 4 \\ x - y = 2 \end{cases}$ 　　**(b)** $\begin{cases} 3x + 2y = 13 \\ x + y = 5 \end{cases}$ 　　**(c)** $\begin{cases} 2x - 3y = 0 \\ x + y = 5 \end{cases}$

(d) $\begin{cases} 3x + 2y = 13 \\ 2x - 3y = 0 \end{cases}$ (e) $\begin{cases} 4x + 3y = 5 \\ 2x - y = 0 \end{cases}$ (f) $\begin{cases} 3x + 2y = 2 \\ 6x + 4y = 0 \end{cases}$

Solution: Recall that to solve a second-order system by the addition method, you:

1. Multiply both equations by appropriate numbers so that the coefficients of one of the variables are opposite numbers.
2. Add the new equations to get an equation in one variable.
3. Solve the equation for that variable.
4. Substitute the solution obtained in one of the original equations and solve for the other variable.
5. Check the proposed solution in both of the given equations [see Example 5-4].

(a) $\begin{cases} x + y = 4 \\ x - y = 2 \end{cases}$ (b) $\begin{cases} 3x + 2y = 13 \\ x + y = 5 \end{cases}$ (c) $\begin{cases} 2x - 3y = 0 \\ x + y = 5 \end{cases}$

$$\begin{array}{r} x + y = 4 \\ x - y = 2 \\ \hline 2x \quad\quad = 6 \end{array}$$

$$x = 3$$
$$x + y = 4$$
$$3 + y = 4$$
$$y = 1$$

$$\begin{array}{r} x + y = 4 \\ \hline 3 + 1 \mid 4 \end{array}$$
$$4 = 4$$

$$\begin{array}{r} x - y = 2 \\ \hline 3 - 1 \mid 2 \end{array}$$
$$2 = 2$$

(3, 1) is the solution.

$$\begin{array}{r} 3x + 2y = 13 \\ -2x - 2y = -10 \\ \hline x \quad\quad = 3 \end{array}$$

$$x + y = 5$$
$$3 + y = 5$$
$$y = 2$$

$$\begin{array}{r} 3x + 2y = 13 \\ \hline 3(3) + 2(2) \mid 13 \\ 9 + 4 \mid 13 \end{array}$$
$$13 = 13$$

$$\begin{array}{r} x + y = 5 \\ \hline 3 + 2 \mid 5 \end{array}$$
$$5 = 5$$

(3, 2) is the solution.

$$\begin{array}{r} 2x - 3y = 0 \\ -2x - 2y = -10 \\ \hline -5y = -10 \end{array}$$

$$y = 2$$
$$x + y = 5$$
$$x + 2 = 5$$
$$x = 3$$

$$\begin{array}{r} 2x - 3y = 0 \\ \hline 2(3) - 3(2) \mid 0 \\ 6 - 6 \mid 0 \end{array}$$
$$0 = 0$$

$$\begin{array}{r} x + y = 5 \\ \hline 3 + 2 \mid 5 \end{array}$$
$$5 = 5$$

(3, 2) is the solution.

(d) $\begin{cases} 3x + 2y = 13 \\ 2x - 3y = 0 \end{cases}$ (e) $\begin{cases} 4x + 3y = 5 \\ 2x - y = 0 \end{cases}$ (f) $\begin{cases} 3x + 2y = 2 \\ 6x + 4y = 0 \end{cases}$

$$\begin{array}{r} 9x + 6y = 39 \\ 4x - 6y = 0 \\ \hline 13x \quad\quad = 39 \end{array}$$

$$x \quad\quad = 3$$
$$2x - 3y = 0$$
$$2(3) - 3y = 0$$
$$6 - 3y = 0$$
$$y = 2$$

$$\begin{array}{r} 4x + 3y = 5 \\ 6x - 3y = 0 \\ \hline 10x \quad\quad = 5 \end{array}$$

$$x \quad\quad = \tfrac{1}{2}$$
$$2x - y = 0$$
$$2(\tfrac{1}{2}) - y = 0$$
$$1 - y = 0$$
$$y = 1$$

$$\begin{array}{r} -6x - 4y = -4 \\ 6x + 4y = 0 \\ \hline 0 = -4 \end{array}$$

The system has no solution and is inconsistent.

$$\begin{array}{r} 3x + 2y = 13 \\ \hline 3(3) + 2(2) \mid 13 \\ 9 + 4 \mid 13 \end{array}$$
$$13 = 13$$

$$\begin{array}{r} 4x + 3y = 5 \\ \hline 4(\tfrac{1}{2}) + 3(1) \mid 5 \\ 2 + 3 \mid 5 \end{array}$$
$$5 = 5$$

$$\begin{array}{c|c} 2x - 3y = 0 \\ \hline 2(3) - 3(2) & 0 \\ 6 - 6 & 0 \\ \end{array}$$

$$0 = 0$$

(3, 2) is the solution.

$$\begin{array}{c|c} 2x - y = 0 \\ \hline 2(\frac{1}{2}) - 1 & 0 \\ 1 - 1 & 0 \\ \end{array}$$

$$0 = 0$$

$(\frac{1}{2}, 1)$ is the solution.

PROBLEM 5-5 Determine whether the given ordered triple is a solution of the given third-order system:

(a) $(1, 1, 1)$, $\begin{cases} x - y - z = -1 \\ x + y - z = 1 \\ x - 2y + z = 0 \end{cases}$
(b) $(1, -1, 1)$, $\begin{cases} 2x - y + 3z = 3 \\ x - 2y + 3z = 3 \\ 2x + y + z = 1 \end{cases}$

Solution: Recall that to determine whether a given ordered triple is a solution of a third-order system, you must determine that it is a solution of all three equations of the system. [see Example 5-5].

(a)
$$\begin{array}{c|c} x - y - z = -1 \\ \hline 1 - 1 - 1 & -1 \\ \end{array} \qquad \begin{array}{c|c} x + y - z = 1 \\ \hline 1 + 1 - 1 & 1 \\ \end{array} \qquad \begin{array}{c|c} x - 2y + z = 0 \\ \hline 1 - 2(1) + 1 & 0 \\ \end{array}$$

$$-1 = -1 \qquad\qquad 1 = 1 \qquad\qquad 0 = 0$$

$(1, 1, 1)$ is a solution of the system.

(b)
$$\begin{array}{c|c} 2x - y + 3z = 3 \\ \hline 2(1) - (-1) + 3(1) & 3 \\ 2 + 1 + 3 & 3 \\ \end{array} \qquad \begin{array}{c|c} x - 2y + 3z = 3 \\ \hline 1 - 2(-1) + 3(1) & 3 \\ 1 + 2 + 3 & 3 \\ \end{array} \qquad \begin{array}{c|c} 2x + y + z = 1 \\ \hline 2(1) + (-1) + 1 & 1 \\ 2 - 1 + 1 & 1 \\ \end{array}$$

$$6 \neq 3 \qquad\qquad 6 \neq 3 \qquad\qquad 2 \neq 1$$

$(1, -1, 1)$ is not a solution of the system.

PROBLEM 5-6 Solve the following th d-order systems using the addition method:

(a) $\begin{cases} x + y + z = 3 \\ x - y + z = 1 \\ x + y - z = 1 \end{cases}$
(b) $\begin{cases} 3x + 2y + z = 4 \\ 4x - 3y + 2z = 11 \\ x - 5y - 4z = 7 \end{cases}$
(c) $\begin{cases} x + y + z = 2 \\ x - 2y - z = 2 \\ 2x + 2y + z = 1 \end{cases}$

(d) $\begin{cases} x - 2y - 6z = 3 \\ 2x + 4y - 12z = 6 \\ 5x + 6y - 30z = 13 \end{cases}$
(e) $\begin{cases} 2x + 3y + 4z = 3 \\ 2x - 3y - 4z = -1 \\ x + 6y + 2z = 3 \end{cases}$
(f) $\begin{cases} x - 3y + z = -3 \\ 2x + 2y - z = 2 \\ 3x + 4y - 3z = 4 \end{cases}$

Solution: Recall that to solve a third-order system by the addition method, you:

1. Eliminate a variable from any two equations, in the manner used in the second-order system, to get an equation in two variables.
2. Eliminate the same variable from the remaining equation and either of the first two equations to get another equation in two variables.
3. Use the addition method to solve the new second-order system obtained in steps 1 and 2.
4. Substitute the solutions obtained in step 3 in one of the original equations and solve it for the third variable.
5. Check the preposed solution in each of the three equations [see Example 5-6].

(a)
$$\begin{array}{l} x + y + z = 3 \\ x - y + z = 1 \\ \hline 2x \quad + 2z = 4 \end{array}$$

$$\begin{array}{l} x - y + z = 1 \\ x + y - z = 1 \\ \hline 2x \quad\quad = 2 \end{array}$$

$$x = 1$$

(b)
$$\begin{array}{l} -6x - 4y - 2z = -8 \\ 4x - 3y + 2z = 11 \\ \hline -2x - 7y \quad = 3 \end{array}$$

$$\begin{array}{l} 12x + 8y + 4z = 16 \\ x - 5y - 4z = 7 \\ \hline 13x + 3y \quad = 23 \end{array}$$

(c)
$$\begin{array}{l} x + y + z = 2 \\ x - 2y - z = 2 \\ \hline 2x - y \quad = 4 \end{array}$$

$$\begin{array}{l} x - 2y - z = 2 \\ 2x + 2y + z = 1 \\ \hline 3x \quad\quad = 3 \end{array}$$

$$x = 1$$

$$2x + 2z = 4$$

$$2(1) + 2z = 4$$
$$2z = 2$$
$$z = 1$$
$$x + y + z = 3$$
$$1 + y + 1 = 3$$
$$y = 1$$

$$\frac{x + y + z = 3}{1 + 1 + 1 \ |\ 3}$$
$$3 = 3$$

$$\frac{x - y + z = 1}{1 - 1 + 1 \ |\ 1}$$
$$1 = 1$$

$$\frac{x + y - z = 1}{1 + 1 - 1 \ |\ 1}$$
$$1 = 1$$

(1, 1, 1) is the solution.

$$-26x - 91y = 39$$
$$\underline{26x + 6y = 46}$$
$$-85y = 85$$
$$y = -1$$
$$-2x - 7y = 3$$
$$-2x - 7(-1) = 3$$
$$-2x = -4$$
$$x = 2$$
$$3x + 2y + z = 4$$
$$3(2) + 2(-1) + z = 4$$
$$z = 0$$

$$\frac{3x + 2y + z = 4}{3(2) + 2(-1) + 0 \ |\ 4}$$
$$4 = 4$$

$$\frac{4x - 3y + 2z = 11}{4(2) - 3(-1) + 2(0) \ |\ 11}$$
$$11 = 11$$

$$\frac{x - 5y - 4z = 7}{2 - 5(-1) - 4(0) \ |\ 7}$$
$$7 = 7$$

(2, −1, 0) is the solution.

$$2x - y = 4$$

$$2(1) - y = 4$$
$$-y = 2$$
$$y = -2$$
$$x + y + z = 2$$
$$1 + (-2) + z = 2$$
$$z = 3$$

$$\frac{x + y + z = 2}{1 + (-2) + 3 \ |\ 2}$$
$$2 = 2$$

$$\frac{x - 2y - z = 2}{1 - 2(-2) - 3 \ |\ 2}$$
$$2 = 2$$

$$\frac{2x - 2y + z = 1}{2(1) + 2(-2) + 3 \ |\ 1}$$
$$1 = 1$$

(1, −2, 3) is the solution.

(d)
$$2x - 4y - 12z = 6$$
$$\underline{2x + 4y - 12z = 6}$$
$$4x \qquad - 24z = 12$$

$$3x - 6y - 18z = 9$$
$$\underline{5x + 6y - 30z = 13}$$
$$8x \qquad - 48z = 22$$

$$-8x + 48z = -24$$
$$\underline{8x - 48z = 22}$$
$$0 = -2$$

This system has no solution, and is inconsistent.

(e)
$$2x + 3y + 4z = 3$$
$$\underline{2x - 3y - 4z = -1}$$
$$4x \qquad\qquad = 2$$
$$x = \tfrac{1}{2}$$

$$2x - 3y - 4z = -1$$
$$\underline{2x + 12y + 4z = 6}$$
$$4x + 9y \qquad = 5$$

$$4x + 9y = 5$$
$$4(\tfrac{1}{2}) + 9y = 5$$
$$2 + 9y = 5$$
$$9y = 3$$
$$y = \tfrac{1}{3}$$
$$2x + 3y + 4z = 3$$
$$2(\tfrac{1}{2}) + 3(\tfrac{1}{3}) + 4z = 3$$
$$1 + 1 + 4z = 3$$
$$4z = 1$$
$$z = \tfrac{1}{4}$$

$(\tfrac{1}{2}, \tfrac{1}{3}, \tfrac{1}{4})$ is the solution.

(f)
$$x - 3y + z = -3$$
$$\underline{2x + 2y - z = 2}$$
$$3x - y \qquad = -1$$

$$3x - 9y + 3z = -9$$
$$\underline{3x + 4y - 3z = 4}$$
$$6x - 5y \qquad = -5$$

$$-6x + 2y = 2$$
$$\underline{6x - 5y = -5}$$
$$-3y = -3$$
$$y = 1$$

$$3x - y = -1$$
$$3x - 1 = -1$$
$$3x = 0$$
$$x = 0$$

$$x - 3y + z = -3$$
$$0 - 3(1) + z = -3$$
$$z = 0$$

(0, 1, 0) is the solution.

PROBLEM 5-7 Evaluate the following second-order determinants:

(a) $\begin{vmatrix} 2 & 3 \\ 1 & 4 \end{vmatrix}$ (b) $\begin{vmatrix} 3 & -1 \\ 2 & 3 \end{vmatrix}$ (c) $\begin{vmatrix} -2 & 3 \\ -2 & -1 \end{vmatrix}$ (d) $\begin{vmatrix} 2 & 1 \\ -2 & -3 \end{vmatrix}$

(e) $\begin{vmatrix} 0 & 2 \\ -3 & -1 \end{vmatrix}$ (f) $\begin{vmatrix} -3 & -1 \\ 0 & -1 \end{vmatrix}$

Solution: Recall that to evaluate a second-order determinant, you find the number $a_1b_2 - a_2b_1$ [see Example 5-7].

(a) $\begin{vmatrix} 2 & 3 \\ 1 & 4 \end{vmatrix}$ (b) $\begin{vmatrix} 3 & -1 \\ 2 & 3 \end{vmatrix}$ (c) $\begin{vmatrix} -2 & 3 \\ -2 & -1 \end{vmatrix}$

$= 2(4) - 1(3)$ $= 3(3) - 2(-1)$ $= -2(-1) - (-2)3$

$= 5$ $= 11$ $= 8$

(d) $\begin{vmatrix} 2 & 1 \\ -2 & -3 \end{vmatrix}$ (e) $\begin{vmatrix} 0 & 2 \\ -3 & -1 \end{vmatrix}$ (f) $\begin{vmatrix} -3 & -1 \\ 0 & -1 \end{vmatrix}$

$= 2(-3) - (-2)(1)$ $= 0(-1) - (-3)(2)$ $= -3(-1) - 0(-1)$

$= -4$ $= 6$ $= 3$

PROBLEM 5-8 Expand the following third-order determinants about the given row or column:

(a) $\begin{vmatrix} 1 & -2 & -1 \\ 3 & 0 & -2 \\ -1 & 2 & 3 \end{vmatrix}$, row 2   (b) $\begin{vmatrix} 3 & 2 & 1 \\ -1 & 0 & 2 \\ 1 & -2 & -3 \end{vmatrix}$, column 2

Solution: Recall that to expand a third-order determinant about a row or column, you form the product of each element of the row or column with its corresponding minor, then add or subtract the products according to the sign array [see Example 5-8].

(a) $\begin{vmatrix} 1 & -2 & -1 \\ 3 & 0 & -2 \\ -1 & 2 & 3 \end{vmatrix} = 3\begin{vmatrix} -2 & -1 \\ 2 & 3 \end{vmatrix} \quad 0\begin{vmatrix} 1 & -1 \\ -1 & 3 \end{vmatrix} \quad (-2)\begin{vmatrix} 1 & -2 \\ -1 & 2 \end{vmatrix}$

$= -3\begin{vmatrix} -2 & -1 \\ 2 & 3 \end{vmatrix} + 0\begin{vmatrix} 1 & -1 \\ -1 & 3 \end{vmatrix} - (-2)\begin{vmatrix} 1 & -2 \\ -1 & 2 \end{vmatrix}$

(b) $\begin{vmatrix} 3 & 2 & 1 \\ -1 & 0 & 2 \\ 1 & -2 & -3 \end{vmatrix} = 2\begin{vmatrix} -1 & 2 \\ 1 & -3 \end{vmatrix} \quad 0\begin{vmatrix} 3 & 1 \\ 1 & -3 \end{vmatrix} \quad (-2)\begin{vmatrix} 3 & 1 \\ -1 & 2 \end{vmatrix}$

$= -2\begin{vmatrix} -1 & 2 \\ 1 & -3 \end{vmatrix} + 0\begin{vmatrix} 3 & 1 \\ 1 & -3 \end{vmatrix} - (-2)\begin{vmatrix} 3 & 1 \\ -1 & 2 \end{vmatrix}$

PROBLEM 5-9 Evaluate the following third-order determinants:

(a) $\begin{vmatrix} 4 & -1 & -2 \\ 0 & 2 & -1 \\ 2 & 1 & 3 \end{vmatrix}$ (b) $\begin{vmatrix} 1 & -1 & -2 \\ 2 & -3 & 1 \\ 4 & 2 & 1 \end{vmatrix}$ (c) $\begin{vmatrix} 1 & -1 & 1 \\ -2 & 0 & 2 \\ -3 & 1 & 3 \end{vmatrix}$

(d) $\begin{vmatrix} 2 & -1 & 3 \\ 3 & 0 & -2 \\ 1 & -3 & 0 \end{vmatrix}$ (e) $\begin{vmatrix} 3 & -2 & 1 \\ -1 & 2 & 0 \\ 2 & 0 & 1 \end{vmatrix}$ (f) $\begin{vmatrix} -2 & 0 & 3 \\ 1 & 2 & -1 \\ -3 & 0 & 2 \end{vmatrix}$

Solution: Recall that to evaluate a third-order determinant, you expand it, then evaluate the resulting expression [see Example 5-9].

(a) $\begin{vmatrix} 4 & -1 & -2 \\ 0 & 2 & -1 \\ 2 & 1 & 3 \end{vmatrix}$

$= +4 \begin{vmatrix} 2 & -1 \\ 1 & 3 \end{vmatrix} - 0 \begin{vmatrix} -1 & -2 \\ 1 & 3 \end{vmatrix} + 2 \begin{vmatrix} -1 & -2 \\ 2 & -1 \end{vmatrix}$

$= 4(6 + 1) - 0(-3 + 2) + 2(1 + 4)$

$= 38$

(b) $\begin{vmatrix} 1 & -1 & -2 \\ 2 & -3 & 1 \\ 4 & 2 & 1 \end{vmatrix}$

$= +1 \begin{vmatrix} -3 & 1 \\ 2 & 1 \end{vmatrix} - 2 \begin{vmatrix} -1 & -2 \\ 2 & 1 \end{vmatrix} + 4 \begin{vmatrix} -1 & -2 \\ -3 & 1 \end{vmatrix}$

$= 1(-3 - 2) - 2(-1 + 4) + 4(-1 - 6)$

$= -39$

(c) $\begin{vmatrix} 1 & -1 & 1 \\ -2 & 0 & 2 \\ -3 & 1 & 3 \end{vmatrix}$

$= -(-1) \begin{vmatrix} -2 & 2 \\ -3 & 3 \end{vmatrix} + 0 \begin{vmatrix} 1 & 1 \\ -3 & 3 \end{vmatrix} - 1 \begin{vmatrix} 1 & 1 \\ -2 & 2 \end{vmatrix}$

$= 1(-6 + 6) + 0(3 + 3) - 1(2 + 2)$

$= -4$

(d) $\begin{vmatrix} 2 & -1 & 3 \\ 3 & 0 & -2 \\ 1 & -3 & 0 \end{vmatrix}$

$= +1 \begin{vmatrix} -1 & 3 \\ 0 & -2 \end{vmatrix} - (-3) \begin{vmatrix} 2 & 3 \\ 3 & -2 \end{vmatrix} + 0 \begin{vmatrix} 2 & -1 \\ 3 & 0 \end{vmatrix}$

$= 1(2 - 0) + 3(-4 - 9) + 0(0 + 3)$

$= -37$

(e) $\begin{vmatrix} 3 & -2 & 1 \\ -1 & 2 & 0 \\ 2 & 0 & 1 \end{vmatrix}$

$= +1 \begin{vmatrix} -1 & 2 \\ 2 & 0 \end{vmatrix} - 0 \begin{vmatrix} 3 & -2 \\ 2 & 0 \end{vmatrix} + 1 \begin{vmatrix} 3 & -2 \\ -1 & 2 \end{vmatrix}$

$= 1(0 - 4) - 0(0 + 4) + 1(6 - 2)$

$= 0$

(f) $\begin{vmatrix} -2 & 0 & 3 \\ 1 & 2 & -1 \\ -3 & 0 & 2 \end{vmatrix}$

$= -0 \begin{vmatrix} 1 & -1 \\ -3 & 2 \end{vmatrix} + 2 \begin{vmatrix} -2 & 3 \\ -3 & 2 \end{vmatrix} - 0 \begin{vmatrix} -2 & 3 \\ 1 & -1 \end{vmatrix}$

$= 0(2 - 3) + 2(-4 + 9) - 0(2 - 3)$

$= 10$

PROBLEM 5-10 Solve the following second-order systems using Cramer's rule:

(a) $\begin{cases} x + y = 3 \\ x + 3y = 1 \end{cases}$
(b) $\begin{cases} 3x + 2y = 12 \\ x + y = 5 \end{cases}$
(c) $\begin{cases} 3x + 2y = 0 \\ x + y = -1 \end{cases}$

(d) $\begin{cases} 3x + 2y = 2 \\ 6x - 4y = 0 \end{cases}$
(e) $\begin{cases} x + 2y = 5 \\ 2x + 4y = 10 \end{cases}$
(f) $\begin{cases} x - 2y = 4 \\ 2x - 4y = 4 \end{cases}$

Solution: Recall that to solve a second-order system using Cramer's rule, you set

$$x = \frac{D_x}{D} \quad \text{and} \quad y = \frac{D_y}{D}$$

and simplify the expressions [see Example 5-10].

(a) $\begin{cases} x + y = 3 \\ x + 3y = 1 \end{cases}$

$D = \begin{vmatrix} 1 & 1 \\ 1 & 3 \end{vmatrix}$

$= 3 - 1 = 2$

$D_x = \begin{vmatrix} 3 & 1 \\ 1 & 3 \end{vmatrix}$

$= 9 - 1 = 8$

$D_y = \begin{vmatrix} 1 & 3 \\ 1 & 1 \end{vmatrix}$

$= 1 - 3 = -2$

(b) $\begin{cases} 3x + 2y = 12 \\ x + y = 5 \end{cases}$

$D = \begin{vmatrix} 3 & 2 \\ 1 & 1 \end{vmatrix}$

$= 3 - 2 = 1$

$D_x = \begin{vmatrix} 12 & 2 \\ 5 & 1 \end{vmatrix}$

$= 12 - 10 = 2$

$D_y = \begin{vmatrix} 3 & 12 \\ 1 & 5 \end{vmatrix}$

$= 15 - 12 = 3$

(c) $\begin{cases} 3x + 2y = 0 \\ x + y = -1 \end{cases}$

$D = \begin{vmatrix} 3 & 2 \\ 1 & 1 \end{vmatrix}$

$= 3 - 2 = 1$

$D_x = \begin{vmatrix} 0 & 2 \\ -1 & 1 \end{vmatrix}$

$= 0 + 2 = 2$

$D_y = \begin{vmatrix} 3 & 0 \\ 1 & -1 \end{vmatrix}$

$= -3 - 0 = -3$

$$x = \frac{D_x}{D} = \frac{8}{2} = 4 \qquad x = \frac{D_x}{D} = \frac{2}{1} = 2 \qquad x = \frac{D_x}{D} = \frac{2}{1} = 2$$

$$y = \frac{D_y}{D} = \frac{-2}{2} = -1 \qquad y = \frac{D_y}{D} = \frac{3}{1} = 3 \qquad y = \frac{D_y}{D} = \frac{-3}{1} = -3$$

$(4, -1)$ is the solution. $\qquad$ $(2, 3)$ is the solution. $\qquad$ $(2, -3)$ is the solution.

(d) $\begin{cases} 3x + 2y = 1 \\ 6x - 4y = 0 \end{cases}$ $\qquad$ (e) $\begin{cases} x + 2y = 5 \\ 2x + 4y = 10 \end{cases}$ $\qquad$ (f) $\begin{cases} x - 2y = 4 \\ 2x - 4y = 4 \end{cases}$

$$D = \begin{vmatrix} 3 & 2 \\ 6 & -4 \end{vmatrix} \qquad D = \begin{vmatrix} 1 & 2 \\ 2 & 4 \end{vmatrix} \qquad D = \begin{vmatrix} 1 & -2 \\ 2 & -4 \end{vmatrix}$$

$$= -12 - 12 = -24 \qquad = 4 - 4 = 0 \qquad = -4 + 4 = 0$$

$$D_x = \begin{vmatrix} 2 & 2 \\ 0 & -4 \end{vmatrix} \qquad D_x = \begin{vmatrix} 5 & 2 \\ 10 & 4 \end{vmatrix} \qquad D_x = \begin{vmatrix} 4 & -2 \\ 4 & -4 \end{vmatrix}$$

$$= -8 - 0 = -8 \qquad = 20 - 20 = 0 \qquad = -16 + 8 = -8$$

$$D_y = \begin{vmatrix} 3 & 2 \\ 6 & 0 \end{vmatrix} \qquad D_y = \begin{vmatrix} 1 & 5 \\ 2 & 10 \end{vmatrix} \qquad D_y = \begin{vmatrix} 1 & 4 \\ 2 & 4 \end{vmatrix}$$

$$= 0 - 12 = -12 \qquad = 10 - 10 = 0 \qquad = 4 - 8 = -4$$

$$x = \frac{D_x}{D} = \frac{-8}{-24} = \frac{1}{3}$$

Because all the determinants are 0, the system is a dependent system and has many solutions.

Because $D = 0$ and at least one of the other determinants is not 0, the system is inconsistent and has no solution.

$$y = \frac{D_y}{D} = \frac{-12}{-24} = \frac{1}{2}$$

$(\frac{1}{3}, \frac{1}{2})$ is the solution.

PROBLEM 5-11 Solve the following third-order systems using Cramer's rule:

(a) $\begin{cases} 3x - y + z = 3 \\ 2x + y - z = 2 \\ x - 2y + 2z = 1 \end{cases}$ $\qquad$ (b) $\begin{cases} x - y - 2z = 1 \\ x - 2y - z = 3 \\ 2x + 3y + z = -8 \end{cases}$

Solution: Recall that to solve a third-order system using Cramer's rule, you use D, D_x, D_y, and D_z in a manner similar to solving second-order systems to find x, y, and z [see Example 5-11].

(a) $\begin{cases} 3x - y + z = 3 \\ 2x + y - z = 2 \\ x - 2y + 2z = 1 \end{cases}$ $\qquad$ (b) $\begin{cases} x - y - 2z = 1 \\ x - 2y - z = 3 \\ 2x + 3y + z = -8 \end{cases}$

$$D = \begin{vmatrix} 3 & -1 & 1 \\ 2 & 1 & -1 \\ 1 & -2 & 2 \end{vmatrix} \qquad D = \begin{vmatrix} 1 & -1 & -2 \\ 1 & -2 & -1 \\ 2 & 3 & 1 \end{vmatrix}$$

$$= +3 \begin{vmatrix} 1 & -1 \\ -2 & 2 \end{vmatrix} - 2 \begin{vmatrix} -1 & 1 \\ -2 & 2 \end{vmatrix} \qquad = +1 \begin{vmatrix} -2 & -1 \\ 3 & 1 \end{vmatrix} - 1 \begin{vmatrix} -1 & -2 \\ 3 & 1 \end{vmatrix}$$

$$+ 1 \begin{vmatrix} -1 & 1 \\ 1 & -1 \end{vmatrix} \qquad + 2 \begin{vmatrix} -1 & -2 \\ -2 & -1 \end{vmatrix}$$

$$= 3(2 - 2) - 2(-2 + 2) + 1(1 - 1) \qquad = 1(-2 + 3) - 1(-1 + 6) + 2(1 - 4)$$

$$= 0 \qquad\qquad = -10$$

$$D_x = \begin{vmatrix} 3 & -1 & 1 \\ 2 & 1 & -1 \\ 1 & -2 & 2 \end{vmatrix}$$

$$= 3\begin{vmatrix} 1 & -1 \\ -2 & 2 \end{vmatrix} - 2\begin{vmatrix} -1 & 1 \\ -2 & 1 \end{vmatrix} + 1\begin{vmatrix} -1 & 1 \\ 1 & -1 \end{vmatrix}$$

$$= 3(2 - 2) - 2(-2 + 2) + 1(1 - 1)$$

$$= 0$$

$$D_y = \begin{vmatrix} 3 & 3 & 1 \\ 2 & 2 & -1 \\ 1 & 1 & 2 \end{vmatrix}$$

$$= 3\begin{vmatrix} 2 & -1 \\ 1 & 2 \end{vmatrix} - 2\begin{vmatrix} 3 & 1 \\ 1 & 2 \end{vmatrix} + 1\begin{vmatrix} 3 & 1 \\ 2 & -1 \end{vmatrix}$$

$$= 3(4 + 1) - 2(6 - 1) + 1(-3 - 2)$$

$$= 0$$

$$D_z = \begin{vmatrix} 3 & -1 & 3 \\ 2 & 1 & 2 \\ 1 & -2 & 1 \end{vmatrix}$$

$$= 3\begin{vmatrix} 1 & 2 \\ -2 & 1 \end{vmatrix} - 2\begin{vmatrix} -1 & 3 \\ -2 & 1 \end{vmatrix} + 1\begin{vmatrix} -1 & 3 \\ 1 & 2 \end{vmatrix}$$

$$= 3(1 + 4) - 2(-1 + 6) + 1(-2 - 3)$$

$$= 0$$

Because all the determinants are 0, the system is dependent and has infinite solutions.

$$D_x = \begin{vmatrix} 1 & -1 & -2 \\ 3 & -2 & -1 \\ -8 & 3 & 1 \end{vmatrix}$$

$$= 1\begin{vmatrix} -2 & -1 \\ 3 & 1 \end{vmatrix} - 3\begin{vmatrix} -1 & -2 \\ 3 & 1 \end{vmatrix} + 8\begin{vmatrix} -1 & -2 \\ -2 & -1 \end{vmatrix}$$

$$= 1(-2 + 3) - 3(-1 + 6) - 8(1 - 4)$$

$$= 10$$

$$D_y = \begin{vmatrix} 1 & 1 & -2 \\ 1 & 3 & -1 \\ 2 & -8 & 1 \end{vmatrix}$$

$$= 1\begin{vmatrix} 3 & -1 \\ -8 & 1 \end{vmatrix} - 1\begin{vmatrix} 1 & -2 \\ -8 & 1 \end{vmatrix} + 2\begin{vmatrix} 1 & -2 \\ 3 & -1 \end{vmatrix}$$

$$= 1(3 - 8) - 1(1 - 16) + 2(-1 + 6)$$

$$= 20$$

$$D_z = \begin{vmatrix} 1 & -1 & 1 \\ 1 & -2 & 3 \\ 2 & 3 & -8 \end{vmatrix}$$

$$= 1\begin{vmatrix} -2 & 3 \\ 3 & -8 \end{vmatrix} - 1\begin{vmatrix} -1 & 1 \\ 3 & -8 \end{vmatrix} + 2\begin{vmatrix} -1 & 1 \\ -2 & 3 \end{vmatrix}$$

$$= 1(16 - 9) - 1(8 - 3) + 2(-3 + 2)$$

$$= 0$$

$$x = \frac{D_x}{D} = \frac{10}{-10} = -1$$

$$y = \frac{D_y}{D} = \frac{20}{-10} = -2$$

$$z = \frac{D_z}{D} = \frac{0}{-10} = 0$$

$(-1, -2, 0)$ is the solution.

PROBLEM 5-12 Solve each age problem using a system:

(a) Gary is 5 years older than Denise. The sum of their ages is 37. How old is each?

(b) Diane is 6 years older than Nelson. In 5 years, Diane will be three times as old as Nelson was 3 years ago. How old will Diane be 5 years from now? How old was Nelson 3 years ago?

Solution: Recall that to solve an age problem using a linear equation, you:

1. *Read* the problem very carefully, several times.
2. *Identify* the unknown ages.
3. *Decide* how to represent the unknown current ages using two variables.
4. *Make a table* to help represent any unknown ages in the future or past.
5. *Translate* the problem to a system.
6. *Solve* the system.
7. *Interpret* the solutions of the system with respect to each represented unknown age to find the proposed solutions of the original problem.

8. *Check* to see if the proposed solutions satisfy all the conditions of the original problem [see Example 5-12].

(a) *Identify:*   The unknown ages are $\begin{cases} \text{the current age of Gary} \\ \text{the current age of Denise} \end{cases}$.

Decide: Let g = the current age of Gary

and d = the current age of Denise [see the following note].

Translate: Gary is 5 years older than Denise.

$$g = 5 + d$$

The sum of their ages is 37.

$$g + d = 37$$

Solve: $g = d + 5$ ⟵ solved equation

$g + d = 37$ ⟵ other system equation

$(d + 5) + d = 37$ Substitute to eliminate a variable.

$d + 5 + d = 37$ Solve for d.

$2d + 5 = 37$

$2d = 32$

$d = 16$ ⟵ the current age of Denise

$g = d + 5$ ⟵ solved equation

$g = (16) + 5$ Substitute for d and then solve for g.

$g = 21$ ⟵ the current age of Gary [check as before]

Note: Because there are no future or past ages to represent, you do not need to make a table in Problem 5-15(a).

(b) *Identify:*   The unknown ages are $\begin{cases} \text{the current age of Diane} \\ \text{the current age of Nelson} \\ \text{their ages 5 years from now} \\ \text{their ages 3 years ago} \end{cases}$.

Decide: Let d = the current age of Diane

and n = the current age of Nelson.

Make a table:

	now	5 years from now	3 years ago
Diane	d	$d + 5$	$d - 3$
Nelson	n	$n + 5$	$n - 3$

Translate: Diane is 6 years older than Nelson.

$$d = 6 + n$$

In 5 years, Diane will be three times as old as Nelson was 3 years ago.

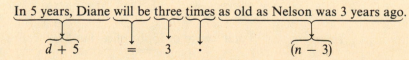

$$d + 5 = 3 \cdot (n - 3)$$

Solve:

$$d = 6 + n \longleftarrow \text{solved equation}$$

$$d + 5 = 3(n - 3) \longleftarrow \text{other equation}$$

$$(6 + n) + 5 = 3(n - 3) \qquad \text{Substitute to eliminate a variable.}$$

$$6 + n + 5 = 3n - 9 \qquad \text{Solve for } n.$$

$$n + 11 = 3n - 9$$

$$-2n + 11 = -9$$

$$-2n = -20$$

$$n = 10 \longleftarrow \text{the current age of Nelson}$$

$$d = 6 + n \longleftarrow \text{solved equation}$$

$$d = 6 + (10) \qquad \text{Substitute for } n \text{ and then solve for } d.$$

$$d = 16 \longleftarrow \text{the current age of Diane}$$

$$d + 5 = (16) + 5 = 21 \longleftarrow \text{the age Diane will be 5 years from now}$$

$$n - 3 = (10) - 3 = 7 \longleftarrow \text{the age Nelson was 3 years ago [check as before]}$$

PROBLEM 5-13 Solve each mixture problem using a system:

(a) The owner of a coffee store wants to make a blend of coffee that is worth \$5.40 per pound (\$5.40/lb) from coffee worth \$5 per pound and coffee worth \$7 per pound. How many pounds of each kind of coffee must be used to make 100 pounds of the blend?

(b) How many gallons of water must be added to 12 gallons of a 25% alcohol solution (25% alcohol and 75% water) in order to have a 15% alcohol solution?

Solution: Recall that to solve a mixture problem using a system, you:

1. *Read* the problem very carefully, several times.
2. *Identify* the unknown base amounts.
3. *Decide* how to represent the unknown base amounts using two variables.
4. *Make a table* to help represent the unknown amount of ingredient in each base amount.
5. *Translate* the problem to a system.
6. *Solve* the system.
7. *Interpret* the solutions of the system with respect to each represented unknown base amount to find the proposed solutions of the original problem.
8. *Check* to see if the proposed solutions satisfy all the conditions of the original problem [see Example 5-13].

(a) *Identify:* The unknown base amounts are $\begin{cases} \text{the number of pounds of \$5/lb coffee} \\ \text{the number of pounds of \$7/lb coffee} \end{cases}$.

Decide: Let x = the number of pounds of \$5/lb coffee

and y = the number of pounds of \$7/lb coffee.

Make a table:

	cost per pound [in dollars]	base amount [in pounds]	value of base amount [in dollars]
\$5/lb coffee	5	x	$5x$
\$7/lb coffee	7	y	$7y$
\$5.40/lb blend	5.40	100	5.40(100) or 540

Translate: Total weight of \$5/lb coffee plus \$7/lb coffee equals 100 pounds.

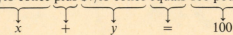

$$x \qquad + \qquad y \qquad = \qquad 100$$

Total value of $5/lb coffee plus $7/lb coffee equals $540.

$$5x + 7y = 540$$

Solve:

$$x + y = 100 \xrightarrow{\text{multiply by } -5} -5x - 5y = -500$$
$$5x + 7y = 540 \xrightarrow{\text{same}} \underline{5x + 7y = 540}$$
$$0 + 2y = 40 \qquad \text{Add to eliminate a variable}$$

$$2y = 40 \qquad \text{Solve for } y.$$
$$y = 20 \text{ (lb)} \longleftarrow \text{number of pounds of } \$7/lb \text{ coffee}$$

$$x + y = 100 \longleftarrow \text{one of the system equations}$$

$$x + (20) = 100 \qquad \text{Substitute for } y \text{ and then solve for } x.$$

$$x = 80 \text{ (lb)} \longleftarrow \text{number of pounds of } \$5/lb \text{ coffee [check as before]}$$

(b) *Identify:* The unknown base amounts are $\begin{cases} \text{the amount of water to be added} \\ \text{the amount of 15\%-alcohol solution} \end{cases}$.

Decide: Let x = the amount of water to be added

and y = the amount of 15%-alcohol solution

Make a table:

	percent [of alcohol]	base amount [in gallons]	amount of alcohol [in gallons]
25% alcohol solution	25%	12	25%(12) or 3
added water	0%	x	0%x or 0
15% alcohol solution	15%	y	15%y or 0.15y

Translate: Total amount of 25% solution plus added water equals 15% solution.

$$12 + x = y$$

Amount of alcohol in 25% solution plus added water equals 15% solution.

$$3 + 0 = 0.15y$$

Solve:

$$3 + 0 = 0.15y \longleftarrow \text{one system equation}$$

$$3 = 0.15y \qquad \text{Solve for } y.$$

$$y = 20 \text{ (gallons)} \longleftarrow \text{the amount of 15\% alcohol solution}$$

$$12 + x = y \longleftarrow \text{other system equation}$$

$$12 + x = (20) \qquad \text{Substitute for } y \text{ and then solve for } x.$$

$$x = 8 \text{ (gallons)} \longleftarrow \text{the amount of water to be added [check as before]}$$

PROBLEM 5-14 Solve each uniform motion problem involving opposite rates using a system:

(a) With uniform effort Greg can row downstream 18 miles in 2 hours. With the same uniform effort, he takes 6 hours to make the return trip up stream. How fast can Greg row in still water with the same uniform effort. What is the rate of the current? What is the landspeed of the boat downstream?

(b) At a constant airspeed of 175 mph, a plane can fly 600 miles against a headwind in the same amount of time that it takes to fly 800 miles with a tailwind of the same velocity. How long is the amount of time? What is the velocity of the wind? What is the groundspeed of the plane with the tailwind?

Solution: Recall that to solve a uniform motion problem involving opposite wind rates using a system, you:

1. *Read* the problem very carefully, several times.
2. *Identify* the unknown rates.
3. *Decide* how to represent the unknown air rates using two variables.
4. *Make a table* when appropriate to help represent the true rates.
5. *Translate* the problem to a system.
6. *Solve* the system.
7. *Interpret* the solutions of the system with respect to each unknown constant rate and true speed to find the proposed solutions of the problem.
8. *Check* to see if the proposed solutions satisfy all the conditions of the original problem [see Examples 5-14 and 5-15].

(a) *Identify:*   The unknown rates are $\begin{cases} \text{the constant waterspeed of the boat} \\ \text{the constant current rate} \\ \text{the landspeed of the boat downstream} \end{cases}$.

Decide: Let $x =$ the constant waterspeed of the boat

and $y =$ the constant current rate.

Make a table:

	distance (d) [in miles]	rate (r) [landspeed in mph]	time (t) [in hours]
downstream	18	$x + y$	2
upstream	18	$x - y$	6

Translate: Rowing downstream, the distance equals the rate times the time.

$$18 = (x + y) \cdot 2$$

Rowing upstream, the distance equals the rate times the time.

$$18 = (x - y) \cdot 6$$

Solve:
$$18 = (x + y)2 \xrightarrow{\text{divide by 2}} x + y = 9 \;\;\leftarrow$$
$$18 = (x - y)6 \xrightarrow[\text{divide by 6}]{} x - y = 3 \;\;\leftarrow \quad \text{standard form}$$
$$\overline{2x + 0 = 12} \quad\quad \text{Add to eliminate a variable.}$$

$$2x = 12 \quad\quad \text{Solve for } x.$$

$$x = 6 \,(\text{mph}) \;\longleftarrow\; \text{waterspeed of the boat}$$

$$x + y = 9 \;\longleftarrow\; \text{one of the system equations}$$

$$(6) + y = 9 \quad\quad \text{Substitute for } x \text{ and then solve for } y.$$

$$y = 3 \,(\text{mph}) \;\longleftarrow\; \text{the current rate}$$

$$x + y = (6) + (3) = 9 \,(\text{mph}) \;\longleftarrow\; \text{the landspeed of the boat downstream [check as before]}$$

(b) *Identify:*   The unknown rates are $\begin{cases} \text{the constant velocity of the wind} \\ \text{the amount of time} \\ \text{the groundspeed with the wind} \end{cases}$.

Decide: Let $w =$ the velocity of the wind

and $t =$ the amount of time

Make a table:

	distance (d) [in miles]	rate (r) [groundspeed in mph]	time (t) [in hours]
against the wind	600	$175 - w$	t
with the wind	800	$175 + w$	t

Translate: Flying against the wind, the distance equals the rate times the time.

$$600 \overset{?}{=} (175 - w) \cdot t$$

Flying with the wind, the distance equals the rate times the time.

$$800 = (175 + w) \cdot t$$

Solve:

$$600 = (175 - w)t \xrightarrow{\text{write in standard form}} 175t - wt = 600$$
$$800 = (175 + w)t \xrightarrow{\text{write in standard form}} 175t + wt = 800$$
$$\overline{\,350t + 0 = 1400}$$ Add to eliminate a variable.

$$350t = 1400$$ Solve for t.

$$t = 4 \text{ (hours)} \longleftarrow \text{the amount of time}$$

$$175t + wt = 800 \longleftarrow \text{one of the system equations}$$

$$175(4) + w(4) = 800$$ Substitute for t.

$$700 + 4w = 800$$ Solve for w.

$$4w = 100$$

$$w = 25 \text{ (mph)} \longleftarrow \text{the constant wind velocity}$$

$$175 + w = 175 + (25)$$ Substitute 25 for w.

$$= 200 \text{ (mph)} \longleftarrow \text{the groundspeed with the wind [check as before]}$$

Supplementary Exercises

PROBLEM 5-15 Determine whether the given ordered pair is a solution of the given second-order system:

(a) $(1, 3)$, $\begin{cases} 4x - y = 1 \\ x + 2y = 7 \end{cases}$ **(b)** $(-3, 0)$, $\begin{cases} x + 3y = -3 \\ 2x - 3y = 0 \end{cases}$ **(c)** $(0, -2)$, $\begin{cases} 4x - y = 2 \\ 2x - 3y = 6 \end{cases}$

(d) $(-2, 1)$, $\begin{cases} 2x + 5y = 1 \\ 3x - 3y = -6 \end{cases}$ **(e)** $(-1, -2)$, $\begin{cases} 2x - 3y = 4 \\ 3x + 2y = 1 \end{cases}$

PROBLEM 5-16 Solve the following system of equations by graphing:

(a) $\begin{cases} 3x + 2y = 8 \\ 2x - 3y = 1 \end{cases}$ **(b)** $\begin{cases} x + y = 5 \\ 3x - 2y = 0 \end{cases}$ **(c)** $\begin{cases} 3x + 2y = 0 \\ 2x - 5y = 19 \end{cases}$

(d) $\begin{cases} 4x - y = 10 \\ x + y = 0 \end{cases}$ (e) $\begin{cases} 2x - 3y = 0 \\ x = -3 \end{cases}$ (f) $\begin{cases} y = 3 \\ 2x - y = -3 \end{cases}$

PROBLEM 5-17 Solve the following systems of equations by the substitution method:

(a) $\begin{cases} x - y = 3 \\ x + y = 5 \end{cases}$ (b) $\begin{cases} x + y = 4 \\ 2x + 3y = 9 \end{cases}$ (c) $\begin{cases} x + y = 1 \\ 3x + 2y = 1 \end{cases}$

(d) $\begin{cases} 3x + y = 1 \\ 2x - 3y = 8 \end{cases}$ (e) $\begin{cases} 2x - 3y = -2 \\ 3x + y = -3 \end{cases}$ (f) $\begin{cases} 3x - 2y = 4 \\ 2x + 3y = -6 \end{cases}$

PROBLEM 5-18 Solve the following systems of equations by the addition method:

(a) $\begin{cases} x - y = 3 \\ x - 3y = 1 \end{cases}$ (b) $\begin{cases} 3x + 2y = 12 \\ x + y = 5 \end{cases}$ (c) $\begin{cases} 3x - 2y = 0 \\ x - y = -1 \end{cases}$

(d) $\begin{cases} 3x + 2y = 2 \\ 6x - 4y = 0 \end{cases}$ (e) $\begin{cases} x + 2y = 5 \\ 2x + 4y = 10 \end{cases}$ (f) $\begin{cases} x - 2y = 4 \\ 2x - 4y = 4 \end{cases}$

PROBLEM 5-19 Determine whether the given ordered-triple is a solution of the given third-order system:

(a) $(1, 1, 1)$, $\begin{cases} x - y - z = -1 \\ x + y - z = 1 \\ x - 2y + z = 0 \end{cases}$ (b) $(-1, 2, 1)$, $\begin{cases} x + y + z = 2 \\ x + 2y - 3z = 0 \\ x + y + 3z = 0 \end{cases}$

(c) $(1, -1, 1)$, $\begin{cases} 2x - y + 3z = 3 \\ x - 2y + 3z = 3 \\ 2x + y + z = 1 \end{cases}$ (d) $(0, -1, 2)$, $\begin{cases} 3x - y + z = 3 \\ 2x + y + 2z = 3 \\ x - 2y - z = 0 \end{cases}$

(e) $(1, 2, -3)$, $\begin{cases} x + y + z = 0 \\ 2x - y - z = 3 \\ 3x - y + z = 2 \end{cases}$

PROBLEM 5-20 Solve the following third-order systems:

(a) $\begin{cases} x + y + z = 2 \\ x - y + z = 0 \\ x + y - z = 4 \end{cases}$ (b) $\begin{cases} 2x + y - z = -2 \\ 3x - y + 2z = -2 \\ x - 2y + z = -2 \end{cases}$ (c) $\begin{cases} x + y + z = 4 \\ x - 2y + z = 7 \\ 2x - y + 2z = 11 \end{cases}$

(d) $\begin{cases} 4x + 2y + z = 2 \\ 2x - y + z = 0 \\ 6x + 3y + z = 3 \end{cases}$ (e) $\begin{cases} x + y - z = 3 \\ 2x + 2y - 2z = 3 \\ x - y + 3z = 3 \end{cases}$ (f) $\begin{cases} 3x + 2y + z = 1 \\ x - 4y - z = -2 \\ 2x + 4y - z = 2 \end{cases}$

PROBLEM 5-21 Evaluate the following second-order determinants:

(a) $\begin{vmatrix} 2 & 5 \\ 1 & 4 \end{vmatrix}$ (b) $\begin{vmatrix} 3 & 4 \\ 2 & -3 \end{vmatrix}$ (c) $\begin{vmatrix} 2 & -1 \\ -2 & -3 \end{vmatrix}$ (d) $\begin{vmatrix} 2 & 0 \\ -3 & -1 \end{vmatrix}$

(e) $\begin{vmatrix} 2 & -3 \\ -2 & -2 \end{vmatrix}$ (f) $\begin{vmatrix} -3 & -1 \\ 0 & 3 \end{vmatrix}$ (g) $\begin{vmatrix} 2 & 0 \\ 1 & 0 \end{vmatrix}$

PROBLEM 5-22 Expand the following third-order determinants about the given row or column:

(a) $\begin{vmatrix} 2 & 4 & 1 \\ 3 & 0 & 1 \\ 1 & 2 & 2 \end{vmatrix}$, row 1 (b) $\begin{vmatrix} 1 & 3 & -1 \\ -2 & 0 & 2 \\ -1 & -2 & 3 \end{vmatrix}$, row 2 (c) $\begin{vmatrix} 3 & 4 & 0 \\ 2 & 1 & 3 \\ -1 & 2 & -1 \end{vmatrix}$, row 3

(d) $\begin{vmatrix} -1 & 4 & 1 \\ 0 & 1 & 0 \\ 3 & 2 & 2 \end{vmatrix}$, column 1   (e) $\begin{vmatrix} 3 & -1 & -3 \\ 2 & 0 & -2 \\ 1 & 2 & 1 \end{vmatrix}$, column 2   (f) $\begin{vmatrix} 4 & 0 & -2 \\ 1 & -3 & -1 \\ 2 & -1 & 4 \end{vmatrix}$, column 3

PROBLEM 5-23 Evaluate the following third-order determinants:

(a) $\begin{vmatrix} 2 & -1 & -2 \\ 1 & 2 & -1 \\ 3 & 0 & 4 \end{vmatrix}$ (b) $\begin{vmatrix} 4 & 2 & 1 \\ 2 & -3 & -1 \\ 1 & 1 & -2 \end{vmatrix}$ (c) $\begin{vmatrix} -3 & -2 & 1 \\ 1 & 0 & -1 \\ 3 & 2 & 1 \end{vmatrix}$

(d) $\begin{vmatrix} 2 & 3 & 1 \\ -1 & 0 & -3 \\ 3 & -2 & 0 \end{vmatrix}$ (e) $\begin{vmatrix} 3 & -1 & 2 \\ -2 & 2 & 0 \\ 1 & 0 & 1 \end{vmatrix}$ (f) $\begin{vmatrix} -2 & 1 & -3 \\ 0 & 2 & 0 \\ 3 & -1 & 2 \end{vmatrix}$

PROBLEM 5-24 Solve the following second-order systems using Cramer's rule:

(a) $\begin{cases} x + y = 4 \\ x - y = 2 \end{cases}$ (b) $\begin{cases} 3x + 2y = 12 \\ x + y = 5 \end{cases}$ (c) $\begin{cases} 2x - 3y = 0 \\ x + y = 5 \end{cases}$

(d) $\begin{cases} 3x - 2y = 12 \\ 2x - 3y = 13 \end{cases}$ (e) $\begin{cases} 4x + 3y = 5 \\ 2x - y = 0 \end{cases}$ (f) $\begin{cases} 3x + 2y = 2 \\ 6x + 4y = 0 \end{cases}$

PROBLEM 5-25 Solve the following third-order system using Cramer's rule:

(a) $\begin{cases} x + y + z = 4 \\ 3x - y + 2z = 14 \\ 2x - 2y - z = 6 \end{cases}$ (b) $\begin{cases} 2x + y - 6z = 5 \\ x - 2y + 6z = 0 \\ 3x + 2y - 3z = 8 \end{cases}$ (c) $\begin{cases} 2x + 4y - z = -2 \\ x + y - 2z = 3 \\ x - y + z = -1 \end{cases}$

(d) $\begin{cases} x - y + z = 4 \\ 2x - 3y + z = 7 \\ x + y + 3z = 6 \end{cases}$ (e) $\begin{cases} x - y + z = 3 \\ 2x - 2y + 2z = 6 \\ x + 2y - z = 3 \end{cases}$ (f) $\begin{cases} x + 2y - z = -2 \\ 2x - y + z = 2 \\ y - 2z = -4 \end{cases}$

PROBLEM 5-26 Solve each age problem using a system:

(a) Myrle is 5 years older than Evelyn. The sum of their ages is 47. How old is each?

(b) Arthur is 10 years older than Carol. Nine years ago, his age was three times her age. How old will each be 4 years from now?

(c) Five years ago, Chuck was twice as old as Jeanne will be 6 years from now. He is now one year older than 4 times her age. How old was Chuck 5 years ago? How old will Jeanne be 6 years from now?

(d) Greg is one-third as old as his father who is two years older than his mother. When Greg was born the sum of the ages of his father and mother was 46. How old is Greg now?

PROBLEM 5-27 Solve each mixture problem using a system:

(a) A nurse wants to strengthen 7 gallons of a 20%-alcohol solution to a 40%-alcohol solution. How much alcohol must be added?

(b) A janitor wants to dilute 7 gallons of 40%-cleaner solution (40% cleaner and 60% water) to a 20%-cleaner solution. How much water must she add?

(c) Tincture of arnica is 20% arnica and 80% alcohol. How much alcohol must be added to a pint of tincture of arnica to get a 5%-arnica solution?

(d) How much water must be evaporated from 100 gallons of a 4%-salt solution (4% salt and 96% water) to get a 5%-salt solution?

(e) A grocer wants 100 pounds of a nut mixture to sell for $6.20/lb. She makes the mixture from nuts that sell for $5/lb and $8/lb. How many pounds of each does it take to make the nut mixture?

(f) Ring gold is usually 14 carat (14K) and the rest copper. Coin gold is usually 90% gold and the rest copper. How many ounces of pure gold (24K) must be added to 24 ounces of ring gold to make an alloy of coin gold? (Hint: nK gold is $\frac{n}{24} \cdot 100\%$ gold.)

PROBLEM 5-28 Solve each uniform motion problem involving opposite rates using a system:

(a) An airplane flying at a constant airspeed with the wind flew 340 miles in 2 hours. At the same constant airspeed, the airplane made the return trip against the same wind in $2\frac{1}{2}$ hours. What were the constant airspeed and the wind velocity? What is the groundspeed when flying with a tailwind?

(b) An airplane flying at a constant airspeed can travel 240 km against a headwind in $1\frac{1}{2}$ hours. At the same constant airspeed the return trip takes one hour and twelve minutes, with a tailwind of the same velocity as the headwind. What are the constant airspeed and the wind velocity? What is the groundspeed when flying against the wind?

(c) A motor boat can travel 36 miles downstream in 3 hours. If it takes twice as long to make the return trip, what are the waterspeed of the boat and the current rate?

(d) A boat can travel downstream 6 miles in 45 minutes. The return trip takes $1\frac{1}{2}$ hours. What are the waterspeed of the boat and the current rate?

(e) At a constant airspeed of 180 km/h, an airplane can fly 500 km with a tailwind in the same amount of time that it takes to fly 400 km against a headwind of the same velocity. What is that amount of time? What is the velocity of the wind? What is the groundspeed of the plane against a headwind?

(f) A boat can travel 8 miles upstream in the same time that it takes to travel 12 miles downstream. The waterspeed of the boat is 20 mph. How long is the period of time? What are the current rate and the landspeed downstream?

Answers to Supplementary Exercises

(5-15) (a) yes   (b) no   (c) yes   (d) no   (e) no

(5-16) (a) (2, 1)   (b) (2, 3)   (c) (2, −3)   (d) (2, −2)   (e) (−3, −2)   (f) (0, 3)

(5-17) (a) (4, 1)   (b) (3, 1)   (c) (−1, 2)   (d) (1, −2)   (e) (−1, 0)   (f) (0, −2)

(5-18) (a) (4, 1)   (b) (2, 3)   (c) (2, 3)   (d) $(\frac{1}{3}, \frac{1}{2})$
      (e) infinite solutions (dependent system)   (f) no solution (inconsistent system)

(5-19) (a) yes   (b) no   (c) no   (d) yes   (e) no

(5-20) (a) (2, 1, −1)   (b) (−1, 1, 1)   (c) (2, −1, 3)   (d) $(\frac{1}{4}, \frac{1}{2}, 0)$
      (e) no solution (inconsistent system)   (f) $(0, \frac{1}{2}, 0)$

(5-21) (a) 3   (b) −17   (c) −8   (d) −2   (e) −10   (f) −9   (g) 0

(5-22) (a) $+2\begin{vmatrix} 0 & 1 \\ 2 & 2 \end{vmatrix} - 4\begin{vmatrix} 3 & 1 \\ 1 & 2 \end{vmatrix} + 1\begin{vmatrix} 3 & 0 \\ 1 & 2 \end{vmatrix}$ (b) $-(-2)\begin{vmatrix} 3 & -1 \\ -2 & 3 \end{vmatrix} + 0\begin{vmatrix} 1 & -1 \\ -1 & 3 \end{vmatrix} - 2\begin{vmatrix} 1 & 3 \\ -1 & -2 \end{vmatrix}$

(c) $+(-1)\begin{vmatrix} 4 & 0 \\ 1 & 3 \end{vmatrix} - 2\begin{vmatrix} 3 & 0 \\ 2 & 3 \end{vmatrix} + (-1)\begin{vmatrix} 3 & 4 \\ 2 & 1 \end{vmatrix}$ (d) $+(-1)\begin{vmatrix} 1 & 0 \\ 2 & 2 \end{vmatrix} - 0\begin{vmatrix} 4 & 1 \\ 2 & 2 \end{vmatrix} + 3\begin{vmatrix} 4 & 1 \\ 1 & 0 \end{vmatrix}$

(e) $-(-1)\begin{vmatrix} 2 & -2 \\ 1 & 1 \end{vmatrix} + 0\begin{vmatrix} 3 & -3 \\ 1 & 1 \end{vmatrix} - 2\begin{vmatrix} 3 & -3 \\ 2 & -2 \end{vmatrix}$

(f) $+(-2)\begin{vmatrix} 1 & -3 \\ 2 & -1 \end{vmatrix} - (-1)\begin{vmatrix} 4 & 0 \\ 2 & -1 \end{vmatrix} + 4\begin{vmatrix} 4 & 0 \\ 1 & -3 \end{vmatrix}$

(5-23) **(a)** 35   **(b)** 39   **(c)** 4   **(d)** -37 **(e)** 0   **(f)** 10

(5-24) **(a)** (3, 1)   **(b)** (2, 3)   **(c)** (3, 2)   **(d)** $(2, -3)$ **(e)** $(\frac{1}{2}, 1)$
       (f) no solution (inconsistent system)

(5-25) **(a)** $(3, -1, 2)$ **(b)** (2, 1, 0)   **(c)** $(0, -1, -2)$ **(d)** $(1, -1, 2)$
       (e) infinite solutions (dependent system)   **(f)** (0, 0, 2)

(5-26) **(a)** 26 years, 21 years   **(b)** 28 years, 18 years   **(c)** 28 years, 14 years
       (d) 12 years

(5-27) **(a)** $2\frac{1}{3}$ gallons   **(b)** 7 gallons   **(c)** 3 pints   **(d)** 20 gallons
       (e) 60 pounds of $5/lb, 40 pounds of $8/lb   **(f)** 76 ounces

(5-28) **(a)** 153 mph, 17 mph, 170 mph   **(b)** 180 km/h, 20 km/h, 160 km/h
       (c) 9 mph, 3 mph   **(d)** 6 mph, 2 mph   **(e)** $2\frac{1}{2}$ hours, 20 km/h, 160 km/h
       (f) 30 minutes, 4 mph, 24 mph

6 POLYNOMIALS

THIS CHAPTER IS ABOUT

☑ Properties of Polynomials
☑ Adding or Subtracting Polynomials
☑ Multiplying Polynomials
☑ Computing Special Products
☑ Dividing Polynomials
☑ Solving Problems by Evaluating Polynomials

6-1. Properties of Polynomials

A. Defining integer exponents

Consider the following lists.

$$3^4 = 81 \longrightarrow 81 \div 3$$
$$3^3 = 27 \longleftarrow$$
$$3^2 = 9$$
$$3^1 = 3$$
$$3^0 = \ ?$$

$$10^4 = 10{,}000 \longrightarrow 10{,}000 \div 10$$
$$10^3 = \ 1{,}000 \longleftarrow$$
$$10^2 = \ \ \ 100$$
$$10^1 = \ \ \ \ 10$$
$$10^0 = \ \ \ \ ?$$

Notice that each power divided by its base produces the power in the row below it ($81 \div 3 = 27$). If this pattern is to continue, then 3^0 must equal $3 \div 3$ or 1, and 10^0 must equal $10 \div 10$ or 1. For example:

    (a) $5^0 = 1$ **(b)** $(-4)^0 = 1$ **(c)** $12.3^0 = 1$ **(d)** $(18 + 2 \cdot 3^5 \div 7)^0 = 1$

These observations lead to the following definition:

For any nonzero real number a, $a^0 = 1$

Caution: a^0 does not mean $a \cdot 0$. If $a \neq 0$, then a^0 means 1.

The previous lists can be extended to include negative integers as exponents.

$$3^4 \ = 81$$
$$3^3 \ = 27$$
$$3^2 \ = 9$$
$$3^1 \ = 3$$
$$3^0 \ = 1$$
$$3^{-1} = \ ?$$
$$3^{-2} = \ ?$$

$$10^4 \ = 10{,}000$$
$$10^3 \ = \ 1{,}000$$
$$10^2 \ = \ \ \ 100$$
$$10^1 \ = \ \ \ \ 10$$
$$10^0 \ = \ \ \ \ \ 1$$
$$10^{-1} = \ ?$$
$$10^{-2} = \ ?$$

If we continue the pattern that each power divided by its base produces the power in the row below, then we will reach the following conclusions.

$$3^{-1} \text{ must equal } 1 \div 3 = \frac{1}{3^1} \qquad\qquad 10^{-1} \text{ must equal } 1 \div 10 = \frac{1}{10^1}$$

$$3^{-2} \text{ must equal } \frac{1}{3} \div 3 = \frac{1}{3} \cdot \frac{1}{3} = \frac{1}{3^2} \qquad 10^{-2} \text{ must equal } \frac{1}{10} \div 10 = \frac{1}{10} \cdot \frac{1}{10} = \frac{1}{10^2}$$

These observations lead to the following definition:

If n is any integer, and if a is any nonzero real number, then $a^{-n} = \dfrac{1}{a^n}$

Some examples of this principle are

(a) $2^{-3} = \dfrac{1}{2^3}$ (b) $(-3)^{-4} = \dfrac{1}{(-3)^4}$ (c) $\dfrac{1}{3^{-2}} = 1 \div (3^{-2}) = 1 \div \dfrac{1}{3^2} = 1 \cdot 3^2 = 3^2$

Caution: A negative exponent does not necessarily mean that the exponential expression is negative. For example:

$$4^{-2} = \frac{1}{4^2} = \frac{1}{16}$$

B. Defining polynomials

An expression of the form ax^n, where a is a constant, and n is a whole number, is called a **monomial** in the variable x. The number a is called the **numerical coefficient of the monomial.** An expression of the form $ax^m y^n$, where m and n are whole numbers, is called a monomial in the variables x and y. A monomial in any number of variables is defined in a similar way. Some examples of monomials are:

(a) $3x^2$ (b) $-17xy^3$ (c) $2.4x^3 w^4 z$ (d) 8

Note: The constant 8 is a monomial since $8 = 8 \cdot 1 = 8x^0$.

In a monomial a variable cannot appear as an exponent, in the denominator, or with an exponent that is not a whole number. For example:

(a) x^y is not a monomial because the variable y appears as an exponent.
(b) $3x^{-2}$ is not a monomial because the variable x appears with an exponent which is not a whole number.
(c) $\dfrac{4}{y^3}$ is not a monomial because the variable y appears in the denominator.

Recall: A term is defined as a number, a variable, or the product or quotient of numbers and variables. The expression $4x^2 - 3x + x^{-1}$ which can be written as $4x^2 + (-3x) + x^{-1}$ has three terms, $4x^2$, $-3x$, and x^{-1} [see Chapter 1].

A **polynomial** is an expression that can be written so that it contains only terms that are monomials. For example:

(a) $4x - 3$ (b) $x^2 + 7x - 5$ (c) $5x^2 y^3$ (d) $4w^3 - 3wy + 7y^2$ (e) 3

Any expression that contains a term which is not a monomial cannot be a polynomial. Some examples of nonpolynomials are:

(a) $5xy + y^{-3}$ is not a polynomial, because y^{-3} is not a monomial.
(b) $x^{1/2} + 5$ is not a polynomial, because $x^{1/2}$ is not a monomial.
(c) $x^3 - y^w$ is not a polynomial, because y^w is not a monomial.
(d) $\dfrac{1}{y} + y^2 - 5y^3$ is not a polynomial, because $\dfrac{1}{y}$ is not a monomial.

Caution: In a polynomial, each term MUST be a monomial.

C. Classifying polynomials

A polynomial is said to be a **simplified polynomial** if all of its like terms have been combined. A simplified polynomial consisting of exactly

(a) one term is classified as a monomial.
(b) two terms is classified as a **binomial.**
(c) three terms is classified as a **trinomial.**

Simplified polynomials consisting of four or more terms are not generally referred to by a special name.

EXAMPLE 6-1: Classify $4x^2 - 5xy + 7xy + 5y^2$ as a monomial, a binomial, or a trinomial.

Solution: $4x^2 - 5xy + 7xy + 5y^2$ is a trinomial because it simplifies to $4x^2 + 2xy + 5y^2$, which has exactly three terms.

D. Finding the degree of a polynomial

In a monomial of the form ax^m, where $a \neq 0$, the whole-number exponent m is called the **degree of the monomial.**

EXAMPLE 6-2: Find the degree of each of the following monomials:

(a) $5x^2$ **(b)** $-7y^3$ **(c)** 3   **(d)** $-w$

Solution: **(a)** $5x^2$ is a monomial of degree 2.
   (b) $-7y^3$ is a monomial of degree 3.
   (c) 3 is a monomial of degree 0, since $3 = 3x^0$.
   (d) $-w$ is a monomial of degree 1, since $-w = -w^1$.

Note: The monomial 0 is not assigned a degree.

The **degree of a monomial in more than one variable** is the sum of the exponents of the variables.

EXAMPLE 6-3: Find the degree of **(a)** $5x^2y^3$ **(b)** $-xy$ **(c)** $4.7uv^2$ **(d)** $2u^3vw^2$

Solution: **(a)** $5x^2y^3$ is a monomial of degree 5 ($2 + 3 = 5$).
   (b) $-xy$ is a monomial of degree 2 ($1 + 1 = 2$).
   (c) $4.7uv^2$ is a monomial of degree 3 ($1 + 2 = 3$).
   (d) $2u^3vw^2$ is a monomial of degree 6 ($3 + 1 + 2 = 6$).

The **degree of a polynomial** is the largest degree of any of its monomial terms (provided like terms have been combined).

EXAMPLE 6-4: Find the degree of $4xy^2 - 3x^2y^3 + 5xy^3 + 3x^2y^3$.

Solution: $4xy^2 - 3x^2y^3 + 5xy^3 + 3x^2y^3 = 4xy^2 + 5xy^3$.   First combine like terms.
   $4xy^2$ is a monomial of degree 3 ($1 + 2 = 3$).   Find the degree of each monomial.
   $5xy^3$ is a monomial of degree 4 ($1 + 3 = 4$).
   $4 > 3$, so the polynomial is of degree 4.

E. Writing polynomials in standard form

A polynomial is in standard form if (1) the variables are written in alphabetical order in each term, (2) the polynomial is simplified, and (3) the terms are arranged in descending powers of the first variable.

EXAMPLE 6-5: Write $4x^2y - 7y^3 + 5y^2x + 3x^3 - yx^2$ in standard form.

Solution:

$$4x^2y - 7y^3 + 5y^2x + 3x^3 - yx^2 = 4x^2y - 7y^3 + 5xy^2 + 3x^3 - x^2y$$

Alphabetize the variables in each term.

$$= 3x^2y - 7y^3 + 5xy^2 + 3x^3$$

Combine like terms.

$$= 3x^3 + 3x^2y + 5xy^2 - 7y^3$$

Arrange in descending powers of the first variable.

6-2. Adding or Subtracting Polynomials

To add polynomials we combine like terms.

EXAMPLE 6-6: Find $(3x^2 + 7x) + (2x^2 - 5x)$.

Solution:

$$(3x^2 + 7x) + (2x^2 - 5x) = 3x^2 + 2x^2 + 7x - 5x$$

Rearrange and group like terms.

$$= (3x^2 + 2x^2) + (7x - 5x)$$

$$= (3 + 2)x^2 + (7 - 5)x$$

This use of the distributive property is often performed mentally.

$$= 5x^2 + 2x$$

Sometimes it is convenient to add polynomials by arranging like terms in columns.

EXAMPLE 6-7: Add $(3x^3 + 4x^2 - 5)$, $(2x^3 + 5x^2 - 3x + 8)$, and $(-x^2 - 3x + 1)$ using a vertical format.

Solution:
$$
\begin{array}{r}
3x^3 + 4x^2 \qquad - 5 \\
2x^3 + 5x^2 - 3x + 8 \\
-x^2 - 3x + 1 \\
\hline
5x^3 + 8x^2 - 6x + 4
\end{array}
$$

Arrange like terms in columns.

Combine like terms.

The **additive inverse of a polynomial** is the sum of the additive inverses of each term of the polynomial. To form the additive inverse of a polynomial, just write the additive inverse of each term.

EXAMPLE 6-8: Find the additive inverse of $3x^2 - 7x + 11$.

Solution: The additive inverse of $3x^2 - 7x + 11$ is $-3x^2 + 7x - 11$.

To subtract a polynomial, you add its additive inverse.

EXAMPLE 6-9: Find $(5y^2 - 7y + 2) - (3y^2 - 4y + 1)$.

change to addition

Solution: $(5y^2 - 7y + 2) - (3y^2 - 4y + 1) = (5y^2 - 7y + 2) + (-3y^2 + 4y - 1)$

write the additive inverse

$$= 5y^2 - 7y + 2 - 3y^2 + 4y - 1 \qquad \text{Add as before.}$$

$$= 5y^2 - 3y^2 - 7y + 4y + 2 - 1$$

$$= 2y^2 - 3y + 1$$

You can also subtract polynomials using a vertical format.

EXAMPLE 6-10: Find $(4x^3 + 5x^2 - 7) - (3x^3 + x^2 - 5x - 4)$ using a vertical format.

Solution:
$$\begin{array}{l} 4x^3 + 5x^2 \quad\quad - 7 \\ -(3x^3 + \ x^2 - 5x - 4) \end{array}$$ Arrange like terms in columns.

$$\begin{array}{l} 4x^3 + 5x^2 \quad\quad - 7 \\ +(-3x^3 - \ x^2 + 5x + 4) \\ \hline x^3 + 4x^2 + 5x - 3 \end{array}$$ Add the additive inverse.

6-3. Multiplying Polynomials

A. Multiplying monomials

Simplify the product of exponential expressions with the same base by using the associative property of multiplication, or by adding the exponents.

$$a^2 \cdot a^3 = (a \cdot a) \cdot (a \cdot a \cdot a) = (a \cdot a \cdot a \cdot a \cdot a) = a^5$$
$$a^2 \cdot a^3 = a^{2+3} = a^5$$

Product Rule for Exponents

If m and n are integers, then: $a^m \cdot a^n = a^{m+n}$.

The product rule for exponents states that we can multiply factors with the same base in exponential notation by simply adding the exponents.

EXAMPLE 6-11: Find the product of each of the following: **(a)** $x^{10} \cdot x^3$ **(b)** $y^7 \cdot y^{11}$ **(c)** $z^0 \cdot z^5$

Solution: Use the product rule for exponents.

(a) $x^{10} \cdot x^3 = x^{10+3} = x^{13}$ **(b)** $y^7 \cdot y^{11} = y^{7+11} = y^{18}$ **(c)** $z^0 \cdot z^5 = z^{0+5} = z^5$

Caution: $x^3 \cdot y^4$ cannot be simplified by the product rule for exponents because the bases are not the same.

To multiply monomials, we use the commutative and associative properties of multiplication to rearrange and group factors.

EXAMPLE 6-12: Find the product of each of the following:

(a) $(2x)(3x^4)$ **(b)** $(-5a^2)(4a^3)$ **(c)** $(-12xy^2)(-3x^2y^5)$
(d) $(3m^2n^3)(-2mn^5)(7mn^4)$ **(e)** $(4xy^2)(-\frac{1}{4}xy^3)$ **(f)** $(0.3p^2q^3)(1.4p^4q^0)$

Solution:

(a) $(2x)(3x^4) = (2 \cdot 3)(x \cdot x^4)$ **(b)** $(-5a^2)(4a^3) = (-5 \cdot 4)(a^2a^3)$

$ = 6x^5$ $ = -20a^5$

(c) $(-12xy^2)(-3x^2y^5) = [(-12)(-3)](x \cdot x^2)(y^2 \cdot y^5)$

$ = 36x^3y^7$

(d) $(3m^2n^3)(-2mn^5)(7mn^4) = [3(-2)7](m^2mm)(n^3n^5n^4)$

$ = -42m^4n^{12}$

(e) $(4xy^2)(-\frac{1}{4}xy^3) = [4(-\frac{1}{4})](xx)(y^2y^3)$ **(f)** $(0.3p^2q^3)(1.4p^4q^0) = [(0.3)(1.4)](p^2p^4)(q^3q^0)$

$\phantom{(e)(4xy^2)(-\frac{1}{4}xy^3)} = -x^2y^5$ $ = 0.42p^6q^3$

To simplify a monomial of the form $(a^m)^n$ you use the definition of exponential notation and the product rule for exponents. For example, $(3^2)^4$ means $(3^2)(3^2)(3^2)(3^2) = 3^8$. It is important to observe that this result could have been obtained by simply multiplying the exponents: $(3^2)^4 = 3^{2 \cdot 4} = 3^8$.

The following rule is a generalization of this observation.

Power Rule for Exponents

If m and n are integers, then $(a^m)^n = a^{mn}$.

The power rule for exponents states that you can find the power of a power by multiplying the exponents.

EXAMPLE 6-13: Find the following powers: (a) $(2^2)^3$ (b) $(x^{-3})^{-4}$ (c) $(y^{-3})^2$

Solution: Use the power rule for exponents.

(a) $(2^2)^3 = 2^{2 \cdot 3}$ (b) $(x^{-3})^{-4} = x^{(-3)(-4)}$ (c) $(y^{-3})^2 = y^{(-3)2}$

$\quad = 2^6$ or 64 $\quad = x^{12}$ $\quad = y^{-6}$ or $\dfrac{1}{y^6}$

The power rule for exponents can be extended to simplify powers of expressions which contain products.

Extended Power Rule for Exponents

If m, n, and p are integers, then: $(a^m b^n)^p = a^{mp} b^{np}$.

EXAMPLE 6-14: Simplify the following powers: (a) $(3x^2)^3$ (b) $(-2x^3 y)^2$

Solution: Use the extended power rule for exponents.

(a) $(3x^2)^3 = 3^{1 \cdot 3} x^{2 \cdot 3}$ (b) $(-2x^3 y)^2 = (-2)^{1 \cdot 2} x^{3 \cdot 2} y^{1 \cdot 2}$

$\quad = 3^3 x^6$ $\quad = (-2)^2 x^6 y^2$

$\quad = 27x^6$ $\quad = 4x^6 y^2$

B. Multiplying a polynomial by a monomial

To multiply a polynomial by a monomial, we use the distributive property to multiply each term of the polynomial by the monomial.

EXAMPLE 6-15: Find the product of each of the following:

(a) $2y(3y + 5)$ (b) $3w(2w^2 - 4w - 5)$ (c) $a^n(a^n + a^3 + 5)$

Solution: (a) $2y(3y + 5) = (2y)(3y) + (2y)(5)$

$\qquad = 6y^2 + 10y$

(b) $3w(2w^2 - 4w - 5) = (3w)(2w^2) + (3w)(-4w) + (3w)(-5)$

$\qquad = 6w^3 + (-12w^2) + (-15w)$

$\qquad = 6w^3 - 12w^2 - 15w$

(c) $a^n(a^n + a^3 + 5) = (a^n)(a^n) + (a^n)(a^3) + (a^n)5$

$\qquad = a^{n+n} + a^{n+3} + 5a^n$

$\qquad = a^{2n} + a^{n+3} + 5a^n$

C. Multiplying two polynomials

To multiply a polynomial by a polynomial, we use the distributive property to multiply each term of the first polynomial by each term of the second polynomial, then combine any resulting like terms.

EXAMPLE 6-16: Find $(3x + 5)(2x + 7)$.

Solution:

$$
\begin{aligned}
(3x + 5)(2x + 7) &= (3x + 5)2x + (3x + 5)7 && \text{Distribute } (3x + 5) \text{ over } 2x + 7. \\
&= (3x)(2x) + 5(2x) + (3x)7 + 5 \cdot 7 && \text{Distribute } 2x \text{ over } 3x + 5 \text{ and} \\
& && \text{also distribute } 7 \text{ over } 3x + 5. \\
&= 6x^2 + 10x + 21x + 35 \\
&= 6x^2 + 31x + 35 && \text{Combine like terms.}
\end{aligned}
$$

It is often convenient to multiply two polynomials using a vertical format.

EXAMPLE 6-17: Use a vertical format to compute the following products:

(a) $(y^2 - 4y + 3)(y + 5)$ **(b)** $(5a^3 + 4a^2 - 2)(3a - 5)$

Solution: **(a)**

$$
\begin{array}{r}
y^2 - 4y + 3 \\
y + 5 \\
\hline
5y^2 - 20y + 15 \quad \longleftarrow \; 5(y^2 - 4y + 3) \\
y^3 - 4y^2 + 3y \qquad\quad \longleftarrow \; y(y^2 - 4y + 3) \\
\hline
y^3 + y^2 - 17y + 15
\end{array}
$$

(b)

$$
\begin{array}{r}
5a^3 + 4a^2 - 2 \\
3a - 5 \\
\hline
-25a^3 - 20a^2 \qquad\quad + 10 \quad \longleftarrow \; (-5)(5a^3 + 4a^2 - 2) \\
15a^4 + 12a^3 \qquad\qquad - 6a \qquad\quad \longleftarrow \; (3a)(5a^3 + 4a^2 - 2) \\
\hline
15a^4 - 13a^3 - 20a^2 - 6a + 10
\end{array}
$$

Note: Only like terms should be placed in each column.

6-4. Computing Special Products

Although we can find the product of two binomials by the methods shown in the previous section, the following method will produce the product quickly without the need of writing intermediate steps.

A. Multiplying two binomials by the FOIL method

If the terms of two binomials are labeled as in Figure 6-1, then the product of the binomials can be formed by the **FOIL method.**

Figure 6-1. $(a + b)(c + d)$

To find the product of two binomials by the FOIL method:

1. Find the product of the **F**irst terms
                     Outside terms
                     Inside terms
                     Last terms
2. Form the sum of the above products. Any like terms should be combined.

EXAMPLE 6-18: Find $(3x + 5)(2x - 7)$.

Solution: Use the FOIL method.

$$(3x + 5)(2x - 7) = 6x^2 + (-21x) + 10x + (-35)$$
$$= 6x^2 - 11x - 35$$

First terms: $(3x)(2x) = 6x^2$
Outside terms: $(3x)(-7) = -21x$
Inside terms: $(5)(2x) = 10x$
Last terms: $(5)(-7) = -35$

In the previous example some intermediate steps have been shown. However, with practice you can compute and combine the F, O, I, L products mentally.

B. Computing the square of a binomial

The problem of computing the square of a binomial occurs so often in mathematics that you should memorize its method of calculation and the form of its product. The examples below show that the FOIL method may be used to compute the square of a binomial.

(a) $(r + s)^2 = ?$

$(r + s)^2 = (r + s)(r + s)$

$$= rr + rs + sr + ss$$

$$= r^2 + 2rs + s^2$$

(b) $(r - s)^2 = ?$

$(r - s)^2 = (r - s)(r - s)$

$$= rr + r(-s) + (-s)r + (-s)(-s)$$

$$= r^2 - 2rs + s^2$$

Both of these examples show that you may also find the square of a binomial by finding (1) the square of the first term, (2) twice the product of both terms, and (3) the square of the last term.

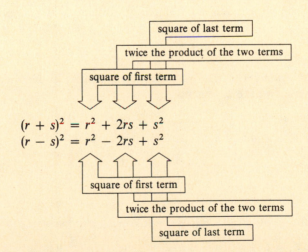

$$(r + s)^2 = r^2 + 2rs + s^2$$
$$(r - s)^2 = r^2 - 2rs + s^2$$

EXAMPLE 6-19: Compute **(a)** $(3x + 5)^2$ and **(b)** $(2y - 7)^2$.

Solution: Use the formulae $(r + s)^2 = r^2 + 2rs + s^2$ or $(r - s)^2 = r^2 - 2rs + s^2$.

(a) $(3x + 5)^2 = (3x)^2 + 2(3x)(5) + (5)^2$

$$= 9x^2 + 30x + 25$$

(b) $(2y - 7)^2 = (2y)^2 - 2(2y)(7) + 7^2$

$$= 4y^2 - 28y + 49$$

Caution: $(r + s)^2 \neq r^2 + s^2$. For example, $(2 + 5)^2 = 7^2$, or 49, while $2^2 + 5^2 = 4 + 25$, or 29.

C. Multiplying the sum and difference of the same two expressions

If r and s are expressions, then the product of the sum and the difference of those expressions is the difference of their squares.

$$\overset{\overset{\text{F}\qquad\text{O}\qquad\quad\text{I}\qquad\quad\text{L}}{}}{(r + s)(r - s)} = rr + (-rs) + rs + s(-s)$$

$$= r^2 + 0 + (-s^2)$$

$$= r^2 - s^2$$

EXAMPLE 6-20: Find the products of the following expressions:
(a) $(3x + 5)(3x - 5)$ **(b)** $(5y + 2z)(5y - 2z)$ **(c)** $(9ab + 1)(9ab - 1)$

Solution: Use the formula $(r + s)(r - s) = r^2 - s^2$.

(a) $(3x + 5)(3x - 5) = (3x)^2 - 5^2$ **(b)** $(5y + 2z)(5y - 2z) = (5y)^2 - (2z)^2$

$$= 9x^2 - 25 \qquad\qquad\qquad\qquad = 25y^2 - 4z^2$$

(c) $(9ab + 1)(9ab - 1) = (9ab)^2 - 1^2$

$$= 81a^2b^2 - 1$$

6-5. Dividing Polynomials

A. Dividing monomials

Consider the following division:

$$\frac{2^5}{2^2} = \frac{2 \cdot 2 \cdot 2 \cdot 2 \cdot 2}{2 \cdot 2} = \frac{\cancel{2} \cdot \cancel{2} \cdot 2 \cdot 2 \cdot 2}{\cancel{2} \cdot \cancel{2}} = 2^3$$

You should observe that the same result is obtained by subtracting the exponents:

$$\frac{2^5}{2^2} = 2^{5-2} = 2^3$$

The **quotient rule for exponents** states that you can divide factors in exponential notation with the same base by simply subtracting the exponents.

Quotient Rule for Exponents

If m and n are integers, and if a is any nonzero real number, then

$$\frac{a^m}{a^n} = a^{m-n}$$

EXAMPLE 6-21: Find the quotient of each of the following expressions:

(a) $\dfrac{x^8}{x^5}$ **(b)** $\dfrac{y^{11}}{y^7}$ **(c)** $\dfrac{z^4}{z^0}$ **(d)** $\dfrac{s^3}{s^4}$ **(e)** $\dfrac{t^5}{t^5}$ **(f)** $\dfrac{u^2}{u^5}$ **(g)** $\dfrac{v^4}{v^{-3}}$ **(h)** $\dfrac{r^{-2}}{r^4}$ **(i)** $\dfrac{a^{-3}}{a^{-5}}$

Solution: Use the quotient rule for exponents.

(a) $\dfrac{x^8}{x^5} = x^{8-5} = x^3$ **(b)** $\dfrac{y^{11}}{y^7} = y^{11-7} = y^4$ **(c)** $\dfrac{z^4}{z^0} = z^{4-0} = z^4$

(d) $\dfrac{s^3}{s^4} = s^{3-4} = s^{-1} = \dfrac{1}{s}$ **(e)** $\dfrac{t^5}{t^5} = t^{5-5} = t^0 = 1$ **(f)** $\dfrac{u^2}{u^5} = u^{2-5} = u^{-3} = \dfrac{1}{u^3}$

(g) $\dfrac{v^4}{v^{-3}} = v^{4-(-3)} = v^7$ **(h)** $\dfrac{r^{-2}}{r^4} = r^{-2-4} = r^{-6} = \dfrac{1}{r^6}$ **(i)** $\dfrac{a^{-3}}{a^{-5}} = a^{-3-(-5)} = a^2$

Caution: x^4/y^3 cannot be simplified by the quotient rule for exponents because the bases are not the same.

To divide monomials you compute the quotient of their numerical coefficients and the quotient of their literal parts.

EXAMPLE 6-22: Find the indicated quotients:

(a) $\dfrac{6x^5}{3x^2}$ (b) $\dfrac{-12a^9}{4a^5}$ (c) $\dfrac{20x^2y^6}{-2xy^8}$ (d) $\dfrac{-48a^3b^7c^2}{-6ab^4c^5}$

Solution: (a) $\dfrac{6x^5}{3x^2} = \dfrac{6}{3}\cdot\dfrac{x^5}{x^2} = 2x^3$ (b) $\dfrac{-12a^9}{4a^5} = \dfrac{-12}{4}\cdot\dfrac{a^9}{a^5} = -3a^4$

(c) $\dfrac{20x^2y^6}{-2xy^8} = \dfrac{20}{-2}\cdot\dfrac{x^2}{x}\cdot\dfrac{y^6}{y^8} = -10xy^{-2} = \dfrac{-10x}{y^2}$ or $-\dfrac{10x}{y^2}$

(d) $\dfrac{-48a^3b^7c^2}{-6ab^4c^5} = \dfrac{-48}{-6}\cdot\dfrac{a^3}{a}\cdot\dfrac{b^7}{b^4}\cdot\dfrac{c^2}{c^5} = 8a^2b^3c^{-3} = \dfrac{8a^2b^3}{c^3}$

B. Dividing a polynomial by a monomial

To divide a polynomial by a monomial, we divide each term of the polynomial by the monomial, then form the sum of the resulting quotients.

EXAMPLE 6-23: Find the indicated quotients:

(a) $\dfrac{4y^2 + 12y}{4y}$ (b) $\dfrac{6w^3 + 12w^2 - 18w + 24}{6}$ (c) $\dfrac{14x^3y^5 - 35x^6y^4}{7x^2y^2}$

Solution: Divide each term of the following polynomials by the monomial, then form the sum of the resulting quotients:

(a) $\dfrac{4y^2 + 12y}{4y} = \dfrac{4y^2}{4y} + \dfrac{12y}{4y} = y + 3$

(b) $\dfrac{6w^4 + 12w^3 - 18w^2 + 24w}{6w} = \dfrac{6w^4}{6w} + \dfrac{12w^3}{6w} - \dfrac{18w^2}{6w} + \dfrac{24w}{6w} = w^3 + 2w^2 - 3w + 4$

(c) $\dfrac{14x^3y^5 - 35x^6y^4}{7x^2y^2} = \dfrac{14x^3y^5}{7x^2y^2} + \dfrac{-35x^6y^4}{7x^2y^2} = 2xy^3 - 5x^4y^2$

C. Dividing a polynomial by a polynomial

The method used to divide a polynomial by a polynomial is similar to the long division algorithm (method) used for natural numbers.

To divide a polynomial by a polynomial:

1. Write in division-box form. Arrange both the divisor and the dividend in standard form. Insert 0 for any missing terms.
2. Divide the first term of the divisor into the first term of the dividend to obtain the first term of the quotient.
3. Multiply the first term of the quotient by each term in the divisor. Write this product under the dividend, aligning like terms.
4. Subtract like terms and bring down one or more terms as needed.
5. If the remainder in step 4 is 0 or of less degree than the divisor, then the division is complete. Otherwise, use the result in step 4 as a new dividend and repeat steps 2–4, until the remainder is 0 or of lower degree than the divisor.

EXAMPLE 6-24: Divide $x^2 - 7 + 5x$ by $x - 3$.

Solution: Use the steps outlined in the previous procedure.

1. $x - 3 \overline{\smash{\big)}\, x^2 + 5x - 7}$

   Write in division-box form. Arrange both the divisor and the dividend in standard form.

2. $\dfrac{x}{x - 3 \overline{\smash{\big)}\, x^2 + 5x - 7}}$

   Divide the first term of the divisor (x) into the first term of the dividend (x^2) to obtain the first term of the quotient (x).

3.
$$
\begin{array}{r}
x \\
x - 3 \overline{\smash{\big)}\, x^2 + 5x - 7} \\
x^2 - 3x
\end{array}
$$

   Multiply the first term of the quotient (x) by each term in the divisor $(x - 3)$. Write this product $(x^2 - 3x)$ under the dividend, aligning like terms.

4.
$$
\begin{array}{r}
x \\
x - 3 \overline{\smash{\big)}\, x^2 + 5x - 7} \\
\underline{x^2 - 3x} \\
8x - 7
\end{array}
$$

   Subtract like terms and bring down one or more terms as needed. Since the remainder $(8x - 7)$ is the same degree as the divisor, repeat steps 2–4.

5.
$$
\begin{array}{r}
x + 8 \\
x - 3 \overline{\smash{\big)}\, x^2 + 5x - 7} \\
\underline{x^2 - 3x} \\
8x - 7 \\
\underline{8x - 24} \\
17
\end{array}
$$

   Repeat steps 2–4, using $(8x - 7)$ as the new dividend. Since the remainder (17) is of lower degree than the divisor, the division is complete.

$$(x^2 - 7 + 5x) \div (x - 3) = \underbrace{x + 8}_{\text{quotient}} + \dfrac{17 \leftarrow \text{remainder}}{x - 3 \leftarrow \text{divisor}}$$

To check the division we multiply the quotient by the divisor and to this product add the remainder. The result should equal the dividend.

Check:
$$
\begin{array}{r}
x + 8 \leftarrow \text{quotient} \\
\text{Multiply:} \quad x - 3 \leftarrow \text{divisor} \\
\hline
-3x - 24 \\
x^2 + 8x \\
\hline
x^2 + 5x - 24 \leftarrow \text{product of the quotient and the divisor} \\
\text{Add:} \qquad\qquad + 17 \leftarrow \text{remainder} \\
\hline
x^2 + 5x - 7
\end{array}
$$

This result equals the divided, so the division is correct.

D. Dividing polynomials using synthetic division

There is a special method for dividing a polynomial by a binomial of the form $x - c$. This special method, which only requires three rows, is called **synthetic division**.

To perform synthetic division, we follow these steps:

1. Write the constant c, followed by the coefficients of the dividend polynomial, which should be arranged in standard form with 0 inserted for any missing terms.
2. Bring down the leading coefficient of the dividend as the first number in the third (bottom) row.
3. Compute the product of c and the number just written in the bottom row. Write this product in the next column to the right in the middle row.
4. Add the numbers in the column from step 3, and write the sum in the bottom row.
5. Alternate between steps 3 and 4 until all columns are filled.
6. Identify the quotient and remainder from the bottom row.

Caution: Synthetic division can only be used if the divisor is a binomial of the form $x - c$.

EXAMPLE 6-25: Divide $3x^3 - x^2 - 11$ by $x - 2$.

Solution: Use the synthetic division method.

Step 1 $\underline{2}\,|\,3 \quad -1 \quad 0 \quad -11$ Write c followed by the coefficients of the dividend. Zero has been inserted for the missing x term.

Step 2 $\underline{2}\,|\,3 \quad -1 \quad 0 \quad -11$ Bring down the leading coefficient of the dividend.

$$3$$

Step 3 $\underline{2}\,|\,3 \quad -1 \quad 0 \quad -11$ Compute the product of c and the number in the bottom row.

$$6$$
$$3$$

Step 4 $\underline{2}\,|\,3 \quad -1 \quad 0 \quad -11$ Add.

$$6$$
$$3 \quad 5$$

Step 5 $\underline{2}\,|\,3 \quad -1 \quad 0 \quad -11$ Alternate between Steps 3 and 4.

$$6 \quad 10 \quad 20$$
$$3 \quad 5 \quad 10 \quad 9$$

Step 6 $\underline{2}\,|\,3 \quad -1 \quad 0 \quad -11$ Identify the quotient and remainder.

$$6 \quad 10 \quad 20$$
$$3 \quad 5 \quad 10 \qquad 9$$

the coefficients of the quotient   the remainder

$$\frac{3x^3 - x^2 - 11}{x - 2} = 3x^2 + 5x + 10 + \frac{9}{x - 2}$$

6-6. Solving Problems by Evaluating Polynomials

To **evaluate a polynomial** in one variable for a given value of that variable, you first substitute the given value of the variable in the polynomial and then compute.

EXAMPLE 6-26: If the sum of all the counting numbers from 1 to n $[1 + 2 + 3 + \cdots + n]$ is represented by the polynomial $\frac{1}{2}n^2 + \frac{1}{2}n$, then evaluate $\frac{1}{2}n^2 + \frac{1}{2}n$ to find the sum of the first ten counting numbers.

Solution: $\dfrac{1}{2}n^2 + \dfrac{1}{2}n = \dfrac{1}{2}(10)^2 + \dfrac{1}{2}(10)$ Substitute 10 for n.

$$= \frac{1}{2}(100) + \frac{1}{2}(10) \qquad \text{Compute.}$$

$$= 50 + 5$$

$$= 55 \qquad \text{sum of the first ten counting numbers}$$

Check: Is the sum $1 + 2 + 3 + \cdots + 10$ equal to 55? Yes:

$$1 + 2 + 3 + 4 + 5 + 6 + 7 + 8 + 9 + 10 = (1 + 9) + (2 + 8) + (3 + 7) + (4 + 6) + 10 + 5$$

$$= 10 + 10 + 10 + 10 + 10 + 5$$

$$= 55$$

RAISE YOUR GRADES

Can you . . . ?

☑ identify polynomials
☑ classify a polynomial as either a monomial, a binomial, or a trinomial
☑ determine the degree of a polynomial
☑ write a polynomial in standard form
☑ find the sum, difference, product, or quotient of two polynomials
☑ use the FOIL method to find the product of two binomials
☑ square a binomial using the formula $(r + s)^2 = r^2 + 2rs + s^2$ or $(r - s)^2 = r^2 - 2rs + s^2$
☑ find the product of the sum and difference of the same two expressions using the formula $(r + s)(r - s) = r^2 - s^2$
☑ use synthetic division to divide a polynomial by a binomial
☑ solve problems by evaluating polynomials

SOLVED PROBLEMS

PROBLEM 6-1 Classify each of the following polynomials as either a monomial, a binomial, or a trinomial: **(a)** $4x + 7$ **(b)** $6x^2y$ **(c)** $5x^2 + 4x - 9$ **(d)** $6x^2 - 8xy + 12y^2$ **(e)** 8 **(f)** $4xy + y^2 + xy$ **(g)** $11x^2 - 2x + 7 - 11x^2$

Solution: Recall that a simplified polynomial consisting of exactly one term is called a monomial, two terms is called a binomial, three terms is called a trinomial [see Example 6-1].

(a) $4x + 7$ is a binomial because it has exactly two terms.
(b) $6x^2y$ is a monomial because it has exactly one term.
(c) $5x^2 + 4x - 9$ is a trinomial because it has exactly three terms.
(d) $6x^2 - 8xy + 12y^2$ is a trinomial because it has exactly three terms.
(e) 8 is a monomial because it has exactly one term.
(f) $4xy + y^2 + xy$ is a binomial because it simplifies to $5xy + y^2$, which has exactly two terms.
(g) $11x^2 - 2x + 7 - 11x^2$ is a binomial because it simplifies to $-2x + 7$, which has exactly two terms.

PROBLEM 6-2 Find the degree of each of the following monomials:

(a) $5x$ **(b)** $-5uv^3$ **(c)** $7xy$ **(d)** $-18mnp^3$ **(e)** $\frac{3}{4}r^2s^5t$

Solution: Recall that in a monomial of the form ax^m, where $a \neq 0$, the whole number exponent m is called the degree of the monomial. The degree of a monomial in more than one variable is the sum of the exponents of the variables [see Examples 6-2 and 6-3].

(a) $5x$ is a monomial of degree 1. $(5x = 5x^1)$
(b) $-5uv^3$ is a monomial of degree 4. The sum of the exponents is: $1 + 3 = 4$.
(c) $7xy$ is a monomial of degree 2. $(1 + 1 = 2)$
(d) $-18mnp^3$ is a monomial of degree 5. $(1 + 1 + 3 = 5)$
(e) $\frac{3}{4}r^2s^5$ is a monomial of degree 8. $(2 + 5 + 1 = 8)$

PROBLEM 6-3 Find the degree of each of the following polynomials:

(a) $3x^4 + 5$ **(b)** $3x^2 + 5x - 4$ **(c)** $2xy^3 + xy^2$ **(d)** $5x^4 - 2x^2 + 7 - x^5$
(e) $6x^2 + 2x - 5x^2 - x^2$

Solution: Recall that the degree of a polynomial is the largest degree of any of its monomial terms (provided like terms have been combined) [see Example 6-4].

(a) $3x^4 + 5$ is a polynomial of degree 4 because the largest degree term $3x^4$ has degree 4.
(b) $3x^2 + 5x - 4$ is a polynomial of degree 2 because the largest degree term $3x^2$ has degree 2.
(c) $2xy^3 + xy^2$ is a polynomial of degree 4 because the largest degree term $2xy^3$ has degree 4.
(d) $5x^4 - 2x^2 + 7 - x^5$ is a polynomial of degree 5 because the largest degree term $-x^5$ has degree 5.
(e) $6x^2 + 2x - 5x^2 - x^2$ is a polynomial of degree one because it simplifies to $2x$, which is a monomial of degree one.

PROBLEM 6-4 Write each of the following polynomials in standard form:

(a) $2x + 5x^2 - 17$ (b) $-2 + x^3$ (c) $4xy + 5x^2 - 7y^2$ (d) $5p^2q - 19q^3 + 8q^2p + 7p^3$
(e) $5x + 7 - x$

Solution: Recall that a polynomial is in standard form if (1) the variables are written in alphabetical order in each term, (2) the polynomial is simplified, and (3) the terms are arranged in descending powers of the first variable [see Example 6-5].

(a) $2x + 5x^2 - 17$ in standard form is written as $5x^2 + 2x - 17$.
(b) $-2 + x^3$ in standard form is written as $x^3 - 2$.
(c) $4xy + 5x^2 - 7y^2$ in standard form is written as $5x^2 + 4xy - 7y^2$.
(d) $5p^2q - 19q^3 + 8q^2p + 7p^3$ in standard form is written as $7p^3 + 5p^2q + 8pq^2 - 19q^3$.
(e) $5x + 7 - x$ in standard form is written as $4x + 7$.

PROBLEM 6-5 Find the indicated sum of each of the following polynomials:

(a) $(4x^2 + 5x + 7) + (3x^2 - 9x + 12)$ (b) $(17x^2y - 9xy^2) + (2x^2y + 10xy^2)$
(c) $(w^3 + 2w - 8) + (-2w - 9 + 4w^3)$

Solution: Recall that to add polynomials, you combine like terms [see Examples 6-6 and 6-7].

(a) $(4x^2 + 5x + 7) + (3x^2 - 9x + 12) = (4x^2 + 3x^2) + (5x - 9x) + (7 + 12)$
$$= (4 + 3)x^2 + (5 - 9)x + (7 + 12)$$
$$= 7x^2 + (-4)x + 19$$
$$= 7x^2 - 4x + 19$$

(b) $(17x^2y - 9xy^2) + (2x^2y + 10xy^2) = (17x^2y + 2x^2y) + (-9xy^2 + 10xy^2)$
$$= (17 + 2)x^2y + (-9 + 10)xy^2$$
$$= 19x^2y + xy^2$$

(c) $(w^3 + 2w - 8) + (-2w - 9 + 4w^3) = (1w^3 + 4w^3) + (2w + (-2w)) + (-8 + (-9))$
$$= (1 + 4)w^3 + (2 + (-2))w + (-8 + (-9))$$
$$= 5w^3 + 0w + (-17)$$
$$= 5w^3 - 17$$

PROBLEM 6-6 Find the additive inverse of the following polynomials:

(a) $2x - 3$ (b) $5y^2 - 2y + 8$ (c) $-a^3 + 2a^2 - 5a + 8$

Solution: Recall that to form the additive inverse of a polynomial, you write the additive inverse of each term of the polynomial [see Example 6-8].

(a) The additive inverse of $2x - 3$ is $-2x + 3$.
(b) The additive inverse of $5y^2 - 2y + 8$ is $-5y^2 + 2y - 8$.
(c) The additive inverse of $-a^3 + 2a^2 - 5a + 8$ is $a^3 - 2a^2 + 5a - 8$.

PROBLEM 6-7 Find the indicated difference of each of the following polynomials:

(a) $(3x - 2) - (2x + 5)$ (b) $(3x^2 - 7x + 2) - (5x^2 - 8x - 11)$
(c) $(w^4 - 3w^2 + 5w - 1) - (2w^3 - 8w + 7)$

Solution: Recall that to subtract a polynomial, you add its additive inverse [see Examples 6-9 and 6-10].

$$\text{(a) } (3x - 2) - (2x + 5) = (3x - 2) + (-2x - 5)$$
$$= x - 7$$

$$\text{(b) } (3x^2 - 7x + 2) - (5x^2 - 8x - 11) = (3x^2 - 7x + 2) + (-5x^2 + 8x + 11)$$
$$= -2x^2 + x + 13$$

$$\text{(c) } (w^4 - 3w^2 + 5w - 1) - (2w^3 - 8w + 7) = (w^4 - 3w^2 + 5w - 1) + (-2w^3 + 8w - 7)$$
$$= w^4 - 2w^3 - 3w^2 + 13w - 8$$

PROBLEM 6-8 Find the indicated product of the following monomials:

(a) $x^4 \cdot x^2$ (b) $y^3 \cdot y^4 \cdot y$ (c) $(3a^2bc^3)(-4ab^3c^5)$

Solution: Recall that to multiply monomials, you multiply their numerical coefficients and attach the product of their literal parts [see Examples 6-11 and 6-12].

$$\text{(a) } x^4 \cdot x^2 = x^{4+2} \qquad \text{(b) } y^3 \cdot y^4 \cdot y = y^{3+4+1}$$
$$= x^6 \qquad\qquad\qquad\qquad = y^8$$

$$\text{(c) } (3a^2bc^3)(-4ab^3c^5) = [3(-4)](a^2a)(bb^3)(c^3c^5)$$
$$= -12a^{2+1}b^{1+3}c^{3+5}$$
$$= -12a^3b^4c^8$$

PROBLEM 6-9 Find the indicated power of each of the following monomials:

(a) $(a^3b)^2$ (b) $(-3xy^4)^3$ (c) $(-mn^2)^4$

Solution: Recall the power rule $(a^m)^n = a^{mn}$, and the extended power rule $(a^mb^n)^p = a^{mp}b^{np}$ [see Examples 6-13 and 6-14].

$$\text{(a) } (a^3b)^2 = a^{3 \cdot 2}b^{1 \cdot 2} \qquad \text{(b) } (-3xy^4)^3 = (-3)^{1 \cdot 3}x^{1 \cdot 3}y^{4 \cdot 3} \qquad \text{(c) } (-mn^2)^4 = (-m)^{1 \cdot 4}n^{2 \cdot 4}$$
$$= a^6b^2 \qquad\qquad\qquad = (-3)^3x^3y^{12} \qquad\qquad\qquad = (-m)^4n^8$$
$$= -27x^3y^{12} \qquad\qquad\qquad = m^4n^8$$

PROBLEM 6-10 Find the indicated product of each of the following polynomials:

(a) $2x(3x + 1)$ (b) $3m^2(2m^2 + 5m - 1)$ (c) $(b^3 - 4b^2 - 3b + 2)b$

Solution: Recall that to multiply a polynomial by a monomial, you use the distributive property to multiply each term of the polynomial by the monomial [see Example 6-15].

$$\text{(a) } 2x(3x + 1) = (2x)(3x) + (2x)1$$
$$= 6x^2 + 2x$$

$$\text{(b) } 3m^2(2m^2 + 5m - 1) = (3m^2)(2m^2) + (3m^2)(5m) - (3m^2)1$$
$$= 6m^4 + 15m^3 - 3m^2$$

$$\text{(c) } (b^3 - 4b^2 - 3b + 2)b = (b^3)b - (4b^2)b - (3b)b + 2b$$
$$= b^4 - 4b^3 - 3b^2 + 2b$$

PROBLEM 6-11 Find $(3x^2 + 2x - 5)(2x + 3)$.

Solution: Use the distributive property to multiply each term of the second polynomial by each term of the first polynomial, then combine any like terms [see Examples 6-16 and 6-17].

Horizontal format:

$(3x^2 + 2x - 5)(2x + 3) = (3x^2 + 2x - 5)2x + (3x^2 + 2x - 5)3$ Distribute $(3x^2 + 2x - 5)$ over $2x + 3$.

$$= (3x^2)(2x) + (2x)(2x) - 5(2x)$$
$$+ (3x^2)3 + (2x)3 - 5 \cdot 3$$

Distribute $2x$ over $(3x^2 + 2x - 5)$ and 3 over $(3x^2 + 2x - 5)$.

$$= 6x^3 + 4x^2 - 10x + 9x^2 + 6x - 15$$

$$= 6x^3 + 13x^2 - 4x - 15$$ Combine like terms.

Vertical format:

$$
\begin{array}{r}
3x^2 + 2x - 5 \\
2x + 3 \\
\hline
9x^2 + 6x - 15 \\
6x^3 + 4x^2 - 10x \\
\hline
6x^3 + 13x^2 - 4x - 15
\end{array}
$$

PROBLEM 6-12 Find each of the following indicated products of two binomials:

(a) $(2x + 7)(3x + 11)$ **(b)** $(4x - 1)(3x + 7)$ **(c)** $(2m - 5n)(4m - 3n)$

Solution: Recall that to find the product of two binomials you may use the FOIL method [see Example 6-18].

(a) $(2x + 7)(3x + 11) = 6x^2 + 22x + 21x + 77$
$$= 6x^2 + 43x + 77$$

(b) $(4x - 1)(3x + 7) = 12x^2 + 28x - 3x - 7$
$$= 12x^2 + 25x - 7$$

(c) $(2m - 5n)(4m - 3n) = 8m^2 - 6mn - 20mn + 15n^2$
$$= 8m^2 - 26mn + 15n^2$$

PROBLEM 6-13 Find the indicated squares of each of the following binomials:

(a) $(3x + 5)^2$ **(b)** $(4y - 3)^2$

Solution: Recall the formulae $(r + s)^2 = r^2 + 2rs + s^2$ and $(r - s)^2 = r^2 - 2rs + s^2$ [see Example 6-19].

(a) $(3x + 5)^2 = (3x)^2 + 2(3x)5 + 5^2$ **(b)** $(4y - 3)^2 = (4y)^2 - 2(4y)3 + 3^2$
$$= 9x^2 + 30x + 25 \qquad\qquad = 16y^2 - 24y + 9$$

PROBLEM 6-14 Find the indicated products of the sum and the difference of the same two expressions:

(a) $(5x + 7)(5x - 7)$ **(b)** $(11a + 2b)(11a - 2b)$

Solution: Recall that $(r + s)(r - s) = r^2 - s^2$ [see Example 6-20].

(a) $(5x + 7)(5x - 7) = (5x)^2 - 7^2$ **(b)** $(11a + 2b)(11a - 2b) = (11a)^2 - (2b)^2$
$$= 25x^2 - 49 \qquad\qquad = 121a^2 - 4b^2$$

PROBLEM 6-15 Find the indicated quotients of each of the following monomials:

(a) $\dfrac{4a^5}{2a^2}$ (b) $\dfrac{-21x^2y^3z}{7xy^3}$ (c) $\dfrac{45m^2n^3}{-9mn}$

Solution: Recall that to divide monomials, you compute the quotient of their numerical coefficients and attach the quotient of their literal parts [see Example 6-22].

(a) $\dfrac{4a^5}{2a^2} = \dfrac{4}{2} \cdot \dfrac{a^5}{a^2}$ (b) $\dfrac{-21x^2y^3z}{7xy^3} = \dfrac{-21}{7} \cdot \dfrac{x^2}{x} \cdot \dfrac{y^3}{y^3} \cdot z$ (c) $\dfrac{45m^2n^3}{-9mn} = \dfrac{45}{-9} \cdot \dfrac{m^2}{m} \cdot \dfrac{n^3}{n}$

$\quad = 2a^3$ $\quad = -3xy^0z$ $\quad = -5mn^2$

$\qquad\qquad\qquad\qquad = -3xz$

PROBLEM 6-16 Find the indicated quotients of each of the following polynomials:

(a) $\dfrac{4x^3 - 8x^4y^3}{2x}$ (b) $\dfrac{16a^3b^4 + 32a^2b^5 - 48ab^6}{16ab^4}$

Solution: Recall that to divide a polynomial by a monomial, you divide each term of the polynomial by the monomial, then form the sum of the resulting quotients [see Example 6-23].

(a) $\dfrac{4x^3 - 8x^4y^3}{2x} = \dfrac{4x^3}{2x} - \dfrac{8x^4y^3}{2x}$ (b) $\dfrac{16a^3b^4 + 32a^2b^5 - 48ab^6}{16ab^4} = \dfrac{16a^3b^4}{16ab^4} + \dfrac{32a^2b^5}{16ab^4} - \dfrac{48ab^6}{16ab^4}$

$\quad = 2x^2 - 4x^3y^3$ $\quad = a^2 + 2ab - 3b^2$

PROBLEM 6-17 Find (a) $(2x^2 + 4x^3 - 7) \div (2x - 3)$ and (b) $(a^3 + 8) \div (a + 2)$.

Solution: Follow the method for division of polynomials [see Example 6-24].

(a)
$$\begin{array}{r}
2x^2 + 4x + 6 \phantom{{}+0} \\
2x - 3 \overline{\smash{\big)}\ 4x^3 + 2x^2 + 0x - 7} \\
\underline{4x^3 - 6x^2 \phantom{{}+0x-7}} \\
8x^2 + 0x \phantom{{}-7} \\
\underline{8x^2 - 12x \phantom{{}-7}} \\
12x - 7 \\
\underline{12x - 18} \\
11
\end{array}$$

(b)
$$\begin{array}{r}
a^2 - 2a + 4 \phantom{{}+0} \\
a + 2 \overline{\smash{\big)}\ a^3 + 0a^2 + 0a + 8} \\
\underline{a^3 + 2a^2 \phantom{{}+0a+8}} \\
-2a^2 + 0a \phantom{{}+8} \\
\underline{-2a^2 - 4a \phantom{{}+8}} \\
4a + 8 \\
\underline{4a + 8} \\
0
\end{array}$$

$(2x^2 + 4x^3 - 7) \div (2x - 3) = 2x^2 + 4x + 6 + \dfrac{11}{2x - 3}$ $(a^3 + 8) \div (a + 2) = a^2 - 2a + 4$

PROBLEM 6-18 Perform each of the following divisions using the synthetic division method:

(a) $(x^2 + 5x - 14) \div (x - 2)$ (b) $(5x^3 - 4x^2 + 7x + 5) \div (x + 1)$
(c) $(x^3 - 8x + 2) \div (x - 3)$ (d) $(2x^4 + 20x^2 + 13x^3 - 5) \div (x + 4)$

Solution: Follow the method for synthetic division [see Example 6-25].

(a) $\begin{array}{r|rrr}
2 & 1 & 5 & -14 \\
 & & 2 & 14 \\
\hline
 & 1 & 7 & 0
\end{array}$

$(x^2 + 5x - 14) \div (x - 2) = x + 7$

(b) $\begin{array}{r|rrrr}
-1 & 5 & -4 & 7 & +5 \\
 & & -5 & 9 & -16 \\
\hline
 & 5 & -9 & 16 & -11
\end{array}$ $\begin{array}{l} x + 1 = x - (-1), \text{ which is in} \\ x - c \text{ form with } c = -1 \end{array}$

$(5x^3 - 4x^2 + 7x + 5) \div (x + 1) = 5x^2 - 9x + 16 + \dfrac{-11}{x + 1}$

(c) $3 \underline{| 1 \quad 0 \quad -8 \quad 2}$
$ \quad 3 \quad 9 \quad 3$
$\overline{1 \quad 3 \quad 1 \quad 5}$

$$(x^3 - 8x + 2) \div (x - 3) = x^2 + 3x + 1 + \frac{5}{x - 3}$$

(d) $-4 \underline{| 2 \quad 13 \quad +20 \quad 0 \quad -5}$
$ -8 \quad -20 \quad 0 \quad 0$
$\overline{2 \quad 5 \quad \quad 0 \quad 0 \quad -5}$

$$\frac{2x^4 + 20x^2 + 13x^3 - 5}{x + 4} = 2x^3 + 5x^2 + \frac{-5}{x + 4}$$

PROBLEM 6-19 Given that the number of single connections possible between n different things is represented by $\frac{1}{2}n^2 - \frac{1}{2}n$, find the number of different handshakes that are possible between four people.

Solution: Recall that to evaluate a polynomial in one variable for a given value of that variable, you first substitute the given value of the variable in the polynomial, then compute its value [see Example 6-26].

$$\frac{1}{2}n^2 - \frac{1}{2}n = \frac{1}{2}(4)^2 - \frac{1}{2}(4) \qquad \text{Substitute 4 for } n.$$

$$= \frac{1}{2}(16) - \frac{1}{2}(4) \qquad \text{Compute.}$$

$$= 8 - 2$$

$$= 6 \longleftarrow \text{number of different handshakes possible between four people}$$

Check: Draw a picture:

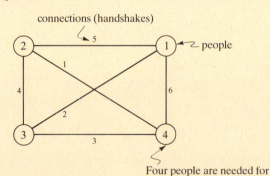

Four people are needed for six different handshakes.

Supplementary Exercises

PROBLEM 6-20 Classify each of the following polynomials as either a monomial, a binomial, or a trinomial: (a) $8x - 5$ (b) $19x^3y$ (c) $16y^2 - 54y + 2$ (d) $8a - 5b$
(e) $15x^2 - 19xy + 37y^2$ (f) $x^3 - 1$ (g) $4x - 5 + 7x$ (h) $3y^2 + 2y - 3y^2$

PROBLEM 6-21 Find the degree of each of the following monomials: (a) $7y$ (b) $-8yw^2$
(c) $32x^3y^7$ (d) $-\frac{1}{2}m^2np^3$ (e) $\frac{2}{5}rst$ (f) $82x^0y^0$ (g) $3x^{11}y$ (h) $4xy$

PROBLEM 6-22 Find the degree of each of the following polynomials: (a) $5x^3 - 7$
(b) $7x^3 - 4x + 8$ (c) $-2xy^5 + 11xy^3$ (d) $14x^4 - 3x^3 + 9x^6$
(e) $-4p^2q - 3q^3 + 2q^2p + 5p^3$ (f) $7x + 8$

PROBLEM 6-23 Write each of the following polynomials in standard form: (a) $6y + 8 - 17y^2$
(b) $-5 + x^5 - 6x$ (c) $3xy + 6y^2 - 2x^2$ (d) $6ab^2 + 5a^2b - 3b^3 + 4a^3$ (e) $16 + 5y$
(f) $32y^2x - 5x^2y$

PROBLEM 6-24 Find the indicated sum of the following polynomials:

(a) $(3x^2 + 4x - 5) + (2x^2 - 5x - 8)$ (b) $(6y - 11) + (2y + 5)$
(c) $(4x^2 + 7 - 8x) + (9x - 5x^2 - 1)$ (d) $(2w^4 - 3w^3 + 2w^2 + 9) + (4w^3 - 7w^2 + w^4 - 11)$
(e) $(2x^2 + 3xy + 5y^2) + (3x^2 - 7xy + 11y^2)$

PROBLEM 6-25 Find the indicated difference of the following polynomials:

(a) $(2x - 7) - (3x + 4)$ (b) $(6x^2 + 7x - 8) - (2x^2 - 8x + 5)$
(c) $(2y^2 + 3y - 5) - (8y + 7 - y^2)$ (d) $(3w^2 - 5w^3 + 6w^4 - 8) - (w^4 + 5w^3 - 3)$
(e) $(5a^2 + 17ab - 8b^2) - (6a^2 - 2ab + b^2)$

PROBLEM 6-26 Find the indicated product of each of the following monomials: (a) $x^5 \cdot x^3$
(b) $y^8 \cdot y^2$ (c) $w^4 \cdot w^3 \cdot w^8$ (d) $z^2 \cdot z$ (e) $x^0 \cdot x^3$ (f) $(3a^2)(4a^5)$ (g) $(-2b^3)(7b^8)$
(h) $(4m^3n^2)(-3mn^4)(5mn)$ (i) $(3x^2y)(-\frac{1}{3}x^2y)$ (j) $(2.5r^2s)(-3rs^3)$

PROBLEM 6-27 Find the indicated power of each of the following monomials: (a) $(2b)^3$
(b) $(3a^2)^4$ (c) $(-5xy^2)^2$ (d) $(-7mn^3)^3$

PROBLEM 6-28 Find the indicated product of each of the following polynomials: (a) $3x(5x + 2)$
(b) $4w^2(w^2 - 3w + 5)$ (c) $(2a^3 + 5a^2 - 4)(2a)$ (d) $(4x + 3)(2x + 5)$
(e) $(2m + 3n)(5m + n)$ (f) $(4k + 5)(3k - 2)$ (g) $(8b - 3)(5b - 7)$ (h) $(9z + 1)(9z - 1)$
(i) $(4q - 3r)(q + 5r)$ (j) $(3a + 2b)(3a + 2b)$ (k) $(15m - 8)(15m + 3)$
(l) $(16p + 13)(12p - 11)$ (m) $(q + 1)(q - 1)$ (n) $(r - 3)(r + 3)$
(o) $(x + 3)(2x^2 - 5x + 7)$ (p) $(3y - 5)(2y^3 + y^2 - 11)$ (q) $(a^2 + 2a - 5)(a^2 - 3a + 7)$
(r) $(2w + 3)(4w^3 + 5w^2 - 4w - 1)$

PROBLEM 6-29 Find the indicated squares of each of the following binomials: (a) $(x + 2)^2$
(b) $(2y + 3)^2$ (c) $(z - 1)^2$ (d) $(5k + 7)^2$ (e) $(4m - 5)^2$ (f) $(11j - 8)^2$
(g) $(x + 3y)^2$ (h) $(15s - 13t)^2$ (i) $(-3k + 5)^2$ (j) $(-3x + 2y)^2$ (k) $(-3t - 5v)^2$
(l) $(2ab^2 + 3c)^2$

PROBLEM 6-30 Find the indicated product of the sum and difference of the same two expressions:

(a) $(3a + 2)(3a - 2)$ (b) $(4x + 5)(4x - 5)$ (c) $(2m + 11)(2m - 11)$ (d) $(5k - 3)(5k + 3)$
(e) $(7j + 1)(7j - 1)$ (f) $(15p + 2)(15p - 2)$ (g) $(8q - 7)(8q + 7)$ (h) $(w^2 + 5)(w^2 - 5)$
(i) $(3m + 2n)(3m - 2n)$ (j) $(17t + 3v)(17t - 3v)$ (k) $(15e - 11f)(15e + 11f)$
(l) $(2r^3 + 5)(2r^3 - 5)$ (m) $(3p^2 + 5q^3)(3p^2 - 5q^3)$ (n) $(8t - rs)(8t + rs)$

PROBLEM 6-31 Find the indicated quotients of each of the following monomials:

(a) $\dfrac{a^6}{a^2}$ (b) $\dfrac{x^{11}}{x^3}$ (c) $\dfrac{6m^8}{2m^7}$ (d) $\dfrac{u^5v^2}{uv^3}$ (e) $\dfrac{14xy^2z^3}{-7xy^3z}$ (f) $(15mp^2q^5) \div (-3m^2p^3q)$

(g) $\dfrac{-28m^3n^7}{-4mn^7}$ (h) $\dfrac{-50e^3f^3}{25e^3f^3}$ (i) $\dfrac{117v^3w^{10}}{-3w^8}$ (j) $\dfrac{-2.4pqr}{-0.3pr}$ (k) $-\dfrac{21f^{20}q^8}{3f^{11}q^{13}h}$ (l) $\dfrac{-51s^5t^4}{-3s^5t}$

PROBLEM 6-32 Find the indicated quotient of each of the following polynomials:

(a) $\dfrac{5a^2 - 7a}{a}$ (b) $\dfrac{6d^2 + 12d}{3d}$ (c) $\dfrac{15p^2q + 10pq^2}{5pq}$ (d) $\dfrac{8r^2s - 12rs^2}{-4rs}$

(e) $\dfrac{16a^4 - 32a^3 + 8a^2 + 64a}{4a}$ **(f)** $\dfrac{14x^4y^2 - 8x^3y^3 + 20x^2y^4}{2xy}$ **(g)** $\dfrac{12y^3 - 30y^2}{-2y^2}$

(h) $\dfrac{144c^5d^3 - 36c^4d^4 - 72c^3d^5}{9c^3d^3}$ **(i)** $\dfrac{1.5x^3y^2 - 2.1x^2y^3}{-0.3xy^2}$ **(j)** $(15p^3q^3r^3 + 24pq^4r^2) \div (3pqr)$

PROBLEM 6-33 Find the quotient of each of the following polynomials:

(a) $\dfrac{x^2 + 7x + 10}{x + 2}$ **(b)** $\dfrac{y^2 - 8y + 12}{y - 2}$ **(c)** $\dfrac{w^3 + 3w^2 - 4w + 9}{w + 1}$ **(d)** $\dfrac{m^2 + 7m + 13}{m - 3}$

(e) $\dfrac{2p^2 + 5p + 7}{2p + 1}$ **(f)** $(25r^2 - 5r + 8) \div (5r + 2)$ **(g)** $(9w^2 + 3w + 1) \div (3w + 2)$

(h) $\dfrac{t^3 - 5t^2 + t}{t - 1}$ **(i)** $\dfrac{3a + 5a^2 - 9}{a - 2}$ **(j)** $\dfrac{4x^3 + 5x^2 - 7 - x}{x^2 - 5}$ **(k)** $\dfrac{4m^3 - 5m^2 + 6m - 8}{m^2 + m + 1}$

(l) $\dfrac{6n^3 - n - 3n^2 + 7}{n^2 + 1 - 2n}$ **(m)** $4x + 1 \overline{\smash{\big)}\, 16x^2 + 8x + 1}$ **(n)** $p - 1 \overline{\smash{\big)}\, p^4 - 5p^3 + 8}$

PROBLEM 6-34 Use synthetic division to find the quotient of each of the following polynomials:

(a) $(x^2 + 11x + 24) \div (x + 3)$ **(b)** $(w^2 + 8w - 20) \div (w - 2)$

(c) $(t^3 + 3t^2 - 4t + 5) \div (t - 1)$ **(d)** $\dfrac{r^2 + 25r + 101}{r + 2}$ **(e)** $\dfrac{4a^2 - 9}{a + 3}$ **(f)** $\dfrac{p^3 - 64}{p - 4}$

(g) $\dfrac{x^2 - 7 - 2x}{x - 3}$ **(h)** $\dfrac{y^5 + 7y^4 + 8y^3 + 10y^2 - 2y + 7}{y - 1}$ **(i)** $\dfrac{n^3 + 125}{n + 5}$ **(j)** $\dfrac{m^3 + 125}{m - 5}$

(k) $\dfrac{2s^2 + 7s - 240}{s + 12}$ **(l)** $(u^4 - 6u^2 + 8) \div (u + 3)$ **(m)** $(b^3 - 8b^2 + 7) \div (b - 1)$

(n) $(t^4 - 1) \div (t - 1)$ **(o)** $(v^6 - 4v^3 + 10) \div (v - 3)$ **(p)** $(z^4 + 2z^2 + 6z^3 + 1) \div (z + 4)$

PROBLEM 6-35 Given that the number of diagonals possible for an n-sided figure is represented by $\frac{1}{2}n^2 - \frac{3}{2}n$, find the number of diagonals possible for a figure with **(a)** 3 sides [triangle]
(b) 4 sides [quadrilateral]   **(c)** 5 sides [pentagon]   **(d)** 6 sides [hexagon]
(e) 7 sides [septagon]   **(f)** 8 sides [octagon]   **(g)** 10 sides [decagon]
(h) 100 sides [centagon].

Evaluate $\frac{1}{2}n^2 + \frac{1}{2}n$ to find the sum of all the counting numbers from 1 to **(i)** 2   **(j)** 3
(k) 5   **(l)** 8   **(m)** 20   **(n)** 50   **(o)** 100   **(p)** 1000.

Evaluate $\frac{1}{2}n^2 - \frac{1}{2}n$ to find the number of different single connections possible through a switchboard connected to **(q)** 2 phones   **(r)** 3 phones   **(s)** 5 phones   **(t)** 6 phones   **(u)** 8 phones
(v) 10 phones   **(w)** 20 phones   **(x)** 50 phones   **(y)** 100 phones
(z) 1000 phones [see Problem 6-19].

Answers to Supplementary Exercises

(6-20) (a) binomial   **(b)** monomial   **(c)** trinomial   **(d)** binomial   **(e)** trinomial
(f) binomial   **(g)** binomial   **(h)** monomial

(6-21) (a) 1   **(b)** 3   **(c)** 10   **(d)** 6   **(e)** 3   **(f)** 0   **(g)** 12   **(h)** 2

(6-22) (a) 3   **(b)** 3   **(c)** 6   **(d)** 6   **(e)** 3   **(f)** 1

(6-23) (a) $-17y^2 + 6y + 8$ (b) $x^5 - 6x - 5$ (c) $-2x^2 + 3xy + 6y^2$
(d) $4a^3 + 5a^2b + 6ab^2 - 3b^3$ (e) $5y + 16$ (f) $-5x^2y + 32xy^2$

(6-24) (a) $5x^2 - x - 13$ (b) $8y - 6$ (c) $-x^2 + x + 6$ (d) $3w^4 + w^3 - 5w^2 - 2$
(e) $5x^2 - 4xy + 16y^2$

(6-25) (a) $-x - 11$ (b) $4x^2 + 15x - 13$ (c) $3y^2 - 5y - 12$ (d) $5w^4 - 10w^3 + 3w^2 - 5$
(e) $-a^2 + 19ab - 9b^2$

(6-26) (a) x^8 (b) y^{10} (c) w^{15} (d) z^3 (e) x^3 (f) $12a^7$ (g) $-14b^{11}$
(h) $-60m^5n^7$ (i) $-x^4y^2$ (j) $-7.5r^3s^4$

(6-27) (a) $8b^3$ (b) $81a^8$ (c) $25x^2y^4$ (d) $-343m^3n^9$

(6-28) (a) $15x^2 + 6x$ (b) $4w^4 - 12w^3 + 20w^2$ (c) $4a^4 + 10a^3 - 8a$ (d) $8x^2 + 26x + 15$
(e) $10m^2 + 17mn + 3n^2$ (f) $12k^2 + 7k - 10$ (g) $40b^2 - 71b + 21$ (h) $81z^2 - 1$
(i) $4q^2 + 17qr - 15r^2$ (j) $9a^2 + 12ab + 4b^2$ (k) $225m^2 - 75m - 24$
(l) $192p^2 - 20p - 143$ (m) $q^2 - 1$ (n) $r^2 - 9$ (o) $2x^3 + x^2 - 8x + 21$
(p) $6y^4 - 7y^3 - 5y^2 - 33y + 55$ (q) $a^4 - a^3 - 4a^2 + 29a - 35$
(r) $8w^4 + 22w^3 + 7w^2 - 14w - 3$

(6-29) (a) $x^2 + 4x + 4$ (b) $4y^2 + 12y + 9$ (c) $z^2 - 2z + 1$ (d) $25k^2 + 70k + 49$
(e) $16m^2 - 40m + 25$ (f) $121j^2 - 176j + 64$ (g) $x^2 + 6xy + 9y^2$
(h) $225s^2 - 390st + 169t^2$ (i) $9k^2 - 30k + 25$ (j) $9x^2 - 12xy + 4y^2$
(k) $9t^2 + 30tv + 25v^2$ (l) $4a^2b^4 + 12ab^2c + 9c^2$

(6-30) (a) $9a^2 - 4$ (b) $16x^2 - 25$ (c) $4m^2 - 121$ (d) $25k^2 - 9$ (e) $49j^2 - 1$
(f) $225p^2 - 4$ (g) $64q^2 - 49$ (h) $w^4 - 25$ (i) $9m^2 - 4n^2$ (j) $289t^2 - 9v^2$
(k) $225e^2 - 121f^2$ (l) $4r^6 - 25$ (m) $9p^4 - 25q^6$ (n) $64t^2 - r^2s^2$

(6-31) (a) a^4 (b) x^8 (c) $3m$ (d) $\dfrac{u^4}{v}$ (e) $-\dfrac{2z^2}{y}$ (f) $-\dfrac{5q^4}{mp}$ (g) $7m^2$ (h) -2

(i) $-39v^3w^2$ (j) $8q$ (k) $-\dfrac{7f^9}{q^5h}$ (l) $17t^3$

(6-32) (a) $5a - 7$ (b) $2d + 4$ (c) $3p + 2q$ (d) $-2r + 3s$ (e) $4a^3 - 8a^2 + 2a + 16$
(f) $7x^3y - 4x^2y^2 + 10xy^3$ (g) $-6y + 15$ (h) $16c^2 - 4cd - 8d^2$
(i) $-5x^2 + 7xy$ (j) $5p^2q^2r^2 + 8q^3r$

(6-33) (a) $x + 5$ (b) $y - 6$ (c) $w^2 + 2w - 6 + \dfrac{15}{w + 1}$ (d) $m + 10 + \dfrac{43}{m - 3}$

(e) $p + 2 + \dfrac{5}{2p + 1}$ (f) $5r - 3 + \dfrac{14}{5r + 2}$ (g) $3w - 1 + \dfrac{3}{3w + 2}$

(h) $t^2 - 4t - 3 + \dfrac{-3}{t - 1}$ (i) $5a + 13 + \dfrac{17}{a - 2}$ (j) $4x + 5 + \dfrac{19x + 18}{x^2 - 5}$

(k) $4m - 9 + \dfrac{11m + 1}{m^2 + m + 1}$ (l) $6n + 9 + \dfrac{11n - 2}{n^2 - 2n + 1}$ (m) $4x + 1$

(n) $p^3 - 4p^2 - 4p - 4 + \dfrac{4}{p - 1}$

(6-34) (a) $x + 8$ (b) $w + 10$ (c) $t^2 + 4t + \dfrac{5}{t - 1}$ (d) $r + 23 + \dfrac{55}{r + 2}$

(e) $4a - 12 + \dfrac{27}{a + 3}$ (f) $p^2 + 4p + 16$ (g) $x + 1 + \dfrac{-4}{x - 3}$

(h) $y^4 + 8y^3 + 16y^2 + 26y + 24 + \dfrac{31}{y - 1}$ (i) $n^2 - 5n + 25$

(j) $m^2 + 5m + 25 + \dfrac{250}{m - 5}$ (k) $2s - 17 + \dfrac{-36}{s + 12}$

(l) $u^2 - 3u^2 + 3u - 9 + \dfrac{35}{u + 3}$ (m) $b^2 - 7b - 7$ (n) $t^3 + t^2 + t + 1$

(o) $v^5 + 3v^4 + 9v^3 + 23v^2 + 69v + 207 + \dfrac{631}{v - 3}$ (p) $z^3 + 2z^2 - 6z + 24 + \dfrac{-95}{z + 4}$

(6-35) (a) 0   (b) 2   (c) 5   (d) 9   (e) 14   (f) 20   (g) 35   (h) 4850   (i) 3
(j) 6   (k) 15   (l) 36   (m) 210   (n) 1275   (o) 5050   (p) 500,500   (q) 1
(r) 3   (s) 10   (t) 15   (u) 28   (v) 45   (w) 190   (x) 1225   (y) 4950
(z) 499,500

7 FACTORING POLYNOMIALS

THIS CHAPTER IS ABOUT

- ☑ Factoring Integers
- ☑ Factoring the Greatest Common Factor from a Polynomial
- ☑ Factoring Trinomials
- ☑ Factoring the Difference of Two Squares
- ☑ Factoring Perfect Square Trinomials
- ☑ Factoring the Sum or Difference of Two Cubes
- ☑ Factoring by Grouping
- ☑ Factoring Trinomials Using the *ac* Method
- ☑ Factoring Using the General Strategy
- ☑ Solving Problems by Factoring Polynomials

7-1. Factoring Integers

When we described the operation of multiplication in Chapter 1, we called the numbers to be multiplied the **factors,** and the result of the multiplication process the **product.** The operation of **factoring** is in a sense the opposite of the operation of multiplication; it is a kind of division in which we start with a product and reduce it to its original factors.

An integer is a factor of another integer if it divides that other integer evenly (with no remainder). For example, the integer factors of 15 are ± 1, ± 3, ± 5, and ± 15.

Note: The symbol "$\pm$" is read "plus or minus." In this case it means that either the positive or the negative values of the integers can be factors: $5 \cdot 3 = 15$, and $-5 \cdot -3 = 15$.

A **prime number** is a natural number greater than 1 that has no natural number factors other than itself and 1. The ten smallest prime numbers are 2, 3, 5, 7, 11, 13, 17, 19, 23, and 29.

Note: The number 1 is not considered to be a prime number.

A **composite number** is a natural number greater than 1 that is not a prime number. The ten smallest composite numbers are 4, 6, 8, 9, 10, 12, 14, 15, 16, and 18.

Each composite number may be factored into prime numbers. The factorization of a composite number is complete when it has been factored into nothing but prime numbers. For example, 15 may be factored into the prime numbers $3 \cdot 5$, and 56 may be factored into $2 \cdot 2 \cdot 2 \cdot 7$. This process is called the **prime factorization** of a composite number.

Note: For convenience, $2 \cdot 2 \cdot 2 \cdot 7$ is often written in exponential form as $2^3 \cdot 7$.

The **Fundamental Theorem of Arithmetic** states that each composite number has one and only one prime factorization, if we disregard the order of the factors. (In other words, $3 \cdot 5$ and $5 \cdot 3$ are considered to be the same prime factorization of 15.)

EXAMPLE 7-1: Find the prime factorization for each of the following integers:
(a) 75 **(b)** 256 **(c)** 391

Solution: (a) $75 = 3 \cdot 25$ (b) $256 = 2 \cdot 128$ (c) $391 = 17 \cdot 23$

$$= 3 \cdot 5 \cdot 5 \qquad\qquad = 2 \cdot 2 \cdot 64$$

$$= 3 \cdot 5^2 \qquad\qquad = 2 \cdot 2 \cdot 2 \cdot 32$$

$$= 2 \cdot 2 \cdot 2 \cdot 2 \cdot 16$$

$$= 2 \cdot 2 \cdot 2 \cdot 2 \cdot 2 \cdot 8$$

$$= 2 \cdot 2 \cdot 2 \cdot 2 \cdot 2 \cdot 2 \cdot 4$$

$$= 2 \cdot 2 \cdot 2 \cdot 2 \cdot 2 \cdot 2 \cdot 2 \cdot 2$$

$$= 2^8$$

7-2. Factoring the Greatest Common Factor from a Polynomial

A polynomial whose numerical coefficients and constant term are integers is called an **integral polynomial.** In this chapter we will discuss the factoring of integral polynomials. To factor a polynomial, we rewrite it as the product of other polynomials.

A. Finding the greatest common factor of monomials

To determine the **greatest common factor (GCF)** of exponential expressions with the same base, take the expression with the smallest exponent. For instance, the GCF of 5^3, 5^2, and 5^6 is 5^2, and the GCF of x^4, x^5, x, and x^2 is simply x.

To find the GCF of monomials, form the product of the GCF of each common base.

EXAMPLE 7-2: Find the GCF of $24a^3b$, $40a^2b^3$, and $56a^3b$.

Solution: Write the prime factorization of each numerical coefficient, followed by the literal parts. The product of the GCF of each common base is the GCF of the monomials.

$$24a^3b = 2^3 \cdot 3 \quad a^3 \quad b$$
$$40a^2b^3 = 2^3 \cdot 5 \quad a^2 \quad b^3$$
$$56a^3b = 2^3 \cdot 7 \quad a^3 \quad b$$

$2^3 \quad \cdot \quad a^2 \cdot b \longleftarrow$ the product of the GCF of each common base

The GCF of $24a^3b$, $40a^2b$, and $56a^3b$ is $8a^2b$: $(8a^2b)(3a) = 24a^3b$

$$(8a^2b)(5b^2) = 40a^2b^3$$

$$(8a^2b)(7a) = 56a^3b$$

The **GCF of a polynomial** is the GCF of the terms of the polynomial. We use the distributive property to **factor the GCF from a polynomial.**

EXAMPLE 7-3: Factor the GCF from $3a^3b + 6a^2b^2 + 12ab^3$.

Solution: Find the GCF of the terms, then use the distributive property to write the polynomial as a product of polynomials.

The GCF of $3a^2b$, $6a^2b^2$, and $12ab^3$ is $3ab$.

$3a^3b + 6a^2b^2 + 12ab^3 = 3ab(a^2) + 3ab(2ab) + 3ab(4b)^2$ Rename each term as factors involving the GCF.

$= 3ab(a^2 + 2ab + 4b^2)$ Use the distributive property to write the polynomial as a product.

Check: Does $a^2 + 2ab + 4b^2$ have any common factors other than 1? No.
Does $3ab(a^2 + 2ab + 4b^2)$ equal the original polynomial? Yes.

7-3. Factoring Trinomials

A. Factoring trinomials of the form $x^2 + bx + c$

Trinomials of the form $x^2 + bx + c$ often result when we multiply two binomials. For example:

$$\overset{\text{F} \quad \text{O} \quad \text{I} \quad \text{L}}{(x + 2)(x + 5)} = x^2 + 5x + 2x + 10$$

$$= x^2 + 7x + 10$$

It is important to observe that the coefficient 7 is the SUM of 2 and 5, and the constant term 10 is the PRODUCT of 2 and 5. In general:

$$\overset{\text{F} \quad \text{O} \quad \text{I} \quad \text{L}}{(x + d)(x + e)} = x^2 + ex + dx + de$$

$$= x^2 + (e + d)x + de$$

Observe that the expression $x^2 + (e + d)x + de$ is in $x^2 + bx + c$ form, with $b = e + d$ and $c = de$.

To factor $x^2 + bx + c$ means to write the trinomial as a product of two binomials of the form $(x + d)$ and $(x + e)$, where $de = c$ and $e + d = b$. Finding d and e is a process of **trial and error.**

EXAMPLE 7-4: Factor $x^2 + 4x - 21$.

Solution: Observe that $x^2 + 4x - 21$ is in $x^2 + bx + c$ form, with $b = 4$ and $c = -21$. Use the trial-and-error process to find factors of -21 whose sum is 4.

Factors of -21	Sum of the factors
$1, -21$	-20
$-1, 21$	20
$3, -7$	-4
$-3, 7$	4 ← this is the correct coefficient of x, so let $d = -3$ and $e = 7$

$\underset{d \; e}{\uparrow \uparrow} \qquad \underset{c}{\uparrow}$

$$x^2 + 4x - 21 = (x + d)(x + e)$$

$$= [x + (-3)](x + 7) \qquad \text{Substituting } -3 \text{ for } d \text{ and } 7 \text{ for } e.$$

$$= (x - 3)(x + 7) \qquad \text{Simplify.}$$

Check: $(x - 3)(x + 7) = x^2 + 7x - 3x - 21$

$$= x^2 + 4x - 21 \longleftarrow \text{ the original polynomial}$$

We can reduce the number of factors under consideration if we use relationships between the signs of the terms to determine the signs of the terms in the binomial factors. If the constant term of the binomial is positive, then the constant terms of the binomial factors are (a) both positive if the coefficient of x is positive, or (b) both negative if the coefficient of x is negative. If the constant term of the trinomial is negative, then the constant terms of the binomials have opposite signs.

To visualize these relationships, examine the following cases, where B and C are natural numbers.

	Trinomial	**Sign pattern of the binomial factors**	
Case 1:	$x^2 + Bx + C$	$= (x + \square)(x + \square)$	[See Example 7-5(a).]
Case 2:	$x^2 - Bx + C$	$= (x - \square)(x - \square)$	[See Example 7-5(b).]
Case 3:	$x^2 + Bx - C$	$= (x + \square)(x - \square)$	[See Example 7-5(c).]
Case 4:	$x^2 - Bx - C$	$= (x + \square)(x - \square)$	[See Example 7-5(d).]

EXAMPLE 7-5: Factor the following trinomials:

(a) $x^2 + 8x + 15$ (b) $x^2 - 7x + 12$ (c) $x^2 + 5x - 14$ (d) $x^2 - 9x - 10$
(e) $y^4 + 3y^2 - 10$

Solution: Factor by trial and error, making use of the sign patterns to reduce the number of factors under consideration.

(a) $x^2 + 8x + 15 = (x + \Box)(x + \Box)$ Since the constant term is positive and the coefficient of x is also positive, you know the constant terms of the binomial factors are both positive.

Positive factors of 15	Sum of the factors
1, 15	16
3, 5	8

The positive factors of 15 that have a sum of 8 are 3 and 5.

$x^2 + 8x + 15 = (x + 3)(x + 5)$ Check as before.

(b) $x^2 - 7x + 12 = (x - \Box)(x - \Box)$ The sign pattern indicates that you need to find two negative factors of 12 whose sum is -7.

Negative factors of 12	Sum of the factors
$-1, -12$	-13
$-2, -6$	-8
$-3, -4$	-7

The factors of 12 that have a sum of -7 are -3 and -4.

$x^2 - 7x + 12 = (x - 3)(x - 4)$ Check as before.

Note: Positive factors of 12 were not considered because they could not have a sum of -7.

(c) $x^2 + 5x - 14 = (x + \Box)(x - \Box)$ You need to find two factors of -14 whose sum is 5. Since their sum is positive, the positive factor must have a larger absolute value than the negative factor.

Factors of -14	Sum of the factors
$-1, 14$	13
$-2, 7$	5

The factors of -14 that have a sum of 5 are -2 and 7.

$x^2 + 5x - 14 = (x + 7)(x - 2)$ Check as before.

(d) $x^2 - 9x - 10 = (x + \Box)(x - \Box)$ You need to find two factors of -10 whose sum is -9. Since their sum is negative, the negative factor must have a larger absolute value than the positive factor.

Factors of -10	Sum of the factors
1, -10	-9

The factors of -10 that have a sum of -9 are 1 and -10.

$x^2 - 9x - 10 = (x + 1)(x - 10)$ Check as before.

(e) $y^4 + 3y^2 - 10 = (y^2)^2 + 3y^2 - 10$ $y^4 + 3y^2 - 10$ is in $x^2 + bx + c$ form, with $b = 3$, $c = -10$, and $x = y^2$.

$= (y^2 + \square)(y^2 - \square)$ You need to find two factors of -10 whose sum is 3. This means the positive factor must be the factor which has the larger absolute value.

Factors of -10	Sum of the factors
$-1, 10$	9
$-2, 5$	3

The desired factors are -2 and 5.

$y^4 + 3y^2 - 10 = (y^2 + 5)(y^2 - 2)$ Check as before.

Some trinomials cannot be factored using only integers. For example, $x^2 + 5x + 3$ cannot be factored using only integers. The only positive integer factors of 3 are 1 and 3, and their sum does not equal 5. A polynomial which cannot be factored using only integers is said to be **irreducible over the integers.**

B. Factoring trinomials of the form $ax^2 + bx + c$

To factor trinomials of the form $ax^2 + bx + c$, we use trial binomial factors whose first terms are factors of a and whose last terms are factors of c. We then check the trial binomial factors to see if their product produces the correct middle term.

EXAMPLE 7-6: Factor $3x^2 + x - 10$.

Solution:

Positive factors of 3	Factors of -10
1, 3	1, -10
	$-1, 10$
	2, -5
	$-2, 5$

Trial factor	Check middle term	
$(1x + 1)(3x - 10)$	$-10x + 3x = -7x$	Use the outer and inner products
$(1x - 10)(3x + 1)$	$1x - 30x = -29x$	of the FOIL method to determine
$(1x - 1)(3x + 10)$	$10x - 3x = 7x$	the middle term of the product of
$(1x + 10)(3x - 1)$	$-1x + 30x = 29x$	the trial factors.
$(1x + 2)(3x - 5)$	$-5x + 6x = 1x$	Stop. $1x$ is the correct middle term.

$3x^2 + x - 10 = (x + 2)(3x - 5)$ Check as before.

Note: Another factorization of $3x^2 + x - 10$ is $(-x - 2)(-3x + 5)$. However, we prefer the first coefficients of the binomial factors to be positive whenever possible. This is the reason we started by considering only positive factors of 3 in this example.

To reduce the number of trial factors to be considered you should:

1. Use relationships between the signs of the terms of the trinomial to determine the signs of the terms in the binomial factors.
2. Observe that if a trinomial does not have a factor common to all of its terms, then neither of its binomial factors will have a common factor.

EXAMPLE 7-7: Factor (a) $10x^2 + 47x + 9$ and (b) $6w^4 + 5w^2 - 14$.

Solution: (a) Observe that $10x^2 + 47x + 9$ is in $ax^2 + bx + c$ form with $a = 10$, $b = 47$, and $c = 9$. Also notice that the terms $10x^2$, $47x$, and 9 do not have a common factor.

Factors of 10	Factors of 9	
1, 10	1, 9	Since all the signs of the terms of the trinomials are
2, 5	3, 3	positive, you only need to consider positive factors.

Trial factors	Check middle term	
$(1x + 1)(10x + 9)$	$9x + 10x = 19x$	
$(1x + 9)(10x + 1)$	$1x + 27x = 28x$	
$(1x + 3)(3x + 3)$	No: $(3x + 3)$ may be factored further	
$(2x + 1)(5x + 9)$	$18x + 5x = 23x$	
$(2x + 9)(5x + 1)$	$2x + 45x = 47x$	Stop. $47x$ is the correct middle term.

$$10x^2 + 47x + 9 = (2x + 9)(5x + 1) \qquad \text{Check as before.}$$

(b) Observe that $6w^4 + 5w^2 - 14$ is in $ax^2 + bx + c$ form with $a = 6$, $b = 5$, $c = -14$, and $x = w^2$. Also notice that the terms $6w^4$, $5w^2$, and -14 do not have a common factor.

Factors of 6	Factors of -14
1, 6	1, -14
2, 3	-1, 14
	2, -7
	-2, 7

Trial factors	Check middle term	
$(1w^2 + 1)(6w^2 - 14)$	No: $(6w^2 - 14)$ may be factored further	
$(1w^2 - 14)(6w^2 + 1)$	$1w^2 - 84w^2 = -83w^2$	
$(1w^2 - 1)(6w^2 + 14)$	No: $(6w^2 + 14)$ may be factored further	
$(1w^2 + 14)(6w^2 - 1)$	$-1w^2 + 84w^2 = 83w^2$	
$(1w^2 + 2)(6w^2 - 7)$	$-7w^2 + 12w^2 = 5w^2$	Stop. $5w^2$ is the correct middle term.

$$6w^4 + 5w^2 - 14 = (w^2 + 2)(6w^2 - 7) \qquad \text{Check as before.}$$

7-4. Factoring the Difference of Two Squares

The binomial $x^2 - y^2$ is **the difference of two squares.** To factor the difference of two squares, we use the product formula

$$(r + s)(r - s) = r^2 - s^2$$

in reverse.

Caution: The **sum of two squares,** $r^2 + s^2$, cannot be factored using integer coefficients.

EXAMPLE 7-8: Factor **(a)** $x^2 - 16$ and **(b)** $9m^4 - 25n^2$.

Solution: **(a)** $x^2 - 16 = x^2 - 4^2$ Rename the polynomial in $r^2 - s^2$ form ($r = x$, $s = 4$).

$$= (x + 4)(x - 4) \qquad \text{Use the product formula } r^2 - s^2 = (r + s)(r - s).$$

(b) $9m^4 - 25n^2 = (3m^2)^2 - (5n)^2$

$$= (3m^2 + 5n)(3m^2 - 5n)$$

7-5. Factoring Perfect Square Trinomials

We call trinomial a **perfect square trinomial** (PST) if it is the square of a binomial. To factor a perfect square trinomial you use the following special product formulas in reverse.

$$(r + s)^2 = r^2 + 2rs + s^2$$

$$(r - s)^2 = r^2 - 2rs + s^2$$

You can factor every PST by the trial-and-error method, but it is generally less time-consuming to be familiar enough with the special product formulas to be able to identify the PST form $r^2 \pm 2rs + s^2$, then write it in its factored form as $(r \pm s)^2$.

EXAMPLE 7-9: Factor (a) $x^2 + 6x + 9$ and (b) $4u^2 - 20uv + 25v^2$.

Solution: (a) $x^2 + 6x + 9 = x^2 + 2(x)(3) + 3^2$ Identify the PST form $r^2 + 2rs + s^2$.

$$= (x + 3)^2 \qquad \text{Use } r^2 + 2rs + s^2 = (r + s)^2.$$

(b) $4u^2 - 20uv + 25v^2 = (2u)^2 - 2(2u)(5v) + (5v)^2$ Identify the $r^2 - 2rs + s^2$ form.

$$= (2u - 5v)^2 \qquad \text{Use } r^2 - 2rs + s^2 = (r - s)^2.$$

7-6. Factoring the Sum or Difference of Two Cubes

The exponential expression a^3 is the **cube** of a. The **cube root** of a^3 is a.

The following products are special because the first is the **sum of two cubes** and the second is **the difference of two cubes.**

$$
\begin{array}{rl}
r^2 - rs + s^2 \\
r + s \\
\hline
r^2s - rs^2 + s^3 \\
r^3 - r^2s + rs^2 \\
\hline
r^3 \qquad\qquad + s^3
\end{array}
\quad\leftarrow \text{ sum of two cubes}
\qquad
\begin{array}{rl}
r^2 + rs + s^2 \\
r - s \\
\hline
- r^2s - rs^2 - s^3 \\
r^3 + r^2s + rs^2 \\
\hline
r^3 \qquad\qquad - s^3
\end{array}
\quad\leftarrow \text{ difference of two cubes}
$$

To factor the sum or difference of two cubes we use the above products in reverse.

$$r^3 + s^3 = (r + s)(r^2 - rs + s^2)$$
$$r^3 - s^3 = (r - s)(r^2 + rs + s^2)$$

In practice we often factor the sum or difference of two cubes using the following procedure:

1. Identify the polynomial as a sum or difference of two cubes.
2. Write the binomial factor; it is the sum or difference of the two cube roots.
3. Write the trinomial factor:
   (a) The first term of the trinomial factor is the square of the first term of the binomial factor.
   (b) The second term of the trinomial factor is the opposite of the product of the two terms of the binomial factor.
   (c) The third term of the trinomial factor is the square of the second term of the binomial factor.

EXAMPLE 7-10: Factor (a) $y^3 + 64$ and (b) $8m^3 - 125n^3$.

Solution: (a) $y^3 + 64 = y^3 + 4^3$ Identify the $r^3 + s^3$ form ($r = y$ and $s = 4$).

$$= (y + 4)(?\ -\ ?\ +\ ?) \qquad \text{Write the binomial factor.}$$
$$= (y + 4)(y^2 - 4y + 16) \qquad \text{Write the trinomial factor.}$$

square of the first term (y) ⎤
opposite of the product of ⎱
the two terms (y) and (4) ⎰
square of the last term (4) ⎦

(b) $8m^3 - 125n^3 = (2m)^3 - (5n)^3$ Identify the $r^3 - s^3$ form

$$= (2m - 5n)(?\ +\ ?\ +\ ?) \qquad (r = 2m \text{ and } s = 5n).$$
$$= (2m - 5n)(4m^2 + 10mn + 25n^2)$$

7-7. Factoring by Grouping

Some polynomials can be factored by **grouping** the terms in pairs.

It is sometimes necessary to rearrange the terms of a polynomial to be able to factor by grouping. It is not always possible to determine which grouping may lead to a factorization without some experimentation. As a guide, however, it is best to group terms that have a common factor or terms that form one of the special polynomials introduced in this chapter.

EXAMPLE 7-11: Factor **(a)** $x^3 + 2x^2 + 3x + 6$ and **(b)** $3ac + 4bc - 6ad - 8bd$.

Solution:

(a)
$$x^3 + 2x^2 + 3x + 6 = (x^3 + 2x^2) + (3x + 6) \quad \text{Regroup as the sum of two binomials}$$
$$= x^2(x + 2) + 3(x + 2) \quad \text{Factor the GCF from each binomial.}$$
$$= (x + 2)(x^2 + 3) \quad \text{Factor the common binomial factor from each term.}$$

(b)
$$3ac + 4bc - 6ad - 8bd = (3ac + 4bc) + (-6ad - 8bd)$$
$$= c(3a + 4b) + (-2d)(3a + 4b)$$
$$= (3a + 4b)[c + (-2d)]$$
$$= (3a + 4b)(c - 2d)$$

Some polynomials can be factored by grouping the terms of the polynomial into groups which form special polynomials.

EXAMPLE 7-12: Factor **(a)** $x^2 + 6xy + 9y^2 - 25$ and **(b)** $a^3 - b^3 + a^2 + ab + b^2$.

Solution:

(a)
$$x^2 + 6xy + 9y^2 - 25 = (x^2 + 6xy + 9y^2) - 25 \quad \text{Group the terms of the PST.}$$
$$= (x + 3y)^2 - 25 \quad \text{Factor the PST.}$$
$$= (x + 3y)^2 - 5^2 \quad \text{Identity the } r^2 - s^2 \text{ form.}$$
$$= [(x + 3y) + 5][(x + 3y) - 5]$$
$$= (x + 3y + 5)(x + 3y - 5)$$

(b)
$$a^3 - b^3 + a^2 + ab + b^2 = (a^3 - b^3) + (a^2 + ab + b^2)$$
$$= (a - b)(a^2 + ab + b^2) + (a^2 + ab + b^2)(1)$$
$$= (a^2 + ab + b^2)[(a - b) + (1)]$$
$$= (a^2 + ab + b^2)(a - b + 1)$$

7-8. Factoring Trinomials Using the *ac* Method

If an expression of the form $ax^2 + bx + c$ can be factored as the product of two binomials using integers, then you can factor it by grouping when you find the integers m and n so that $mn = ac$ and $m + n = b$. This result leads to the ***ac* method of factoring trinomials** of the form $ax^2 + bx + c$.

To factor $ax^2 + bx + c$ using the *ac* method:

1. Find integers m and n so that $mn = ac$ and $m + n = b$.
2. Substitute $mx + nx$ for bx to produce a polynomial of the form $ax^2 + mx + nx + c$.
3. Factor $ax^2 + mx + nx + c$ by grouping (see Section 7-7).
4. Check the result of step 3 by finding the product of the factors and comparing it with the original trinomial.

EXAMPLE 7-13: Factor $6x^2 + 19x - 20$ using the *ac* method.

Solution: In $6x^2 + 19x - 20$, $a = 6$, $b = 19$, and $c = -20$.

$$\underline{ac = 6(-20) = -120} \qquad\qquad \underline{b = 19}$$

$$
\begin{array}{ll}
(-1)(120) & -1 + 120 = 119 \\
(-2)(60) & -2 + 60 = 58 \\
(-3)(40) & -3 + 40 = 37 \\
(-4)(30) & -4 + 30 = 26 \\
(-5)(24) & -5 + 24 = 19
\end{array}
$$

Let $m = -5$ and $n = 24$.

$$6x^2 + 19x - 20 = 6x^2 - 5x + 24x - 20 \qquad \text{Substitute } -5x + 24x \text{ for } 19x.$$

$$= (6x^2 - 5x) + (24x - 20) \qquad \text{Factor by grouping.}$$

$$= x(6x - 5) + 4(6x - 5)$$

$$= (6x - 5)(x + 4)$$

Check: $(6x - 5)(x + 4) = 6x^2 + 24x - 5x - 20 \qquad$ Use the FOIL method.

$$= 6x^2 + 19x - 20 \longleftarrow \text{ the original trinomial}$$

Note: If the values of m and n are reversed, you will still get the same result. Try it.

7-9. Factoring Using the General Strategy

A polynomial is **completely factored** if it is written as a product of polynomials each of which is irreducible over the integers. To completely factor polynomials, we use the following **general strategy.**

To completely factor a polynomial:

1. Factor the GCF from each term.
2. If the polynomial is a binomial, then try to factor it using the appropriate special product formulas:
   (a) the difference of two squares: $r^2 - s^2 = (r + s)(r - s)$
   (b) the sum of two cubes: $r^3 + s^3 = (r + s)(r^2 - 2rs + s^2)$
   (c) the difference of two cubes: $r^3 - s^3 = (r - s)(r^2 + 2rs + s^2)$
3. If a polynomial is a trinomial, then try to factor it
   (a) as a PST: $r^2 \pm 2rs + s^2 = (r \pm s)^2$
   (b) using the trial-and-error method or the *ac* method.
4. Try to factor the polynomial by grouping.
5. After each factorization, examine each new factor to determine if it too can be factored.

EXAMPLE 7-14: Factor (a) $162a^4 - 2$ (b) $2x^3y + 14x^2y + 24xy$
(c) $12a^2cx - 42a^2x + 30abcx - 105abx$.

Solution: Use the general strategy.

(a) $162a^4 - 2 = 2(81a^4 - 1) \qquad$ Factor the GCF from each term.

$$= 2(9a^2 + 1)(9a^2 - 1) \qquad \text{Factor using the difference of two squares.}$$

$$= 2(9a^2 + 1)(3a + 1)(3a - 1) \qquad \text{Factor using the difference of two squares.}$$

(b) $2x^3 + 14x^2y + 24xy = 2xy(x^2 + 7x + 12) \qquad$ Factor the GCF from each term.

$$= 2xy(x + 3)(x + 4) \qquad \text{Factor using trial-and error.}$$

(c) $12a^2cx - 42a^2x + 30abcx - 105abx = 3ax(4ac - 14a + 10bc - 35b) \qquad$ Factor the GCF.

$$= 3ax[(4ac - 14a) + (10bc - 35b)] \qquad \text{Factor by}$$
$$= 3ax[2a(2c - 7) + 5b(2c - 7)] \qquad \text{grouping.}$$

$$= 3ax[(2c - 7)(2a + 5b)]$$

$$= 3ax(2c - 7)(2a + 5b)$$

7-10. Solving Problems by Factoring Polynomials

To solve a problem that can be represented by an equation where one member is zero and the other member is a second-degree polynomial that can be factored as the product of two first-degree polynomials, you first factor the second-degree polynomial, then use the zero-product property.

Recall: The sum of all the counting numbers from 1 to n $[1 + 2 + 3 + \cdots + n]$ is represented by the second-degree polynomial $\frac{1}{2}n^2 + \frac{1}{2}n$ [see Section 6-6].

EXAMPLE 7-15: Use $\frac{1}{2}n^2 + \frac{1}{2}n$ to find how many of the first n counting numbers are needed to make a sum of 55.

Solution: The sum of the first n counting numbers is 55.

$$\frac{1}{2}n^2 + \frac{1}{2}n = 55$$

$$\frac{1}{2}n^2 + \frac{1}{2}n - 55 = 0 \qquad \text{Make one member zero.}$$

$$2 \cdot \tfrac{1}{2}n^2 + 2 \cdot \tfrac{1}{2}n - 2(55) = 2(0) \qquad \text{Clear fractions.}$$

$$n^2 + n - 110 = 0 \longleftarrow \text{fractions cleared}$$

$$(n + 11)(n - 10) = 0 \qquad \begin{array}{l}\text{Factor the second-degree} \\ \text{polynomial over the integers.}\end{array}$$

$$n + 11 = 0 \quad \text{or } n - 10 = 0 \qquad \begin{array}{l}\text{Use the zero-product} \\ \text{property: } ab = 0 \text{ means} \\ a = 0 \text{ or } b = 0.\end{array}$$

$$\text{proposed solutions} \longrightarrow n = -11 \text{ or} \qquad n = 10$$

$n = -11$ cannot be a solution of the original problem because it is not possible to have a negative number of counting numbers.

$n = 10$ means the sum of the first ten counting numbers is 55.

Check: Is the sum $1 + 2 + 3 + \cdots + 10$ equal to 55? Yes:

$$1 + 2 + 3 + 4 + 5 + 6 + 7 + 8 + 9 + 10 = (1 + 9) + (2 + 8) + (3 + 7) + (4 + 6) + 10 + 5$$

$$= 10 + 10 + 10 + 10 + 10 + 5$$

$$= 55$$

RAISE YOUR GRADES

Can you ... ?

☑ factor integers
☑ find the GCF of a polynomial
☑ factor the GCF from a polynomial
☑ factor trinomials of the form $x^2 + bx + c$
☑ factor trinomials of the form $ax^2 + bx + c$
☑ factor the difference of two squares
☑ factor perfect square trinomials
☑ factor the sum or difference of two cubes
☑ factor by grouping
☑ factor trinomials by the ac method
☑ factor using the general strategy
☑ solve problems by factoring

SOLVED PROBLEMS

PROBLEM 7-1 Find the GCF of $30x^3y^2z^4$, $75x^3y^3$, and $90x^3y^4z^2$.

Solution: Recall that to find the GCF of polynomial terms, you write each term in its prime factored form and then form the product of the GCF of each common base [see Example 7-2].

$$30x^3y^2z^4 = 2 \cdot \boxed{3} \cdot \boxed{5} \boxed{x^3} \boxed{y^2} z^4$$
$$75x^3y^3 = \boxed{3} \cdot \boxed{5^2} \boxed{x^3} \boxed{y^3}$$
$$90x^3y^4z^2 = 2 \cdot \boxed{3^2} \cdot \boxed{5} \boxed{x^3} \boxed{y^4} z^2$$
$$\qquad\qquad\quad 3 \quad \cdot \quad 5 \quad x^3 \quad y^2$$

The GCF of $30x^3y^2z^4$, $75x^3y^3$, and $90x^3y^4z^2$ is $15x^3y^2$.

PROBLEM 7-2 Factor the GCF from each of the following polynomials:

(a) $4x + 10$ **(b)** $3x^2 + 6x + 21$ **(c)** $-2x^2 - 12x - 2$

Solution: Recall that to factor the GCF from a polynomial, you write each term as a product which involves the GCF, and then use the distributive property to write the polynomial as a product [see Example 7-3].

(a) $4x + 10 = 2(2x) + 2 \cdot 5$ **(b)** $3x^2 + 6x + 21 = 3 \cdot x^2 + 3 \cdot (2x) + 3 \cdot 7$

$\qquad\qquad\quad = 2(2x + 5)$ $\qquad\qquad\qquad\qquad\quad = 3(x^2 + 2x + 7)$

(c) $-2x^2 - 12x - 2 = -2 \cdot x^2 + (-2) \cdot 6x + (-2) \cdot 1$

$\qquad\qquad\qquad\qquad = -2(x^2 + 6x + 1)$

PROBLEM 7-3 Factor the following polynomials:

(a) $x^2 + 9x + 14$ **(b)** $x^2 - x - 12$ **(c)** $y^2 + 4y - 21$ **(d)** $z^4 - 11z + 30$
(e) $w^2 + 8w + 1$

Solution: Recall that to factor a trinomial of the form $x^2 + Bx + C$, you use the trial-and-error method [see Examples 7-4 and 7-5].

(a) $x^2 + 9x + 14 = (x + \square)(x + \square)$ Find two positive factors of 14 whose sum is 9.

$\qquad\qquad\qquad = (x + 2)(x + 7)$ $\quad 2 \cdot 7 = 14$, and $2 + 7 = 9$.

(b) $x^2 - x - 12 = (x + \square)(x - \square)$ Find two factors of -12 whose sum is -1.

$\qquad\qquad\qquad = (x + 3)(x - 4)$ $\quad 3(-4) = -12$, and $3 + (-4) = -1$.

(c) $y^2 + 4y - 21 = (y + \square)(y - \square)$ Find two factors of -21 whose sum is 4.

$\qquad\qquad\qquad = (y + 7)(y - 3)$ $\quad 7(-3) = -21$, and $7 + (-3) = 4$.

(d) $z^4 - 11z + 30 = (z^2 - \square)(z^2 - \square)$ Find two negative factors of 30 whose sum is -11.

$\qquad\qquad\qquad = (z^2 - 5)(z^2 - 6)$ $\quad (-5)(-6) = 30$, and $(-5) + (-6) = -11$.

(e) $w^2 + 8w + 1$ is irreducible over the integers because there are no integers whose product is 1 and whose sum is 8.

PROBLEM 7-4 Factor the following polynomials:

(a) $4x^2 + 23x + 15$ **(b)** $2y^2 + y - 15$ **(c)** $3a^4 - 31a^2 + 10$ **(d)** $6z^2 + 5z - 4$

Solution: Recall that to factor a trinomial of the form $ax^2 + bx + c$, you use the trial-and-error method or the *ac* method [see Examples 7-6 and 7-7].

(a) $4x^2 + 23x + 15 = (\square x + \square)(\square x + \square)$

Factors of 4	Factors of 15
1, 4	1, 15
2, 2	3, 5

Trial factors	Check middle term
$(1x + 1)(4x + 15)$	$15x + 4x = 19x$
$(1x + 15)(4x + 1)$	$1x + 60x = 61x$
$(1x + 3)(4x + 5)$	$5x + 12x = 17x$
$(1x + 5)(4x + 3)$	$3x + 20x = 23x$ ⟵ the correct middle term

$$4x^2 + 23x + 15 = (x + 5)(4x + 3)$$

(b) $2y^2 - y - 15 = (\Box y + \Box)(\Box y - \Box)$

Factors of 2	Factors of −15
1, 2	−1, 15
	1, −15
	−3, 5
	3, −5

Trial factors	Check middle term
$(1y - 1)(2y + 15)$	$15y - 2y = 13y$
$(1y + 15)(2y - 1)$	$-1y + 30y = 29y$
$(1y + 1)(2y - 15)$	$-15y + 2y = -13y$
$(1y - 15)(2y + 1)$	$1y - 30y = -29y$
$(1y - 3)(2y + 5)$	$5y - 6y = -1y$ ⟵ the correct middle term

$$2y^2 - y - 15 = (y - 3)(2y + 5)$$

(c) $3a^4 - 31a^2 + 10 = (\Box a^2 - \Box)(\Box a^2 - \Box)$

Factors of 3	Factors of 10
1, 3	−1, −10
	−2, −5

Trial factors	Check middle term
$(1a^2 - 1)(3a^2 - 10)$	$-10a^2 - 3a^2 = -13a^2$
$(1a^2 - 10)(3a - 1)$	$-1a^2 - 30a^2 = -31a^2$ ⟵ the correct middle term

$$3a^4 - 31a^2 + 10 = (a^2 - 10)(3a^2 - 1)$$

(d) $6z^2 + 5z - 4 = (\Box z + \Box)(\Box z - \Box)$

Factors of 6	Factors of −4
1, 6	−1, 4
2, 3	1, −4
	2, −2

Trial factors	Check middle term
$(1z - 1)(6z + 4)$	No: $(6z + 4)$ may be factored further
$(1z + 4)(6z - 1)$	$-1z + 24z = 23z$
$(1z + 1)(6z - 4)$	No: $(6z - 4)$ may be factored further
$(1z - 4)(6z + 1)$	$1z - 24z = -23z$
$(1z + 2)(6z - 2)$	No: $(6z - 2)$ may be factored further
$(1z - 2)(6z + 2)$	No: $(6z + 2)$ may be factored further
$(2z - 1)(3z + 4)$	$8z - 3z = 5z$ ⟵ the correct middle term

$$6z^2 + 5z - 4 = (2z - 1)(3z + 4)$$

PROBLEM 7-5 Factor the following polynomials: **(a)** $y^2 - 9$ **(b)** $16m^2 - 25$ **(c)** $81p^2 - 1$

Solution: Recall that $(r^2 - s^2) = (r + s)(r - s)$ [see Example 7-8].

(a) $y^2 - 9 = y^2 - 3^2$

$\qquad = (y + 3)(y - 3)$

(b) $16m^2 - 25 = (4m)^2 - 5^2$

$\qquad = (4m + 5)(4m - 5)$

(c) $81p^4 - 1 = (9p)^2 - 1^2$

$\qquad = (9p + 1)(9p - 1)$

PROBLEM 7-6 Factor **(a)** $m^2 + 10m + 25$ and **(b)** $16z^2 - 24z + 9$.

Solution: Recall that $r^2 + 2rs + s^2 = (r + s)^2$ and $r^2 - 2rs + s^2 = (r - s)^2$ [see Example 7-9].

(a) $m^2 + 10m + 25 = m^2 + 2 \cdot 5m + 5^2$

$\qquad = (m + 5)^2$

(b) $16z^2 - 24z + 9 = (4z)^2 - 2 \cdot 12z + 3^2$

$\qquad = (4z - 3)^2$

PROBLEM 7-7 Factor the following polynomials: **(a)** $x^3 + 64$ **(b)** $27p^3 - 8q^3$ **(c)** $n^6 + 1$

Solution: Recall that $r^3 + s^3 = (r + s)(r^2 - rs + s^2)$ and $r^3 - s^3 = (r - s)(r^2 + rs + s^2)$ [see Example 7-10].

(a) $x^3 + 64 = x^3 + 4^3$

$\qquad = (x + 4)(? - ? + ?)$

$\qquad = (x + 4)(x^2 - 4x + 16)$

(b) $27p^3 - 8q^3 = (3p)^3 - (2q)^3$

$\qquad = (3p - 2q)(? + ? + ?)$

$\qquad = (3p - 2q)(9p^2 + 6pq + 4q^2)$

(c) $n^6 + 1 = (n^2)^3 + 1^3$

$\qquad = (n^2 + 1)(n^4 - n^2 + 1)$

PROBLEM 7-8 Factor the following polynomials:

(a) $y^3 + 4y^2 + 2y + 8$ **(b)** $3x^3 + 4x^2 - 9x - 12$ **(c)** $u^3 + v^3 + 2u^2 - 2uv + 2v^2$

Solution: Recall that to factor a polynomial with more than three terms, you should try grouping terms [see Examples 7-11 and 7-12].

(a) $y^3 + 4y^2 + 2y + 8 = (y^3 + 4y^2) + (2y + 8)$

$\qquad = y^2(y + 4) + 2(y + 4)$

$\qquad = (y + 4)(y^2 + 2)$

(b) $3x^3 + 4x^2 - 9x - 12 = (3x^3 + 4x^2) + (-9x - 12)$

$\qquad = x^2(3x + 4) + (-3)(3x + 4)$

$\qquad = (3x + 4)[x^2 + (-3)]$

$\qquad = (3x + 4)(x^2 - 3)$

(c) $u^3 + v^3 + 2u^2 - 2uv + 2v^2 = (u^3 + v^3) + (2u^2 - 2uv + 2v^2)$

$\qquad = (u + v)(u^2 - uv + v^2) + 2(u^2 - uv + v^2)$

$\qquad = (u^2 - uv + v^2)[(u + v) + 2]$

$\qquad = (u^2 - uv + v^2)(u + v + 2)$

PROBLEM 7-9 Factor each trinomial using the *ac* method:

(a) $12x^2 + 40x + 25$ **(b)** $14y^2 - 83y - 6$

Solution [see Example 7-13].

(a) In $12x^2 + 40x + 25$, $a = 12$, $b = 40$, and $c = 25$.

$ac = 12 \cdot 25 = 300$	$b = 40$	
$1 \cdot 300 = 300$	$1 + 300 = 301$	
$2 \cdot 150 = 300$	$2 + 150 = 152$	
$3 \cdot 100 = 300$	$3 + 100 = 103$	
$4 \cdot 75 \ = 300$	$4 + \ 75 = \ \ 79$	
$5 \cdot 60 \ = 300$	$5 + \ 60 = \ \ 65$	
$6 \cdot 50 \ = 300$	$6 + \ 50 = \ \ 56$	
$10 \cdot 30 \ \ = 300$	$10 + \ 30 = \ \ 40$	Stop.
	$\quad \uparrow \qquad \uparrow$	
	$\quad m \qquad n$	

$$12x^2 + 40x + 25 = 12x^2 + 10x + 30x + 25$$
$$= (12x^2 + 10x) + (30x + 25)$$
$$= 2x(6x + 5) + 5(6x + 5)$$
$$= (6x + 5)(2x + 5) \qquad \text{Check as before.}$$

(b) In $14y^2 - 83y - 6$, $a = 14$, $b = -83$, and $c = -6$.

$ac = 14 \cdot (-6) = -84$	$b = -83$	
$1 \cdot (-84) = -84$	$1 + (-84) = -83$	Stop.
	$\ \uparrow \qquad \uparrow$	
	$\ m \qquad n$	

$$14y^2 - 83y - 6 = 14y^2 + 1y + (-84)y - 6$$
$$= (14y^2 + 1y) + [(-84y) - 6]$$
$$= y(14y + 1) + (-6)(14y + 1)$$
$$= (14y + 1)(y + (-6))$$
$$= (14y + 1)(y - 6) \qquad \text{Check as before.}$$

PROBLEM 7-10 Factor completely:

(a) $30x^2y + 14xy - 4y$ **(b)** $32x^3y - 108y$ **(c)** $48w^4 - 3$

Solution: Recall that to factor a polynomial completely, you use the general strategy [see Example 7-14].

(a) $30x^2y + 14xy - 4y = 2y(15x^2 + 7x - 2)$
$$= 2y(3x + 2)(5x - 1)$$

(b) $32x^3y - 108y = 4y(8x^3 - 27)$
$$= 4y(2x - 3)(4x^2 + 6x + 9)$$

(c) $48w^4 - 3 = 3(16w^4 - 1)$
$$= 3(4w^2 + 1)(4w^2 - 1)$$
$$= 3(4w^2 + 1)(2w + 1)(2w - 1)$$

PROBLEM 7-11 Given that the number of single connections possible between n different things is represented by $\frac{1}{2}n^2 - \frac{1}{2}n$, find the number of people that are needed to create 6 different handshakes.

Solution: Recall that to solve a problem that can be represented by an equation where one member is zero and the other member is second-degree polynomial that factors as the product of two first-degree polynomials, you first factor the second-degree polynomial, then use the zero-product property [see Example 7-15].

The number of single handshakes possible between n different people is 6.

$$\tfrac{1}{2}n^2 - \tfrac{1}{2}n \qquad\qquad = 6$$

$$\tfrac{1}{2}n^2 - \tfrac{1}{2}n - 6 = 0$$

$$2 \cdot \tfrac{1}{2}n^2 - 2 \cdot \tfrac{1}{2}n - 2(6) = 2(0)$$

$$n^2 - n - 12 = 0$$

$$(n + 3)(n - 4) = 0$$

$$n + 3 = 0 \text{ or } n - 4 = 0$$

$$n = \cancel{-3} \qquad n = 4 \longleftarrow \text{ number of people needed for 6 different handshakes}$$

Check: Draw a picture:

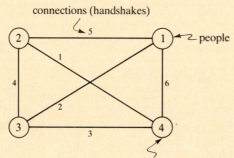

connections (handshakes)

people

Four people are needed for
six different handshakes.

Supplementary Exercises

PROBLEM 7-12 Find the GCF of the given monomials:

(a) $2x, 10$ (b) $4y^2, 24y$ (c) $6z^2, 12z^3, 15z^4$ (d) $40x^6, 25x^3$ (e) $w^4v^3, -8w^3v^5, 6wv^4$
(f) $60m^4n^3p, 36m^3n^4p, 84m^3n^3p$

PROBLEM 7-13 Factor the GCF from each polynomial:

(a) $3x + 6$ (b) $15x + 5$ (c) $2y - 2$ (d) $6m^2 + 12m + 18$
(e) $32x^3 - 8x^2 + 16x + 24$ (f) $4xy^2 + 7y$ (g) $51x^3w^2 - 6x^2w^3$
(h) $15x^3y - 5x^2y + 10xy + 25y$ (i) $49p^5 - 14p^6$ (j) $125w^7 - 225w^4v$

PROBLEM 7-14 Factor each trinomial of the form $x^2 + bx + c$:

(a) $x^2 + 4x + 3$ (b) $y^2 - 9y + 14$ (c) $w^2 + 5w - 24$ (d) $p^2 + 17p + 60$
(e) $q^2 + 2q + 1$ (f) $r^2 - 2r - 3$ (g) $x^2 + 7x + 3$ (h) $x^2 - 5x - 6$
(i) $x^4 + 7x^2 + 12$ (j) $w^4 - 19w^2 + 34$ (k) $u^4 + u^2 - 6$ (l) $z^4 - 4z^2 - 21$

PROBLEM 7-15 Factor each trinomial of the form $ax^2 + bx + c$:

(a) $6x^2 + 5x + 1$ (b) $10w^2 + 39w + 35$ (c) $3z^2 - 19z - 14$ (d) $6y^2 + 7y - 5$
(e) $8p^2 + 9p + 1$ (f) $33q^2 + 49q - 10$ (g) $2t^4 + 17t^2 + 21$ (h) $10m^2 + 21m - 10$
(i) $15u^2 + 91u + 6$ (j) $10x^2 + 21xy + 9y^2$ (k) $6x^4 + 25x^2 + 4$ (l) $45y^2 + 81y + 10$
(m) $8k^2 - 14kj - 15j^2$ (n) $30q^2 - 29qr - 35r^2$ (o) $6t^2 + 37t + 6$

PROBLEM 7-16 Factor each difference of two squares:

(a) $x^2 - 9$ (b) $y^2 - 1$ (c) $w^2 - 64$ (d) $z^2 - 121$ (e) $4x^2 - 9$ (f) $9y^2 - 1$
(g) $25u^2 - 169$ (h) $100x^2 - 9y^2$ (i) $81z^2 - 16v^2$ (j) $9m^2 - 49n^2$ (k) $p^4 - 4$
(l) $16q^4 - 9r^2$ (m) $900w^2 - 121$ (n) $x^8 - 25$ (o) $p^{10} - 1$

PROBLEM 7-17 Factor each PST:

(a) $w^2 + 6w + 9$ (b) $m^2 + 14m + 49$ (c) $n^2 - 8n + 16$ (d) $x^2 - 24x + 144$
(e) $9p^2 + 30p + 25$ (f) $25t^2 + 70t + 49$ (g) $4q^2 - 12q + 9$ (h) $u^4 + 10u^2 + 25$
(i) $m^2 + 18m + 81$ (j) $n^2 - 18n + 81$ (k) $121r^2 + 66r + 9$ (l) $25p^2 + 70pq + 49q^2$
(m) $b^2 - 6bc + 9c^2$ (n) $64u^2 + 80uv + 25v^2$ (o) $x^4 + 2x^2y^2 + y^4$

PROBLEM 7-18 Factor each sum or difference of two cubes:

(a) $y^3 + 1$ (b) $x^3 - 1$ (c) $z^3 - 64$ (d) $p^3 + 125$ (e) $8t^3 + 1$ (f) $216v^3 + 125w^3$
(g) $8w^3 - 729x^3$ (h) $x^6 + 1$ (i) $x^6 - 8$ (j) $x^6 + y^6$ (k) $m^{15} - 27$ (l) $1 - v^3$
(m) $1331k^3 - 27$ (n) $512p^3 - 1$ (o) $729q^6 + 8$

PROBLEM 7-19 Factor each polynomial by grouping:

(a) $y^3 + 5y^2 + 2y + 10$ (b) $3x^2 + xy + 3xy^2 + y^3$ (c) $2ab + 5bc + 4ad + 10cd$
(d) $3wx + 5wy + 9ux + 15uy$ (e) $3ab + bm - 6ay - 2my$ (f) $x^3 + 3x^2 - 3x - 9$
(g) $a^3 - 5a^2 - 6a + 30$ (h) $4x^2 + 12xy + 9y^2 - 25$ (i) $9x^2 - 30xy + 25y^2 - 49$
(j) $m^3 - n^3 + m^2 + mn + n^2$ (k) $a^3 + b^3 + a^2 - ab + b^2$ (l) $9m^2 - 6mn + n^2 - 25$
(m) $49 - 4p^2 - 20pq - 25q^2$ (n) $64 - 9t^2 + 6ts - s^2$ (o) $x^2 - y^2 - 2yz - z^2$

PROBLEM 7-20 Factor each trinomial by the *ac* method:

(a) $6x^2 + 11x + 3$ (b) $4y^2 + 11y + 6$ (c) $-6w^2 - w + 15$ (d) $6a^2 - 13a + 5$
(e) $15b^2 + 19b - 8$ (f) $10x^2 - 19xy - 15y^2$ (g) $20m^2 + 41m + 20$
(h) $6m^2 - 25m + 14$ (i) $8a^2 + 37a - 15$ (j) $8a^2 - 43a + 15$ (k) $14b^2 + 19b - 3$
(l) $2x^2 + 3x - 5$ (m) $8a^2 + 3a - 5$ (n) $20a^2 - 79a - 4$ (o) $12p^2 + 7pq - 12q^2$

PROBLEM 7-21 Completely factor each polynomial using the general strategy:

(a) $2a^2 - 2b^2$ (b) $3x^3 + 30x^2 + 75x$ (c) $9c^3d + 15c^2d^2 + 6cd^3$ (d) $5a^3 - 40b^3$
(e) $54x^3 + 2y^3$ (f) $x^4 - 1$ (g) $y^8 - 1$ (h) $2m^3 + 5m^2n - 7mn^2$
(i) $(3k + 5)(a^2 + ab) + (3k + 5)(ab + b^2)$ (j) $x^3y^3 + x^6y^6$ (k) $10m^5 - 53m^4 + 15m^3$
(l) $a^3 - a^3b^3$ (m) $x^6 - 8x^4 - 9x^2$ (n) $4a^2x^2 - 4a^2 - x^2 + 1$
(o) $9x^4 - 9x^2y^2 - x^2 + y^2$ (p) $x^{2n+1} - 10x^{n+1} + 25x$ (q) $6y^{n+2} + y^{n+1} - 15y^n$
(r) $a^3(a - 1) - a + 1$

PROBLEM 7-22 Given that the number of diagonals possible for a *n*-sided figure is represented by $\frac{1}{2}n^2 - \frac{3}{2}n$, find the kind of figure that has: (a) 0 diagonals   (b) 1 diagonal   (c) 2 diagonals
(d) 3 diagonals   (e) 4 diagonals   (f) 5 diagonals   (g) 9 diagonals   (h) 10 diagonals
(i) 14 diagonals   (j) 20 diagonals   (k) 35 diagonals   (l) 4850 diagonals

Use $\frac{1}{2}n^2 + \frac{1}{2}n$ to find how many of the first *n* counting numbers are needed to make a sum of:
(m) 3   (n) 6   (o) 15   (p) 36   (q) 210.

Use $\frac{1}{2}n^2 - \frac{1}{2}n$ to find the number of different phones, going through a switchboard, that are needed to create: (r) 3 single connections   (s) 10 single connections   (t) 15 single connections
(u) 28 single connections   (v) 45 single connections   (w) 190 single connections
(x) 1225 single connections   (y) 4950 single connections   (z) 499,500 single connections

Answers to Supplementary Exercises

(7-12) **(a)** 2 **(b)** $4y$ **(c)** $3z^2$ **(d)** $5x^3$ **(e)** wv^3 **(f)** $12m^3n^3p$

(7-13) **(a)** $3(x + 2)$ **(b)** $5(3x + 1)$ **(c)** $2(y - 1)$ **(d)** $6(m^2 + 2m + 3)$
(e) $8(4x^3 - x^2 + 2x + 3)$ **(f)** $y(4xy + 7)$ **(g)** $3x^2w^2(17x - 2w)$
(h) $5y(3x^3 - x^2 + 2x + 5)$ **(i)** $7p^5(7 - 2p)$ **(j)** $25w^4(5w^3 - 9v)$

(7-14) **(a)** $(x + 1)(x + 3)$ **(b)** $(y - 2)(y - 7)$ **(c)** $(w + 8)(w - 3)$ **(d)** $(p + 5)(p + 12)$
(e) $(q + 1)(q + 1)$ or $(q + 1)^2$ **(f)** $(r + 1)(r - 3)$ **(g)** irreducible over the integers
(h) $(x + 1)(x - 6)$ **(i)** $(x^2 + 3)(x^2 + 4)$ **(j)** $(w^2 - 2)(w^2 - 17)$
(k) $(u^2 - 2)(u^2 + 3)$ **(l)** $(z^2 + 3)(z^2 - 7)$

(7-15) **(a)** $(3x + 1)(2x + 1)$ **(b)** $(5w + 7)(2w + 5)$ **(c)** $(3z + 2)(z - 7)$
(d) $(3y + 5)(2y - 1)$ **(e)** $(8p + 1)(p + 1)$ **(f)** $(11q - 2)(3q + 5)$
(g) $(2t^2 + 3)(t^2 + 7)$ **(h)** $(5m - 2)(2m + 5)$ **(i)** $(15u + 1)(u + 6)$
(j) $(5x + 3y)(2x + 3y)$ **(k)** $(6x^2 + 1)(x^2 + 4)$ **(l)** $(15y + 2)(3y + 5)$
(m) $(4k + 3j)(2k - 5j)$ **(n)** $(10q + 7r)(3q - 5r)$ **(o)** $(6t + 1)(t + 6)$

(7-16) **(a)** $(x + 3)(x - 3)$ **(b)** $(y + 1)(y - 1)$ **(c)** $(w + 8)(w - 8)$ **(d)** $(z + 11)(z - 11)$
(e) $(2x + 3)(2x - 3)$ **(f)** $(3y + 1)(3y - 1)$ **(g)** $(5u + 13)(5u - 13)$
(h) $(10x + 3y)(10x - 3y)$ **(i)** $(9z + 4v)(9z - 4v)$ **(j)** $(3m + 7n)(3m - 7n)$
(k) $(p^2 + 2)(p^2 - 2)$ **(l)** $(4q^2 + 3r)(4q^2 - 3r)$ **(m)** $(30w + 11)(30w - 11)$
(n) $(x^4 + 5)(x^4 - 5)$ **(o)** $(p^5 + 1)(p^5 - 1)$

(7-17) **(a)** $(w + 3)^2$ **(b)** $(m + 7)^2$ **(c)** $(n - 4)^2$ **(d)** $(x - 12)^2$ **(e)** $(3p + 5)^2$
(f) $(5t + 7)^2$ **(g)** $(2q - 3)^2$ **(h)** $(u^2 + 5)^2$ **(i)** $(m + 9)^2$ **(j)** $(n - 9)^2$
(k) $(11r + 3)^2$ **(l)** $(5p + 7q)^2$ **(m)** $(b - 3c)^2$ **(n)** $(8u + 5v)^2$ **(o)** $(x^2 + y^2)^2$

(7-18) **(a)** $(y + 1)(y^2 - y + 1)$ **(b)** $(x - 1)(x^2 + x + 1)$ **(c)** $(z - 4)(z^2 + 4z + 16)$
(d) $(p + 5)(p^2 - 5p + 25)$ **(e)** $(2t + 1)(4t^2 - 2t + 1)$
(f) $(6v + 5w)(36v^2 - 30vw + 25w^2)$ **(g)** $(2w - 9x)(4w^2 + 18wx + 81x^2)$
(h) $(x^2 + 1)(x^4 - x^2 + 1)$ **(i)** $(x^2 - 2)(x^4 + 2x^2 + 4)$ **(j)** $(x^2 + y^2)(x^4 - x^2y^2 + y^4)$
(k) $(m^5 - 3)(m^{10} + 3m^5 + 9)$ **(l)** $(1 - v)(1 + v + v^2)$
(m) $(11k - 3)(121k^2 + 33k + 9)$ **(n)** $(8p - 1)(64p^2 + 8p + 1)$
(o) $(9q^2 + 2)(81q^4 - 18q^2 + 4)$

(7-19) **(a)** $(y + 5)(y^2 + 2)$ **(b)** $(3x + y)(x + y^2)$ **(c)** $(2a + 5c)(b + 2d)$
(d) $(3x + 5y)(w + 3u)$ **(e)** $(3a + m)(b - 2y)$ **(f)** $(x + 3)(x^2 - 3)$
(g) $(a - 5)(a^2 - 6)$ **(h)** $(2x + 3y + 5)(2x + 3y - 5)$ **(i)** $(3x - 5y + 7)(3x - 5y - 7)$
(j) $(m^2 + mn + n^2)(m - n + 1)$ **(k)** $(a^2 - ab + b^2)(a + b + 1)$
(l) $(3m - n + 5)(3m - n - 5)$ **(m)** $(7 + 2p + 5q)(7 - 2p - 5q)$
(n) $(8 + 3t - s)(8 - 3t + s)$ **(o)** $(x + y + z)(x - y - z)$

(7-20) **(a)** $(2x + 3)(3x + 1)$ **(b)** $(4y + 3)(y + 2)$ **(c)** $(3w + 5)(-2w + 3)$
(d) $(3a - 5)(2a - 1)$ **(e)** $(5b + 8)(3b - 1)$ **(f)** $(5x + 3y)(2x - 5y)$
(g) $(5m + 4)(4m + 5)$ **(h)** $(3m - 2)(2m - 7)$ **(i)** $(8a - 3)(a + 5)$
(j) $(8a - 3)(a - 5)$ **(k)** $(7b - 1)(2b + 3)$ **(l)** $(x - 1)(2x + 5)$
(m) $(8a - 5)(a + 1)$ **(n)** $(20a + 1)(a - 4)$ **(o)** $(4p - 3q)(3p + 4q)$

(7-21) **(a)** $2(a + b)(a - b)$ **(b)** $3x(x + 5)^2$ **(c)** $3cd(3c + 2d)(c + d)$
(d) $5(a - 2b)(a^2 + 2ab + 4b^2)$ **(e)** $2(3x + y)(9x^2 - 3xy + y^2)$
(f) $(x^2 + 1)(x + 1)(x - 1)$ **(g)** $(y^4 + 1)(y^2 + 1)(y + 1)(y - 1)$ **(h)** $m(2m + 7n)(m - n)$
(i) $(3k + 5)(a + b)^2$ **(j)** $x^3y^3(1 + xy)(1 - xy + x^2y^2)$ **(k)** $m^3(10m - 3)(m - 5)$
(l) $a^3(1 - b)(1 + b + b^2)$ **(m)** $x^2(x^2 + 1)(x + 3)(x - 3)$
(n) $(x + 1)(x - 1)(2a + 1)(2a - 1)$ **(o)** $(3x + 1)(3x - 1)(x + y)(x - y)$
(p) $x(x^n - 5)^2$ **(q)** $y^n(3y + 5)(2y - 3)$ **(r)** $(a - 1)^2(a^2 + a + 1)$

(7-22) **(a)** 3 sides [triangle]   **(b)** not possible for a figure to have 1 diagonal
(c) 4 sides [quadrilateral]   **(d)** not possible for a figure to have 3 diagonals
(e) not possible for a figure to have 4 diagonals   **(f)** 5 sides [pentagon]
(g) 6 sides [hexagon]   **(h)** not possible for a figure to have 10 diagonals
(i) 7 sides [septagon]   **(j)** 8 sides [octagon]   **(k)** 10 sides [decagon]
(l) 100 sides [centagon]   **(m)** 2   **(n)** 3   **(o)** 5   **(p)** 8   **(q)** 20   **(r)** 3
(s) 5   **(t)** 6   **(u)** 8   **(v)** 10   **(w)** 20   **(x)** 50   **(y)** 100   **(z)** 1000

RATIONAL EXPRESSIONS AND EQUATIONS

8-1. Renaming Rational Expressions in Equivalent Forms

A fraction that has a polynomial for both its numerator and its denominator is called a **rational expression.** Some examples of rational expressions are

$$\frac{2}{3} \qquad \frac{y}{y-1} \qquad \frac{x^2 - 2xy + y^2}{x^2 - y^2} \qquad \frac{w^2 - 7w + 12}{w^2 + 2w - 15}$$

Every rational expression represents a quotient of two polynomials. Since division by 0 is not allowed, the denominator of any rational expression cannot equal zero. Every numerical replacement of the variable(s) that causes the denominator of a rational expression to equal zero is called an **excluded value of the rational expression.**

EXAMPLE 8-1: Find the excluded value(s) of: **(a)** $\dfrac{3y+1}{2y+5}$ **(b)** $\dfrac{5x}{x-y}$ **(c)** $\dfrac{4x}{x^2-9}$

Solution: Set the denominator equal to 0 and solve the resulting equation.

(a) $2y + 5 = 0$

$\qquad 2y = -5$

$\qquad y = -\dfrac{5}{2}$ ⟵ the excluded value of $\dfrac{3y+1}{2y+5}$

(b) $x - y = 0$

$\qquad x = y$ ⟵ there are an infinite number of excluded values because the denominator will equal 0 whenever $x = y$

(c) $\qquad x^2 - 9 = 0$

$\qquad (x + 3)(x - 3) = 0$ Factor.

$\qquad x + 3 = 0 \text{ or } x - 3 = 0$ Use the zero product property.

$\qquad x = -3 \text{ or } \qquad x = 3$ ⟵ the excluded values of $\dfrac{4x}{x^2-9}$

Some rational expressions do not have any excluded values. For example, $3x/(x^2 + 1)$ does not have an excluded value because the denominator $x^2 + 1$ is never equal to 0.

When we write a rational expression, it is understood that the value(s) of the variable(s) that make the denominator equal to 0 are excluded. Sometimes the necessary restriction(s) are listed to the right of the rational expression:

$$\underbrace{\frac{2}{3x - 4}}_{\text{rational expression}}, \qquad x \neq \frac{4}{3} \longleftarrow \text{restriction}$$

We can rename rational expressions in equivalent forms by using the following rule.

The Fundamental Rule of Rational Expressions:

if p, q, and r are polynomials, then $\dfrac{pr}{qr} = \dfrac{p}{q}$, provided $q \neq 0$ and $r \neq 0$.

This rule states that if a nonzero polynomial is a common factor in both the numerator and the denominator of a rational expression, then the rational expression is equal to the rational expression formed by eliminating the common polynomial factor.

A rational expression is in **lowest terms** if the numerator and the denominator have no common factor other than 1 or -1.

To reduce a rational expression to lowest terms:

1. Factor the numerator and the denominator completely.
2. Use the Fundamental Rule of Rational Expressions to eliminate all nonzero factors common to both the numerator and the denominator.

EXAMPLE 8-2: Reduce $\dfrac{x^2 - 2x - 15}{2x^2 + 20x + 42}$ to lowest terms.

Solution: $\dfrac{x^2 - 2x - 15}{2x^2 + 20x + 42} = \dfrac{(x + 3)(x - 5)}{2(x + 3)(x + 7)}$ Factor.

$$= \dfrac{\cancel{(x + 3)}(x - 5)}{2\cancel{(x + 3)}(x + 7)}$$ Eliminate the common factor(s).

$$= \dfrac{x - 5}{2(x + 7)} \text{ or } \dfrac{x - 5}{2x + 14}$$

Because the expressions $x - y$ and $y - x$ are opposites of each other, their quotient is -1. This result can sometimes be used to reduce rational expressions.

EXAMPLE 8-3: Reduce $\dfrac{a^2 - b^2}{3a + 3b}$ to lowest terms.

Solution: $\dfrac{a^2 - b^2}{3b - 3a} = \dfrac{(a + b)(a - b)}{3(b - a)}$

$$= -\dfrac{a + b}{3}$$ Use $\dfrac{(a - b)}{(b - a)} = -1$.

Note: Since $-\dfrac{p}{q} = \dfrac{-p}{q} = \dfrac{p}{-q}$, the answer above, $-\dfrac{a + b}{3}$, may be written in the following equivalent

forms: $-\dfrac{a + b}{3} = \dfrac{-(a + b)}{3} = \dfrac{-a - b}{3} = \dfrac{a + b}{-3}$.

8-2. Operations on Rational Expressions

A. Multiplying rational expressions

To multiply two rational expressions:

1. Factor each numerator and each denominator.
2. Eliminate all common nonzero factors which occur in both the numerator and the denominator of the rational expressions.
3. Write the indicated product of all remaining factors in the numerators over the indicated product of all remaining factors in the denominators.

EXAMPLE 8-4: Find $\dfrac{7x - 14}{x + 3} \cdot \dfrac{x^2 - x - 12}{x^2 - 7x + 10}$.

Solution: $\dfrac{7x - 14}{x + 3} \cdot \dfrac{x^2 - x - 12}{x^2 - 7x + 10} = \dfrac{7(x - 2)}{x + 3} \cdot \dfrac{(x + 3)(x - 4)}{(x - 2)(x - 5)}$ Factor.

$$= \frac{7(x - 2)}{(x + 3)} \cdot \frac{(x + 3)(x - 4)}{(x - 2)(x - 5)}$$ Eliminate the common factors.

$$= \frac{7(x - 4)}{x - 5}$$

Note: It is common practice not to multiply the factors which remain. Thus the answer $\dfrac{7(x - 4)}{x - 5}$ is generally preferred over the equivalent form $\dfrac{7x - 28}{x - 5}$.

B. Dividing rational expressions

The **reciprocal of a rational expression,** like any fraction, is the rational expression with the numerator and the denominator interchanged. For example, the reciprocal of $(3x^2 + 4)/(x^2 - 9)$ is $(x^2 - 9)/(3x^2 + 4)$. To divide two rational expressions, we multiply by the reciprocal of the divisor.

EXAMPLE 8-5: Find $\dfrac{3a}{a + 3b} \div \dfrac{6ab}{5a + 15b}$.

Solution: $\dfrac{3a}{a + 3b} \div \dfrac{6ab}{5a + 15b} = \dfrac{3a}{a + 3b} \cdot \dfrac{5a + 15b}{6ab}$ Multiply by the reciprocal of the divisor.

$$= \frac{3a}{a + 3b} \cdot \frac{5(a + 3b)}{2 \cdot 3ab}$$ Factor.

$$= \frac{3a}{a + 3b} \cdot \frac{5(a + 3b)}{2 \cdot 3ab}$$ Reduce.

$$= \frac{5}{2b}$$

C. Combining (adding and subtracting) rational expressions

To combine (add or subtract) two rational expressions with common denominators, we add or subtract their numerators and place the result over the common denominator. It may be possible to reduce the result further.

EXAMPLE 8-6: Find **(a)** $\dfrac{3a - 5}{2a^2 - 7a - 15} + \dfrac{a + 11}{2a^2 - 7a - 15}$ and **(b)** $\dfrac{4x + 1}{x^2 - 9} - \dfrac{3x - 2}{x^2 - 9}$.

Solution:

(a) $\dfrac{3a - 5}{2a^2 - 7a - 15} + \dfrac{a + 11}{2a^2 - 7a - 15} = \dfrac{3a - 5 + a + 11}{2a^2 - 7a - 15}$
    Add the numerators and place the result over the common denominator.

$\qquad\qquad = \dfrac{4a + 6}{2a^2 - 7a - 15}$
    Simplify.

$\qquad\qquad = \dfrac{2(\cancel{2a + 3})}{(\cancel{2a + 3})(a - 5)}$
    Reduce (if possible).

$\qquad\qquad = \dfrac{2}{a - 5}$

(b) $\dfrac{4x + 1}{x^2 - 9} - \dfrac{3x - 2}{x^2 - 9} = \dfrac{(4x + 1) - (3x - 2)}{x^2 - 9}$
    Subtract the numerators and place the result over the common denominator.

$\qquad\qquad = \dfrac{4x + 1 - 3x + 2}{x^2 - 9}$

$\qquad\qquad = \dfrac{(\cancel{x + 3})}{(\cancel{x + 3})(x - 3)}$
    Reduce.

$\qquad\qquad = \dfrac{1}{x - 3}$

It is important to observe the use of parentheses in the above subtraction. A common mistake is to write

$$\dfrac{4x + 1}{x^2 - 9} - \dfrac{3x - 2}{x^2 - 9} \quad \text{as} \quad \dfrac{4x + 1 - 3x \overset{\text{error}}{-} 2}{x^2 - 9}$$

The entire numerator, $3x - 2$, is to be subtracted. Using parentheses around the numerator to be subtracted prevents this mistake.

To add or subtract rational expressions which have different denominators, you must first express each rational expression as an equivalent rational expression, such that they all have a common denominator. As with rational numbers, you could use any common denominator; however it will generally be easier to use the **least common denominator (LCD) of the rational expressions.**

To find the LCD of two or more rational expressions:

1. Factor each denominator completely. Express repeated factors in exponential form.
2. Identify the largest power of each factor.
3. The LCD is the product of the powers identified in step 2.

EXAMPLE 8-7: Find the LCD of $\dfrac{5y}{9y^2 - 36}$ and $\dfrac{-2y + 3}{3y^2 + 12y + 12}$.

Solution:

$9y^2 - 36 = 9(y^2 - 4) \qquad$ and $\quad 3y^2 + 12y + 12 = 3(y^2 + 4y + 4) \qquad$ Factor each denominator.

$\qquad\quad = 3^2(y + 2)(y - 2) \qquad\qquad\qquad\qquad\quad = 3(y + 2)^2$

In the above factorizations: 3^2 is the largest power of 3,
$\qquad\qquad\qquad\qquad\qquad\quad (y + 2)^2$ is the largest power of $y + 2$, and
$\qquad\qquad\qquad\qquad\qquad\quad (y - 2)^1$ is the largest power of $y - 2$.

The LCD of $\dfrac{5y}{9y^2 - 36}$ and $\dfrac{-2y + 3}{3y^2 + 12y + 12}$ is $3^2(y + 2)^2(y - 2)$.

Note: The powers 3^2 and $(y + 2)^2$ could be written as 9 and $y^2 + 4y + 4$. However, it is generally preferred to leave them in their exponential form.

To rename rational expressions in terms of the LCD of the rational expressions, you multiply the numerator and denominator of each rational expression by those factor(s) whose product with the denominator results in the LCD.

EXAMPLE 8-8: Rename the rational expressions $\dfrac{x + 5}{x^2 + 3x}$ and $\dfrac{4x}{2x + 6}$ in terms of the LCD of the rational expressions.

Solution: The LCD of $\dfrac{x + 5}{x^2 + 3x}$ and $\dfrac{4x}{2x + 6}$ is $2x(x + 3)$.

$$\frac{x + 5}{x^2 + 3x} = \frac{(x + 5)}{x(x + 3)} \cdot \frac{2}{2} = \frac{2x + 10}{2x(x + 3)}$$

The factor whose product with $x^2 + 3x$ will result in the LCD is 2. Multiply by 1 in the form of $\dfrac{2}{2}$.

$$\frac{4x}{2x + 6} = \frac{4x}{2(x + 3)} \cdot \frac{x}{x} = \frac{4x^2}{2x(x + 3)}$$

The factor whose product with $2x + 6$ will result in the LCD is x. Multiply by 1 in the form of $\dfrac{x}{x}$.

To combine rational expressions which have different denominators:

1. Determine the LCD of the rational expressions.
2. Rename each rational expression as an equivalent rational expression whose denominator is the LCD from step 1.
3. Add or subtract the rational expressions developed in step 2, using the procedure for adding or subtracting rational expressions with a common denominator.
4. Reduce, if possible.

EXAMPLE 8-9: Find **(a)** $\dfrac{2w}{3w + 4} + \dfrac{5}{w - 3}$ and **(b)** $\dfrac{z}{z^2 - 25} - \dfrac{2}{z^2 - z - 20}$.

Solution: **(a)** The LCD of $\dfrac{2w}{3w + 4}$ and $\dfrac{5}{w - 3}$ is $(3w + 4)(w - 3)$.

$$\frac{2w}{3w + 4} + \frac{5}{w - 3} = \frac{2w \cdot (w - 3)}{(3w + 4)(w - 3)} + \frac{5 \cdot (3w + 4)}{(w - 3)(3w + 4)}$$

$$= \frac{2w(w - 3) + 5(3w + 4)}{(3w + 4)(w - 3)}$$

$$= \frac{2w^2 - 6w + 15w + 20}{(3w + 4)(w - 3)}$$

$$= \frac{2w^2 + 9w + 20}{(3w + 4)(w - 3)}$$

(b) The LCD of $\dfrac{z}{z^2 - 25}$ and $\dfrac{2}{z^2 - z - 20}$ is $(z + 4)(z - 5)(z + 5)$.

$$\frac{z}{z^2 - 25} - \frac{2}{z^2 - z - 20} = \frac{z}{(z + 5)(z - 5)} - \frac{2}{(z + 4)(z - 5)}$$

$$= \frac{z \cdot (z + 4)}{(z + 5)(z - 5)(z + 4)} - \frac{2 \cdot (z + 5)}{(z + 4)(z - 5)(z + 5)}$$

$$= \frac{z^2 + 4z}{(z + 5)(z - 5)(z + 4)} - \frac{2z + 10}{(z + 4)(z - 5)(z + 5)}$$

$$= \frac{(z^2 + 4z) - (2z + 10)}{(z + 4)(z - 5)(z + 5)}$$

$$= \frac{z^2 + 2z - 10}{(z + 4)(z - 5)(z + 5)}$$

8-3. Simplifying Complex Fractions

A **complex fraction** is a fraction that has a fraction in its numerator, its denominator, or both. Some examples of complex fractions are

$$\frac{\dfrac{3}{4}}{\dfrac{7}{11}} \qquad \frac{3 - \dfrac{2}{x}}{4x + 1} \qquad \frac{\dfrac{x+y}{x-y} - 5}{\dfrac{3x+4y}{x-2y}}$$

If a complex fracton is written using bar notation, the longest fraction bar is called the **main fraction bar.** The main fraction bar separates the complex fraction into two parts. The part above the main fraction bar is called the **numerator of the complex fraction,** and the part below the main fraction bar is called the **denominator of the complex fraction.**

To simplify a complex fraction:

1. Simplify both the numerator and the denominator of the complex fraction to single fractions.
2. Divide the single fraction in the numerator by the single fraction in the denominator.
3. Reduce, if possible.

EXAMPLE 8-10: Simplify $\dfrac{\dfrac{1}{x} - x}{1 + \dfrac{1}{x}}$.

Solution: $\dfrac{\dfrac{1}{x} - x}{1 + \dfrac{1}{x}} = \dfrac{\dfrac{1}{x} - \dfrac{x \cdot x}{1 \cdot x}}{\dfrac{x}{x} + \dfrac{1}{x}}$ Simplify both the numerator and the denominator of the complex fraction.

$$= \frac{\dfrac{1 - x^2}{x}}{\dfrac{x + 1}{x}}$$

$$= \frac{1 - x^2}{x} \div \frac{x + 1}{x}$$

$$= \frac{1 - x^2}{x} \cdot \frac{x}{x + 1} \qquad \text{Multiply by the reciprocal of the divisor.}$$

$$= \frac{1 - x^2}{x + 1}$$

$$= \frac{(1 + x)(1 - x)}{x + 1}$$

$$= 1 - x$$

Complex fractions that do not contain another complex fraction in either their numerator or denominator can often be simplified by the **LCD method.**

To simplify a complex fraction by the LCD method:

1. Find the LCD of all the fractions in the numerator and the denominator of the complex fraction.
2. Multiply both the numerator and the denominator of the complex fraction by the LCD from step 1.
3. Reduce, if possible.

EXAMPLE 8-11: Use the LCD method to simplify $\dfrac{\frac{1}{a} - \frac{1}{b}}{\frac{a}{b} + \frac{b}{a}}$.

Solution: The LCD of $\dfrac{1}{a}$, $-\dfrac{1}{b}$, $\dfrac{a}{b}$, and $\dfrac{b}{a}$ is ab.

$$\frac{\frac{1}{a} - \frac{1}{b}}{\frac{a}{b} + \frac{b}{a}} = \frac{\frac{1}{a} - \frac{1}{b}}{\frac{a}{b} + \frac{b}{a}} \cdot \frac{ab}{ab}$$

$$= \frac{\frac{1}{a} \cdot ab - \frac{1}{b} \cdot ab}{\frac{a}{b} \cdot ab + \frac{b}{a} \cdot ab}$$

$$= \frac{b - a}{a^2 + b^2}$$

8-4. Solving Rational Equations

In Section 2-2, you solved fractional equations containing constants in the denominators. You cleared the equation of fractions by multiplying both members of the equation by the LCD of all the denominators. If an equation contains rational expressions, you may be able to solve the equation by clearing fractions, provided that you restrict the variable(s) from any excluded value(s).

To solve an equation containing rational expressions:

1. Determine the excluded value(s).
2. Multiply both members of the equation by the LCD of all the denominators.
3. Solve the resulting equation. The solution(s) of this equation are called proposed solution(s).
4. Compare the proposed solution(s) with the excluded value(s). Reject any proposed solution that is also an excluded value, and check any proposed solution that is not an excluded value in the original equation.

EXAMPLE 8-12: Solve **(a)** $\dfrac{2}{x - 4} - \dfrac{3}{x + 1} = 0$ and **(b)** $\dfrac{5x}{x - 1} + 4 = \dfrac{2x + 3}{x - 1}$.

Solution:

(a) The excluded values of $\dfrac{2}{x - 4} - \dfrac{3}{x + 1} = 0$ are 4 and -1.

$$(x - 4)(x + 1)\left[\frac{2}{x - 4} - \frac{3}{x + 1}\right] = (x - 4)(x + 6)0 \qquad \text{Multiply both members by the LCD: } (x - 4)(x + 6).$$

$$(x - 4)(x + 1)\frac{2}{x - 4} - (x - 4)(x + 1)\frac{3}{x + 1} = (x - 4)(x + 6)0$$

$$\cancel{(x - 4)}(x + 1)\frac{2}{\cancel{x - 4}} - (x - 4)\cancel{(x + 1)}\frac{3}{\cancel{x + 1}} = 0 \qquad \text{Simplify.}$$

$$2x + 2 - 3x + 12 = 0$$

$$-x + 14 = 0$$

$$-x = -14$$

$$x = 14 \ \longleftarrow \ \text{a proposed solution}$$

Check: The proposed solution is not equal to either of the excluded values, so it must be checked in the original equation.

$$\frac{2}{x-4} - \frac{3}{x+1} = 0$$

$$\frac{2}{14-4} - \frac{3}{14+1} \mid 0$$

$$\frac{2}{10} - \frac{3}{15} \mid 0$$

$$\frac{1}{5} - \frac{1}{5} \mid 0$$

$$0 \mid 0 \longleftarrow 14 \text{ checks}$$

(b) The excluded value of $\dfrac{5x}{x-1} + 4 = \dfrac{2x+3}{x-1}$ is 1.

$$(x-1)\left[\frac{5x}{x-1} + 4\right] = (x-1)\frac{(2x+3)}{(x-1)}$$

$$\cancel{(x-1)}\frac{5x}{\cancel{x-1}} + (x-1)4 = \cancel{(x-1)}\frac{(2x+3)}{\cancel{(x-1)}}$$

$$5x + 4x - 4 = 2x + 3$$

$$7x = 7$$

$$x = 1 \longleftarrow \text{a proposed solution}$$

1 is not a solution of the original equation because 1 is an excluded value. The original equation $\dfrac{5x}{x-1} + 4 = \dfrac{2x+3}{x-1}$ has no solution.

8-5. Solving Rational Formulas

A formula containing only rational expressions is called a **rational formula.** To solve a rational formula for a given variable, you first clear fractions and then isolate the given variable in one member as shown in Section 2-5.

EXAMPLE 8-13: Solve the electricity formula $I = \dfrac{E}{R + \dfrac{r}{n}}$ for n.

Solution:

$$I = \frac{E}{R + \dfrac{r}{n}} \longleftarrow \text{given formula}$$

$$\left(R + \frac{r}{n}\right)I = \left(\cancel{R + \frac{r}{n}}\right) \cdot \frac{E}{\cancel{R + \dfrac{r}{n}}} \qquad \text{Clear fractions.}$$

$$RI + \frac{r}{n} \cdot I = E$$

$$n(RI) + \cancel{n} \cdot \frac{rI}{\cancel{n}} = n(E)$$

$$nRI + rI = nE \longleftarrow \text{fractions cleared}$$

$$nRI - nRI + rI = nE - nRI \qquad \text{Isolate the given variable } n$$
$$\qquad\qquad\qquad\qquad\qquad\qquad\qquad \text{in one member.}$$

$$rI = n(E - RI)$$

$$\text{solution} \longrightarrow \frac{rI}{E - RI} = n \longleftarrow n \text{ is isolated}$$

Caution: $\dfrac{rI}{E-RI} \neq \dfrac{rI}{E} - \dfrac{rI}{RI}$ $\left[\text{e.g., } 1 = \dfrac{2}{2} = \dfrac{2}{4-2} \neq \dfrac{2}{4} - \dfrac{2}{2} = \dfrac{1}{2} - 1 = -\dfrac{1}{2}\right].$

8-6. Solve Problems Using Rational Equations

Agreement: In this chapter the words "rational equation" will mean *a rational equation in one variable.*

To **solve a problem using a rational equation,** you:

1. *Read* the problem very carefully, several times.
2. *Identify* the unknowns.
3. *Decide* how to represent the unknowns using one variable.
4. *Make a table* to help represent the unknowns when appropriate.
5. *Translate* the problem to a rational equation.
6. *Solve* the rational equation.
7. *Interpret* the solution of the rational equation with respect to each represented unknown to find the proposed solutions of the original problem.
8. *Check* to see if the proposed solutions satisfy all the conditions of the original problem.

A. Solving number problems using rational equations

EXAMPLE 8-14: Solve this number problem using a rational equation.

1. *Read:*   The larger of two numbers is twice the smaller number. The sum of their reciprocals is 1. What are the numbers?

2. *Identify:*   The unknown numbers are $\left\{\begin{array}{l}\text{the smaller number}\\ \text{the larger number}\end{array}\right\}$.

3.   Let n = the smaller number

  then $2n$ = the larger number.

4. *Translate:* $\underbrace{\text{The sum of their reciprocals}}$ is 1.

$$\dfrac{1}{n} + \dfrac{1}{2n} \stackrel{?}{=} 1$$

5. *Solve:*   The LCD of $\dfrac{1}{n}, \dfrac{1}{2n},$ and $\dfrac{1}{1}$ is $2n$.

$$2n\left(\dfrac{1}{n} + \dfrac{1}{2n}\right) = 2n(1)$$

$$2n\left(\dfrac{1}{n}\right) + 2n\left(\dfrac{1}{2n}\right) = 2n(1)$$

$$2 + 1 = 2n$$

$$3 = 2n$$

$$\dfrac{3}{2} = n$$

6. *Interpret:* $n = \frac{3}{2}$ means the smaller number is $\frac{3}{2}$.

  $2n = 2(\frac{3}{2}) = 3$ means the larger number is 3.

7. *Check:*   Did you find two numbers? Yes: $\frac{3}{2}$ and 3.

  Is the larger of the two numbers twice the smaller number? Yes: $3 = 2(\frac{3}{2})$.

  Is the sum of their reciprocals 1? Yes: $(1/\frac{3}{2}) + \frac{1}{3} = \frac{2}{3} + \frac{1}{3} = 1$.

B. Solving work problems using rational equations

The **unit rate** for a person who completes a certain job (1 job) in n units of time (hours, days, etc.) is $1/n$ job per unit. The unit rate indicates the *fractional* part of a job that can be completed in *one* unit of the given time period.

EXAMPLE 8-15: Find the unit rate for a person who completes a certain job in:

(a) 4 hours **(b)** 4 days **(c)** 4 weeks

Solution: **(a)** The unit rate for 1 job in 4 hours is: $\dfrac{1 \text{ job}}{4 \text{ hours}} = \dfrac{1 \text{ job}}{4 \text{ hour}}$ or $\dfrac{1}{4}$ job per hour.

(b) The unit rate for 1 job in 4 days is: $\dfrac{1 \text{ job}}{4 \text{ days}} = \dfrac{1 \text{ job}}{4 \text{ day}}$ or $\dfrac{1}{4}$ job per day.

(c) The unit rate for 1 job in 4 weeks is: $\dfrac{1 \text{ job}}{4 \text{ weeks}} = \dfrac{1 \text{ job}}{4 \text{ week}}$ or $\dfrac{1}{4}$ job per week.

To find the amount of **work** completed with respect to the whole job, you evaluate the following **work formula:** work = (unit rate)(time period) or $w = rt$.

EXAMPLE 8-16: If a given person can complete a certain job in 8 hours, then how much of the job can that person complete in 6 hours?

Solution: The unit rate (r) for 1 job [the whole job] in 8 hours is: $r = \frac{1}{8}$ (job per hour).
The time period (t) spent towards completing the whole 8-hour job is: $t = 6$ (hours).
The amount of work (w) completed in 6 hours with respect to the whole 8-hour job is:

work = (unit rate)(time period) or $w = rt$

$$= \left(\frac{1}{8}\,\frac{\text{job}}{\text{hours}}\right)(6 \text{ hours}) \qquad\qquad = \left(\frac{1}{8}\right)(6)$$

$$= \frac{6}{8}\text{job} \qquad\qquad\qquad\qquad = \frac{6}{8}$$

$$= \frac{3}{4}\text{job} \qquad\qquad\qquad\qquad = \frac{3}{4} \text{ [of the whole 8-hour job]}$$

If a given person can complete a certain job in 8 hours, then in 6 hours that same person can complete $\frac{3}{4}$ of the same job.

EXAMPLE 8-17: Solve this work problem using a rational equation.

1. *Read:* Shari can paint a certain house in 3 hours. Shari and Gary working together can paint the same house in 2 hours. How long would it take Gary working alone to paint the house?

2. *Identify:* The unknown time period is the time needed for Gary to paint the house alone.

3. *Decide:* Let x = the time needed for Gary to paint the house alone.

4. *Make a table:*

	unit rate (r) [per hour]	time period (t) [in hours]	work ($w = rt$) [portion of whole job]	
Shari	$\dfrac{1}{3}$	2	$\dfrac{1}{3}(2)$ or $\dfrac{2}{3}$	[see Note 1.]
Gary	$\dfrac{1}{x}$	2	$\dfrac{1}{x}(2)$ or $\dfrac{2}{x}$	

5. *Translate:* Shari's work plus Gary's work equals the total work [see the following Note 2].

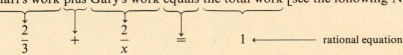

$$\frac{2}{3} \qquad + \qquad \frac{2}{x} \qquad = \qquad 1 \longleftarrow \text{rational equation}$$

6. *Solve:* The LCD of $\frac{2}{3}, \frac{2}{x}$, and $\frac{1}{1}$ is $3x$.

$$3x\left(\frac{2}{3} + \frac{2}{x}\right) = 3x(1)$$

$$\cancel{3}x\left(\frac{2}{\cancel{3}}\right) + 3\cancel{x}\left(\frac{2}{\cancel{x}}\right) = 3x(1)$$

$$2x + 6 = 3x$$

$$6 = x$$

7. *Interpret:* $x = 6$ means the time needed for Gary to paint the house alone is 6 hours.

8. *Check:* Does Shari's unit rate [per hour] plus Gary's unit rate equal their unit rate together? Yes: $\frac{1}{3} + \frac{1}{6} = \frac{2}{6} + \frac{1}{6} = \frac{3}{6} = \frac{1}{2}$.

Note 1: In Example 8-17, Shari's portion of the total work is known (2/3 of the whole job) while Gary's portion of the total work is unknown (2/x of the whole job).

Note 2: The **total work** is the whole completed job (painting the house) or: total work = 1 (complete job).

C. Solving uniform motion problems using rational equations

To represent an unknown time (t) given the distance (d) and the rate (r), you solve the distance formula ($d = rt$) for t to get the **time formula:** $t = d/r$.

To represent an unknown rate (r) given the distance (d) and the time (t), you solve the distance formula ($d = rt$) for r to get the **rate formula:** $r = d/t$.

EXAMPLE 8-18: Solve this uniform motion problem using a rational equation.

1. *Read:* It takes a boat the same amount of time to go 30 km up a certain river as it does to go 50 km down the river. The speed of the boat in still water (waterspeed) is 20 km/h. What is the rate of the current? What was the rate of the boat downriver? How long did it take to go upriver (or downriver)?

2. *Identify:* The unknown rates and times are
$\begin{cases} \text{the rate of the current} \\ \text{the rate of the boat upriver} \\ \text{the rate of the boat downriver} \\ \text{the time to go upriver} \\ \text{the time to return downriver} \end{cases}$.

3. *Decide:* Let $x =$ the rate of the current

then $20 - x =$ the rate of the boat upriver

and $20 + x =$ the rate of the boat downriver.

4. *Make a table:*

	distance (d) [in kilometers]	rate (r) [km/h]	time $t = d/r$ [in hours]
upriver	30	$20 - x$	$\dfrac{30}{20 - x}$
downriver	50	$20 + x$	$\dfrac{50}{20 + x}$

5. *Translate:* The time to go upriver equals the time to go downriver.

$$\frac{30}{20 - x} = \frac{50}{20 + x}$$

6. *Solve:* The LCD of $\dfrac{30}{20-x}$ and $\dfrac{50}{20+x}$ is $(20-x)(20+x)$.

$$(20-x)(20+x) \cdot \frac{30}{20-x} = (20-x)(20+x) \cdot \frac{50}{20+x}$$

$$(20+x)30 = (20-x)50$$

$$(20+x)3 = (20-x)5$$

$$60 + 3x = 100 - 5x$$

$$8x = 40$$

$$x = 5$$

7. *Interpret:* $x = 5$ means the rate of the current is 5 km/h.
$20 - x = 20 - 5 = 15$ means the rate of the boat upriver is 15 km/h.
$20 + x = 20 + 5 = 25$ means the rate of the boat downriver is 25 km/h.
$\dfrac{30}{20-x} = \dfrac{30}{15} = 2$ means the time to go upriver is 2 hours.

8. *Check:* Does it take 2 hours to go 30 km upriver or 50 km downriver? Yes:
$\dfrac{50}{20+x} = \dfrac{50}{25} = 2$ means the time to go downriver is also 2 hours.

RAISE YOUR GRADES

Can you . . . ?

- ☑ find the excluded value(s) of a rational expression
- ☑ use the Fundamental Rule of Rational Expressions to rename a rational expression in equivalent forms
- ☑ find the LCD of two or more rational expressions
- ☑ add, subtract, multiply, and divide with rational expressions
- ☑ simplify a complex fraction
- ☑ solve rational equations
- ☑ solve rational formulas
- ☑ solve problems using rational expressions
- ☑ solve a rational formula for a given variable
- ☑ solve a number problem using a rational equation
- ☑ solve a work problem using a rational equation
- ☑ solve a uniform motion problem using a rational equation

SOLVED PROBLEMS

PROBLEM 8-1 Find the excluded value(s): **(a)** $\dfrac{1}{2x-3}$ **(b)** $\dfrac{5y}{6-2y}$ **(c)** $\dfrac{7z}{z^2+6}$

Solution: Recall that to find the excluded values of a rational expression, you set the denominator equal to zero and solve for the variable(s) [see Example 8-1].

(a) $2x - 3 = 0$

$2x = 3$

$x = \dfrac{3}{2}$ ⟵ the excluded value of $\dfrac{1}{2x-3}$

(b) $6 - 2y = 0$

$$- 2y = -6$$

$$y = 3 \longleftarrow \text{the excluded value of } \frac{5y}{6 - 2y}$$

(c) $\dfrac{7z}{z^2 + 6}$ has no excluded values because $z^2 + 6$ is never equal to 0.

PROBLEM 8-2 Reduce (a) $\dfrac{3a^2 + 3a - 36}{2a^2 - a - 15}$ and (b) $\dfrac{2x^3 - 2y^3}{4x^2 - 8xy + 4y^2}$ to lowest terms.

Solution: Recall that to reduce a rational expression to lowest terms, you factor the numerator and the denominator and then use the Fundamental Rule of Rational Expressions to eliminate common factors [see Examples 8-2 and 8-3].

(a) $\dfrac{3a^2 + 3a - 36}{2a^2 - a - 15} = \dfrac{3(a - 3)(a + 4)}{(2a + 5)(a - 3)} = \dfrac{3\cancel{(a - 3)}(a + 4)}{(2a + 5)\cancel{(a - 3)}} = \dfrac{3(a + 4)}{2a + 5}$

(b) $\dfrac{2x^3 - 2y^3}{4x^2 - 8xy + 4y^2} = \dfrac{2\cancel{(x - y)}(x^2 + xy + y^2)}{2 \cdot 2\cancel{(x - y)}(x - y)} = \dfrac{x^2 + xy + y^2}{2(x - y)}$

PROBLEM 8-3 Find the indicated products:

(a) $\dfrac{m + 3}{2m - 10} \cdot \dfrac{m - 5}{m + 3}$ (b) $\dfrac{3x + 6}{2x + 8} \cdot \dfrac{x - 5}{2x + 4}$ (c) $\dfrac{u^2 - v^2}{u^2 + 2uv + v^2} \cdot \dfrac{3u + 3v}{u - v}$

(d) $\dfrac{a^2 - 2a - 15}{a^2 + 10a + 16} \cdot \dfrac{a^2 - 5a - 14}{a^2 + 7a + 12}$ (e) $\dfrac{x^2 - 9}{5 - 3x} \cdot \dfrac{3x - 5}{x + 3}$

Solution: Recall that to multiply two rational expressions, you factor each numerator and each denominator, eliminate all common nonzero factors that occur in both a numerator and a denominator, and write the indicated product of all remaining factors in the numerators over the indicated product of all remaining factors in the denominators [see Example 8-4].

(a) $\dfrac{m + 3}{2m - 10} \cdot \dfrac{m - 5}{m + 3} = \dfrac{\cancel{(m + 3)}}{2\cancel{(m - 5)}} \cdot \dfrac{\cancel{(m - 5)}}{\cancel{(m + 3)}} = \dfrac{1}{2}$

(b) $\dfrac{3x + 6}{2x + 8} \cdot \dfrac{x - 5}{2x + 4} = \dfrac{3\cancel{(x + 2)}}{2(x + 4)} \cdot \dfrac{(x - 5)}{2\cancel{(x + 2)}} = \dfrac{3(x - 5)}{4(x + 4)}$

(c) $\dfrac{u^2 - v^2}{u^2 + 2uv + v^2} \cdot \dfrac{3u + 3v}{u - v} = \dfrac{(u + v)\cancel{(u - v)}}{\cancel{(u + v)}(u + v)} \cdot \dfrac{3\cancel{(u + v)}}{\cancel{(u - v)}} = 3$

(d) $\dfrac{a^2 - 2a - 15}{a^2 + 10a + 16} \cdot \dfrac{a^2 - 5a - 14}{a^2 + 7a + 12} = \dfrac{(a - 5)\cancel{(a + 3)}}{\cancel{(a + 2)}(a + 8)} \cdot \dfrac{\cancel{(a + 2)}(a - 7)}{\cancel{(a + 3)}(a + 4)} = \dfrac{(a - 5)(a - 7)}{(a + 8)(a + 4)}$

(e) $\dfrac{x^2 - 9}{5 - 3x} \cdot \dfrac{3x - 5}{x + 3} = \dfrac{\cancel{(x + 3)}(x - 3)}{(5 - 3x)} \cdot \dfrac{(3x - 5)}{\cancel{(x + 3)}} = \dfrac{(x - 3)(3x - 5)}{(5 - 3x)} = -(x - 3)$

PROBLEM 8-4 Find the indicated quotients:

(a) $\dfrac{6ab}{6a + 3b} \div \dfrac{2a}{2a + b}$ (b) $\dfrac{x + 3}{x^2 - 4} \div \dfrac{x^2 + 7x + 12}{x^2 + 7x + 10}$ (c) $\dfrac{a^2 - b^2}{a + b} \div \dfrac{b - a}{a^2}$

(d) $\dfrac{6w^2 + w - 12}{6w^2 - w - 12} \div \dfrac{6w^2 + 19w + 15}{6w^2 - 19w + 15}$

Solution: Recall that to divide two rational expressions, you multiply by the reciprocal of the divisor [see Example 8-5].

(a) $\dfrac{6ab}{6a + 3b} \div \dfrac{2a}{2a + b} = \dfrac{6ab}{6a + 3b} \cdot \dfrac{2a + b}{2a}$

$$= \dfrac{2 \cdot 3ab}{3(2a + b)} \cdot \dfrac{(2a + b)}{2a}$$

$$= b$$

(b) $\dfrac{x + 3}{x^2 - 4} \div \dfrac{x^2 + 7x + 12}{x^2 + 7x + 10} = \dfrac{x + 3}{x^2 - 4} \cdot \dfrac{x^2 + 7x + 10}{x^2 + 7x + 12}$

$$= \dfrac{(x + 3)}{(x + 2)(x - 2)} \cdot \dfrac{(x + 2)(x + 5)}{(x + 3)(x + 4)}$$

$$= \dfrac{x + 5}{(x - 2)(x + 4)}$$

(c) $\dfrac{a^2 - b^2}{a + b} \div \dfrac{b - a}{a^2} = \dfrac{a^2 - b^2}{a + b} \cdot \dfrac{a^2}{(b - a)}$

$$= \dfrac{(a + b)(a - b)}{(a + b)} \cdot \dfrac{a^2}{(b - a)}$$

$$= a^2$$

(d) $\dfrac{6w^2 + w - 12}{6w^2 - w - 12} \div \dfrac{6w^2 + 19w + 15}{6w^2 - 19w + 15} = \dfrac{6w^2 + w - 12}{6w^2 - w - 12} \cdot \dfrac{6w^2 - 19w + 15}{6w^2 + 19w + 15}$

$$= \dfrac{(2w + 3)(3w - 4)}{(2w - 3)(3w + 4)} \cdot \dfrac{(2w - 3)(3w - 5)}{(2w + 3)(3w + 5)}$$

$$= \dfrac{(3w - 4)(3w - 5)}{(3w + 4)(3w + 5)}$$

PROBLEM 8-5 Find the indicated sums or differences:

(a) $\dfrac{3}{x} + \dfrac{4}{x}$ (b) $\dfrac{a}{b} - \dfrac{c}{b}$ (c) $\dfrac{5a + 1}{2a^2 + 7a + 9} + \dfrac{3a - 8}{2a^2 + 7a + 9}$ (d) $\dfrac{6w - 2}{w^2 + 4} - \dfrac{2w - 5}{w^2 + 4}$

Solution: Recall that to combine two rational expressions which have common denominators, you add or subtract their numerators and place the result over the common denominator. This result may be reduced further [see Example 8-6].

(a) $\dfrac{3}{x} + \dfrac{4}{x} = \dfrac{3 + 4}{x} = \dfrac{7}{x}$ (b) $\dfrac{a}{b} - \dfrac{c}{b} = \dfrac{a - c}{b}$

(c) $\dfrac{5a + 1}{2a^2 + 7a + 9} + \dfrac{3a - 8}{2a^2 + 7a + 9} = \dfrac{8a - 7}{2a^2 + 7a + 9}$

(d) $\dfrac{6w - 2}{w^2 + 4} - \dfrac{2w - 5}{w^2 + 4} = \dfrac{(6w - 2) - (2w - 5)}{w^2 + 4} = \dfrac{6w - 2 - 2w + 5}{w^2 + 4} = \dfrac{4w + 3}{w^2 + 4}$

PROBLEM 8-6 Find the LCD of the following pairs of rational expressions:

(a) $\dfrac{5}{x}, \dfrac{2}{y}$ (b) $\dfrac{3x}{4x^2 - 4y^2}, \dfrac{2y}{x^2 + 2xy + y^2}$ (c) $\dfrac{6a + 5b}{24(a - b)^3}, \dfrac{2a - 7b}{20(a - b)^2(a + b)}$

Solution: Recall that to find the LCD of two or more rational expressions, you:

1. Factor each denominator completely. Express repeated factors in exponential form.
2. Identify the largest power of each factor.
3. Take the LCD from the product of the powers identified in step 2 [see Example 8-7].

(a) The denominators of $\dfrac{5}{x}$ and $\dfrac{2}{y}$ do not have a common factor, so the LCD of $\dfrac{5}{x}$ and $\dfrac{2}{y}$ is xy.

(b) $4x^2 - 4y^2 = 4(x^2 - y^2)$ and $x^2 + 2xy + y^2 = (x + y)^2$
$$= 2^2(x + y)(x - y)$$

In the above factorizations: 2^2 is the largest power of 2,
$(x + y)^2$ is the largest power of $x + y$, and
$(x - y)^1$ is the largest power of $x - y$.

The LCD of $\dfrac{3x}{4x^2 - 4y^2}$ and $\dfrac{2y}{x^2 + 2xy + y^2}$ is $2^2(x + y)^2(x - y)$.

(c) $24(a - b)^3 = 2^3 \cdot 3(a - b)^3$ and $20(a - b)^2(a + b) = 2^2 \cdot 5(a - b)^2(a + b)$

In the above factorizations: 2^3 is the largest power of 2,
3^1 is the largest power of 3,
5^1 is the largest power of 5,
$(a - b)^3$ is the largest power of $a - b$, and
$(a + b)^1$ is the largest power of $a + b$.

The LCD of $\dfrac{6a + 5b}{24(a - b)^3}$ and $\dfrac{2a - 7b}{20(a - b)^2(a + b)}$ is $2^3 \cdot 3 \cdot 5(a - b)^3(a + b)$.

PROBLEM 8-7 Rename these rational expressions in terms of their LCD:

(a) $\dfrac{3}{a}, \dfrac{5}{b}$ **(b)** $\dfrac{4}{m}, \dfrac{5m}{m - 3}$ **(c)** $\dfrac{5x}{x^2 - y^2}, \dfrac{2x}{x^2 + 2xy + y^2}$

Solution: Recall that to rename rational expressions in terms of the LCD of the rational expressions, you multiply the numerator and the denominator of each rational expression by the factor(s) whose product with the denominator is the LCD [see Example 8-8].

(a) The LCD of $\dfrac{3}{2}$ and $\dfrac{5}{b}$ is ab.

$$\frac{3}{a} = \frac{3}{a} \cdot \frac{b}{b} = \frac{3b}{ab} \quad \text{and} \quad \frac{5}{b} = \frac{5}{b} \cdot \frac{a}{a} = \frac{5a}{ab}$$

(b) The LCD of $\dfrac{4}{m}$ and $\dfrac{5m}{m - 3}$ is $m(m - 3)$.

$$\frac{4}{m} = \frac{4}{m} \cdot \frac{(m - 3)}{(m - 3)} = \frac{4m - 12}{m(m - 3)} \quad \text{and} \quad \frac{5m}{m - 3} = \frac{5m}{(m - 3)} \cdot \frac{m}{m} = \frac{5m^2}{m(m - 3)}$$

(c) The LCD of $\dfrac{5x}{x^2 - y^2}$ and $\dfrac{2x}{x^2 + 2xy + y^2}$ is $(x + y)^2(x - y)$.

$$\frac{5x}{x^2 - y^2} = \frac{5x}{(x + y)(x - y)} = \frac{5x}{(x + y)(x - y)} \cdot \frac{(x + y)}{(x + y)} = \frac{5x^2 + 5xy}{(x + y)^2(x - y)}$$

$$\frac{2x}{x^2 + 2xy + y^2} = \frac{2x}{(x + y)^2} = \frac{2x}{(x + y)^2} \cdot \frac{(x - y)}{(x - y)} = \frac{2x^2 - 2xy}{(x + y)^2(x - y)}$$

PROBLEM 8-8 Find the indicated sums or differences:

(a) $\dfrac{3}{x} + \dfrac{5}{y}$ **(b)** $\dfrac{2m}{m + 5} - \dfrac{m}{m - 3}$ **(c)** $\dfrac{4x}{x^2 - 9} + \dfrac{5}{x^2 + 5x + 6}$

(d) $\dfrac{2z}{z^2 + 7z + 12} - \dfrac{3z}{z^2 - 2z - 15}$

Solution: Recall that to combine rational expressions that have different denominators you:
   1. Determine the LCD of the rational expressions.
   2. Rename each rational expression as an equivalent expression whose denominator is the LCD from step 1.

3. Add or subtract the rational expressions developed in step 2, using the procedure for adding or subtracting rational expressions with a common denominator.

4. Reduce if possible [see Example 8-9].

(a) The LCD of $\dfrac{3}{x}$ and $\dfrac{5}{y}$ is xy.

$$\frac{3}{x} + \frac{5}{y} = \frac{3y}{xy} + \frac{5x}{yx}$$

$$= \frac{3y + 5x}{xy}$$

(b) The LCD of $\dfrac{2m}{m + 5}$ and $\dfrac{m}{m - 3}$ is $(m + 5)(m - 3)$.

$$\frac{2m}{m + 5} - \frac{m}{m - 3} = \frac{2m(m - 3)}{(m + 5)(m - 3)} - \frac{m(m + 5)}{(m - 3)(m + 5)}$$

$$= \frac{2m(m - 3) - [m(m + 5)]}{(m + 5)(m - 3)}$$

$$= \frac{2m^2 - 6m - m^2 - 5m}{(m + 5)(m - 3)}$$

$$= \frac{m^2 - 11m}{(m + 5)(m - 3)} \quad \text{or} \quad \frac{m(m - 11)}{(m + 5)(m - 3)}$$

(c) The LCD of $\dfrac{4x}{x^2 - 9}$ and $\dfrac{5}{x^2 + 5x + 6}$ is $(x + 3)(x - 3)(x + 2)$.

$$\frac{4x}{x^2 - 9} + \frac{5}{x^2 + 5x + 6} = \frac{4x}{(x + 3)(x - 3)} + \frac{5}{(x + 2)(x + 3)}$$

$$= \frac{4x(x + 2)}{(x + 3)(x - 3)(x + 2)} + \frac{5(x - 3)}{(x + 2)(x + 3)(x - 3)}$$

$$= \frac{4x(x + 2) + 5(x - 3)}{(x + 3)(x - 3)(x + 2)}$$

$$= \frac{4x^2 + 8x + 5x - 15}{(x + 3)(x - 3)(x + 2)}$$

$$= \frac{4x^2 + 13x - 15}{(x + 3)(x - 3)(x + 2)}$$

(d) The LCD of $\dfrac{2z}{z^2 + 7z + 12}$ and $\dfrac{3z}{z^2 - 2z - 15}$ is $(z + 3)(z + 4)(z - 5)$.

$$\frac{2z}{z^2 + 7z + 12} - \frac{3z}{z^2 - 2z - 15} = \frac{2z}{(z + 3)(z + 4)} - \frac{3z}{(z + 3)(z - 5)}$$

$$= \frac{2z(z - 5)}{(z + 3)(z + 4)(z - 5)} - \frac{3z(z + 4)}{(z + 3)(z - 5)(z + 4)}$$

$$= \frac{2z(z - 5) - [3z(z + 4)]}{(z + 3)(z + 4)(z - 5)}$$

$$= \frac{2z^2 - 10z - 3z^2 - 12z}{(z + 3)(z + 4)(z - 5)}$$

$$= \frac{-z^2 - 22z}{(z + 3)(z + 4)(z - 5)} \quad \text{or} \quad \frac{-z(z + 22)}{(z + 3)(z + 4)(z - 5)}$$

PROBLEM 8-9 Simplify: **(a)** $\dfrac{\dfrac{a}{b}}{\dfrac{a}{c}}$ **(b)** $\dfrac{1 + \dfrac{1}{x}}{2 - \dfrac{1}{x}}$ **(c)** $\dfrac{\dfrac{a-b}{a+b} - 1}{\dfrac{2}{a+b}}$

Solution: Recall that to simplify a complex fraction, you:

1. Simplify both the numerator and the denominator of the complex fraction to single fractions.
2. Divide the single fraction in the numerator by the single fraction in the denominator.
3. Reduce, if possible [see Examples 8-10 and 8-11].

(a) $\dfrac{\dfrac{a}{b}}{\dfrac{a}{c}} = \dfrac{a}{b} \div \dfrac{a}{c} = \dfrac{a}{b} \cdot \dfrac{c}{a} = \dfrac{\cancel{a}}{b} \cdot \dfrac{c}{\cancel{a}} = \dfrac{c}{b}$

(b) $\dfrac{1 + \dfrac{1}{x}}{2 - \dfrac{1}{x}} = \dfrac{\dfrac{x}{x} + \dfrac{1}{x}}{\dfrac{2x}{x} - \dfrac{1}{x}} = \dfrac{\dfrac{x+1}{x}}{\dfrac{2x-1}{x}} = \dfrac{x+1}{x} \div \dfrac{2x-1}{x} = \dfrac{x+1}{\cancel{x}} \cdot \dfrac{\cancel{x}}{2x-1} = \dfrac{x+1}{2x-1}$

(c) $\dfrac{\dfrac{a-b}{a+b} - 1}{\dfrac{2}{a+b}} = \dfrac{\dfrac{a-b}{a+b} - \dfrac{a+b}{a+b}}{\dfrac{2}{a+b}} = \dfrac{\dfrac{-2b}{a+b}}{\dfrac{2}{a+b}} = \dfrac{-2b}{a+b} \div \dfrac{2}{a+b} = -b$

PROBLEM 8-10 Solve each rational formula for the indicated variable:

(a) Finance formula, $T = \dfrac{24I}{B(n+1)}$, for n **(b)** Optics formula, $\dfrac{1}{f} = \dfrac{1}{a} + \dfrac{1}{b}$, for a

Solution: Recall that to solve a rational formula for a given variable, you first clear fractions and then isolate the given variable in one member as shown in Section 2-6 [see Example 8-13].

(a) $T = \dfrac{24I}{B(n+1)}$ ⟵ given formula

$B(n+1) \cdot T = \cancel{B(n+1)} \cdot \dfrac{24I}{\cancel{B(n+1)}}$ Clear the fractions.

$BnT + BT = 24I$ ⟵ fractions cleared

$BnT = 24I - BT$ Isolate the given variable n in one member.

$n = \dfrac{24I - BT}{BT}$ or $\dfrac{24I}{BT} - 1$ ⟵ solution

(b) $\dfrac{1}{f} = \dfrac{1}{a} + \dfrac{1}{b}$ ⟵ given formula

$\cancel{f}ab \cdot \dfrac{1}{\cancel{f}} = f\cancel{a}b \cdot \dfrac{1}{\cancel{a}} + fa\cancel{b} \cdot \dfrac{1}{\cancel{b}}$ Clear fractions (the LCD is fab).

$ab = fb + fa$ ⟵ fractions cleared

$ab - fa = fb$ Isolate the given variable a in one member.

$a(b - f) = fb$

$a = \dfrac{fb}{b - f}$ ⟵ solution

PROBLEM 8-11 Solve each number problem using a rational equation:

(a) The denominator of a certain fraction is 1 less than 3 times the numerator. If 6 is added to both the numerator and denominator, the result equals $\frac{1}{2}$. What is the fraction?

(b) One number is 3 times another number. The product of their reciprocals equals the reciprocal of their sum. Find the numbers.

Solution: Recall that to solve a number problem using a rational equation, you:

1. *Read* the problem very carefully, several times.
2. *Identify* the unknown numbers.
3. *Decide* how to represent the unknown numbers using one variable.
4. *Translate* the problem to a rational equation using key words.
5. *Solve* the rational equation.
6. *Interpret* the solution of the rational equation with respect to each represented unknown number to find the proposed solutions of the original problem.
7. *Check* to see if the proposed solutions satisfy all the conditions of the original problem [see Example 8-14].

(a) *Identify:*   The unknown numbers are $\begin{cases} \text{the numerator of the fraction} \\ \text{the denominator of the fraction} \end{cases}$.

  Decide: Let x = the numerator of the fraction

     then $3x - 1$ = the denominator of the fraction,

     because $3x - 1$ is 1 less than 3 times x.

  Translate: If 6 is added to the numerator and denominator, the result equals $\frac{1}{2}$.

$$\frac{x + (6)}{3x - 1 + (6)} = \frac{1}{2}$$

  Solve: The LCD of $\dfrac{x + 6}{3x + 5}$ and $\dfrac{1}{2}$ is $2(3x + 5)$.

$$2(3x + 5) \cdot \frac{x + 6}{3x + 5} = 2(3x + 5) \cdot \frac{1}{2}$$

$$2(x + 6) = 3x + 5$$

$$2x + 12 = 3x + 5$$

$$7 = x$$

  Interpret: $x = 7$ means the numerator of the fraction is 7.
     $3x - 1 = 3(7) - 1 = 20$ means the denominator of the fraction is 20.
     The numerator is 7 and the denominator is 20 means the fraction is $\frac{7}{20}$.

  Check: Is the denominator 1 less than 3 times the numerator? Yes: $20 = 3(7) - 1$.
     If 6 is added to both the numerator and denominator of $\frac{7}{20}$, is the result equal to $\frac{1}{2}$? Yes: $\dfrac{7 + 6}{20 + 6} = \dfrac{13}{26} = \dfrac{1(13)}{2(13)} = \dfrac{1}{2}$.

(b) *Identify:*   The unknown numbers are $\begin{cases} \text{the smaller number} \\ \text{the larger number} \end{cases}$.

  Decide: Let n = the smaller number

     then $3n$ = the larger number.

  Translate: The product of their reciprocals equals the reciprocal of their sum.

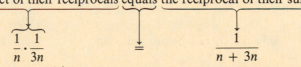

$$\frac{1}{n} \cdot \frac{1}{3n} = \frac{1}{n + 3n}$$

Solve: The LCD of $\dfrac{1}{3n^2}$ and $\dfrac{1}{4n}$ is $12n^2$

$$4(3n^2) \cdot \frac{1}{3n^2} = 3n(4n) \cdot \frac{1}{4n}$$

$$4 = 3n$$

$$\frac{4}{3} = n$$

Interpret: $n = \frac{4}{3}$ means the smaller number is $\frac{4}{3}$.
$3n = 3 \cdot \frac{4}{3} = 4$ means the larger number is 4.

Check: Is 4 three times $\frac{4}{3}$? Yes: $4 = 3(\frac{4}{3})$
Does the product of their reciprocals equal the reciprocal of their sum?
Yes: $\dfrac{1}{\frac{4}{3}} \cdot \dfrac{1}{4} = \dfrac{3}{4} \cdot \dfrac{1}{4} = \dfrac{3}{16}$ and $\dfrac{1}{4 + \frac{4}{3}} = \dfrac{1}{\frac{12}{3} + \frac{4}{3}} = \dfrac{1}{\frac{16}{3}} = \dfrac{3}{16}$.

PROBLEM 8-12 Find the unit rate for a person who completes a certain job in:

(a) 3 hours   **(b)** 5 days   **(c)** 2 weeks   **(d)** 10 minutes

Solution: Recall that the unit rate for a person who completes a certain job (1 job) in n hours is $1/n$ job per hour [see Example 8-15].

(a) The unit rate for 1 job in 3 hours is: $\dfrac{1 \text{ job}}{3 \text{ hours}} = \dfrac{1}{3}\dfrac{\text{job}}{\text{hour}}$ or $\dfrac{1}{3}$ job per hour.

(b) The unit rate for 1 job in 5 days is: $\dfrac{1 \text{ job}}{5 \text{ days}} = \dfrac{1}{5}\dfrac{\text{job}}{\text{day}}$ or $\dfrac{1}{5}$ job per day.

(c) The unit rate for 1 job in 2 weeks is: $\dfrac{1 \text{ job}}{2 \text{ weeks}} = \dfrac{1}{2}\dfrac{\text{job}}{\text{week}}$ or $\dfrac{1}{2}$ job per week.

(d) The unit rate for 1 job in 10 minutes is: $\dfrac{1 \text{ job}}{10 \text{ minutes}} = \dfrac{1}{10}\dfrac{\text{job}}{\text{minute}}$ or $\dfrac{1}{10}$ job per minute.

PROBLEM 8-13 If a person can complete a job in 6 days, then how much of the job can that person complete in 4 days?

Solution: Recall that to find the amount of work completed with respect to the whole job, you evaluate the work formula: work = (unit rate)(time period), or $w = rt$ [see Example 8-16].

The unit rate (r) for 1 job in 6 days is: $r = \frac{1}{6}$(job per day).
The time period (t) spent towards completing the whole 6-day job is: $t = 4$(days).
The amount of work (w) completed in 4 days with respect to the whole 6-day job is:

$$w = rt$$

$$= (\tfrac{1}{6})(4)$$

$$= \tfrac{4}{6}$$

$$= \tfrac{2}{3} \text{ (of the whole job)}$$

PROBLEM 8-14 Solve each work problem using a rational equation:

(a) Kristina can paint a certain room with a brush in 8 hours. Stephanie can paint the same room with a roller in 2 hours. How long will it take for them to paint the room together?

(b) An outlet pipe can empty a pool in 24 hours. An inlet pipe can fill the same pool in 16 hours. How long will it take to fill the pool using the inlet pipe if the outlet pipe is unintentionally left open?

Solution: Recall that to solve a work problem using a rational equation, you:

1. *Read* the problem very carefully, several times.
2. *Identify* the unknown time periods.
3. *Decide* how to represent the unknown time periods using one variable.
4. *Make a table* to help represent individual amounts of work using the formula work = (unit rate)(time period), or $w = rt$.
5. *Translate* the problem to a rational equation.
6. *Solve* the rational equation.
7. *Interpret* the solution of the rational equation with respect to each unknown time periods to find the proposed solutions of the original problem.
8. *Check* to see if the proposed solutions satisfy all the conditions of the original problem [see Example 8-17].

(a) *Identify:* The unknown time period is the time needed to paint the room together.

Decide: Let x = the time needed to paint the room together.

Make a table:

	unit rate (r) [per hour]	time period (t) [in hours]	work $[w = rt]$ [portion of whole job]
Kristina	$\frac{1}{8}$	x	$\frac{1}{8}x$ or $\frac{x}{8}$
Stephanie	$\frac{1}{2}$	x	$\frac{1}{2}x$ or $\frac{x}{2}$

Translate: Kristina's work plus Stephanie's work equals the total work.

$$\frac{x}{8} + \frac{x}{2} = 1$$

Solve: The LCD of $\frac{x}{8}, \frac{x}{2}$, and $\frac{1}{1}$ is 8.

$$8 \cdot \frac{x}{8} + 8\frac{x}{2} = 8(1)$$

$$x + 4x = 8$$

$$5x = 8$$

$$x = \frac{8}{5} \text{ or } 1\frac{3}{5}$$

Interpret: $x = 1\frac{3}{5}$ means Kristina and Stephanie can paint the room together in $1\frac{3}{5}$ hours.

Check: If Kristina works $\frac{8}{5}$ hours, then she does $\frac{1}{8} \cdot \frac{8}{5} = \frac{1}{5}$ of the job. Likewise, Stephanie does $\frac{1}{2} \cdot \frac{8}{5} = \frac{4}{5}$ of the job, and $\frac{1}{5} + \frac{4}{5} = 1$ (the entire job).

(b) *Identify:* The unknown time period is the time to fill the pool with both pipes open.

Decide: Let x = the time to fill the pool with both pipes open.

Make a table:

	unit rate (r) [per hour]	time period (t) [in hours]	work $[w = rt]$ [portion of whole job]
outlet	$\frac{1}{24}$	x	$\frac{1}{24}x = \frac{x}{24}$
inlet	$\frac{1}{16}$	x	$\frac{1}{16}x = \frac{x}{16}$

Translate: Water from inlet pipe minus water from outlet pipe equals total work.

$$\frac{x}{16} \quad - \quad \frac{x}{24} \quad = \quad 1$$

Solve: The LCD of $\frac{x}{16}$, $\frac{x}{24}$, and $\frac{1}{1}$ is 48.

$$3(16) \cdot \frac{x}{16} - 2(24) \cdot \frac{x}{24} = 48(1)$$

$$3x - 2x = 48$$

$$x = 48$$

Interpret: $x = 48$ means the pool can be filled with both pipes open in 48 hours.

Check: Does the unit rate [per hour] of the inlet pipe minus the unit rate of the outlet pipe equal their unit rate together?
Yes: $\frac{1}{16} - \frac{1}{24} = \frac{3}{48} - \frac{2}{48} = \frac{1}{48}$.

PROBLEM 8-15 Solve each uniform motion problem using a rational equation:

(a) One car is traveling at 60 mph. Another car is traveling at 55 mph. How long will it take the faster car to catch the lower car if the faster car is 12 miles behind the slower car?

(b) An airplane has a 7-hour fuel supply. If the pilot flies out as far as possible at 150 mph and then straight back again at 200 mph using all the fuel, how long did each part of the trip take?

Solution: Recall that to solve a uniform motion problem using a rational equation, you:

1. *Read* the problem very carefully, several times.
2. *Identify* the unknown distances, rates, and times.
3. *Decide* how to represent the unknown distances, rates, and or times using one variable.
4. *Make a table* to help represent the unknown rates (or times) using $r = d/t$ (or $t = d/r$).
5. *Translate* the problem to a rational equation.
6. *Solve* the rational equation.
7. *Interpret* the solution of the rational equation with respect to each unknown distance, rate, and time to find the proposed solutions of the original problem.
8. *Check* to see if the proposed solutions satisfy all the conditions of the original problem [see Example 8-18].

(a) *Identify:* The unknown time and distances are $\begin{cases} \text{the distance for the slower car} \\ \text{the distance for the faster car} \\ \text{the time for the faster car} \end{cases}$.

Decide: Let d = the distance for the slower car
then $d + 12$ = the distance for the faster car.

Make a table:

	distance (d) [in miles]	rate (r) [mph]	time ($t = d/r$) [in hours]
faster car	$d + 12$	60	$\frac{d + 12}{60}$
slower car	d	55	$\frac{d}{55}$

Translate: The time for slower car equals the time for the faster car.

$$\frac{d}{55} \quad = \quad \frac{d + 12}{60}$$

Solve: The LCD for $\dfrac{d}{55}$ and $\dfrac{d + 12}{60}$ is 660.

$$12(\cancel{55}) \cdot \frac{d}{\cancel{55}} = 11(\cancel{60}) \cdot \frac{d + 12}{\cancel{60}}$$

$$12(d) = 11(d + 12)$$

$$12d = 11d + 132$$

$$d = 132$$

Interpret: $d = 132$ means the distance for the slower car is 132 miles.

$d + 12 = 132 + 12 = 144$ means the distance for the faster car is 144 miles.

$\dfrac{d + 12}{60} = \dfrac{144}{60} = 2.4$ means the time for the faster car is 2.4 hours.

Check: Is the time for the slower car to travel 132 miles at 55 mph equal to the 2.4 hours it took the faster car to travel 144 miles at 60 mph? Yes:

$$t = \frac{d}{r} = \frac{132}{55} = 2.4 \text{ (hours)}.$$

(b) *Identify:* The unknown distance and times are $\left\{\begin{array}{l}\text{the distance out (or back)}\\ \text{the time to fly out}\\ \text{the time to fly back}\end{array}\right\}$.

Decide: Let $d =$ the distance out (or back).

Make a table:

	distance (d) [in miles]	rate (r) [mph]	time ($t = d/r$) [in hours]
out	d	150	$\dfrac{d}{150}$
back	d	200	$\dfrac{d}{200}$

Translate:

The time to fly out plus the time to fly back is 7 hours.

$$\frac{d}{150} \quad + \quad \frac{d}{200} \quad = \quad 7$$

Solve: The LCD for $\dfrac{d}{150}, \dfrac{d}{200},$ and $\dfrac{7}{1}$ is 600.

$$4(\cancel{150}) \cdot \frac{d}{\cancel{150}} + 3(\cancel{200}) \cdot \frac{d}{\cancel{200}} = 600(7)$$

$$4d + 3d = 4200$$

$$7d = 4200$$

$$d = 600$$

Interpret: $d = 600$ means the distance out (or back) is 600 miles.

$\dfrac{d}{150} = \dfrac{600}{150} = 4$ means the time to fly out is 4 hours.

$\dfrac{d}{200} = \dfrac{600}{200} = 3$ means the time to fly back is 3 hours.

Check: Is the time for the round trip 7 hours? Yes: $4 + 3 = 7$.

Supplementary Exercises

PROBLEM 8-16 Find the excluded value(s):

(a) $\dfrac{3}{x-5}$ **(b)** $\dfrac{5y}{3y-8}$ **(c)** $\dfrac{2a}{2a+7}$ **(d)** $\dfrac{x+4}{x}$

PROBLEM 8-17 Reduce each rational expression to lowest terms:

(a) $\dfrac{x-4}{4-x}$ **(b)** $\dfrac{x^2-y^2}{x^3-y^3}$ **(c)** $\dfrac{x^2-4x-21}{x^2-9x+14}$ **(d)** $\dfrac{4x^2+20x+25}{6x^2+x-35}$ **(e)** $\dfrac{18x^2+3x-3}{36x^2-9}$

(f) $\dfrac{a-b}{2b-2a}$

PROBLEM 8-18 Find the indicated products:

(a) $\dfrac{a+7}{a-2}\cdot\dfrac{a-2}{a+3}$ **(b)** $\dfrac{3}{y}\cdot\dfrac{y-5}{6}$ **(c)** $\dfrac{w^2-16}{2w+8}\cdot\dfrac{2}{w-4}$

(d) $\dfrac{x^2+10x+25}{3x^2+8x-35}\cdot\dfrac{9x^2-42x+49}{2x^2+9x-5}$ **(e)** $\dfrac{x+y}{x^2-y^2}\cdot x$ **(f)** $\dfrac{x-2y}{4x+5}\cdot\dfrac{4}{2y-x}$

PROBLEM 8-19 Find the indicated quotients:

     x **(b)** $\dfrac{a+3}{a-2}\div\dfrac{a+5}{a-2}$ **(c)** $\dfrac{x^2-4y^2}{3x+5}\div\dfrac{x+2y}{3x+5}$ **(d)** $\dfrac{m^3-n^3}{m^3+n^3}\div\dfrac{m^2+mn+n^2}{m^2-mn+n^2}$

(e) $\dfrac{18x^2+12x+2}{12x^2+36x+27}\div\dfrac{18x^2-2}{12x^2-27}$ **(f)** $\dfrac{2w^2+w-15}{w^2-4w-21}\div\dfrac{-2w+5}{w^2-6w-7}$

PROBLEM 8-20 Find the indicated sums or differences:

(a) $\dfrac{5}{y}+\dfrac{1}{y}$ **(b)** $\dfrac{m}{n}-\dfrac{p}{n}$ **(c)** $\dfrac{3a+1}{2a+5}+\dfrac{a+4}{2a+5}$ **(d)** $\dfrac{5p-q}{3p+q}-\dfrac{2p-2q}{3p+q}$

(e) $\dfrac{x-5}{x^2+x+1}+\dfrac{x-5}{x^2+x+1}$ **(f)** $\dfrac{2w-5}{w^2-w+1}-\dfrac{w-4}{w^2-w+1}$

(g) $\dfrac{a^2+4a+1}{a^2+5a+1}+\dfrac{a^2-7a+2}{a^2+5a+1}$ **(h)** $\dfrac{b^2-7b-3}{b^2+2b+1}-\dfrac{b^2-8b-4}{b^2+2b+1}$

(i) $\dfrac{c^2+5c+1}{2c^2+14c+2}+\dfrac{c^2+7c+1}{2c^2+14c+2}$ **(j)** $\dfrac{x+3y}{x^2+5y^2}+\dfrac{2(x-3y)}{x^2+5y^2}$

(k) $\dfrac{3v+1}{v^2+5v+7}-\dfrac{3(2v-5)}{v^2+5v+7}$

PROBLEM 8-21 Rename the rational expressions in terms of the LCD of the rational expressions:

(a) $\dfrac{3}{x},\dfrac{4}{x+2}$ **(b)** $\dfrac{2y}{y^2-2y},\dfrac{3y+5}{y-2},\dfrac{5}{y}$ **(c)** $\dfrac{2z}{z^2+6z+9},\dfrac{3z-1}{z^2-z-12}$

PROBLEM 8-22 Find the indicated sums or differences:

(a) $\dfrac{3}{x}+\dfrac{4}{y}$ **(b)** $\dfrac{2}{x}-\dfrac{5}{y}$ **(c)** $\dfrac{3x}{2x+5}+\dfrac{1}{x-4}$ **(d)** $\dfrac{5w}{w-3}-\dfrac{2w}{w+1}$

(e) $\dfrac{y}{y^2 + 2y + 1} + \dfrac{2}{y^2 - 1}$ (f) $\dfrac{z}{z^2 + 4z + 4} - \dfrac{3}{z^2 - 4}$ (g) $\dfrac{x - 3}{x^2 + 7x + 12} + \dfrac{x + 5}{x^2 + 5x + 6}$

(h) $\dfrac{p - 2}{p^2 + 9p + 8} - \dfrac{p - 3}{p^2 + 10p + 16}$ (i) $\dfrac{3r + 2}{2r^2 + 5r + 2} + \dfrac{2r - 5}{r^2 - 3r - 10}$

(j) $\dfrac{4t - 3}{2t^2 + t - 15} - \dfrac{t - 8}{2t^2 + 7t + 3}$

PROBLEM 8-23 Simplify the following complex fractions:

(a) $\dfrac{\dfrac{1}{x}}{1 - \dfrac{1}{x}}$ (b) $\dfrac{1 + \dfrac{1}{y}}{y}$ (c) $\dfrac{1 - \dfrac{1}{z}}{1 + \dfrac{1}{z}}$ (d) $\dfrac{\dfrac{2}{x} - x}{3 + \dfrac{1}{x}}$ (e) $\dfrac{1}{1 - \dfrac{1}{1 + a}}$ (f) $\dfrac{2 + \dfrac{3}{w}}{2 - \dfrac{3}{w}}$

(g) $\dfrac{\dfrac{x + y}{x - y} - 1}{\dfrac{3}{x - y}}$ (h) $\dfrac{a + b}{\dfrac{a}{b} - \dfrac{b}{a}}$

PROBLEM 8-24 Solve the following rational equations:

(a) $\dfrac{2}{x} = \dfrac{7}{8}$ (b) $\dfrac{4}{x + 1} = \dfrac{3}{x - 1}$ (c) $\dfrac{2x}{x + 3} - 4 = \dfrac{5x}{x + 3}$ (d) $\dfrac{2x}{x + 3} = \dfrac{2x}{x - 3}$

(e) $\dfrac{5y}{2y + 1} + 3 = \dfrac{y}{2y + 1}$ (f) $\dfrac{w}{w + 2} + \dfrac{3w}{w + 2} = 7$ (g) $\dfrac{5}{2z + 4} = \dfrac{3}{2z - 5}$

(h) $\dfrac{4x}{(x + 3)^2} = \dfrac{2}{x + 3}$ (i) $\dfrac{4}{x - 2} + \dfrac{5}{2} = \dfrac{2x}{x - 2}$ (j) $\dfrac{3}{x + 2} + \dfrac{4}{x + 3} = \dfrac{3}{x^2 + 5x + 6}$

(k) $\dfrac{-11}{x^2 + 3x - 4} = \dfrac{2}{x + 4} - \dfrac{3}{x - 1}$

PROBLEM 8-25 Solve each rational formula for the indicated variable:

(a) Lever formula: $\dfrac{w}{W} = \dfrac{L}{l}$ for W (b) Acceleration formula: $a = \dfrac{v - v_0}{t}$ for t

(c) Electricity formula: $I = \dfrac{E}{R_1 + R_2}$ for R_1 (d) Series formula: $s = \dfrac{a}{1 - r}$ for r

(e) Electricity formula: $\dfrac{1}{R} = \dfrac{1}{R_1} + \dfrac{1}{R_2}$ for R_1 (f) Finance formula: $P = \dfrac{A}{1 + rt}$ for r

(g) Einstein's velocity formula: $v = \dfrac{v_1 + v_2}{1 + \dfrac{v_1 v_2}{c^2}}$ for c^2

PROBLEM 8-26 Solve each number problem using a rational equation:

(a) The denominator of a fraction is 3 greater than the numerator. If both the numerator and denominator are increased by 1, the resulting fraction equals $\frac{1}{2}$. What is the original fraction?

(b) The numerator of a fraction is 5 less than the denominator. When 1 is added to the numerator and 1 is subtracted from the denominator, the resulting fraction equals $\frac{1}{2}$. Find "the resulting fraction."

(c) When the same number is added to both the numerator and denominator of $\frac{13}{17}$, the resulting fraction equals $\frac{9}{11}$. What is the number?

(d) The sum of the reciprocal of twice a number and twice the number's reciprocal is $\frac{11}{3}$. Find the number.

PROBLEM 8-27 Solve each work problem using a rational equation:

(a) Ryan can complete a certain job in 6 hours. Sean can complete the same job in 8 hours. How long will it take to complete the same job working together?

(b) Mr. Ivanhoe and Mr. Tippy can complete a certain job in 3 days working together. Mr. Ivanhoe can complete the same job in 5 days working alone. How long would it take Mr. Tippy to complete the job working alone?

(c) An inlet pipe can fill a certain tank in 15 hours. An outlet pipe can drain the same tank in 20 hours. How long will it take to fill the tank if both pipes are left open?

(d) A certain sink can be filled in 15 minutes using the faucets. The same sink can be drained in 10 minutes by pulling the plug. How long will it take to empty a full sink by pulling the plug if the faucets are left on?

PROBLEM 8-28 Solve each uniform motion problem using a rational equation:

(a) A certain row boat can travel twice as fast downstream as it can upstream. It takes the boat a total of 3 hours to travel 5 miles downstream and 2 miles upstream. What is the rate of the current? What is the rate of the boat upstream? How long did the 5-mile trip downstream take?

(b) It takes a certain car 2 hours longer to travel 330 km than to travel 220 km at a constant rate. Find the constant rate of the car. Find the time needed to make the entire 550 km trip.

(c) It takes a motor boat twice as long to go 60 km upstream as it does to go 45 km downstream at a constant waterspeed of 30 km/h. What is the rate of the current? What is the landspeed of the boat upstream? How long does the trip downstream take?

(d) Elizabeth can run around a certain track in 1 minute flat. It takes Carlos 20 seconds longer than Elizabeth to complete one lap around the same track. How long will it take Elizabeth to gain one full lap on Carlos if they both start at the same place?

Answers to Supplementary Exercises

(8-16) (a) 5   (b) $\dfrac{8}{3}$ (c) $-\dfrac{7}{2}$ (d) 0

(8-17) (a) -1 (b) $\dfrac{x+y}{x^2+xy+y^2}$ (c) $\dfrac{x+3}{x-2}$ (d) $\dfrac{2x+5}{3x-7}$ (e) $\dfrac{3x-1}{3(2x-1)}$ (f) $-\dfrac{1}{2}$

(8-18) (a) $\dfrac{a+7}{a+3}$ (b) $\dfrac{y-5}{2y}$ (c) 1   (d) $\dfrac{3x-7}{2x-1}$ (e) $\dfrac{x}{x-y}$ (f) $-\dfrac{4}{4x+5}$

(8-19) (a) $\dfrac{z}{y}$ (b) $\dfrac{a+3}{a+5}$ (c) $x-2y$ (d) $\dfrac{m-n}{m+n}$ (e) $\dfrac{(3x+1)(2x-3)}{(2x+3)(3x-1)}$ (f) $-w-1$

(8-20) (a) $\dfrac{6}{y}$ (b) $\dfrac{m-p}{n}$ (c) $\dfrac{4a+5}{2a+5}$ (d) 1   (e) $\dfrac{2x-10}{x^2+x+1}$ (f) $\dfrac{w-1}{w^2-w+1}$

(g) $\dfrac{2a^2-3a+3}{a^2+5a+1}$ (h) $\dfrac{1}{b+1}$ (i) $\dfrac{c^2+6c+1}{c^2+7c+1}$ (j) $\dfrac{3x-3y}{x^2+5y^2}$ (k) $\dfrac{-3v+16}{v^2+5v+7}$

(8-21) (a) $\dfrac{3}{x}=\dfrac{3x+6}{x(x+2)}, \dfrac{4}{x+2}=\dfrac{4x}{x(x+2)}$

(b) $\dfrac{2y}{y^2-2y}=\dfrac{2y}{y(y-2)}, \dfrac{3y+5}{y-2}=\dfrac{3y^2+5y}{y(y-2)}, \dfrac{5}{y}=\dfrac{5y-10}{y(y-2)}$

(c) $\dfrac{2z}{z^2 + 6z + 9} = \dfrac{2z^2 - 8z}{(z + 3)^2(z - 4)}, \dfrac{3z - 1}{z^2 - z - 12} = \dfrac{3z^2 + 8z - 3}{(z + 3)^2(z - 4)}$

(8-22) (a) $\dfrac{3y + 4x}{xy}$ (b) $\dfrac{2y - 5x}{xy}$ (c) $\dfrac{3x^2 - 10x + 5}{(2x + 5)(x - 4)}$ (d) $\dfrac{3w^2 + 11w}{(w - 3)(w + 1)}$

(e) $\dfrac{y^2 + y + 2}{(y + 1)^2(y - 1)}$ (f) $\dfrac{z^2 - 5z - 6}{(z + 2)^2(z - 2)}$ (g) $\dfrac{2x^2 + 8x + 14}{(x + 2)(x + 3)(x + 4)}$

(h) $\dfrac{2p - 1}{(p + 1)(p + 2)(p + 8)}$ (i) $\dfrac{7r^2 - 21r - 15}{(2r + 1)(r + 2)(r - 5)}$ (j) $\dfrac{6t^2 + 19t - 43}{(2t - 5)(t + 3)(2t + 1)}$

(8-23) (a) $\dfrac{1}{x - 1}$ (b) $\dfrac{y + 1}{y^2}$ (c) $\dfrac{z - 1}{z + 1}$ (d) $\dfrac{2 - x^2}{3x + 1}$ (e) $\dfrac{1 + a}{a}$ (f) $\dfrac{2w + 3}{2w - 3}$

(g) $\dfrac{2y}{3}$ (h) $\dfrac{ab}{a - b}$

(8-24) (a) $\frac{16}{7}$ (b) 7 (c) $-\frac{12}{7}$ (d) 0 (e) $-\frac{3}{10}$ (f) $-\frac{14}{3}$ (g) $\frac{37}{4}$ (h) 3
(i) no solution (j) no solution (k) -3

(8-25) (a) $W = \dfrac{wl}{L}$ (b) $t = \dfrac{v - v_0}{a}$ (c) $R_1 = \dfrac{E - IR_2}{I}$ or $\dfrac{E}{I} - R_2$ (d) $r = \dfrac{s - a}{s}$ or $1 - \dfrac{a}{s}$

(e) $R_1 = \dfrac{R_2 R}{R_2 - R}$ (f) $r = \dfrac{A - P}{Pt}$ or $\dfrac{A}{Pt} - \dfrac{1}{t}$ (g) $c^2 = \dfrac{v_1 v_2 v}{v_1 + v_2 - v}$

(8-26) (a) $\frac{2}{5}$ (b) $\frac{3}{6}$ (c) 5 (d) $\frac{15}{22}$

(8-27) (a) $3\frac{3}{7}$ hr (b) $7\frac{1}{2}$ days (c) 60 hr (d) 30 min

(8-28) (a) $\frac{3}{4}$ mph, $1\frac{1}{2}$ mph, $1\frac{2}{3}$ hr (b) 55 km/h, 10 hr (c) 6 km/h, 24 km/h, $1\frac{1}{4}$ hr (d) 4 min.

MIDTERM EXAMINATION

Chapters 1–8

Part 1: Skills and Concepts (40 questions)

1. Graph $\{x \mid x \in W \text{ and } x < 4\}$:

   (a) [number line from −5 to 5] **(b)** [number line from −5 to 5]

   (c) [number line from −5 to 5] **(d)** [number line from −5 to 5]

   (e) none of these

2. Find $\{-4, -3, -1, 0, 2\} \cap \{-3, -2, -1, 1\}$:
   (a) $\{-4, -3, -2, -1, 0, 1, 2\}$ **(b)** $\{-3\}$ **(c)** $\{-3, -1\}$ **(d)** $\{-4, -2, 0, 1, 2\}$
   (e) none of these

3. Evaluate $-3^2 + |-5|$:
   (a) 14 **(b)** −4 **(c)** −14 **(d)** 1 **(e)** none of these

4. Evaluate $(8 - 6 \div 2) + 2 \cdot 3^2$:
   (a) 23 **(b)** 19 **(c)** 41 **(d)** 37 **(e)** none of these

5. Evaluate $a[6b - 4(4a - 3) + 5 \cdot b^2]$ for $a = 3$ and $b = -2$:
   (a) −204 **(b)** −156 **(c)** −25 **(d)** −84 **(e)** none of these

6. The solution of $5 - 4x = 1$ is:
   (a) −1 **(b)** $-\frac{3}{2}$ **(c)** 1 **(d)** $\frac{3}{2}$ **(e)** none of these

7. The solution of $2(x - 3) = -3(2 - 3x)$ is:
   (a) $\frac{3}{7}$ **(b)** 0 **(c)** $-\frac{3}{7}$ **(d)** $-\frac{12}{7}$ **(e)** none of these

8. The solution of $\frac{1}{3} - \frac{1}{2}x = \frac{1}{6}$ is:
   (a) 1 **(b)** $\frac{1}{3}$ **(c)** $-\frac{1}{3}$ **(d)** −1 **(e)** none of these

9. The solution of $3x - 4y = 12$ for y is:
   (a) $y = \frac{3}{4}x - 3$ **(b)** $y = \frac{3}{4}x + 3$ **(c)** $y = -\frac{3}{4}x - 3$ **(d)** $y = -\frac{3}{4}x + 3$ **(e)** none of these

10. The solutions of $|x - 3| = 2$ are:
    (a) −1 and 5 **(b)** 1 and −5 **(c)** −1 and 1 **(d)** 1 and 5 **(e)** none of these

11. The solution set of $3x - 2 < 7$ is:
    (a) $\{x \mid x < 3\}$ **(b)** $\{x \mid x < 5\}$ **(c)** $\{x \mid x > 3\}$ **(d)** $\{x \mid x = 3\}$ **(e)** none of these

12. The solution set of $4 - 3x \geq 5x + 7$ is:
    (a) $\{x \mid x \leq -\frac{3}{8}\}$ **(b)** $\{x \mid x \geq -\frac{3}{8}\}$ **(c)** $\{x \mid x < \frac{3}{8}\}$ **(d)** $\{x \mid x \leq -\frac{3}{2}\}$ **(e)** none of these

13. The solution set of $1 \leq 5 - 4x \leq 9$ is:
    (a) $\{x \mid -1 \leq x < 1\}$ **(b)** $\{x \mid -1 \leq x \leq 1\}$ **(c)** $\{x \mid x \leq 1\}$ **(d)** $\{x \mid -4 \leq x \leq 4\}$
    (e) none of these

14. The solution set of $|2x - 3| \geq 3$ is:
    (a) $\{x \mid x \geq 3 \text{ or } x \leq -3\}$ **(b)** $\{x \mid x \leq 3 \text{ or } x \geq 0\}$ **(c)** $\{x \mid x \leq -3 \text{ or } x \geq 0\}$
    (d) $\{x \mid x \geq 3 \text{ or } x \leq 0\}$ **(e)** none of these

15. The solution set of $|3 - x| \leq 2$ is:
    (a) $\{x \mid x \leq 1\}$ (b) $\{x \mid -5 \leq x \leq -1\}$ (c) $\{x \mid 1 \leq x \leq 5\}$ (d) $\{x \mid x \leq 1\}$
    (e) none of these

16. The slope of the graph of $5x - 2y = 6$ is:
    (a) $-\frac{5}{2}$ (b) $-\frac{2}{5}$ (c) -3 (d) $\frac{5}{2}$ (e) none of these

17. The equation of the line with y-intercept $(0, -2)$ and a slope of 3 is:
    (a) $y = 3x - 2$ (b) $y = 3x + 2$ (c) $y = -3x + 2$ (d) $y = -3x - 2$ (e) none of these

18. The equation of the line that contains $(2, -3)$ and $(-3, 0)$ is:
    (a) $y = -\frac{3}{5}x - \frac{9}{5}$ (b) $y = -\frac{5}{3}x - 3$ (c) $y = \frac{3}{5}x - \frac{9}{5}$ (d) $y = -\frac{3}{5}x + \frac{9}{5}$
    (e) none of these

19. The graph of $3x - 2y \geq 6$ is:

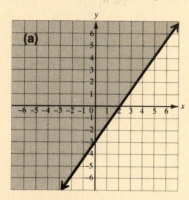 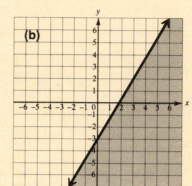 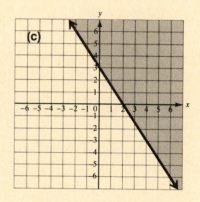

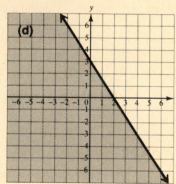 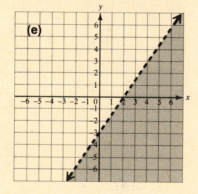

20. The equation of the line that contains $(2, -3)$ and is perpendicular to the graph of $3x - 4y = 12$ is:
    (a) $3x - 4y = 18$ (b) $4x - 3y = 17$ (c) $4x + 3y = -1$ (d) $4x + 3y = 12$
    (e) none of these

21. The solution of $\begin{cases} 3x - 2y = 12 \\ 2x + y = 1 \end{cases}$ obtained by the addition method is:

    (a) $(2, 3)$ (b) $(-2, 3)$ (c) $(-2, -3)$ (d) $(2, -3)$ (e) none of these

22. The solution of $\begin{cases} 2x + y - 3z = 8 \\ x - 3y + z = -1 \\ 3x + 2y - 2z = 7 \end{cases}$ obtained by the addition method is:

    (a) $(1, 0, -2)$ (b) $(0, -1, -3)$ (c) $(1, 0, 2)$ (d) $(1, -2, 0)$ (e) none of these

23. The solution of $\begin{cases} x - 2y + 3z = 7 \\ 2x + y - z = 0 \\ 3x - y + 2z = 7 \end{cases}$ obtained by the addition method is:

    (a) $(0, 1, 2)$ (b) $(-1, 0, 2)$ (c) $(1, 0, -2)$ (d) $(1, 0, 2)$ (e) none of these

24. The value of $\begin{vmatrix} 3 & 2 & 1 \\ -2 & 0 & -3 \\ 0 & -1 & 2 \end{vmatrix}$ is:

    (a) -19 (b) 19 (c) -3 (d) 1 (e) none of these

25. The solution of $\begin{cases} 2x - 5y = 11 \\ 3x + 2y = 7 \end{cases}$ obtained using Cramer's rule is:

    (a) $(3, 1)$ (b) $(-3, -1)$ (c) $(-3, 1)$ (d) $(-1, 3)$ (e) none of these

26. Find $(2x^3 + 4x - 5) + (3x^2 - 2x + 7)$:
    (a) $2x^3 + 3x^2 + 2x + 2$ (b) $5x^3 + 2x + 2$ (c) $2x^3 + 3x^2 + 6x + 12$ (d) $5x^6 + 2x^2 - 2$
    (e) none of these

27. Find $(6x^2 - 2x - 5) - (2x^2 - 8x + 11)$:
    (a) $4x^2 - 6x - 16$ (b) $4x^2 + 6x + 6$ (c) $4x^2 + 6x - 16$ (d) $8x^2 + 10x + 6$
    (e) none of these

28. Find $(3x + 5)(2x - 7)$:
    (a) $6x^2 + 31x - 35$ (b) $6x^2 - 11x - 35$ (c) $6x^2 - 31x - 35$ (d) $5x^2 - 11x - 35$
    (e) none of these

29. Find $(6x^2 - 5x - 7) \div (2x - 1)$:

    (a) $3x - 1 - \dfrac{8}{2x - 1}$ (b) $3x - 4 - \dfrac{3}{2x - 1}$ (c) $3x - 4 - \dfrac{11}{2x - 1}$ (d) $3x + 1 + \dfrac{9}{2x - 1}$

    (e) none of these

30. Use synthetic division to find $(y^3 - 3y^2 + 7y - 10) \div (y - 2)$:

    (a) $y^2 - y + 5$ (b) $y^2 - y + 9 + \dfrac{8}{y - 2}$ (c) $y^2 - 5y + 17 - \dfrac{44}{y - 2}$ (d) $y^2 - 5y - 3$

    (e) none of these

31. Factor $x^2 - 5x + 6$:
    (a) $(x + 1)(x - 6)$ (b) $(x - 2)(x - 3)$ (c) $(x + 3)(x - 2)$ (d) $(x - 1)(x + 6)$
    (e) none of these

32. Factor $81x^2 - 49$:
    (a) $(9x - 7)^2$ (b) $(9x + 7)(9x - 7)$ (c) $(9x + 7)^2$ (d) $(27x + 7)(3x - 7)$
    (e) none of these

33. Factor $27w^3 - 125$:
    (a) $(3w - 5)(9w^2 + 30w + 25)$ (b) $(3w - 5)(9w^2 - 30w + 25)$ (c) $(3w - 5)(9w^2 + 15w + 25)$
    (d) $(3w + 5)(9w^2 - 15w + 5)$ (e) none of these

34. Factor $3w^2 - 2wy + 6wz - 4yz$:
    (a) $w(3w - 2y)$ (b) $(w + 2z)(3w + 2y)$ (c) $(3w - 2y)(2w + z)$ (d) $(3w - 2y)(w + 2z)$
    (e) none of these

35. Factor $512m^5 - 162m$:
(a) $2m(16m^2 + 9)(16m^2 - 9)$ (b) $2m(16m^2 + 9)(4m + 3)(4m - 3)$ (c) $2m(4m + 3)^3(4m - 3)$
(d) $2m(4m + 3)^2(4m - 3)^2$ (e) none of these

36. Find the excluded value(s) of $\dfrac{5}{2x^2 - 3x - 14}$:

(a) 0 (b) -2 (c) $-2, 7$ (d) $-2, \frac{7}{2}$ (e) none of the others

37. Reduce $\dfrac{6x^2 + 9x + 3}{6x^2 - 6}$ to lowest terms.

(a) $\dfrac{2x + 1}{x - 1}$ (b) $\dfrac{9x + 1}{-6}$ (c) $\dfrac{2x + 1}{2x - 2}$ (d) $-\dfrac{1}{2}$ (e) none of these

38. Find $\dfrac{3x + 3}{2x^2 - x - 10} \div \dfrac{x^2 - 3x - 4}{x^2 + x - 2}$:

(a) $\dfrac{3(x - 1)}{(2x - 5)(x - 4)}$ (b) $\dfrac{(2x - 5)(x - 4)}{3(x - 1)}$ (c) $\dfrac{3(x + 1)^2(x - 4)}{(x + 2)^2(2x - 5)(x - 1)}$ (d) $\dfrac{-1}{2x^2 - 10x + 20}$

(e) none of these

39. Simplify $\dfrac{\dfrac{x + y}{x - y} + 3}{\dfrac{2}{x - y}}$:

(a) $2x - y$ (b) $\dfrac{4x - 2y}{x - y}$ (c) $2x - y$ (d) $\dfrac{x + y + 3}{2}$ (e) none of these

40. Solve $\dfrac{3}{x - 2} - \dfrac{2}{x + 1} = 0$:

(a) -7 (b) -2 (c) 2 (d) $-\dfrac{7}{5}$ (e) none of these

Part 2: Problem Solving (10 problems)

41. Richard invested $10,000, part at 8% per year, and the rest at 12% per year. His annual return on investment is $1000. How much interest was earned by the 12% investment?

42. Two trains are 460 miles apart and traveling towards each other on parallel tracks. The constant rate of one train is 6 mph slower than the other train. After 5 hours, the two trains passed each other. How far did the slower train travel to the passing point?

43. In 3 years, Richard will be twice as old as Margie is now. Two years ago, the sum of their ages was 20. How old is each now?

44. The sum of the digits of a 2-digit number is 14. When the reversed number is subtracted from the original number, the result is 36. What is the original number?

45. A nurse wants to strengthen 20 cc of a 50%-alcohol solution to an 80%-alcohol solution. How much pure alcohol should be added?

46. At a constant airspeed against a headwind it took Ann 7 hours to fly 1050 miles to San Jose. Flying at the same constant airspeed, with a tailwind of the same velocity, it took her 5 hours to return. What is the wind velocity and her airspeed?

47. A bathtub can be filled with both faucets on in 20 minutes. It takes 30 minutes to drain the bathtub. How long will it take to fill the bathtub with both faucets on if the drain is accidentally left open?

48. An airplane travels 600 km with a tailwind at 25 km/h. On the return trip against the same wind velocity, it takes the plane the same amount of time to travel only 450 km. What is the plane's airspeed? How long is "the same amount of time?"

49. Working alone, Paul can complete a certain job in 4 hours less time than Ernie. Working together, they can complete the same job in 3 hours. Exactly how long would it take each to complete the job alone?

50. A bus traveled 560 km at a constant rate. A truck traveled the same distance 10 km/h faster than the bus in 1 hour less time. Find the constant rate of the bus. How long did it take the truck to travel the 560 km?

Midterm Examination Answers

Part 1

1. a	**9.** a	**17.** a	**25.** e	**33.** c
2. c	**10.** d	**18.** a	**26.** a	**34.** d
3. b	**11.** a	**19.** b	**27.** c	**35.** b
4. a	**12.** a	**20.** c	**28.** b	**36.** d
5. d	**13.** b	**21.** d	**29.** a	**37.** e
6. c	**14.** d	**22.** a	**30.** a	**38.** a
7. b	**15.** c	**23.** d	**31.** b	**39.** a
8. b	**16.** d	**24.** d	**32.** b	**40.** a

Part 2

41. $600.00

42. 215 mi

43. 9 yr, 15 yr

44. 95

45. 30 cc

46. 30 mph (wind), 180 mph (airspeed)

47. 1 hr

48. 175 km/h, 3 hr

49. $(1 + \sqrt{13})$hr, $(5 + \sqrt{13})$hr

50. 70 km/h, 7 hr

9 RADICALS AND EXPONENTS

THIS CHAPTER IS ABOUT

☑ Radicals and Rational Exponents
☑ Operations on Radical Expressions
☑ Solving Equations Involving Radicals

9-1. Radicals and Rational Exponents

A. Evaluating expressions that involve rational exponents

If n is a natural number, and $b^n = a$, then we call b the **nth root** of a. For example:

- 5 is a second root (**square root**) of 25, because $5^2 = 25$
- -5 is a square root of 25, because $(-5)^2 = 25$
- 2 is a third root (**cube root**) of 8, because $2^3 = 8$
- -6 is a cube root of -216, because $(-6)^3 = -216$
- 4 is a fifth root of 1024, because $4^5 = 1024$

Every positive real number has a positive square root and a negative square root. The positive square root is called the **principal square root:** Both 3 and -3 are square roots of 9; however, the principal square root of 9 is 3.

Negative real numbers do not have a real-number square root because there is no real number b such that b^2 is a negative real number.

Roots are often denoted by using **radical notation.** The radical notation $\sqrt[n]{a}$ is sometimes called a **radical expression.** The number a is called the **radicand.** The number n, which may be any natural number, is called the **index** or **order of the radical.**

- If n is even, and $a > 0$, then $\sqrt[n]{a}$ is the positive real number b such that $b^n = a$, and b is called the principal nth root of a.
  $a = 0$, then $\sqrt[n]{a} = 0$.
  $a < 0$, then $\sqrt[n]{a}$ does not represent a real number.

- If n is odd, and $a > 0$, then $\sqrt[n]{a}$ is the positive number b such that $b^n = a$.
  $a = 0$, then $\sqrt[n]{a} = 0$.
  $a < 0$, then $\sqrt[n]{a}$ is the negative real number b such that $b^n = a$.

Note 1: If $n = 2$ then $\sqrt[2]{a}$ is written as simply $\sqrt{a}$.

Note 2: Radical expressions that have an even index do not represent real numbers when their radicands are negative numbers. For this reason, restrictions are sometimes placed on the variables which appear in the radicand of a radical with an even index so that the radicand will represent a nonnegative real number.

To represent the negative square root of a real number, we write a negative sign IN FRONT OF the radical sign. For example, $\sqrt{25} = 5$, while $-\sqrt{25} = -5$.

You might be tempted to think $\sqrt{x^2} = x$, but this is not always true:

$$\text{If } x = 4, \text{ then } \sqrt{x^2} = \sqrt{4^2} = \sqrt{16} = 4 = x$$

$$\text{If } x = -4, \text{ then } \sqrt{x^2} = \sqrt{(-4)^2} = \sqrt{16} = 4 = -x$$

$$\text{If } x = 0, \text{ then } \sqrt{x^2} = \sqrt{0^2} = \sqrt{0} = 0 = x$$

This example shows that in every case, if $x \geq 0$, then $\sqrt{x^2} = x$, while if $x < 0$, then $\sqrt{x^2} = -x$. The above result can be stated more concisely using absolute value notation:

$$\text{For any real number } x, \sqrt{x^2} = |x|.$$

The following definitions establish the relationship between radicals and rational exponents.

Definition 9-1:

If n is a natural number, then $a^{\frac{1}{n}} = \sqrt[n]{a}$. If n is an even number, a is restricted to nonnegative real numbers.

EXAMPLE 9-1: Evaluate each of the following:

(a) $9^{\frac{1}{2}}$ (b) $(-8)^{\frac{1}{3}}$ (c) $16^{\frac{1}{4}}$ (d) $-25^{\frac{1}{2}}$

Solution: Use $a^{\frac{1}{n}} = \sqrt[n]{a}$ and the definition of radical notation.

(a) $9^{\frac{1}{2}} = \sqrt{9} = 3$ (b) $(-8)^{\frac{1}{3}} = \sqrt[3]{-8} = -2$ (c) $16^{\frac{1}{4}} = \sqrt[4]{16} = 2$
(d) $-25^{\frac{1}{2}} = -\sqrt{25} = -5$

Definition 9-2:

If m is an integer and n is a natural number, then $a^{\frac{m}{n}} = \sqrt[n]{a^m} = (\sqrt[n]{a})^m$. If n is an even number, a is restricted to nonnegative real numbers.

EXAMPLE 9-2: Write each of the following in radical notation:

(a) $4^{\frac{2}{6}}$ (b) $3^{0.5}$ (c) $2^{0.75}$ (d) $-11^{\frac{2}{8}}$

Solution: Use $a^{\frac{m}{n}} = \sqrt[n]{a^m}$.

(a) $4^{\frac{2}{6}} = 4^{\frac{1}{3}} = \sqrt[3]{4^1} = \sqrt[3]{4}$ (b) $3^{0.5} = 3^{\frac{1}{2}} = \sqrt{3^1} = \sqrt{3}$
(c) $2^{0.75} = 2^{\frac{3}{4}} = \sqrt[4]{2^3} = \sqrt[4]{8}$ (d) $-11^{\frac{2}{8}} = -11^{\frac{1}{4}} = -\sqrt[4]{11^1} = -\sqrt[4]{11}$

For computational purposes, it is often more convenient to use $a^{\frac{m}{n}} = (\sqrt[n]{a})^m$.

EXAMPLE 9-3: Evaluate each of the following:

(a) $81^{\frac{3}{4}}$ (b) $27^{\frac{4}{3}}$ (c) $(-27)^{\frac{2}{3}}$ (d) $32^{-\frac{3}{5}}$

Solution: Use $a^{\frac{m}{n}} = (\sqrt[n]{a})^m$.

(a) $81^{\frac{3}{4}} = (\sqrt[4]{81})^3 = 3^3 = 27$ (b) $27^{\frac{4}{3}} = (\sqrt[3]{27})^4 = 3^4 = 81$

(c) $(-27)^{\frac{2}{3}} = (\sqrt[3]{-27})^2 = (-3)^2 = 9$ (d) $32^{-\frac{3}{5}} = (\sqrt[5]{32})^{-3} = 2^{-3} = \frac{1}{2^3} = \frac{1}{8}$

B. Simplifing expressions that involve rational exponents

The following is a restatement of the rules for exponents, which were stated in Chapter 6 and which are now extended to include rational exponents. In this chapter we will assume that the variables that appear as bases are restricted so that every power represents a real number.

Rules for Exponents

If m and n are rational numbers, then

• $a^m \cdot a^n = a^{m+n}$ (The Product Rule for Exponents).

• $\dfrac{a^m}{a^n} = a^{m-n}$ (The Quotient Rule for Exponents).

• $(a^m)^n = a^{mn}$ (The Power Rule for Exponents).

An exponential expression is said to be **simplified** if each different base occurs only once, no power is raised to a power, and each exponent is positive.

EXAMPLE 9-4: Simplify each of the following:

(a) $x^{\frac{1}{3}} \cdot x^{\frac{1}{6}}$ (b) $\dfrac{w^{\frac{2}{3}}}{w^{\frac{1}{4}}}$ (c) $(x^{-1})^{-\frac{1}{2}}$

Solution: Use the rules for exponents.

$$\text{(a)} \quad x^{\frac{1}{3}} \cdot x^{\frac{1}{6}} = x^{\frac{1}{3}+\frac{1}{6}} \qquad \text{Use the product rule for exponents.}$$

$$= x^{\frac{1}{2}}$$

$$\text{(b)} \quad \dfrac{w^{\frac{2}{3}}}{w^{\frac{1}{4}}} = w^{\frac{2}{3}-\frac{1}{4}} \qquad \text{Use the quotient rule for exponents.}$$

$$= w^{\frac{5}{12}}$$

$$\text{(c)} \quad (x^{-1})^{-\frac{1}{2}} = x^{(-1)(-\frac{1}{2})} \qquad \text{Use the power rule for exponents.}$$

$$= x^{\frac{1}{2}}$$

Exponential expressions that involve more than one factor may be simplified by using the following rule.

Extended Power Rule for Exponents

$\left(\dfrac{a^p b^q}{c^r}\right)^s = \dfrac{a^{ps} b^{qs}}{c^{rs}}$, where the exponents are rational numbers and the bases are restricted to avoid a zero factor in the denominator.

EXAMPLE 9-5: Simplify $\left(\dfrac{-6a^2 b^{-4}}{3ab^{-2}}\right)^{-2}$

Solution: $\left(\dfrac{-6a^2 b^{-4}}{3ab^{-2}}\right)^{-2} = (-2a^{2-1} b^{-4-(-2)})^{-2}$ Simplify inside the parentheses.

$$= (-2ab^{-2})^{-2}$$

$$= (-2)^{-2} a^{-2} b^4 \qquad \text{Use the extended power rule.}$$

$$= \dfrac{b^4}{(-2)^2 a^2}$$

$$= \dfrac{b^4}{4a^2}$$

9-2. Operations on Radical Expressions

A. Simplifying radical expressions

You may also use Definition 9-2 to write radical expressions in rational exponential notation.

EXAMPLE 9-6: Write each of the following radicals in rational exponential notation:

(a) $\sqrt[3]{x^2}$ (b) $\sqrt[5]{uv^3}$ (c) $\sqrt{x^3 y}$

Solution: Use Definition 9-2 and the rules for exponents.

(a) $\sqrt[3]{x^2} = x^{\frac{2}{3}}$ (b) $\sqrt[5]{uv^3} = (uv^3)^{\frac{1}{5}} = u^{\frac{1}{5}} v^{\frac{3}{5}}$ (c) $\sqrt{x^3 y} = (x^3 y)^{\frac{1}{2}} = x^{\frac{3}{2}} y^{\frac{1}{2}}$

The rules for rational exponents can be stated in terms of radicals. In radical form they are called the **rules for radicals.**

Rules for Radicals

If a and b are positive real numbers, and n is a natural number, then:

- $\sqrt[n]{ab} = \sqrt[n]{a}\sqrt[n]{b}$ (The Product Rule for Radicals).

- $\sqrt[n]{\dfrac{a}{b}} = \dfrac{\sqrt[n]{a}}{\sqrt[n]{b}}$ (The Quotient Rule for Radicals).

- $\sqrt[n]{a^n} = a$ (The Identity Rule for Radicals).

In this section we will restrict all variables and variable expressions that appear as radicands to positive real numbers, so that we can apply the above rules of radicals.

The rules for radicals are particularly useful for writing radicals in equivalent forms. We often **simplify radicals** by writing them in an equivalent form in which the radical contains only powers less than the index number.

EXAMPLE 9-7: Simplify (a) $\sqrt[3]{24}$ and (b) $\sqrt{20x^3y^6}$.

Solution: (a) $\sqrt[3]{24} = \sqrt[3]{2^3 \cdot 3}$ Write the radicand as powers of prime factors.

$\qquad\qquad\quad = \sqrt[3]{2^3}\sqrt[3]{3}$ Use the product rule for radicals.

$\qquad\qquad\quad = 2\sqrt[3]{3}$ Use the identity rule for radicals.

$\quad$ (b) $\sqrt{20x^3y^6} = \sqrt{2^2 \cdot 5 \cdot x^3 y^6}$

$\qquad\qquad\qquad = \sqrt{2^2 \cdot 5x^2 x(y^3)^2}$ Write the radicand as a product of powers of the index and powers less than the index.

$\qquad\qquad\qquad = \sqrt{2^2 x^2 (y^3)^2}\sqrt{5x}$

$\qquad\qquad\qquad = 2xy^3\sqrt{5x}$

The process of writing a radical in an equivalent radical form with a smaller index is called **reducing the order of the radical.** We often reduce the order of a radical by first writing the radical in exponential form and then applying the rules for exponents.

EXAMPLE 9-8: Reduce the order of each radical: (a) $\sqrt[4]{9}$ (b) $\sqrt[6]{81a^2}$ (c) $\sqrt[8]{64x^2y^6}$

Solution: Write the radical expression in exponential form, then use the rules for exponents.

$\quad$ (a) $\sqrt[4]{9} = 9^{\frac{1}{4}} = (3^2)^{\frac{1}{4}} = 3^{\frac{2}{4}} = 3^{\frac{1}{2}} = \sqrt{3}$

$\quad$ (b) $\sqrt[6]{81a^2} = (81a^2)^{\frac{1}{6}} = (3^4 a^2)^{\frac{1}{6}} = 3^{\frac{4}{6}} a^{\frac{2}{6}} = 3^{\frac{2}{3}} a^{\frac{1}{3}} = (3^2 a^1)^{\frac{1}{3}} = \sqrt[3]{3^2 a} = \sqrt[3]{9a}$

$\quad$ (c) $\sqrt[8]{64x^2y^6} = (64x^2y^6)^{\frac{1}{8}} = (2^6 x^2 y^6)^{\frac{1}{8}} = 2^{\frac{6}{8}} x^{\frac{2}{8}} y^{\frac{6}{8}} = 2^{\frac{3}{4}} x^{\frac{1}{4}} y^{\frac{3}{4}} = (2^3 x^1 y^3)^{\frac{1}{4}} = \sqrt[4]{8xy^3}$

We often find it convenient to write an expression involving radicals in a form where no radical appears in the denominator of the expression. The procedure used to clear a radical from the denominator is called **rationalizing the denominator.**

EXAMPLE 9-9: Rationalize the denominator of each of the following:

(a) $\dfrac{3}{\sqrt{w}}$ (b) $\dfrac{\sqrt[3]{x}}{\sqrt[3]{z}}$ (c) $\dfrac{6}{\sqrt[3]{2y^2}}$

Solution: Multiply both the numerator and the denominator by the same factor so that you can apply the identity rule for radicals ($\sqrt[n]{a^n} = a$) to clear the radical from the denominator.

$\quad$ (a) $\dfrac{3}{\sqrt{w}} = \dfrac{3}{\sqrt{w}} \cdot \dfrac{\sqrt{w}}{\sqrt{w}} = \dfrac{3\sqrt{w}}{\sqrt{w^2}} = \dfrac{3\sqrt{w}}{w}$

$\quad$ (b) $\dfrac{\sqrt[3]{x}}{\sqrt[3]{z}} = \dfrac{\sqrt[3]{x}}{\sqrt[3]{z}} \cdot \dfrac{\sqrt[3]{z^2}}{\sqrt[3]{z^2}} = \dfrac{\sqrt[3]{xz^2}}{\sqrt[3]{z^3}} = \dfrac{\sqrt[3]{xz^2}}{z}$

$\quad$ (c) $\dfrac{6}{\sqrt[3]{2y^2}} = \dfrac{6}{\sqrt[3]{2y^2}} \cdot \dfrac{\sqrt[3]{2^2y}}{\sqrt[3]{2^2y}} = \dfrac{6\sqrt[3]{4y}}{\sqrt[3]{2^3y^3}} = \dfrac{6\sqrt[3]{4y}}{2y} = \dfrac{3\sqrt[3]{4y}}{y}$

A radical expression is in **simplest radical form** if it involves only one radical and

(1) the radicand involves only powers less than the index,
(2) the order of the radical is reduced as far as possible,
(3) no radical appears in a denominator, and
(4) no fraction appears in the radicand.

EXAMPLE 9-10: Write $\sqrt{\dfrac{25b^3}{60a}}$ in simplest radical form.

Solution: $\sqrt{\dfrac{25b^3}{60a}} = \sqrt{\dfrac{5^2 b^3}{2^2 \cdot 3 \cdot 5a}}$ Factor into prime factors.

$\qquad\qquad = \sqrt{\dfrac{5b^3}{2^2 \cdot 3a}}$ Reduce to lowest terms.

$\qquad\qquad = \dfrac{\sqrt{5b^3}}{\sqrt{2^2 \cdot 3a}}$

$\qquad\qquad = \dfrac{b\sqrt{5b}}{2\sqrt{3a}}$ Remove powers of the index.

$\qquad\qquad = \dfrac{b\sqrt{5b}}{2\sqrt{3a}} \cdot \dfrac{\sqrt{3a}}{\sqrt{3a}}$ Rationalize the denominator.

$\qquad\qquad = \dfrac{b\sqrt{15ab}}{6a}$

Some radicals do not represent rational numbers. For example, $\sqrt{2}$ is an irrational number. Its decimal representation is a nonrepeating, nonterminating decimal. We often find it convenient to approximate an irrational number with a rational number. Appendix Table 1 lists approximate square roots and cube roots for the natural numbers from 1 to 100. To approximate the square root or cube root of a natural number greater than 100, you factor the radicand, find exact or approximate roots of the factors, and then find the product of those roots.

EXAMPLE 9-11: Use Appendix Table 1 to approximate **(a)** $\sqrt{539}$ and **(b)** $\sqrt[3]{135}$.

Solution: **(a)** $\sqrt{539} = \sqrt{7^2 \cdot 11}$ Factor the radicand.   **(b)** $\sqrt[3]{135} = \sqrt[3]{3^3 \cdot 5}$

$\qquad\qquad = 7\sqrt{11}$ $= 3\sqrt[3]{5}$

$\qquad\qquad \approx 7(3.317)$ Use Appendix Table 1. $\approx 3(1.710)$

$\qquad\qquad = 23.219$ $= 5.130$

B. Adding or subtracting radical expressions

Radicals with the same index and the same radicand are called **like radicals.** You can add or subtract like radicals by using the distributive property. Just as addition and subtraction of terms are referred to as combining like terms, addition and subtraction of like radicals are referred to as **combining like radicals.**

EXAMPLE 9-12: Combine like radicals:

(a) $3\sqrt{2} + 4\sqrt{2}$ **(b)** $\sqrt[3]{2x} - 4\sqrt[3]{2x}$ **(c)** $3w\sqrt[4]{w} + 2w\sqrt[4]{w} - w\sqrt[4]{w}$

Solution: **(a)** $3\sqrt{2} + 4\sqrt{2} = (3 + 4)\sqrt{2} = 7\sqrt{2}$ **(b)** $\sqrt[3]{2x} - 4\sqrt[3]{2x} = (1 - 4)\sqrt[3]{2x} = -3\sqrt[3]{2x}$
$\qquad\quad$ **(c)** $3w\sqrt[4]{w} + 2w\sqrt[4]{w} - w\sqrt[4]{w} = (3w + 2w - w)\sqrt[4]{w} = 4w\sqrt[4]{w}$

Some radicals can only be combined if we first simplify them.

EXAMPLE 9-13: Simplify and combine like radicals:

(a) $\sqrt{8} + 3\sqrt{2}$ (b) $\sqrt{27y} - 2\sqrt{75y} + 3\sqrt{48y}$ (c) $3a\sqrt[3]{27a} - 2\sqrt[3]{a^4} + \dfrac{1}{a}\sqrt[3]{a^7}$

Solution: (a) $\sqrt{8} + 3\sqrt{2} = \sqrt{2^3} + 3\sqrt{2}$

$$= 2\sqrt{2} + 3\sqrt{2}$$

$$= 5\sqrt{2}$$

(b) $\sqrt{27y} - 2\sqrt{75y} + 3\sqrt{48y} = \sqrt{3^3 y} - 2\sqrt{3 \cdot 5^2 y} + 3\sqrt{2^4 \cdot 3y}$

$$= 3\sqrt{3y} - 10\sqrt{3y} + 12\sqrt{3y}$$

$$= 5\sqrt{3y}$$

(c) $3a\sqrt[3]{27a} - 2\sqrt[3]{a^4} + \dfrac{1}{a}\sqrt[3]{a^7} = 3a\sqrt[3]{3^3 a} - 2\sqrt[3]{a^4} + \dfrac{1}{a}\sqrt[3]{a^7}$

$$= 9a\sqrt[3]{a} - 2a\sqrt[3]{a} + a\sqrt[3]{a}$$

$$= 8a\sqrt[3]{a}$$

C. Multiplying or dividing radical expressions

To multiply radicals which involve the same index, you use the product rule for radicals.

Caution: To use the product rule for radicals, the index MUST be the same for each of the radicals.

EXAMPLE 9-14: Find the indicated products: (a) $\sqrt{2}\sqrt{7}$ (b) $\sqrt[3]{w^2}\sqrt[3]{wy^2}$ (c) $\sqrt{3}\sqrt{30}$

Solution: Use $\sqrt[n]{a}\sqrt[n]{b} = \sqrt[n]{ab}$.

(a) $\sqrt{2}\sqrt{7} = \sqrt{2 \cdot 7}$ (b) $\sqrt[3]{w^2}\sqrt[3]{wy^2} = \sqrt[3]{w^3 y^2}$ (c) $\sqrt{3}\sqrt{30} = \sqrt{3 \cdot 30}$

$\qquad\qquad = \sqrt{14}$ $\qquad\qquad\qquad\qquad = w\sqrt[3]{y^2}$ $\qquad\qquad\qquad\quad = \sqrt{2 \cdot 3^2 \cdot 5}$

$$= 3\sqrt{2 \cdot 5}$$

$$= 3\sqrt{10}$$

Sometimes it is necessary to use the distributive property to multiply radical expressions.

EXAMPLE 9-15: Find the indicated products: (a) $\sqrt{2}(\sqrt{3} - 3\sqrt{5})$ (b) $(2\sqrt{5a} + 7\sqrt{2a})\sqrt{2a}$

Solution:

(a) $\sqrt{2}(\sqrt{3} - 3\sqrt{5}) = \sqrt{2} \cdot \sqrt{3} - \sqrt{2} \cdot 3 \cdot \sqrt{5}$ (b) $(2\sqrt{5a} + 7\sqrt{2a})\sqrt{2a} = 2\sqrt{5a}\sqrt{2a} + 7\sqrt{2a}\sqrt{2a}$

$\qquad\qquad\qquad = \sqrt{6} - 3\sqrt{10}$ $\qquad\qquad\qquad\qquad\qquad = 2\sqrt{10a^2} + 7\sqrt{2a \cdot 2a}$

$$= 2\sqrt{10a^2} + 7\sqrt{2^2 a^2}$$

$$= 2a\sqrt{10} + 14a$$

If the radical expressions are both binomials, you can multiply them using the FOIL method or the special product rules from Chapter 7.

EXAMPLE 9-16: Find the indicated products:

(a) $(2\sqrt{3} - \sqrt{2})(\sqrt{3} + \sqrt{2})$ (b) $(\sqrt{7a} + \sqrt{5})^2$ (c) $(\sqrt{a} + \sqrt{b})(\sqrt{a} - \sqrt{b})$

Solution: (a) $(2\sqrt{3} - \sqrt{2})(\sqrt{3} + \sqrt{2}) = 2\sqrt{3}\sqrt{3} + 2\sqrt{3}\sqrt{2} - \sqrt{2}\sqrt{3} - \sqrt{2}\sqrt{2}$ Use the FOIL method.

$$= 2 \cdot 3 + 2\sqrt{6} - \sqrt{6} - 2$$

$$= 6 + \sqrt{6} - 2$$

$$= 4 + \sqrt{6}$$

(b) $(\sqrt{7a} + \sqrt{5})^2 = (\sqrt{7a})^2 + 2\sqrt{7a}\sqrt{5} + (\sqrt{5})^2$ Use $(r + s)^2 = r^2 + 2rs + s^2$.

$$= 7a + 2\sqrt{35a} + 5$$

(c) $(\sqrt{a} + \sqrt{b})(\sqrt{a} - \sqrt{b}) = (\sqrt{a})^2 - (\sqrt{b})^2$ Use $(r + s)(r - s) = r^2 - s^2$

$$= a - b$$

The above example is worthy of special attention because the product of the radical expressions does not involve a radical. The radical expressions $(\sqrt{a} + \sqrt{b})$ and $(\sqrt{a} - \sqrt{b})$ are called **rationalizing factors** or **conjugates** of each other.

Division by a radical expression generally involves changing the fraction to an equivalent form in which the denominator is clear of radicals and writing the result in simplest radical form. However, we sometimes use the quotient rule of radicals to divide radical expressions with the same index.

EXAMPLE 9-17: Find the indicated quotients:

(a) $\dfrac{\sqrt{192}}{\sqrt{6}}$ **(b)** $\dfrac{3\sqrt{75x}}{\sqrt{3}}$ **(c)** $\dfrac{\sqrt[3]{4a^5b^7c}}{\sqrt[3]{a^2b}}$

Solution: Use the quotient rule of radicals, then write the result in simplest radical form.

(a) $\dfrac{\sqrt{192}}{\sqrt{6}} = \sqrt{\dfrac{192}{6}} = \sqrt{32} = 4\sqrt{2}$

(b) $\dfrac{3\sqrt{75x}}{\sqrt{3}} = 3\sqrt{\dfrac{75x}{3}} = 3\sqrt{25x} = 15\sqrt{x}$

(c) $\dfrac{\sqrt[3]{4a^5b^7c}}{\sqrt[3]{a^2b}} = \sqrt[3]{\dfrac{4a^5b^7c}{a^2b}} = \sqrt[3]{4a^3b^6c} = ab^2\sqrt[3]{4c}$

To rationalize the denominator of an expression which has a binomial denominator, we use the conjugate of the denominator.

EXAMPLE 9-18: Rationalize the denominator of **(a)** $\dfrac{1}{\sqrt{5} - \sqrt{3}}$ and **(b)** $\dfrac{5y}{2 + \sqrt{y}}$.

Solution: **(a)** $\dfrac{1}{\sqrt{5} - \sqrt{3}} = \dfrac{1}{\sqrt{5} - \sqrt{3}} \cdot \dfrac{\sqrt{5} + \sqrt{3}}{\sqrt{5} + \sqrt{3}}$ Multiply both the numerator and the denominator by the conjugate of the denominator.

$$= \dfrac{\sqrt{5} + \sqrt{3}}{5 - 3}$$

$$= \dfrac{\sqrt{5} + \sqrt{3}}{2}$$

(b) $\dfrac{5y}{2 + \sqrt{y}} = \dfrac{5y}{2 + \sqrt{y}} \cdot \dfrac{2 - \sqrt{y}}{2 - \sqrt{y}}$

$$= \dfrac{10y - 5y\sqrt{y}}{2^2 - (\sqrt{y})^2}$$

$$= \dfrac{10y - 5y\sqrt{y}}{4 - y}$$

Note: Because the radicands are assumed to be nonnegative in this section, $(\sqrt{y})^2$ must equal y.

9-3. Solving Equations Involving Radicals

A. Solving equations that involve a single radical term

An equation that has one or more radicals containing a variable in a radicand is called a **radical equation.** To solve a radical equation, we use the following principle to clear the radical(s).

The Power Principle for Equations

If x is a solution of $a = b$, then x is a solution of $a^n = b^n$ for any natural number n.

EXAMPLE 9-19: Solve $\sqrt{7x - 5} + 1 = 5$.

Solution: $\sqrt{7x - 5} + 1 = 5$

$$\sqrt{7x - 5} = 4 \qquad \text{Isolate the radical as one member of the equation.}$$
$$(\sqrt{7x - 5})^2 = 4^2 \qquad \text{Use the power principle of equations with } n = 2.$$
$$7x - 5 = 16 \longleftarrow \text{radical cleared}$$
$$7x = 21$$
$$x = 3 \longleftarrow \text{proposed solution}$$

Check:
$$\frac{\sqrt{7x - 5} + 1 = 5}{\begin{array}{c|c} \sqrt{7 \cdot 3 - 5} + 1 & 5 \\ 4 + 1 & 5 \\ 5 & 5 \longleftarrow \text{3 checks} \end{array}}$$

It is important to note that the converse of the power principle may not always be true. That is, if x is a solution of $a^n = b^n$, x may not be a solution of $a = b$. For example, $x^2 = (-2)^2$ has both 2 and -2 as solutions, but -2 is the only solution of $x = -2$. A solution of $a^n = b^n$ that is not also a solution of $a = b$ is called an **extraneous solution.** Proposed solutions obtained by using the power principle for equations must be checked to determine whether they are actual solutions of the original equation or whether they are extraneous solutions. In this section, to solve means to find proposed solutions and check to eliminate all extraneous solutions.

EXAMPLE 9-20: Solve $x = \sqrt{2x + 2} + 3$.

Solution:
$$x = \sqrt{2x + 2} + 3$$
$$x - 3 = \sqrt{2x + 2} \qquad \text{Add } -3 \text{ to isolate the radical term.}$$
$$(x - 3)^2 = (\sqrt{2x + 2})^2 \qquad \text{Use the power principle for equations.}$$
$$x^2 - 6x + 9 = 2x + 2$$
$$x^2 - 8x + 7 = 0$$
$$(x - 1)(x - 7) = 0$$
$$x - 1 = 0 \text{ or } x - 7 = 0 \qquad \text{Use the zero product rule.}$$
$$x = 1 \text{ or } \qquad x = 7 \longleftarrow \text{proposed solutions}$$

Check: For $x = 1$:
$$\frac{x = \sqrt{2x + 2} + 3}{\begin{array}{c|c} 1 & \sqrt{2(1) + 2} + 3 \\ 1 & 2 + 3 \\ 1 & 5 \longleftarrow \text{1 does not check} \end{array}}$$

Find $x = 7$:
$$\frac{x = \sqrt{2x + 2} + 3}{\begin{array}{c|c} 7 & \sqrt{2(7) + 2} + 3 \\ 7 & 4 + 3 \\ 7 & 7 \longleftarrow \text{7 checks} \end{array}}$$

Since 7 checks but 1 does not check, the only solution is 7.

B. Solving equations that involve more than one radical term

If an equation has more than one radical term, you may need to use the power principle for equations more than once. The general procedure is as follows.

To solve an equation with more than one radical term:

1. Isolate one of the radical terms.
2. Use the power principle for equations.
3. If a radical remains, repeat steps 1 and 2.
4. Solve the resulting equation and check proposed solutions in the original equation.

EXAMPLE 9-21: Solve $\sqrt{4y - 8} - \sqrt{2y + 3} = 1$.

Solution: $\sqrt{4y - 8} - \sqrt{2y + 3} = 1$

$$\sqrt{4y - 8} = \sqrt{2y + 3} + 1 \qquad \text{Add } \sqrt{2y + 3}, \text{ to isolate one of the radical terms.}$$

$$(\sqrt{4y - 8})^2 = (\sqrt{2y + 3} + 1)^2 \qquad \text{Use the power principle for equations.}$$

$$4y - 8 = 2y + 3 + 2\sqrt{2y + 3} + 1$$

$$2y - 12 = 2\sqrt{2y + 3}$$

$$y - 6 = \sqrt{2y + 3}$$

$$(y - 6)^2 = (\sqrt{2y + 3})^2 \qquad \text{Use the power principle again.}$$

$$y^2 - 12y + 36 = 2y + 3$$

$$y^2 - 14y + 33 = 0$$

$$(y - 3)(y - 11) = 0$$

$$y - 3 = 0 \text{ or } y - 11 = 0 \qquad \text{Use the zero product rule.}$$

$$y = 3 \text{ or } \qquad y = 11$$

Check: For $y = 3$:

$$\begin{array}{c|c} \sqrt{4y - 8} - \sqrt{2y + 3} & = 1 \\ \hline \sqrt{4(3) - 8} - \sqrt{2(3) + 3} & 1 \\ \sqrt{4} - \sqrt{9} & 1 \\ -1 & 1 \quad \longleftarrow \text{ 3 does not check} \end{array}$$

For $y = 11$:

$$\begin{array}{c|c} \sqrt{4y - 8} - \sqrt{2y + 3} & = 1 \\ \hline \sqrt{4(11) - 8} - \sqrt{2(11) + 3} & 1 \\ \sqrt{36} - \sqrt{25} & 1 \\ 1 & 1 \quad \longleftarrow \text{ 11 checks} \end{array}$$

The only solution of $\sqrt{4y - 8} - \sqrt{2y + 3} = 1$ is $y = 11$.

Caution: When you square a binomial do not forget the middle term. For example, a common error is to write $(\sqrt{2y + 3} + 1)^2$ as $(\sqrt{2y + 3})^2 + 1^2$ instead of the correct result: $(\sqrt{2y + 3})^2 + 2\sqrt{2y + 3} + 1^2$.

RAISE YOUR GRADES

Can you . . . ?

- ☑ evaluate radical expressions
- ☑ evaluate expressions that involve rational exponents
- ☑ simplify expressions involving rational exponents
- ☑ rename a radical expression in simplest radical form
- ☑ combine like radicals

☑ multiply and divide with radical expressions
☑ rationalize the denominator of a fractional expression
☑ solve radical equations

SOLVED PROBLEMS

PROBLEM 9-1 Evaluate each of the following: (a) $16^{\frac{1}{2}}$ (b) $64^{\frac{1}{3}}$ (c) $-625^{\frac{1}{4}}$

Solution: Use $a^{\frac{1}{n}} = \sqrt[n]{a}$ [see Example 9-1].

(a) $16^{\frac{1}{2}} = \sqrt{16} = 4$ (b) $64^{\frac{1}{3}} = \sqrt[3]{64} = 4$ (c) $-625^{\frac{1}{4}} = -\sqrt[4]{625} = -5$

PROBLEM 9-2 Write each of the following in radical notation:

(a) $5^{\frac{2}{4}}$ (b) $2^{\frac{2}{6}}$ (c) $-7^{\frac{1}{4}}$ (d) $3^{0.75}$ (e) $11^{0.4}$

Solution: Use $a^{\frac{m}{n}} = \sqrt[n]{a^m}$ [see Example 9-2].

(a) $5^{\frac{2}{4}} = 5^{\frac{1}{2}} = \sqrt{5^1} = \sqrt{5}$ (b) $2^{\frac{2}{6}} = 2^{\frac{1}{3}} = \sqrt[3]{2^1} = \sqrt[3]{2}$ (c) $-7^{\frac{1}{4}} = -\sqrt[4]{7^1} = -\sqrt[4]{7}$
(d) $3^{0.75} = 3^{\frac{3}{4}} = \sqrt[4]{3^3} = \sqrt[4]{27}$ (e) $11^{0.4} = 11^{\frac{4}{10}} = 11^{\frac{2}{5}} = \sqrt[5]{11^2} = \sqrt[5]{121}$

PROBLEM 9-3 Evaluate each of the following:

(a) $16^{\frac{3}{4}}$ (b) $-9^{\frac{3}{2}}$ (c) $(-8)^{\frac{2}{3}}$ (d) $64^{\frac{5}{6}}$ (e) $(-32)^{\frac{7}{5}}$

Solution: Use $a^{\frac{m}{n}} = (\sqrt[n]{a})^m$ [see Example 9-3].

(a) $16^{\frac{3}{4}} = (\sqrt[4]{16})^3 = 2^3 = 8$ (b) $-9^{\frac{3}{2}} = -(\sqrt{9})^3 = -(3)^3 = -27$
(c) $(-8)^{\frac{2}{3}} = (\sqrt[3]{-8})^2 = (-2)^2 = 4$ (d) $64^{\frac{5}{6}} = (\sqrt[6]{64})^5 = 2^5 = 32$
(e) $(-32)^{\frac{7}{5}} = (\sqrt[5]{-32})^7 = (-2)^7 = -128$

PROBLEM 9-4 Simplify each of the following: (a) $x^{\frac{1}{4}} \cdot x^{\frac{1}{3}}$ (b) $\dfrac{y}{y^{\frac{1}{4}}}$ (c) $(z^{-2})^{-3}$

Solution: Use the rules of exponents [see Example 9-4].

(a) $x^{\frac{1}{4}} \cdot x^{\frac{1}{3}} = x^{\frac{1}{4}+\frac{1}{3}} = x^{\frac{7}{12}}$ (b) $\dfrac{y}{y^{\frac{1}{4}}} = y^{1-\frac{1}{4}} = y^{\frac{3}{4}}$ (c) $(z^{-2})^{-3} = z^{(-2)(-3)} = z^6$

PROBLEM 9-5 Simplify each of the following: (a) $(2x^{-3}y^2)^{-1}$ (b) $(m^{-6}n^3)^{\frac{1}{3}}$ (c) $\left(\dfrac{32pq^6r}{2r^{-3}}\right)^{\frac{1}{2}}$

Solution: Use the extended power rule [see Example 9-5].

(a) $(2x^{-3}y^2)^{-1} = 2^{-1}x^{(-3)(-1)}y^{2(-1)} = 2^{-1}x^3y^{-2} = \dfrac{x^3}{2y^2}$

(b) $(m^{-6}n^3)^{\frac{1}{3}} = m^{(-6)(\frac{1}{3})}n^{3(\frac{1}{3})} = m^{-2}n^1 = \dfrac{n}{m^2}$

(c) $\left(\dfrac{32pq^6r}{2r^{-3}}\right)^{\frac{1}{2}} = (16pq^6r^4)^{\frac{1}{2}} = 16^{\frac{1}{2}}p^{\frac{1}{2}}q^{6 \cdot \frac{1}{2}}r^{4 \cdot \frac{1}{2}} = 4p^{\frac{1}{2}}q^3r^2$

PROBLEM 9-6 Write each of the following in rational exponential notation:
(a) $y\sqrt{y}$ (b) $\sqrt[3]{uv^2}$ (c) $\sqrt{x^6y^4z}$ (d) $\sqrt[3]{x^{12}}$ (e) $-\sqrt[3]{w^6}$ (f) $\sqrt{a^2 + b^3}$

Solution: Use $\sqrt[n]{a^m} = a^{\frac{m}{n}}$ and the rules of exponents [see Example 9-6].

(a) $y\sqrt{y} = y \cdot y^{\frac{1}{2}} = y^{\frac{3}{2}}$

(b) $\sqrt[3]{uv^2} = (uv^2)^{\frac{1}{3}} = u^{\frac{1}{3}}v^{\frac{2}{3}}$

(c) $\sqrt{x^6y^4z} = (x^6y^4z)^{\frac{1}{2}} = x^3y^2z^{\frac{1}{2}}$

(d) $\sqrt[3]{x^{12}} = x^{\frac{12}{3}} = x^4$

(e) $-\sqrt[3]{w^6} = -w^{\frac{6}{3}} = -w^2$

(f) $\sqrt{a^2 + b^3} = (a^2 + b^3)^{\frac{1}{2}}$

PROBLEM 9-7 Simplify each of the following: (a) $\sqrt{120}$ (b) $\sqrt[3]{40}$ (c) $\sqrt{75x^5y^4}$

Solution: Recall that to simplify a radical means to write it in an equivalent form in which the radical contains only powers less than the index number [see Example 9-7].

(a) $\sqrt{120} = \sqrt{2^3 \cdot 3 \cdot 5} = \sqrt{2^2 \cdot 2^1 \cdot 3 \cdot 5} = \sqrt{2^2}\sqrt{2 \cdot 3 \cdot 5} = 2\sqrt{30}$

(b) $\sqrt[3]{40} = \sqrt[3]{2^3 \cdot 5} = \sqrt[3]{2^3}\sqrt[3]{5} = 2\sqrt[3]{5}$

(c) $\sqrt{75x^5y^4} = \sqrt{3 \cdot 5^2(x^2)^2x(y^2)^2} = \sqrt{5^2(x^2)^2(y^2)^2}\sqrt{3x} = 5x^2y^2\sqrt{3x}$

PROBLEM 9-8 Reduce the order of each of the following: (a) $\sqrt[6]{x^3}$ (b) $\sqrt[4]{x^{12}y^2}$ (c) $\sqrt[8]{64y^2z^4}$

Solution: Write in exponential form and use the rules of exponents [see Example 9-8].

(a) $\sqrt[6]{x^3} = x^{\frac{3}{6}} = x^{\frac{1}{2}} = \sqrt{x}$

(b) $\sqrt[4]{x^{12}y^2} = (x^{12}y^2)^{\frac{1}{4}} = x^{\frac{12}{4}}y^{\frac{2}{4}} = x^3y^{\frac{1}{2}} = x^3\sqrt{y}$

(c) $\sqrt[8]{64y^2z^4} = (2^6y^2z^4)^{\frac{1}{8}} = 2^{\frac{6}{8}}y^{\frac{2}{8}}z^{\frac{4}{8}} = 2^{\frac{3}{4}}y^{\frac{1}{4}}z^{\frac{2}{4}} = \sqrt[4]{2^3yz^2} = \sqrt[4]{8yz^2}$

PROBLEM 9-9 Rationalize the denominator of each of the following:

(a) $\dfrac{3}{\sqrt{5}}$ (b) $\dfrac{5}{\sqrt[3]{4}}$ (c) $\dfrac{x}{\sqrt{xy}}$ (d) $\dfrac{5}{\sqrt{7} - \sqrt{2}}$ (e) $\dfrac{\sqrt{x} - \sqrt{y}}{\sqrt{x} + \sqrt{y}}$ (f) $\dfrac{5w}{3 + \sqrt{w}}$

Solution: Recall that to rationalize the denominator of a radical expression, you multiply both the numerator and the denominator by the same factor, which clears the radical(s) from the denominator [see Examples 9-9 and 9-18].

(a) $\dfrac{3}{\sqrt{5}} = \dfrac{3}{\sqrt{5}} \cdot \dfrac{\sqrt{5}}{\sqrt{5}} = \dfrac{3\sqrt{5}}{5}$

(b) $\dfrac{5}{\sqrt[3]{4}} = \dfrac{5}{\sqrt[3]{2^2}} = \dfrac{5}{\sqrt[3]{2^2}} \cdot \dfrac{\sqrt[3]{2}}{\sqrt[3]{2}} = \dfrac{5\sqrt[3]{2}}{\sqrt[3]{2^3}} = \dfrac{5\sqrt[3]{2}}{2}$

(c) $\dfrac{x}{\sqrt{xy}} = \dfrac{x}{\sqrt{xy}} \cdot \dfrac{\sqrt{xy}}{\sqrt{xy}} = \dfrac{x\sqrt{xy}}{\sqrt{(xy)^2}} = \dfrac{x\sqrt{xy}}{xy} = \dfrac{\sqrt{xy}}{y}$

(d) $\dfrac{5}{\sqrt{7} - \sqrt{2}} = \dfrac{5}{\sqrt{7} - \sqrt{2}} \cdot \dfrac{\sqrt{7} + \sqrt{2}}{\sqrt{7} + \sqrt{2}} = \dfrac{5(\sqrt{7} + \sqrt{2})}{7 - 2} = \dfrac{5(\sqrt{7} + \sqrt{2})}{5} = \sqrt{7} + \sqrt{2}$

(e) $\dfrac{\sqrt{x} - \sqrt{y}}{\sqrt{x} + \sqrt{y}} = \dfrac{\sqrt{x} - \sqrt{y}}{\sqrt{x} + \sqrt{y}} \cdot \dfrac{\sqrt{x} - \sqrt{y}}{\sqrt{x} - \sqrt{y}} = \dfrac{x - 2\sqrt{x}\sqrt{y} + y}{x - y} = \dfrac{x - 2\sqrt{xy} + y}{x - y}$

(f) $\dfrac{5w}{3 + \sqrt{w}} = \dfrac{5w}{3 + \sqrt{w}} \cdot \dfrac{3 - \sqrt{w}}{3 - \sqrt{w}} = \dfrac{15w - 5w\sqrt{w}}{9 - w}$

PROBLEM 9-10 Write each of the following in simplest radical form:

(a) $\dfrac{\sqrt{15b}}{\sqrt{25a^2}}$ (b) $\dfrac{5\sqrt{xy^2}}{\sqrt{5z}}$ (c) $\sqrt[3]{\dfrac{35}{49w^2}}$ (d) $\sqrt[4]{\dfrac{4pq^3}{81p^3q^{-1}}}$

Solution: Recall that a radical expression is in simplest radical form if it involves only one radical, the radicand involves only powers less than the index, the order of the radical is reduced as far as possible, no radicals appear in a denominator, and no fraction appears in the radicand [see Example 9-10].

(a) $\dfrac{\sqrt{15b}}{\sqrt{25a^2}} = \sqrt{\dfrac{15b}{25a^2}} = \sqrt{\dfrac{3b}{5a^2}} = \dfrac{\sqrt{3b}}{\sqrt{5a^2}} = \dfrac{\sqrt{3b}}{a\sqrt{5}} = \dfrac{\sqrt{3b}}{a\sqrt{5}} \cdot \dfrac{\sqrt{5}}{\sqrt{5}} = \dfrac{\sqrt{15b}}{5a}$

(b) $\dfrac{5\sqrt{xy^2}}{\sqrt{5z}} = \dfrac{5y\sqrt{x}}{\sqrt{5z}} = \dfrac{5y\sqrt{x}}{\sqrt{5z}} \cdot \dfrac{\sqrt{5z}}{\sqrt{5z}} = \dfrac{5y\sqrt{5xz}}{5z} = \dfrac{y\sqrt{5xz}}{z}$

(c) $\sqrt[3]{\dfrac{35}{49w^2}} = \sqrt[3]{\dfrac{5}{7w^2}} = \dfrac{\sqrt[3]{5}}{\sqrt[3]{7w^2}} \cdot \dfrac{\sqrt[3]{7^2w}}{\sqrt[3]{7^2w}} = \dfrac{\sqrt[3]{5 \cdot 7^2 w}}{\sqrt[3]{7^3 w^3}} = \dfrac{\sqrt[3]{245w}}{7w}$

(d) $\sqrt[4]{\dfrac{4pq^3}{81p^3q^{-1}}} = \sqrt[4]{\dfrac{2^2q^4}{3^4p^2}} = \left(\dfrac{2^2q^4}{3^4p^2}\right)^{\frac{1}{4}} = \dfrac{2^{\frac{1}{2}}q}{3^1 p^{\frac{1}{2}}} = \dfrac{q\sqrt{2}}{3\sqrt{p}} = \dfrac{q\sqrt{2}}{3\sqrt{p}} \cdot \dfrac{\sqrt{p}}{\sqrt{p}} = \dfrac{q\sqrt{2p}}{3p}$

PROBLEM 9-11 Use Appendix Table 1 to approximate **(a)** $\sqrt{275}$ and **(b)** $\sqrt[3]{648}$.

Solution: Factor the radicand and use Appendix Table 1 [see Example 9-11].

    (a) $\sqrt{275} = \sqrt{5^2 \cdot 11} = \sqrt{5^2}\sqrt{11} \approx 5(3.317) = 16.585$

    (b) $\sqrt[3]{648} = \sqrt[3]{2^3 \cdot 3^4} = \sqrt[3]{2^3}\sqrt[3]{3^3}\sqrt[3]{3} \approx 2 \cdot 3(1.442) = 8.652$

PROBLEM 9-12 Combine like radicals:

(a) $5\sqrt{3} + 7\sqrt{3}$ **(b)** $\sqrt[3]{5y} - 7\sqrt[3]{5y}$ **(c)** $2x\sqrt{xy} + 3x\sqrt{xy} - x\sqrt{xy}$

Solution: Recall that to combine like radicals, you use the distributive property [see Example 9-12].

    (a) $5\sqrt{3} + 7\sqrt{3} = (5 + 7)\sqrt{3} = 12\sqrt{3}$

    (b) $\sqrt[3]{5y} - 7\sqrt[3]{5y} = (1 - 7)\sqrt[3]{5y} = -6\sqrt[3]{5y}$

    (c) $2x\sqrt{xy} + 3x\sqrt{xy} - x\sqrt{xy} = (2x + 3x - x)\sqrt{xy} = 4x\sqrt{xy}$

PROBLEM 9-13 Simplify and combine like radicals:

(a) $3\sqrt{8} + 5\sqrt{2}$ **(b)** $2x\sqrt{80x^5} - 3\sqrt{45x^7}$ **(c)** $5\sqrt{75y} - 2\sqrt{48y} + \sqrt{243y}$

Solution: First simplify each radical and then use the distributive property [see Example 9-13].

    (a) $3\sqrt{8} + 5\sqrt{2} = 3\sqrt{2^3} + 5\sqrt{2} = 6\sqrt{2} + 5\sqrt{2} = 11\sqrt{2}$

    (b) $2x\sqrt{80x^5} - 3\sqrt{45x^7} = 8x^3\sqrt{5x} - 9x^3\sqrt{5x} = -x^3\sqrt{5x}$

    (c) $5\sqrt{75y} - 2\sqrt{48y} + \sqrt{243y} = 25\sqrt{3y} - 8\sqrt{3y} + 9\sqrt{3y} = 26\sqrt{3y}$

PROBLEM 9-14 Find the indicated products:

(a) $\sqrt{3}\sqrt{5}$ **(b)** $\sqrt[3]{x}\sqrt[3]{x^2y}$ **(c)** $\sqrt{3a}(\sqrt{7a} + 5\sqrt{3})$ **(d)** $(3 + 2\sqrt{a})(5 + \sqrt{a})$
(e) $(\sqrt{5x} + \sqrt{3y})(\sqrt{5x} - \sqrt{3y})$

Solution: Recall that to multiply radicals, you use the product rule for radicals [see Examples 9-14, 9-15, and 9-16].

    (a) $\sqrt{3}\sqrt{5} = \sqrt{3 \cdot 5} = \sqrt{15}$ **(b)** $\sqrt[3]{x}\sqrt[3]{x^2y} = \sqrt[3]{x^3y} = x\sqrt[3]{y}$

    (c) $\sqrt{3a}(\sqrt{7a} + 5\sqrt{3}) = \sqrt{21a^2} + 5\sqrt{3^2a} = a\sqrt{21} + 15\sqrt{a}$

    (d) $(3 + 2\sqrt{a})(5 + \sqrt{a}) = 15 + 3\sqrt{a} + 10\sqrt{a} + 2a = 15 + 13\sqrt{a} + 2a$

    (e) $(\sqrt{5x} + \sqrt{3y})(5x - \sqrt{3y}) = (\sqrt{5x})^2 - (\sqrt{3y})^2 = 5x - 3y$

PROBLEM 9-15 Find the indicated quotients: **(a)** $\dfrac{\sqrt{60x^3y}}{\sqrt{12xy}}$ **(b)** $\dfrac{\sqrt[3]{144z}}{\sqrt[3]{40z^2}}$

Solution: Use the quotient rule for radicals and then write the result in simplest radical form [see Example 9-17].

    (a) $\dfrac{\sqrt{60x^3y}}{\sqrt{12xy}} = \sqrt{\dfrac{60x^3y}{12xy}} = \sqrt{5x^2} = x\sqrt{5}$

    (b) $\dfrac{\sqrt[3]{144z}}{\sqrt[3]{40z^2}} = \sqrt[3]{\dfrac{144z}{40z^2}} = \sqrt[3]{\dfrac{18}{5z}} = \dfrac{\sqrt[3]{18}}{\sqrt[3]{5z}} = \dfrac{\sqrt[3]{18}}{\sqrt[3]{5z}} \cdot \dfrac{\sqrt[3]{5^2z^2}}{\sqrt[3]{5^2z^2}} = \dfrac{\sqrt[3]{450z^2}}{\sqrt[3]{5^3z^3}} = \dfrac{\sqrt[3]{450z^2}}{5z}$

PROBLEM 9-16 Solve **(a)** $\sqrt[3]{y^2} + 11 = 3$ and **(b)** $\sqrt{4y - 3} = \sqrt{2y + 2} + 1$.

Solution: Recall that to solve an equation involving radicals, you use the power principle for equations [see Examples 9-19, 9-20, and 9-21].

(a)
$$\sqrt[3]{y^2 + 11} = 3$$
$$(\sqrt[3]{y^2 + 11})^3 = 3^3 \qquad \text{Use the power principle for equations, with } n = 3.$$
$$y^2 + 11 = 27$$
$$y^2 - 16 = 0$$
$$(y + 4)(y - 4) = 0$$
$$y + 4 = 0 \quad \text{or} \quad y - 4 = 0$$
$$y = -4 \text{ or} \qquad y = 4$$

Check: For $y = \pm 4$:

$\sqrt[3]{y^2 + 11} = 3$	
$\sqrt[3]{(\pm 4)^2 + 11}$	3
$\sqrt[3]{27}$	3
3	3 $\longleftarrow$ both -4 and 4 check

The solutions are $y = -4$ and $y = 4$.

(b)
$$\sqrt{4y - 3} = \sqrt{2y + 2} + 1$$
$$(\sqrt{4y - 3})^2 = (\sqrt{2y + 2} + 1)^2$$
$$4y - 3 = 2y + 2 + 2\sqrt{2y + 2} + 1$$
$$2y - 6 = 2\sqrt{2y + 2}$$
$$y - 3 = \sqrt{2y + 2}$$
$$(y - 3)^2 = (\sqrt{2y + 2})^2$$
$$y^2 - 6y + 9 = 2y + 2$$
$$y^2 - 8y + 7 = 0$$
$$(y - 1)(y - 7) = 0$$
$$y - 1 = 0 \text{ or } y - 7 = 0$$
$$y = 1 \text{ or} \qquad y = 7$$

Check: For $y = 1$:

$\sqrt{4y - 3} = \sqrt{2y + 2} + 1$	
$\sqrt{4(1) - 3}$	$\sqrt{2(1) + 2} + 1$
$\sqrt{1}$	$\sqrt{4} + 1$
1	3 $\longleftarrow$ 1 does not check

For $y = 7$:

$\sqrt{4y - 3} = \sqrt{2y + 2} + 1$	
$\sqrt{4(7) - 3}$	$\sqrt{2(7) + 2} + 1$
$\sqrt{25}$	$\sqrt{16} + 1$
5	5 $\longleftarrow$ 7 checks

The only solution of $\sqrt{4y - 3} = \sqrt{2y + 2} + 1$ is 7.

Supplementary Exercises

PROBLEM 9-17 Evaluate each of the following:

(a) $81^{\frac{1}{2}}$ **(b)** $8^{\frac{1}{3}}$ **(c)** $(-125)^{\frac{1}{3}}$ **(d)** $81^{\frac{1}{4}}$ **(e)** $243^{\frac{1}{5}}$ **(f)** $1024^{\frac{1}{10}}$ **(g)** $25^{\frac{3}{2}}$ **(h)** $27^{\frac{2}{3}}$

(i) $(-27)^{\frac{2}{3}}$ **(j)** $-16^{\frac{3}{4}}$ **(k)** $64^{\frac{2}{6}}$ **(l)** $100^{\frac{5}{2}}$

PROBLEM 9-18 Write each of the following in radical notation:

(a) $3^{\frac{2}{3}}$ (b) $4^{\frac{1}{4}}$ (c) $-10^{\frac{1}{2}}$ (d) $(-10)^{\frac{1}{3}}$ (e) $2^{0.75}$ (f) $5^{0.25}$

PROBLEM 9-19 Simplify each of the following:

(a) $p^3 p^{-4}$ (b) $q^{\frac{1}{2}} q^{\frac{1}{5}}$ (c) $\dfrac{x^7}{x^5}$ (d) $\dfrac{y^3}{y^{2\frac{1}{2}}}$ (e) $(m^{-2})^{-3}$ (f) $(n^{-\frac{1}{2}})^{-6}$ (g) $(3x^{-2}y^3)^{-1}$

(h) $(u^{-4}v^2)^{\frac{1}{2}}$ (i) $\left(\dfrac{20s^4t^{-3}}{5s^6t^{-5}}\right)^{\frac{1}{2}}$ (j) $\left(\dfrac{x^{-\frac{1}{2}}y^{\frac{1}{3}}}{z^{\frac{3}{4}}}\right)^{12}$

PROBLEM 9-20 Write each of the following in rational exponential notation:

(a) $\sqrt[3]{x^2}$ (b) $\sqrt[4]{y^2z}$ (c) $x\sqrt{x}$ (d) $-x^2\sqrt[5]{x^{10}y^{15}}$ (e) $\sqrt{p^2-q^2}$

PROBLEM 9-21 Reduce the order of each of the following:

(a) $\sqrt[6]{x^3}$ (b) $\sqrt[8]{y^2z^6}$ (c) $\sqrt[6]{64x^4y^2}$

PROBLEM 9-22 Rationalize the denominator of each of the following:

(a) $\dfrac{5}{\sqrt{6}}$ (b) $\dfrac{3}{\sqrt[3]{9}}$ (c) $\dfrac{w}{\sqrt{wy^2}}$ (d) $\dfrac{5}{\sqrt{7}+\sqrt{2}}$ (e) $\dfrac{m}{m-\sqrt{n}}$ (f) $\dfrac{\sqrt{s}+\sqrt{t}}{\sqrt{s}-\sqrt{t}}$

PROBLEM 9-23 Write each of the following in simplest radical form:

(a) $\sqrt{32}$ (b) $\sqrt{45m^3}$ (c) $\sqrt{\dfrac{15w^4x^{-1}}{60w^5x^3}}$ (d) $\sqrt[3]{\dfrac{12p}{75p^3}}$ (e) $\dfrac{\sqrt[5]{32x}}{\sqrt[5]{-x^3}}$

PROBLEM 9-24 Use Appendix Table 1 to approximate each of the following:

(a) $\sqrt{162}$ (b) $\sqrt{153}$ (c) $\sqrt[3]{200}$ (d) $\sqrt[3]{3000}$ (e) $\sqrt{184}$ (f) $\sqrt[3]{-3773}$

PROBLEM 9-25 Combine like radicals:

(a) $\sqrt{5}-4\sqrt{5}$ (b) $\sqrt[3]{2x}+5\sqrt[3]{2x}$ (c) $x\sqrt[5]{xy^3}+3x\sqrt[5]{xy^3}-7x\sqrt[5]{xy^3}$

PROBLEM 9-26 Simplify and combine like radicals:

(a) $\sqrt{32}-\sqrt{8}$ (b) $5x\sqrt{16x}+3\sqrt{x^3}$ (c) $\sqrt[3]{16y^4}-y\sqrt[3]{54y}$
(d) $\sqrt{153m^2n}+m\sqrt{68n}-\sqrt{17m^2n}$

PROBLEM 9-27 Find the indicated products:

(a) $\sqrt{2}\sqrt{7}$ (b) $\sqrt{x}\sqrt{xy}$ (c) $x\sqrt[3]{xy^2}\sqrt[3]{xy}$ (d) $\sqrt{5x}(\sqrt{7x}-\sqrt{5})$ (e) $(\sqrt{w}+3\sqrt{wx})\sqrt{w}$
(f) $3\sqrt[3]{p}(2\sqrt[3]{p^4}-\sqrt[3]{p^2}+2\sqrt[3]{p})$ (g) $(5+\sqrt{2r})^2$ (h) $(4+\sqrt{3s})(4-\sqrt{3s})$
(i) $(\sqrt{a}-\sqrt{b})^2$ (j) $(\sqrt{a}+\sqrt{b})(\sqrt{a}-\sqrt{b})$

PROBLEM 9-28 Find the indicated quotients:

(a) $\dfrac{\sqrt{125}}{\sqrt{5}}$ (b) $\dfrac{\sqrt{64x^2}}{\sqrt{2x}}$ (c) $\dfrac{\sqrt[3]{64y^2}}{\sqrt[3]{2y^{-1}}}$ (d) $\dfrac{\sqrt[5]{160x}}{\sqrt[5]{5x^3}}$

PROBLEM 9-29 Solve the following equations:

(a) $\sqrt{-5x+21}=6$ (b) $\sqrt{3x+7}+x=21$ (c) $\sqrt[3]{2x-13}=3$
(d) $\sqrt{3x+6}-\sqrt{4x-3}=2$ (e) $\sqrt{x-3}=\sqrt{-x+15}$ (f) $\sqrt{x+4}\sqrt{x}=\sqrt{x^2+10}$
(g) $\sqrt{y-5}+\sqrt{2y-15}=0$

Answers to Supplementary Exercises

(9-17) (a) 9　(b) 2　(c) -5　(d) 3　(e) 3　(f) 2　(g) 125　(h) 9　(i) 9
(j) -8　(k) 4　(l) 100,000

(9-18) (a) $\sqrt[3]{9}$　(b) $\sqrt[4]{4}$　(c) $-\sqrt{10}$　(d) $\sqrt[3]{-10}$　(e) $\sqrt[4]{8}$　(f) $\sqrt[4]{5}$

(9-19) (a) $\dfrac{1}{p}$　(b) $q^{\frac{7}{10}}$　(c) x^2　(d) $y^{\frac{1}{2}}$　(e) m^6　(f) n^3　(g) $\dfrac{x^2}{3y^3}$　(h) $\dfrac{v}{u^2}$

(i) $\dfrac{2t}{s}$　(j) $\dfrac{y^4}{x^6 z^9}$

(9-20) (a) $x^{\frac{2}{3}}$　(b) $y^{\frac{1}{2}}z^{\frac{1}{4}}$　(c) $x^{\frac{3}{2}}$　(d) $-x^4 y^3$　(e) $(p^2-q^2)^{\frac{1}{2}}$

(9-21) (a) $\sqrt{x}$　(b) $\sqrt[4]{yz^3}$　(c) $2\sqrt[3]{x^2 y}$

(9-22) (a) $\dfrac{5\sqrt{6}}{6}$　(b) $\sqrt[3]{3}$　(c) $\dfrac{\sqrt{w}}{y}$　(d) $\sqrt{7}-\sqrt{2}$　(e) $\dfrac{m^2+m\sqrt{n}}{m^2-n}$　(f) $\dfrac{s+2\sqrt{st}+t}{s-t}$

(9-23) (a) $4\sqrt{2}$　(b) $3m\sqrt{5m}$　(c) $\dfrac{\sqrt{w}}{2x^2 w}$　(d) $\dfrac{\sqrt[3]{20p}}{5p}$　(e) $-\dfrac{2\sqrt[5]{x^3}}{x}$

(9-24) (a) 12.726　(b) 12.369　(c) 5.848　(d) 14.42　(e) 13.564　(f) -15.568

(9-25) (a) $-3\sqrt{5}$　(b) $6\sqrt[3]{2x}$　(c) $-3x\sqrt[5]{xy^3}$

(9-26) (a) $2\sqrt{2}$　(b) $23x\sqrt{x}$　(c) $-y\sqrt[3]{2y}$　(d) $4m\sqrt{17n}$

(9-27) (a) $\sqrt{14}$　(b) $x\sqrt{y}$　(c) $xy\sqrt[3]{x^2}$　(d) $x\sqrt{35}-5\sqrt{x}$　(e) $w+3w\sqrt{x}$
(f) $6p\sqrt[3]{p^2}-3p+6\sqrt[3]{p^2}$　(g) $25+10\sqrt{2r}+2r$　(h) $16-3s$　(i) $a-2\sqrt{ab}+b$
(j) $a-b$

(9-28) (a) 5　(b) $4\sqrt{2x}$　(c) $2y\sqrt[3]{4}$　(d) $\dfrac{2\sqrt[5]{x^3}}{x}$

(9-29) (a) -3　(b) 14　(c) 20　(d) 1　(e) 9　(f) $\frac{5}{2}$　(g) no solution

10 COMPLEX NUMBERS

THIS CHAPTER IS ABOUT

☑ Complex Numbers
☑ Adding or Subtracting Complex Numbers
☑ Multiplying Complex Numbers
☑ Dividing Complex Numbers

10-1. Complex Numbers

The equation $x + 5 = 1$ has no solution among the natural numbers, but it does have -4 as a solution if we extend the solutions under consideration to include integers.

The equation $2x = 7$ has no solution among the integers, but it does have $\frac{7}{2}$ as a solution if we extend the solutions under consideration to include rational numbers.

The equation $x^2 = 3$ has no solution among the rational numbers, but it does have $\sqrt{3}$ and $-\sqrt{3}$ as solutions if we extend the solutions under consideration to include irrational numbers.

The equation $x^2 = -1$ has no solution among the real numbers. Recall that $x^2 \geq 0$ for every real number x. If $x^2 = -1$ is to have a solution, then the solutions under consideration must be extended beyond the set of real numbers. This extension was first made during the 16th century by defining a new type of number whose square is a negative number.

Definition 10-1:

The symbol i, called the **imaginary unit,** is a number whose square is -1, that is $i^2 = -1$ and $i = \sqrt{-1}$.

We use the following definition to express the square root of any negative real number as the product of i and a positive real number.

Definition 10-2:

For any positive real number a, $\sqrt{-a} = i\sqrt{a}$.

Note: It is best to write i in front of the radical to avoid confusing $\sqrt{a}i$ with $\sqrt{ai}$.

EXAMPLE 10-1: Write each of the following radicals as a product of a real number and i:

(a) $\sqrt{-9}$ **(b)** $\sqrt{-8}$ **(c)** $\sqrt{-60x^3}$ (assume $x \in R$ and $x \geq 0$)

Solution: Use Definition 10-2, $\sqrt{-a} = i\sqrt{a}$.

$$\textbf{(a)} \ \sqrt{-9} = i\sqrt{9} \qquad \textbf{(b)} \ \sqrt{-8} = i\sqrt{8} \qquad \textbf{(c)} \ \sqrt{-60x^3} = i\sqrt{60x^3}$$
$$= i3 \qquad\qquad\qquad = i2\sqrt{2} \qquad\qquad\qquad = i2x\sqrt{15x}$$
$$= 3i \qquad\qquad\qquad = 2i\sqrt{2} \qquad\qquad\qquad = 2xi\sqrt{15x}$$

Any number of the form bi, where b is a nonzero real number, is called an **imaginary number.** Some examples of imaginary numbers are

$$3i \qquad 4.2i \qquad -7i \qquad i\sqrt{5}$$

249

If a and b are real numbers, the expression $a + bi$ is called a **complex number.** The number a is called the **real part** and the number b is called the **imaginary part.** Some examples of complex numbers are

$$2 + 5i \qquad -4 + i \qquad \tfrac{1}{2} - 5i \qquad 8 \qquad 4i$$

Every real number is a complex number. For example, $7 = 7 + 0 = 7 + 0i$. Every imaginary number is also a complex number. For example, $3i = 0 + 3i$.

10-2. Adding or Subtracting Complex Numbers

To combine two complex numbers:

1. Add (subtract) their real parts. Add (subtract) their imaginary parts.
2. Write the sum of the above results in $a + bi$ form.

EXAMPLE 10-2: Find the indicated sums:

(a) $(5 + 3i) + (-2 + 7i)$ **(b)** $(4 - 3i) + (-1 - 5i)$ **(c)** $9 + (2 - 14i)$

Solution: **(a)** $(5 + 3i) + (-2 + 7i) = [5 + (-2)] + (3 + 7)i$ Add real parts and add
$$= 3 + 10i \longleftarrow a + bi \text{ form} \qquad \text{imaginary parts.}$$

(b) $(4 - 3i) + (-1 - 5i) = [4 + (-1)] + [(-3) + (-5)]i$
$$= 3 + (-8)i$$
$$= 3 - 8i$$

(c) $9 + (2 - 14i) = (9 + 0i) + (2 - 14i)$
$$= (9 + 2) + [0 + (-14)]i$$
$$= 11 + (-14)i$$
$$= 11 - 14i$$

EXAMPLE 10-3: Find the indicated differences:

(a) $(3 + 5i) - (4 - 8i)$ **(b)** $(-6 + 7i) - (-5 + i)$ **(c)** $(-\tfrac{3}{4} + 2i) - (-\tfrac{1}{2} + \tfrac{1}{8}i)$

Solution: **(a)** $(3 + 5i) - (4 - 8i) = (3 - 4) + [5 - (-8)]i$ Subtract real parts and
$$= -1 + 13i \longleftarrow a + bi \text{ form} \qquad \text{subtract imaginary parts.}$$

(b) $(-6 + 7i) - (-5 + i) = [-6 - (-5)] + (7 - 1)i$
$$= -1 + 6i$$

(c) $(-\tfrac{3}{4} + 2i) - (-\tfrac{1}{2} + \tfrac{1}{8}i) = [(-\tfrac{3}{4}) - (-\tfrac{1}{2})] + (2 - \tfrac{1}{8})i$
$$= -\tfrac{1}{4} + \tfrac{15}{8}i$$

10-3. Multiplying Complex Numbers

To multiply a complex number by a real number or an imaginary number we use the distributive property.

EXAMPLE 10-4: Find the indicated products: **(a)** $3(-4 + 5i)$ **(b)** $5i(1 + 2i)$

Solution: **(a)** $3(-4 + 5i) = 3(-4) + 3(5i)$ **(b)** $5i(1 + 2i) = 5i(1 + 2i)$
$$= -12 + 15i \qquad\qquad\qquad\qquad = 5i + 10i^2$$
$$= 5i - 10 \text{ or } -10 + 5i$$

Multiplication of two complex numbers may be accomplished by use of the formula

$$(a + bi)(c + di) = (ac - bd) + (ad + bc)i$$

However, the method shown in Example 10-5 is generally used because it involves the familiar FOIL method and does not require additional memorization.

EXAMPLE 10-5: Find the indicated products: **(a)** $(-3 + 4i)(5 - 2i)$ **(b)** $(4i^2 + i)(3 - \sqrt{-16})$

Solution: Use the FOIL method and the definition $i^2 = -1$.

$$
\begin{aligned}
\textbf{(a)} \;\; (-3 + 4i)(5 - 2i) &= -15 + 6i + 20i - 8i^2 && \text{Use the FOIL method.} \\
&= -15 + 6i + 20i - 8(-1) && \text{Rename } i^2 \text{ as } -1. \\
&= -15 + 6i + 20i + 8 && \text{Simplify.} \\
&= [-15 + 8] + [6i + 20i] \\
&= -7 + 26i \longleftarrow a + bi \text{ form}
\end{aligned}
$$

$$
\begin{aligned}
\textbf{(b)} \;\; (4i^2 + i)(3 - \sqrt{-16}) &= (-4 + i)[3 - (+4i)] && \text{Write each number in } a + bi \\
&= (-4 + i)(3 - 4i) && \text{form.} \\
&= -12 + 16i + 3i - 4i^2 && \text{Use the FOIL method.} \\
&= -8 + 19i
\end{aligned}
$$

The product rule for radicals, $\sqrt{a} \cdot \sqrt{b} = \sqrt{ab}$, does not hold when both radicands are negative numbers. For example, if the product rule for radicals is applied to $\sqrt{-2} \cdot \sqrt{-5}$, we get $\sqrt{(-2)(-5)} = 10$, which is not correct. However, $\sqrt{-2} \cdot \sqrt{-5} = i\sqrt{2} \cdot i\sqrt{5} = -\sqrt{10}$, which is correct. To compute $\sqrt{a} \cdot \sqrt{b}$ when both a and b are negative numbers, always express each radical in terms of i before multiplying.

For any integer n, i^n is either i, -1, $-i$, or 1. The first four powers of i are

$$i^1 = i$$

$$i^2 = -1$$

$$i^3 = i^2 \cdot i = (-1)i = -i$$

$$i^4 = i^2 \cdot i^2 = (-1)(-1) = 1$$

Since $i^4 = 1$ it is true that for any integer n, $(i^4)^n$ will also equal 1, because $1^n = 1$. We often use $(i^4)^n = 1$ to simplify powers of i. For large powers of i it is helpful to divide the exponent of i by 4. For example:

$$
\begin{aligned}
i^{3006} &= (i^4)^{751} \cdot i^2 \\
&= 1 \cdot i^2 \\
&= i^2 \\
&= -1
\end{aligned}
$$

$$4 \overline{)3006} = 751 \qquad 3{,}006 = 4(751) + 2$$

EXAMPLE 10-6: Simplify each of the following:

(a) i^5 **(b)** i^6 **(c)** i^{15} **(d)** i^{41} **(e)** i^{1024}

Solution: Write the exponential expression as a product involving the largest possible power of (i^4) and either 1, i, i^2 or i^3.

(a) $i^5 = i^4 \cdot i = 1 \cdot i = i$
(b) $i^6 = i^4 \cdot i^2 = 1(-1) = -1$
(c) $i^{15} = i^{12} \cdot i^3 = (i^4)^3(i^3) = (1)^3(i^3) = (1)(-i) = -i$
(d) $i^{41} = i^{40} \cdot i = (i^4)^{10} \cdot i = 1^{10} \cdot i = i$
(e) $i^{1024} = (i^4)^{256} \cdot 1 = 1^{256} \cdot 1 = 1$

The **conjugate** of $a + bi$ (denoted by $\overline{a + bi}$) is $a - bi$. The conjugate of $a - bi$ (denoted by $\overline{a - bi}$) is $a + bi$.

EXAMPLE 10-7: Find the indicated conjugates: **(a)** $\overline{-3 + 5i}$ **(b)** $\overline{5}$ **(c)** $\overline{-8i}$

Solution: **(a)** $\overline{-3 + 5i} = -3 - 5i$ **(b)** $\overline{5} = \overline{5 + 0i}$ **(c)** $\overline{-8i} = \overline{0 - 8i}$

$$= 5 - 0i \text{ or } 5 \qquad\qquad = 0 + 8i \text{ or } 8i$$

It should be noted that the product of the complex number $a + bi$ and its conjugate $a - bi$ is the real number $a^2 + b^2$.

$$(a + bi)(a - bi) = a^2 - b^2i^2$$
$$= a^2 - b^2(-1)$$
$$= a^2 + b^2$$

10-4. Dividing Complex Numbers

We often use conjugates to find the quotient of two complex numbers.

EXAMPLE 10-8: Find the indicated quotients: **(a)** $\dfrac{2 - 3i}{1 + 4i}$ **(b)** $\dfrac{4}{3 - 2i}$ **(c)** $\dfrac{1}{i}$

Solution: **(a)** $\dfrac{2 - 3i}{1 + 4i} = \dfrac{2 - 3i}{1 + 4i} \cdot \dfrac{1 - 4i}{1 - 4i}$

Multiply both numerator and denominator by the conjugate of the denominator to produce a real number in the denominator.

$$= \frac{(2 - 3i)(1 - 4i)}{(1 + 4i)(1 - 4i)}$$

$$= \frac{2 - 8i - 3i + 12i^2}{1 - 16i^2}$$

$$= \frac{-10 - 11i}{1 + 16}$$

$$= \frac{-10 - 11i}{17} \text{ or } -\frac{10}{17} - \frac{11}{17}i \longleftarrow a + bi \text{ form}$$

(b) $\dfrac{4}{3 - 2i} = \dfrac{4}{3 - 2i} \cdot \dfrac{3 + 2i}{3 + 2i}$ **(c)** $\dfrac{1}{i} = \dfrac{1}{i} \cdot \dfrac{i}{i}$

$$= \frac{4(3 + 2i)}{(3 - 2i)(3 + 2i)} \qquad\qquad = \frac{i}{i^2}$$

$$= \frac{12 + 8i}{9 - 4i^2} \qquad\qquad = \frac{i}{-1}$$

$$= \frac{12 + 8i}{9 + 4} \qquad\qquad = -i$$

$$= \frac{12 + 8i}{13} \text{ or } \frac{12}{13} + \frac{8}{13}i$$

RAISE YOUR GRADES

Can you . . . ?

☑ identify the real part and the imaginary part of a complex number
☑ find the conjugate of a complex number
☑ add, subtract, multiply, and divide complex numbers
☑ simplify powers of i

SOLVED PROBLEMS

PROBLEM 10-1 Write each of the following radicals as a product of a real number and *i*:

(a) $\sqrt{-16}$ (b) $\sqrt{-12}$ (c) $\sqrt{-4y}$ (d) $\sqrt{-45x^3}$

Solution: Use Definition 10-2: For any positive real number a, $\sqrt{-a} = i\sqrt{a}$ [see Example 10-1].

(a) $\sqrt{-16} = i\sqrt{16}$ (b) $\sqrt{-12} = i\sqrt{12}$ (c) $\sqrt{-4y} = i\sqrt{4y}$ (d) $\sqrt{-45x^2} = i\sqrt{45x^2}$

$\qquad\qquad = i4$ $= i2\sqrt{3}$ $= i2\sqrt{y}$ $= i3x\sqrt{5}$

$\qquad\qquad = 4i$ $= 2i\sqrt{3}$ $= 2i\sqrt{y}$ $= 3xi\sqrt{5}$

PROBLEM 10-2 Find the indicated sums:

(a) $(2 + 3i) + (4 + 5i)$ (b) $(5 - 2i) + (7 + 11i)$ (c) $(-2 - 7i) + (-4 + 7i)$
(d) $(\sqrt{2} + 3i) + (4\sqrt{2} - 9i)$

Solution: Recall that to add complex numbers, add their real parts, then add their imaginary parts [see Example 10-2].

$\qquad$ (a) $(2 + 3i) + (4 + 5i) = (2 + 4) + (3 + 5)i$

$\qquad\qquad\qquad\qquad\qquad\quad = 6 + 8i$

$\qquad$ (b) $(5 - 2i) + (7 + 11i) = (5 + 7) + (-2 + 11)i$

$\qquad\qquad\qquad\qquad\qquad\qquad = 12 + 9i$

$\qquad$ (c) $(-2 - 7i) + (-4 + 7i) = [-2 + (-4)] + [(-7) + 7]i$

$\qquad\qquad\qquad\qquad\qquad\qquad = -6 + 0i$ or -6

$\qquad$ (d) $(\sqrt{2} + 3i) + (4\sqrt{2} - 9i) = (\sqrt{2} + 4\sqrt{2}) + [3 + (-9)]i$

$\qquad\qquad\qquad\qquad\qquad\qquad = 5\sqrt{2} - 6i$

PROBLEM 10-3 Find the indicated differences:

(a) $(2 + 11i) - (5 + 7i)$ (b) $(-3 + 2i) - (8 - 9i)$ (c) $16 - (-2 + 3i)$
(d) $(\frac{1}{2} + \frac{2}{3}i) - (-\frac{1}{4}i)$

Solution: Recall that to subtract complex numbers, subtract their real parts and subtract their imaginary parts [see Example 10-3].

$\qquad$ (a) $(2 + 11i) - (5 + 7i) = (2 - 5) + (11 - 7)i$

$\qquad\qquad\qquad\qquad\qquad\quad = -3 + 4i$

$\qquad$ (b) $(-3 + 2i) - (8 - 9i) = (-3 - 8) + [2 - (-9)]i$

$\qquad\qquad\qquad\qquad\qquad\qquad = -11 + 11i$

$\qquad$ (c) $16 - (-2 + 3i) = (16 + 0i) - (-2 + 3i)$

$\qquad\qquad\qquad\qquad\qquad = [16 - (-2)] + [0 - 3]i$

$\qquad\qquad\qquad\qquad\qquad = 18 - 3i$

$\qquad$ (d) $(\frac{1}{2} + \frac{2}{3}i) - (-\frac{1}{4}i) = (\frac{1}{2} + \frac{2}{3}i) - (0 - \frac{1}{4}i)$

$\qquad\qquad\qquad\qquad\qquad\qquad = (\frac{1}{2} - 0) + [\frac{2}{3} - (-\frac{1}{4})]i$

$\qquad\qquad\qquad\qquad\qquad\qquad = \frac{1}{2} + \frac{11}{12}i$

PROBLEM 10-4 Find the indicated products:

(a) $6(2 + 3i)$ **(b)** $-5(4 - \sqrt{-36})$ **(c)** $-4i(2 - 7i)$ **(d)** $i\sqrt{2}(\sqrt{2} - i\sqrt{3})$

Solution: Use the distributive property and write the result in $a + bi$ form [see Example 10-4].

(a) $6(2 + 3i) = 6 \cdot 2 + 6 \cdot 3i$
$$= 12 + 18i$$

(b) $-5(4 - \sqrt{-36}) = -5(4 - i\sqrt{36})$
$$= -5(4 - 6i)$$
$$= -20 + 30i$$

(c) $-4i(2 - 7i) = (-4i)(2) - (-4i)(7i)$
$$= -8i + 28i^2$$
$$= -8i - 28 \text{ or } -28 - 8i$$

(d) $i\sqrt{2}(\sqrt{2} - i\sqrt{3}) = i(\sqrt{2})^2 - i^2\sqrt{2}\sqrt{3}$
$$= 2i - (-1)\sqrt{6}$$
$$= 2i + \sqrt{6} \text{ or } \sqrt{6} + 2i$$

PROBLEM 10-5 Find the indicated products:

(a) $(2 + 5i)(7 + 11i)$ **(b)** $(-3 + 5i)(7 - 4i)$ **(c)** $(\frac{1}{2} + 3i)(\frac{2}{3} - \frac{1}{4}i)$
(d) $(\sqrt{2} + i\sqrt{5})(\sqrt{3} - i\sqrt{5})$

Solution: Use the FOIL method and the definition $i^2 = -1$ [see Example 10-5].

(a) $(2 + 5i)(7 + 11i) = 2 \cdot 7 + 2(11i) + (5i)7 + (5i)(11i)$
$$= 14 + 22i + 35i + 55i^2$$
$$= 14 + 22i + 35i - 55$$
$$= (14 - 55) + (22i + 35i)$$
$$= -41 + 57i$$

(b) $(-3 + 5i)(7 - 4i) = (-3)(7) + (-3)(-4i) + (5i)7 + (5i)(-4i)$
$$= -21 + 12i + 35i - 20i^2$$
$$= -21 + 12i + 35i + 20$$
$$= (-21 + 20) + (12i + 35i)$$
$$= -1 + 47i$$

(c) $(\frac{1}{2} + 3i)(\frac{2}{3} - \frac{1}{4}i) = \frac{1}{2} \cdot \frac{2}{3} + (\frac{1}{2})(-\frac{1}{4}i) + (3i)(\frac{2}{3}) + (3i)(-\frac{1}{4}i)$
$$= \frac{1}{3} - \frac{1}{8}i + 2i - \frac{3}{4}i^2$$
$$= \frac{1}{3} - \frac{1}{8}i + 2i + \frac{3}{4}$$
$$= (\frac{1}{3} + \frac{3}{4}) + (-\frac{1}{8}i + 2i)$$
$$= \frac{13}{12} + \frac{15}{8}i$$

(d) $(\sqrt{2} + i\sqrt{5})(\sqrt{3} - i\sqrt{5}) = \sqrt{2}\sqrt{3} + (\sqrt{2})(-i\sqrt{5}) + (i\sqrt{5})\sqrt{3} - (i\sqrt{5})(i\sqrt{5})$
$$= \sqrt{6} - i\sqrt{10} + i\sqrt{15} - 5i^2$$
$$= \sqrt{6} - i\sqrt{10} + i\sqrt{15} + 5$$
$$= (\sqrt{6} + 5) + (-i\sqrt{10} + i\sqrt{15})$$
$$= (\sqrt{6} + 5) + (-\sqrt{10} + \sqrt{15})i$$

PROBLEM 10-6 Simplify: **(a)** i^9 **(b)** i^{31} **(c)** i^{-6} **(d)** i^{560} **(e)** i^{2070}

Solution: Write the exponential expression as a product which involves a power of (i^4) [see Example 10-6].

(a) $i^9 = i^8 \cdot i^1 = (i^4)^2(i^1) = 1 \cdot i = i$
(b) $i^{31} = (i^4)^7(i^3) = 1 \cdot i^3 = 1 \cdot (-i) = -i$

(c) $i^{-6} = \dfrac{1}{i^6} = \dfrac{1}{(i^4)(i^2)} = \dfrac{1}{1(i^2)} = \dfrac{1}{1(-1)} = -1$

(d) $i^{560} = (i^4)^{140} = 1$

(e) $i^{2070} = (i^4)^{517}(i^2) = 1(i^2) = -1$

PROBLEM 10-7 Find the conjugates of each of the following complex numbers:

(a) $2 + 7i$ (b) $-5 - 8i$ (c) $6i + 3$ (d) 8 (e) $-7i$ (f) $5 - \sqrt{-36}$

Solution: Recall that the conjugate of $a + bi$ (denoted by $\overline{a + bi}$) is $a - bi$ [see Example 10-7].

(a) $\overline{2 + 7i} = 2 - 7i$ (b) $\overline{-5 - 8i} = -5 + 8i$ (c) $\overline{6i + 3} = \overline{3 + 6i}$

$$= 3 - 6i$$

(d) $\overline{8} = \overline{8 + 0i}$ (e) $\overline{-7i} = \overline{0 - 7i}$ (f) $\overline{5 - \sqrt{-36}} = \overline{5 - i\sqrt{36}}$

$\quad = 8 - 0i$ $\quad = 0 + 7i$ $\quad = \overline{5 - 6i}$

$\quad = 8$ $\quad = 7i$ $\quad = 5 + 6i$

PROBLEM 10-8 Find the indicated quotients: (a) $\dfrac{4 + 3i}{2 - 5i}$ (b) $\dfrac{7}{3 + 4i}$ (c) $\dfrac{6 + i}{-3i}$

(d) $\dfrac{2}{4 - \sqrt{-8}}$

Solution: Multiply both the numerator and the denominator by the conjugate of the denominator, and write in $a + bi$ form [see Example 10-8].

(a) $\dfrac{4 + 3i}{2 - 5i} = \dfrac{4 + 3i}{2 - 5i} \cdot \dfrac{2 + 5i}{2 + 5i}$

$\quad = \dfrac{(4 + 3i)(2 + 5i)}{(2 - 5i)(2 + 5i)}$

$\quad = \dfrac{8 + 20i + 6i + 15i^2}{4 - 25i^2}$

$\quad = \dfrac{-7 + 26i}{4 + 25}$

$\quad = -\dfrac{7}{29} + \dfrac{26}{29}i$

(b) $\dfrac{7}{3 + 4i} = \dfrac{7}{3 + 4i} \cdot \dfrac{3 - 3i}{3 - 4i}$

$\quad = \dfrac{7(3 - 4i)}{(3 + 4i)(3 - 4i)}$

$\quad = \dfrac{21 - 28i}{9 - 16i^2}$

$\quad = \dfrac{21 - 28i}{25}$

$\quad = \dfrac{21}{25} - \dfrac{28}{25}i$

(c) $\dfrac{6 + i}{-3i} = \dfrac{6 + i}{-3i} \cdot \dfrac{3i}{3i}$

$\quad = \dfrac{(6 + i)3i}{(-3i)(3i)}$

$\quad = \dfrac{18i + 3i^2}{-9i^2}$

$\quad = \dfrac{18i - 3}{9}$

$\quad = \dfrac{18i}{9} - \dfrac{3}{9}$

$\quad = 2i - \dfrac{1}{3}$ or $-\dfrac{1}{3} + 2i$

(d) $\dfrac{2}{4 - \sqrt{-8}} = \dfrac{2}{4 - 2i\sqrt{2}}$

$\quad = \dfrac{2}{4 - 2i\sqrt{2}} \cdot \dfrac{4 + 2i\sqrt{2}}{4 + 2i\sqrt{2}}$

$\quad = \dfrac{8 + 4i\sqrt{2}}{16 - 8i^2}$

$\quad = \dfrac{8 + 4i\sqrt{2}}{16 + 8}$

$\quad = \dfrac{8 + 4i\sqrt{2}}{24}$

$\quad = \dfrac{8}{24} + \dfrac{4i\sqrt{2}}{24}$ or $\dfrac{1}{3} + \dfrac{\sqrt{2}}{6}i$

Supplementary Exercises

PROBLEM 10-9 Write each of the following radicals as a product of a real number and i (assume all variables represent positive real numbers):

(a) $\sqrt{-81}$ (b) $\sqrt{-140}$ (c) $\sqrt{-64x}$ (d) $\sqrt{-200xy^2}$

PROBLEM 10-10 Find the indicated sums:

(a) $(3 - 5i) + (7 + 2i)$ (b) $(8 + i) + (9 - 5i)$ (c) $(-3 + i\sqrt{5}) + (3 - 5i\sqrt{5})$
(d) $(\frac{2}{3} + \frac{5}{6}i) + (\frac{1}{2} - \frac{1}{4}i)$ (e) $(6.2 - 4.1i) + (5.3 + 7.8i)$

PROBLEM 10-11 Find the indicated differences:

(a) $(2 - 5i) - (8 - 3i)$ (b) $(19 - 5i) - (6 + i)$ (c) $(8 - i\sqrt{3}) - (8 + i\sqrt{3})$
(d) $(\frac{1}{4} - \frac{1}{3}i) - (\frac{1}{2} + \frac{2}{3}i)$ (e) $(17.4 - 3.8i) - (12.2 - 4.4i)$

PROBLEM 10-12 Find the indicated products:

(a) $4(3 - 5i)$ (b) $5(-2 + 7i)$ (c) $\frac{1}{2}(12 - 6i)$ (d) $\sqrt{2}(3\sqrt{2} - i\sqrt{2})$ (e) $0.6(2.1 + 4.3i)$
(f) $2i(3 - 5i)$ (g) $i(7i - 5)$ (h) $\frac{1}{2}i(16 - 10i)$ (i) $3i(\sqrt{-49} + 2i^2)$ (j) $i\sqrt{3}(\sqrt{3} - \sqrt{-27})$

PROBLEM 10-13 Find the indicated products:

(a) $(3 + 5i)(2 + 7i)$ (b) $(6 - i)(2 + 5i)$ (c) $(8 + 2i)(8 - 2i)$ (d) $(12 + 11i)(2 - 9i)$
(e) $(\frac{1}{2} + 4i)(\frac{3}{4} - i)$ (f) $(2 + 3i)(2 + 3i)$ (g) $(0.4 + 0.5i)(2.1 - 0.3i)$
(h) $(\sqrt{5} + i\sqrt{3})(\sqrt{5} - i\sqrt{3})$

PROBLEM 10-14 Simplify the following powers of i:

(a) i^7 (b) i^8 (c) i^{21} (d) i^{50} (e) i^{200} (f) i^{203} (g) i^{-4} (h) i^{-18} (i) i^{5304}
(j) i^{3221}

PROBLEM 10-15 Find the conjugates of each of the following complex numbers:

(a) $4 - 5i$ (b) $11 + 3i$ (c) $\frac{1}{2} - \frac{2}{5}i$ (d) 19 (e) $31i$ (f) $4 - \sqrt{-49}$ (g) $17i - 2$
(h) $-0.4i - 0.8$ (i) $17 + i\sqrt{3}$ (j) $12i^2$

PROBLEM 10-16 Find the indicated quotients:

(a) $\dfrac{4 - 5i}{3 + 2i}$ (b) $\dfrac{7 + 11i}{8 - i}$ (c) $\dfrac{16}{2 + i}$ (d) $\dfrac{15 + i}{2i}$ (e) $\dfrac{(\frac{1}{2} + 3i)}{-8i}$ (f) $\dfrac{5}{i\sqrt{3}}$ (g) $\dfrac{4 - i}{6i^2 - 5i}$

(h) $\dfrac{3 + i\sqrt{2}}{5 - i\sqrt{2}}$

Answers to Supplementary Exercises

(10-9) (a) $9i$ (b) $2i\sqrt{35}$ (c) $8i\sqrt{x}$ (d) $10iy\sqrt{2x}$

(10-10) (a) $10 - 3i$ (b) $17 - 4i$ (c) $-4i\sqrt{5}$ (d) $\frac{7}{6} + \frac{7}{12}i$ (e) $11.5 + 3.7i$

(10-11) (a) $-6 - 2i$ (b) $13 - 6i$ (c) $-2i\sqrt{3}$ or $0 - 2i\sqrt{3}$ (d) $-\frac{1}{4} - i$ (e) $5.2 + 0.6i$

(10-12) **(a)** $12 - 20i$ **(b)** $-10 + 35i$ **(c)** $6 - 3i$ **(d)** $6 - 2i$ **(e)** $1.26 + 2.58i$
(f) $10 + 6i$ **(g)** $-7 - 5i$ **(h)** $5 + 8i$ **(i)** $-21 - 6i$ **(j)** $9 + 3i$

(10-13) **(a)** $-29 + 31i$ **(b)** $17 + 28i$ **(c)** 68 or $68 + 0i$ **(d)** $123 - 86i$ **(e)** $\frac{35}{8} + \frac{5}{2}i$
(f) $-5 + 12i$ **(g)** $0.99 + 0.93i$ **(h)** 8 or $8 + 0i$

(10-14) **(a)** $-i$ **(b)** 1 **(c)** i **(d)** -1 **(e)** 1 **(f)** $-i$ **(g)** 1 **(h)** -1 **(i)** 1
(j) i

(10-15) **(a)** $4 + 5i$ **(b)** $11 - 3i$ **(c)** $\frac{1}{2} + \frac{2}{5}i$ **(d)** 19 or $19 + 0i$ **(e)** $-31i$ or $0 - 31i$
(f) $4 + 7i$ **(g)** $-2 - 17i$ **(h)** $-0.8 + 0.4i$ **(i)** $17 - i\sqrt{3}$ **(j)** -12

(10-16) **(a)** $\dfrac{2}{13} - \dfrac{23}{13}i$ **(b)** $\dfrac{9}{13} + \dfrac{19}{13}i$ **(c)** $\dfrac{32}{5} - \dfrac{16}{5}i$ **(d)** $\dfrac{1}{2} - \dfrac{15}{2}i$ **(e)** $-\dfrac{3}{8} + \dfrac{1}{16}i$

(f) $0 - \dfrac{5\sqrt{3}}{3}i$ **(g)** $-\dfrac{19}{61} + \dfrac{26}{61}i$ **(h)** $\dfrac{13}{27} + \dfrac{8\sqrt{2}}{27}i$

11 RELATIONS AND FUNCTIONS

THIS CHAPTER IS ABOUT

☑ Understanding Relations and Functions
☑ Inverse Functions
☑ Solving Variation Problems
☑ Solving Applied Variation Problems

11-1. Understanding Relations and Functions

A. Finding the domain and range of a relation

A **relation** is a correspondence between ordered pairs of numbers. In a relation every member of a first set, called the **domain,** is matched with the members of a second set, called the **range.** The correspondence between domain and range is not always a one-to-one correspondence; either set may consist of a single element, while the other set consists of an infinite number of elements.

A set of ordered pairs is often used to specify a relation. For example, the set $\{(1, 3), (2, 5), (2, 7), (4, 11)\}$ specifies a relation in which 1 corresponds with 3, 2 with both 5 and 7, and 4 with 11.

If a relation is specified by a set of ordered pairs, then the set of all first components of the ordered pairs is the domain of the relation, and the set of all second components of the ordered pairs is the range of the relation.

EXAMPLE 11-1: Find the **(a)** domain and **(b)** range of the relation specified by $\{(1, 3), (2, 5), (2, 7), (4, 11)\}$.

Solution:

(a) $\{(1, 3), (2, 5), (2, 7), (4, 11)\}$

$\{1, 2, 4\}$ ⟵ the domain is the set of all first components

(b) $\{(1, 3), (2, 5), (2, 7), (4, 11)\}$

$\{3, 5, 7, 11\}$ ⟵ the range is the set of all second components

Another very concise way to specify a relation is to create an equation that has as its solution set the set of ordered pairs that specify that relation. A third method used to specify a relation is to graph the set of ordered pairs. The following table shows an example of these three methods used to specify the same relation.

Table 11-1. Three Methods of Specifying the Same Relation

Method	Example
A set of ordered pairs	$\{(0, 0), (1, 1), (2, 4)\}$
An equation	$y = x^2$, for $x = 0, 1,$ or 2
A graph	

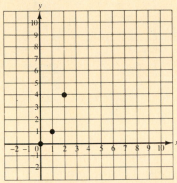

The variable that is used to represent a domain value of a relation is called the **independent variable.** The variable that is used to represent a range value is called the **dependent variable.** In this text, x will be used for the independent variable and y will be used for the dependent variable unless otherwise indicated.

If a relation is specified by an equation and the domain of the relation is not stated, then the domain will include all the real numbers that can be substituted for the independent variable that produce at least one real value of the dependent variable. The range will then consist of all those real numbers that the dependent variable assumes in this substitution process.

EXAMPLE 11-2: Find the (a) domain and (b) range of the relation specified by $y = \sqrt{x + 1} + 3$.

Solution: (a) $y = \sqrt{x + 1} + 3$ is a real number provided $x + 1 \geq 0$. $x + 1 \geq 0$ implies that $x \geq -1$. The domain of the relation specified by $y = \sqrt{x + 1} + 3$ is $\{x \mid x \geq -1\}$.

(b) Because the value of $\sqrt{x + 1}$ must be greater than or equal to 0, the value of $y = \sqrt{x + 1} + 3$ must be greater than or equal to 3. Therefore the range of the relation specified by $y = \sqrt{x + 1} + 3$ is $\{y \mid y \geq 3\}$.

To find the domain of a relation from its graph, we project each point of the graph vertically onto the x-axis. To find the range, we project each point horizontally onto the y-axis.

EXAMPLE 11-3: Find the (a) domain and (b) range of the relation specified by the graph on Grid A:

Grid A:

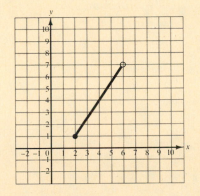

(a) Project the graph vertically onto the *x*-axis.

(b) Project the graph horizontally onto the *y*-axis.

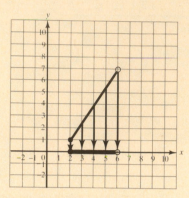

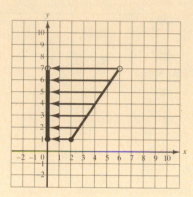

The domain is $\{x \mid 2 \le x < 6\}$.

The range is $\{y \mid 1 \le y < 7\}$.

B. Identifying functions

The concept of a **function** is one of the most important concepts in mathematics.

A function is a relation in which each member in the domain corresponds to one and only one member in the range.

All functions are relations, but some relations are not functions. If a member of the domain corresponds with more than one member of the range, the relation is not a function. Thus, if we use a set of ordered pairs to specify a relation, we see that if any of the first components are repeated (i.e., correspond with more than one member of the range) then the relation is not a function.

EXAMPLE 11-4: Determine which of the following sets of ordered pairs specifies a function:

(a) $\{(-2, 3), (1, -1), (5, 7)\}$ **(b)** $\{(-3, 0), (4, 1), (4, 5), (6, 11)\}$

Solution: **(a)** $\{(-2, 3), (1, -1), (5, 7)\}$ specifies a function because each domain member is paired with one and only one range member.

(b) $\{(-3, 0), (4, 1), (4, 5), (6, 11)\}$ does not specify a function because the domain member 4 is paired with the two range members, 1 and 5.

To determine whether a relation specified by a graph is a function, we use the following test.

The Vertical Line Test

A graph specifies a function if every vertical line intersects the graph in no more than one point (no points or one point).

Observe that a vertical line is the graph of a relation in which *all* the members of the domain correspond to a single member of the range. Therefore, if the graph of a relation is intersected more than once by a vertical line, it *must* have a domain member that corresponds to more than one range member.

EXAMPLE 11-5: Determine which of the following graphs specify a function:

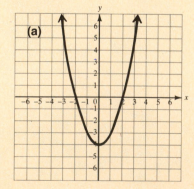

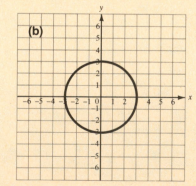

Solution:

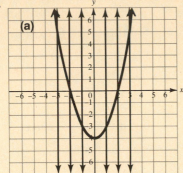

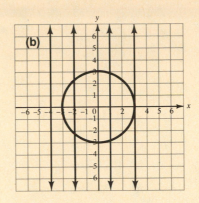

All vertical lines intersect the graph in exactly one point. This graph specifies a function.

Some vertical lines intersect the graph in more than one point. This graph does not specify a function.

C. Evaluating functions

It is often helpful to think of a function as a machine which accepts a member of the domain of the function and produces the corresponding range member as its output.

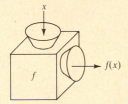

Figure 11-1. The function machine (the equation of the function) has been named f; x is the input (the domain value), and $f(x)$ is the output (the range value).

The functional notation $y = f(x)$, read "y equals f of x," is often used to designate that y is a function of x. The notation $f(a)$ represents the value of the dependent variable y when $x = a$ (provided a is a member of the domain of the function f).

Caution: The symbols f and $f(x)$ do not have the same meaning. The symbol f is not a variable but is used to represent a function; it is in neither the domain nor the range. On the other hand, $f(x)$ is the member of the range that is associated with the member x of the domain.

Since $y = f(x)$, it is often convenient to replace the dependent variable of an equation with the symbol $f(x)$. For example, using functional notation $y = 3x + 1$ is written as $f(x) = 3x + 1$.

If a function f is specified by an equation, then we can evaluate $f(a)$ by substituting a for the independent variable in the equation.

Caution: The notation $f(a)$ does not mean f times a.

EXAMPLE 11-6: Given that $f(x) = x^2 - 5x + 2$ and $g(x) = 5x - 3$, find each of the following:

(a) $f(2)$ **(b)** $g(-3)$ **(c)** $f(h + 1)$ **(d)** $f(0) - g(0)$

Solution:

    (a) $f(x) = x^2 - 5x + 2$

        $f(2) = 2^2 - 5(2) + 2$ Substitute 2 for x.

            $= 4 - 10 + 2$ Simplify.

            $= -4$

(b) $g(-x) = 5x - 3$

$g(-3) = 5(-3) - 3$ Substitute -3 for x.

$= -15 - 3$ Simplify.

$= -18$

(c) $f(x) = x^2 - 5x + 2$

$f(h + 1) = (h + 1)^2 - 5(h + 1) + 2$ Substitute $h + 1$ for x.

$= h^2 + 2h + 1 - 5h - 5 + 2$ Simplify.

$= h^2 - 3h - 2$

(d) $f(x) - g(x) = (x^2 - 5x + 2) - (5x - 3)$

$f(0) - g(0) = (0^2 - 5(0) + 2) - (5(0) - 3)$ Substitute 0 for x in each function.

$= (0 - 0 + 2) - (0 - 3)$ Simplify.

$= 5$

D. Graphing functions

Some functions occur so often in mathematics that you will find it helpful to study them in more detail.

For example, if m and b are real numbers, then the function $f(x) = mx + b$ is called a **linear function;** Recall that an equation of the form $y = mx + b$ graphs as a straight line with slope m and y-intercept b (see Section 4-4).

The function $f(x) = x$ is called the **identity function.** The identity function is the special linear function obtained by letting $m = 1$ and $b = 0$ in the equation $f(x) = mx + b$. The identity function pairs each real number with itself. The graph of the identity function is shown in Figure 11-2.

Figure 11-2

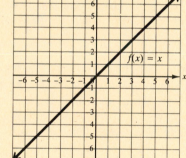

If b is a real number, then the function $f(x) = b$ is called a **constant function.** The graph of a constant function is always a horizontal line (see Section 4-3).

The function $f(x) = |x|$ is called the **absolute value function.** The graph of the absolute value function is shown in Figure 11-3.

Figure 11-3

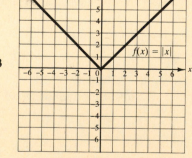

More precise methods of graphing relations will be considered later in this book, and in more advanced mathematics courses. However, the relations in this chapter may be graphed by plotting several ordered pairs which are solutions of the relation and then drawing a smooth curve through these points.

EXAMPLE 11-7: Graph (a) $f(x) = -x^2 + 4$ and (b) $g(x) = |x + 3|$.

Solution:

(a)

x	y
-3	-5
-2	0
-1	3
0	4
1	3
2	0
3	5

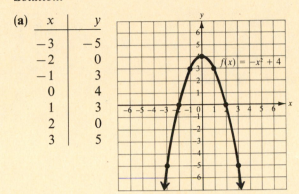

(b)

x	y
-5	2
-4	1
-3	0
-2	1
-1	2
0	3
1	4
2	5

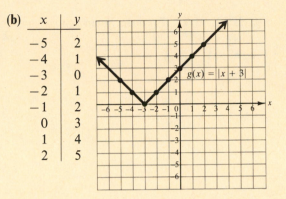

To graph a function specified by an equation containing two or more parts, we graph each part over its indicated domain.

EXAMPLE 11-8: Graph $h(x) = \begin{cases} 3 & \text{if } x \leq -3. \\ -x & \text{if } -3 < x < 0 \\ x & \text{if } 0 \leq x \end{cases}$

Solution:

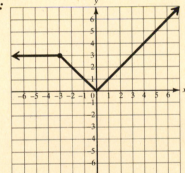

11-2. Inverse Functions

A. Finding the inverse of a function

The **inverse of a function** is a relation in which the pairing of the domain and range elements is reversed. Frequently the symbol f^{-1} is used to denote the inverse of the function f.

If a function f is specified by a set of ordered pairs, then its inverse f^{-1} can be found by interchanging the order of the components in each ordered pair.

EXAMPLE 11-9: Find the inverse of the function specified by $f = \{(1, -5), (2, 3), (4, 9)\}$.

Solution: $f = \{(1, -5), (2, 3), (4, 9)\}$

Interchange the components of each ordered pair.

$f^{-1} = \{(-5, 1), (3, 2), (9, 4)\}$

Note: The domain of f is the range of f^{-1}, and the range of f is the domain of f^{-1}.

Caution: The inverse of a function may not be a function. For example, $\{(1, 3), (2, 3), (5, 7)\}$ specifies a function but its inverse $\{(3, 1), (3, 2), (7, 5)\}$ does not specify a function because the domain member 3 is paired with more than one range member.

To find the inverse of a function which is specified by an equation involving the two variables x and y, we interchange the variables and then solve for y.

EXAMPLE 11-10: Find the inverse of the functions specified by **(a)** $y = 3x + 6$ and **(b)** $f(x) = \dfrac{x}{x + 3}$.

Solution: **(a)**

$$y = 3x + 6$$

$$x = 3y + 6 \qquad\qquad \text{Interchange } x \text{ and } y.$$

$$x - 6 = 3y \qquad\qquad \text{Solve for } y.$$

$$\tfrac{1}{3}x - 2 = y \longleftarrow \text{ the inverse of } y = 3x + 6$$

(b)

$$f(x) = \frac{x}{x + 3}$$

$$y = \frac{x}{x + 3} \qquad\qquad \text{Write the equation in terms of } x \text{ and } y.$$

$$x = \frac{y}{y + 3} \qquad\qquad \text{Interchange } x \text{ and } y.$$

$$x(y + 3) = \frac{y}{(y + 3)} \cdot (y + 3) \qquad \text{Solve for } y.$$

$$xy + 3x = y$$

$$xy - y = -3x$$

$$y(x - 1) = -3x$$

$$y = \frac{-3x}{x - 1}$$

$$f^{-1}(x) = \frac{-3x}{x - 1} \longleftarrow \text{ the inverse of } f(x) = \frac{x}{x + 3}$$

B. Graphing the inverse of a function

The graph of a function f and the graph of its inverse f^{-1} are **reflections** of each other across the line $y = x$ (See Figure 11-4). This is because the graphs of any two ordered pairs (a, b) and (b, a) will be reflections of each other across the line $y = x$.

Figure 11-4

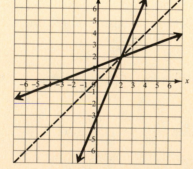

Note: The vertical line test shows that in Figure 11-4, f^{-1} is also a function.

To graph the inverse of a function specified by a graph, you graph its reflection across the line $y = x$.

EXAMPLE 11-11: Graph the inverse g^{-1} of the function g that is graphed on Grid B:

Grid B:

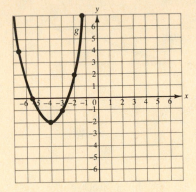

Solution:

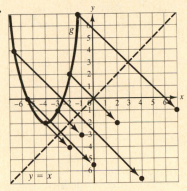

Plot points which are reflections across the line $y = x$ of points on the graph of g.

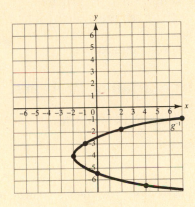

Sketch a smooth curve through the plotted points to obtain the graph of g^{-1}.

Note: The vertical line test shows that the inverse g^{-1} is not a function.

C. Identifying one-to-one functions

A function f is a **one-to-one function** if each member of the range of f is paired with one and only one member of the domain of f. Only one-to-one functions have inverses that are functions.

To identify a one-to-one function from its graph, we use the following test.

The Horizontal Line Test

A function is a one-to-one function if each horizontal line intersects the graph of the function in no more than one point (no points or one point).

EXAMPLE 11-12: Determine which of the following graphs specify a one-to-one function:

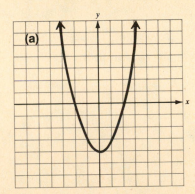

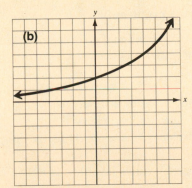

Solution: (a)

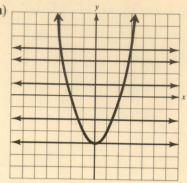

(b)

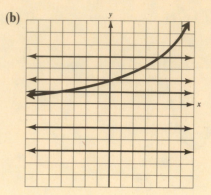

Some horizontal lines intersect the graph in more than one point. This is not the graph of a one-to-one function.

All horizontal lines intersect the graph in at most one point. This is the graph of a one-to-one function.

11-3. Solving Variation Problems

A. Solving direct variation problems

A **variation** is a special type of function. It relates variables by means of multiplication, division, or both.

A variation is called a **direct variation** if it can be represented by an equation of the form $y = kx^n$ where $x > 0, n > 0$, and $k > 0$. The number k is called the **variation constant** or the **constant of proportionality.** Some examples of direct variations are shown below.

- The equation $y = 3x$ means y varies directly as x. The variation constant is 3.
- The equation $C = \pi d$ mean C varies directly as d. The variation constant is π.
- The equation $A = s^2$ means A varies directly as the square of s. The variation constant is 1.

To solve a problem that involves a direct variation, we first write the direct variation equation and then evaluate the variation constant.

EXAMPLE 11-13: If y varies directly as x, and $y = 10$ when $x = 2$, find y when $x = 3$.

Solution:		
$y = kx$		Write the direct variation equation.
$10 = k \cdot 2$		Substitute the given values $y = 10$ and $x = 2$.
$5 = k$		Evaluate the variation constant.
$y = 5x$		Write the direct variation equation with k replaced by its known value.
$y = 5 \cdot 3$		To find y when $x = 3$, substitute 3 for x in the equation and evaluate y.
$y = 15$		

B. Solving inverse variation problems

A variation is called an **inverse variation** if it can be represented by an equation of the form $y = k/x^n$, where $x > 0, n > 0$, and the variation constant $k > 0$. Some examples of inverse variations are shown below.

- The equation $t = \dfrac{10}{r}$ means t varies inversely as r. The variation constant is 10.
- The equation $p = \dfrac{k}{q^3}$ means p varies inversely as the cube of q. The variation constant is k.
- The equation $S = \dfrac{1}{2r}$ means S varies inversely as r. The variation constant is $\frac{1}{2}$.

To solve a problem that involves an inverse variation, we first write the inverse variation equation and then evaluate the variation constant.

EXAMPLE 11-14: If y varies inversely as the square of x, and $y = 10$ when $x = 10$, find y when $x = 4$.

Solution: $y = \dfrac{k}{x^2}$ Write the inverse variation equation.

$10 = \dfrac{k}{10^2}$ Substitute the given values $y = 10$ and $x = 10$.

$1000 = k$ Evaluate the variation constant.

$y = \dfrac{1000}{x^2}$ Write the inverse variation equation with k replaced with its known value.

$y = \dfrac{1000}{4^2}$ Evaluate y.

$y = 62\tfrac{1}{2}$

C. Solving joint variation problems

Joint variations are functions with more than one independent variable, and the independent variables are related by means of multiplication. Some examples of joint variation problems are shown below.

· The equation $y = 3wx$ means y varies jointly as w and x. The variation constant is 3.
· The equation $V = \pi r^2 h$ means V varies jointly as r^2 and h. The variation constant is π.
· The equation $V = lwh$ means V varies jointly as l, w, and h. The variation constant is 1.

Joint variations can be solved by writing the joint variation equation and then evaluating the variation constant.

EXAMPLE 11-15: w varies jointly as p and the square root of q. If $w = 60$ when $p = 4$ and $q = 9$, find w when $p = 10$ and $q = 4$.

Solution: $w = kp\sqrt{q}$ Write the joint variation equation.

$60 = k \cdot 4\sqrt{9}$ Substitute.

$\dfrac{60}{4\sqrt{9}} = k$ Evaluate the variation constant k.

$5 = k$

$w = 5p\sqrt{q}$ Write the joint variation equation with k replaced with its known value.

$w = 5 \cdot 10 \cdot \sqrt{4}$ Substitute.

$w = 100$ Evaluate w.

D. Solving combined variation problems

When direct and inverse variations occur together, they are called **combined variations.** Some examples of combined variation problems are shown below.

· The equation $y = \dfrac{5u}{v}$ means y varies directly as u and inversely as v. The variation constant is 5.

· The equation $s = \dfrac{3.5wd^2}{l^3}$ means s varies directly as the product of w and the square of d, and inversely as the cube of l. The variation constant is 3.5.

• The equation $a = \dfrac{2\sqrt{b}}{cd}$ means a varies directly as the square root of b and inversely as the product of c and d. The variation constant is 2.

Combined variations are also solved by first evaluating the variation constant.

EXAMPLE 11-16: y varies directly as the product of x and w and inversely as the cube of z. If $y = 2$ when $x = 2$, $w = 9$, and $z = 3$, find y when $x = 4$, $w = 14$, and $z = 2$.

Solution: $y = \dfrac{kxw}{z^3}$ Write the combined variation equation.

$2 = \dfrac{k \cdot 2 \cdot 9}{3^3}$ Substitute the given values $y = 2$, $x = 2$, $w = 9$, and $z = 3$.

$\dfrac{2 \cdot 3^3}{2 \cdot 9} = k$ Evaluate the variation constant.

$3 = k$

$y = \dfrac{3xw}{z^3}$ Write the combined variation equation with k replaced with its known value.

$y = \dfrac{3 \cdot 4 \cdot 14}{2^3}$ Evaluate y.

$y = 21$

11-4. Solving Applied Variation Problems

Variation problems have a great many applications in science and engineering.
To **solve an applied variation problem,** we:

1. *Read* the problem carefully, several times.
2. *Identify* each different quantity.
3. *Decide* how to represent the different quantities using different variables.
4. *Translate* to a **general variation formula** (if k is unknown).
5. *Find k* by evaluating the general variation formula.
6. *Replace k* with the constant found in step 5 to find a **specific variation formula.**
7. *Solve* the specific variation formula for the unknown variable.
8. *Interpret* the solution with respect to the unknown quantity to find the proposed solution of the original problem.
9. *Check* to see if the proposed solution satisfies all the conditions of the original problem.

EXAMPLE 11-17: Solve the following applied variation problem:

1. *Read:*   *Ohm's Law:* An electrical current in amperes (A) varies directly as the electromotive force in volts (V) and inversely as the resistance in ohms (Ω). The current flowing is 3 A when a 2 Ω wire is connected to a 6 V battery. What is the resistance of a lamp that draws $\frac{1}{2}$ A when used on a 110 V line?

2. *Identify:*   The different quantities are $\begin{cases} \text{the electrical current} \\ \text{the electromotive force} \\ \text{the resistance} \end{cases}$.

3. *Decide:*   Let I = the electrical current

              and E = the electromotive force

              and R = the resistance [see the following note].

4. *Translate:* I varies directly as E and inversely as R means:

              $I = \dfrac{kE}{R}$ ⟵ general variation formula

5. *Find k:*

$$3 = \frac{k(6)}{2}$$ Substitute: $I = 3$ A when $R = 2\ \Omega$ and $E = 6$ V.

$$3 = 3k$$

$$1 = k \longleftarrow \text{variation constant}$$

6. *Substitute k:*

$$I = \frac{(1)E}{R}$$

$$I = \frac{E}{R} \longleftarrow \text{specific variation formula } [\textbf{Ohm's Law}]$$

7. *Solve:*

$$\frac{1}{2} = \frac{110}{R}$$ Substitute: What is R when $I = \frac{1}{2}$ A and $E = 110$ V?

$$2R \cdot \frac{1}{2} = 2R \cdot \frac{110}{R}$$

$$R = 2(110)$$

$$R = 220$$

8. *Interpret:* $R = 220$ means the resistance of a lamp that draws $\frac{1}{2}$ A when used on a 110 V line is 220 Ω.

9. *Check:*   Does a lamp with a resistance of 220 Ω draw $\frac{1}{2}$ A when used on a 110 V line using Ohm's Law?

Yes: $I = \dfrac{E}{R} = \dfrac{110}{220} = \dfrac{1}{2}$

Note: When representing electrical current, electromotive force, and resistance, the variables I, E, and R are traditionally used.

RAISE YOUR GRADES

Can you . . . ?

☑ find the domain and range of a relation specified by a set of ordered pairs
☑ find the domain and range of a relation specified by an equation
☑ find the domain and range of a relation specified by a graph
☑ determine whether a relation specified by a set of ordered pairs is a function
☑ determine whether a relation specified by a graph is a function
☑ evaluate $f(a)$, where f is a function specified by an equation and a is a member of the domain of f
☑ identify and graph linear functions
☑ identify and graph the identity function
☑ identify and graph constant functions
☑ identify and graph the absolute value function
☑ graph functions specified by an equation
☑ find the inverse of a function specified by a set of ordered pairs
☑ find the inverse of a function specified by an equation
☑ graph the inverse of a function specified by a graph
☑ identify one-to-one functions
☑ solve variation problems
☑ solve applied variation problems

SOLVED PROBLEMS

PROBLEM 11-1 Find the domain and range of each of the following relations specified by a set of ordered pairs:

(a) $\{(1, 5), (1, -3)\}$ **(b)** $\{(-\frac{1}{2}, 0), (0, 0), (9, 3)\}$ **(c)** $\{(1, 1), (2, 2), (3, 3), (4, 4)\}$

(d) $\{(1, -3), (2, -3), (3, -3), (4, -3), (5, -3)\}$ **(e)** $\{(\sqrt{2}, 1), (\sqrt{3}, \pi)\}$

(f) $\{(-\sqrt{7}, -2), (4, -\sqrt{11}), (6, 25), (7, -\sqrt{23})\}$

Solution: Recall that if a relation is specified by a set of ordered pairs, then the set of all first components of the ordered pairs is the domain of relation, and the set of all second components of the ordered pairs is the range of the relation [see Example 11-1].

    (a) $\{(1, 5), (1, -3)\}$ has domain $\{1\}$, and range $\{-3, 5\}$.
    (b) $\{(-\frac{1}{2}, 0), (0, 0), (9, 3)\}$ has domain $\{-\frac{1}{2}, 0, 9\}$, and range $\{0, 3\}$.
    (c) $\{(1, 1), (2, 2), (3, 3), (4, 4)\}$ has domain $\{1, 2, 3, 4\}$ and range $\{1, 2, 3, 4\}$.
    (d) $\{(1, -3), (2, -3), (3, -3), (4, -3), (5, -3)\}$ has domain $\{1, 2, 3, 4, 5\}$ and range $\{-3\}$.
    (e) $\{(\sqrt{2}, 1), (\sqrt{3}, \pi)\}$ has domain $\{\sqrt{2}, \sqrt{3}\}$ and range $\{1, \pi\}$.
    (f) $\{(-\sqrt{7}, -2), (4, -\sqrt{11}), (6, 25), (7, -\sqrt{23})\}$ has domain $\{-\sqrt{7}, 4, 6, 7\}$ and range $\{-\sqrt{23}, -\sqrt{11}, -2, 25\}$.

PROBLEM 11-2 Find the domain and range of each relation specified by the following equations:

(a) $y = 3x + 4$ **(b)** $y = \sqrt{x}$ **(c)** $y = \dfrac{1}{x - 1}$ **(d)** $y = x^2$ for $x \in \{1, 2, 3\}$

(e) $y = \sqrt{x - 3} + 5$ **(f)** $y = |x - 3|$

Solution: Recall that, if a relation is specified by an equation and the domain is not stated, then the domain will include all the real numbers that can be substituted for the independent variable which produce at least one real value of the dependent variable. The range consists of the real numbers that the dependent variable assumes in the above substitution process [see Example 11-2].

    (a) $y = 3x + 4$ is a real number for all real numbers x. Therefore the domain of $y = 3x + 4$ is the set of all real numbers, designated as $\{x | x \in R\}$.
    Because y assumes all real numbers as x assumes all real numbers, the range of $y = 3x + 4$ is $\{y | y \in R\}$.
    (b) $y = \sqrt{x}$ is a real number provided $x \geq 0$. Therefore the domain of $y = \sqrt{x}$ is $\{x | x \geq 0\}$.
    Because $\sqrt{x}$ only represents nonnegative real numbers, the range of $y = \sqrt{x}$ is $\{y | y \geq 0\}$.
    (c) $y = \dfrac{1}{x - 1}$ is a real number provided $x \neq 1$. Therefore the domain of $y = \dfrac{1}{x - 1}$ is $\{x | x \in R \text{ and } x \neq 1\}$.
    Because y assumes all real numbers except 0 as x assumes all real numbers except 1, the range of $y = \dfrac{1}{x - 1}$ is $\{y | y \in R \text{ and } y \neq 0\}$.
    (d) $y = x^2$ for $x \in \{1, 2, 3\}$ means that the domain is $\{1, 2, 3\}$.
    Because y assumes the values 1, 4, and 9 as x assumes the values 1, 2, and 3, the range is $\{1, 4, 9\}$.
    (e) $y = \sqrt{x - 3} + 5$ is a real number provided $x - 3 \geq 0$. $x - 3 \geq 0$ implies that $x \geq 3$. Therefore the domain of $y = \sqrt{x - 3} + 5$ is $\{x | x \geq 3\}$.
    Because $\sqrt{x - 3}$ assumes all values greater than or equal to 0, $y = \sqrt{x - 3} + 5$ assumes all values greater than or equal to 5. The range of $y = \sqrt{x - 3} + 5$ is $\{y | y \geq 5\}$.
    (f) $y = |x - 3|$ is a real number for all real numbers x. Therefore the domain of $y = |x - 3|$ is $\{x | x \in R\}$.
    Because $|x - 3|$ only represents nonnegative real numbers, the range of $y = |x - 3|$ is $\{y | y \geq 0\}$.

PROBLEM 11-3 Find both the domain and range of each relation specified by the following graphs:

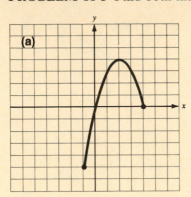

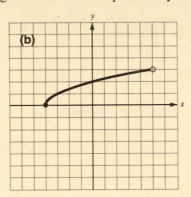

Solution: Recall that, to find the domain of a relation from its graph, you project each point of the graph vertically onto the *x*-axis. To find the range, project each point horizontally onto the *y*-axis [see Example 11-3].

(a) Project the graph vertically onto the *x*-axis.   Project the graph horizontally onto the *y*-axis.

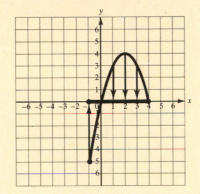

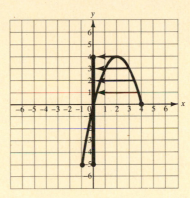

The domain is $\{x \mid -1 \le x \le 4\}$.   The range is $\{y \mid -5 \le y \le 4\}$.

(b) Project the graph vertically onto the *x*-axis.   Project the graph horizontally onto the *y*-axis.

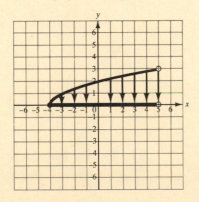

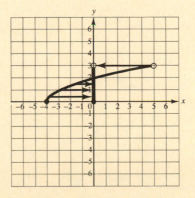

The domain is $\{x \mid -4 \le x < 5\}$.   The range is $\{y \mid 0 \le y < 3\}$.

PROBLEM 11-4 Determine which of the following sets of ordered pairs specify a function:

(a) $\{(-3, 1), (0, 1), (3, 1)\}$ **(b)** $\{(-1, -1), (0, 4), (-1, 1)\}$ **(c)** $\{(-3.5, 2), (0.25, 7), (11, 2)\}$

(d) $\{(\sqrt{2}, \sqrt{2}), (\sqrt{3}, \sqrt{3}), (\sqrt{5}, \sqrt{5}), (\sqrt{6}, \sqrt{6})\}$ **(e)** $\{(-\frac{1}{3}, 2), (-\frac{1}{4}, 3), (-\frac{1}{5}, 4), (-\frac{1}{6}, 5), (-\frac{1}{7}, 6)\}$

(f) $\{(-2, 4), (-1, 1), (0, 0), (1, 1), (2, 4)\}$ **(g)** $\{(0, 0), (1, 1), (2, 4), (3, 9), (2, 3), (4, 16)\}$

Solution: Recall that a set of ordered pairs specifies a function if each domain member is paired with one and only one range member [see Example 11-4].

(a) $\{(-3, 1), (0, 1), (3, 1)\}$ is a function.

(b) $\{(-1, -1), (0, 4), (-1, 1)\}$ is not a function because the domain member -1 is paired with both -1 and 1.

(c) $\{(-3.5, 2), (0.25, 7), (11, 2)\}$ is a function.

(d) $\{(\sqrt{2}, \sqrt{2}), (\sqrt{3}, \sqrt{3}), (\sqrt{5}, \sqrt{5}), (\sqrt{6}, \sqrt{6})\}$ is a function.

(e) $\{(-\frac{1}{3}, 2), (-\frac{1}{4}, 3), (-\frac{1}{5}, 4), (-\frac{1}{6}, 5), (-\frac{1}{7}, 6)\}$ is a function.

(f) $\{(-2, 4), (-1, 1), (0, 0), (1, 1), (2, 4)\}$ is a function.

(g) $\{(0, 0), (1, 1), (2, 4), (3, 9), (2, 3), (4, 16)\}$ is not a function because the domain member 2 is paired with both 3 and 4.

PROBLEM 11-5 Determine which of the following graphs specify a function:

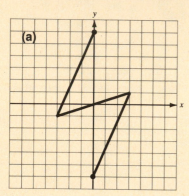

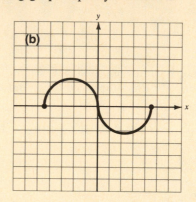

Solution: Recall the vertical line test. A graph specifies a function if every vertical line intersects the graph in no more than one point (no points or one point) [see Example 11-5].

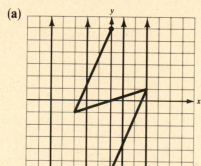

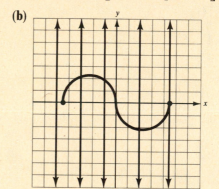

Some vertical lines intersect the graph in more than one point. This graph does not specify a function.

All vertical lines intersect the graph in no more than one point. This graph specifies a function.

PROBLEM 11-6 Given that $f(x) = 2x^2 + 4$, $g(x) = x - 3$, and $h(x) = 4$, find each of the following:

(a) $f(3)$ **(b)** $g(-4)$ **(c)** $h(2)$ **(d)** $f(-3)$ **(e)** $g(8)$ **(f)** $h(20)$ **(g)** $f(0) + g(0)$

(h) $f(8) - h(8)$ **(i)** $\dfrac{f(4)}{g(4)}$ **(j)** $4[g(1) - f(1)]$ **(k)** $h(a) - h(b)$

Solution: Recall that, to evaluate $f(a)$, substitute a for the independent variable in the equation which specifies the function f [see Example 11-6].

(a) $f(x) = 2x^2 + 4$ **(b)** $g(x) = x - 3$ **(c)** $h(x) = 4$ **(d)** $f(x) = 2x^2 + 4$

$f(3) = 2(3)^2 + 4$ $g(-4) = -4 - 3$ $h(2) = 4$ $f(-3) = 2(-3)^2 + 4$

$= 2 \cdot 9 + 4$ $= -7$ $= 2 \cdot 9 + 4$

$= 22$ $= 22$

(e) $g(x) = x - 3$ **(f)** $h(x) = 4$ **(g)** $f(x) + g(x) = (2x^2 + 4) + (x - 3)$

$\qquad g(8) = 8 - 3 \qquad h(20) = 4 \qquad f(0) + g(0) = (2(0)^2 + 4) + (0 - 3)$

$\qquad\qquad = 5 \qquad\qquad\qquad\qquad\qquad\qquad = (0 + 4) + (0 - 3)$

$\qquad\qquad\qquad\qquad\qquad\qquad\qquad\qquad\qquad = 1$

(h) $f(x) - h(x) = (2x^2 + 4) - (4)$ **(i)** $\dfrac{f(x)}{g(x)} = \dfrac{2x^2 + 4}{x - 3}$

$\qquad f(8) - h(7) = (2(8)^2 + 4) - (4) \qquad \dfrac{f(4)}{g(4)} = \dfrac{2(4)^2 + 4}{2 - 3}$

$\qquad\qquad\qquad = (128 + 4) - 4$

$\qquad\qquad\qquad = 128 \qquad\qquad\qquad\qquad\qquad = 36$

(j) $4[g(x) - f(x)] = 4[(x - 3) - (2x^2 + 4)]$ **(k)** $h(x) - h(x) = 4 - 4$

$\qquad 4[g(1) - f(1)] = 4[(1 - 3) - (2(1)^2 + 4)] \qquad\qquad h(a) - h(b) = 4 - 4$

$\qquad\qquad\qquad = 4[(-2) - 6] \qquad\qquad\qquad\qquad\qquad = 0$

$\qquad\qquad\qquad = -32$

PROBLEM 11-7 Graph each of the following functions:

(a) $f(x) = -|x + 1|$ **(b)** $g(x) = -x^2 + 1$ **(c)** $h(x) = \begin{cases} 1 & \text{if } x < -1 \\ 3 & \text{if } x \geq -1 \end{cases}$

(d) $G(x) = \begin{cases} x + 3 & \text{if } x \leq 0 \\ \frac{1}{2}x - 5 & \text{if } x > 0 \end{cases}$

Solution: Recall that the functions in this chapter can be graphed by plotting several solutions and then drawing a smooth curve through these points. To graph a function with two or more parts, graph each part over the indicated domain [see Examples 11-7 and 11-8].

(a)

x	$f(x)$
-5	-4
-4	-3
-3	-2
-2	-1
-1	0
0	-1
1	-2
2	-3
3	-4
4	-5

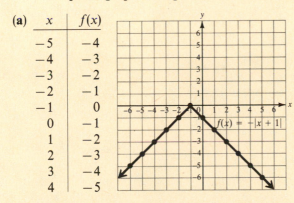

(b)

x	$g(x)$
-2	-3
-1	0
0	1
1	0
2	-3

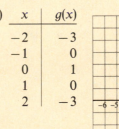

(c) Graph each part of the function.

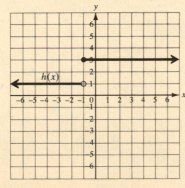

(d) Graph each part of the function.

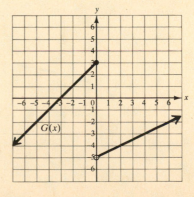

PROBLEM 11-8 Find the inverse of each of the following functions specified by sets of ordered pairs:

(a) $f = \{(3, 0), (4, 1), (5, 4)\}$ **(b)** $g = \{(2, 2), (7, 7)\}$ **(c)** $h = \{(-4, -5), (-3, -1), (1, 2), (4, -5)\}$

Solution: Recall that if a function f is specified by a set of ordered pairs, then its inverse f^{-1} can be formed by interchanging the order of the components in each ordered pair belonging to f [see Example 11-9].

        (a) $f = \{(3, 0), (4, 1), (5, 4)\}$ **(b)** $g = \{(2, 2), (7, 7)\}$

         $f^{-1} = \{(0, 3), (1, 4), (4, 5)\}$ $g^{-1} = \{(2, 2), (7, 7)\}$

        (c) $h = \{(-4, 5), (-3, -1), (1, 2), (4, -5)\}$

         $h^{-1} = \{(5, -4), (-1, -3), (2, 1), (-5, 4)\}$

PROBLEM 11-9 Find the inverse of each of the following functions specified by equations involving the two variables x and y:

(a) $y = \frac{1}{4}x - 3$ **(b)** $x + 2y = 16$ **(c)** $y = x + 1$ **(d)** $y = \dfrac{4x + 3}{x}$

Solution: Recall that to find the inverse of a function specified by an equation involving the two variables x and y, you interchange the variables, then solve for y [see Example 11-10].

   (a) $y = \frac{1}{4}x - 3$ **(b)** $x + 2y = 16$

               $x = \frac{1}{4}y - 3$ $y + 2x = 16$

            $x + 3 = \frac{1}{4}y$ $y = -2x + 16$ ⟵ the inverse of

          $4x + 12 = y$ ⟵ the inverse of $x + 2y = 16$

                $y = \frac{1}{4}x - 3$

   (c) $y = x + 1$ **(d)** $y = \dfrac{4x + 3}{x}$

              $x = y + 1$

           $x - 1 = y$ ⟵ the inverse of $x = \dfrac{4y + 3}{y}$

               $y = x + 1$ $xy = 4y + 3$

                                  $xy - 4y = 3$

                                  $y(x - 4) = 3$

                                  $y = \dfrac{3}{x - 4}$ ⟵ the inverse of

                                          $y = \dfrac{4x + 3}{x}$

PROBLEM 11-10 Graph the inverse of each function specified by the following graphs:

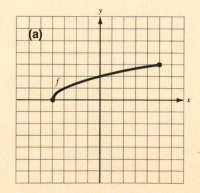
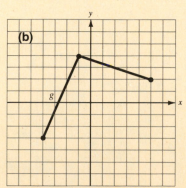

Solution: Recall that to graph the inverse of a function f specified by a graph, you graph its reflection f^{-1} across the line $y = x$ [see Example 11-11].

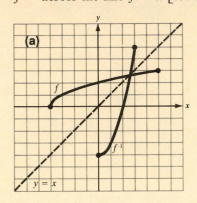
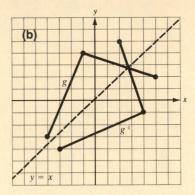

PROBLEM 11-11 Determine which of the following graphs specify a one-to-one function:

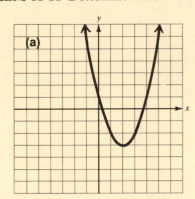
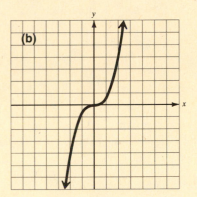

Solution: Recall the horizontal line test. A function is a one-to-one function if each horizontal line in a coordinate system intersects the graph of the function in no more than one point (no point or one point) [see Example 11-12].

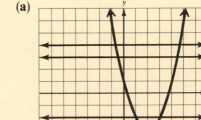
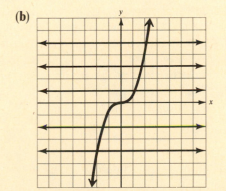

Some horizontal lines intersect the graph in more than one point. This is not the graph of a one-to-one function.

All horizontal lines intersect the graph in at most one point. This is the graph of a one-to-one function.

PROBLEM 11-12 Solve the following variation problems:

(a) If y varies directly as x, and $y = 30$ when $x = 2$, find y when $x = 6$.

(b) If w varies inversely as the cube of x, and $w = 24$ when $x = \frac{1}{2}$, find w when $x = 3$.

(c) p varies jointly as q and the square of r. If $p = 108$ when $q = 2$ and $r = 3$, find p when $q = 4$ and $r = 4$.

(d) s varies directly as w and inversely as the square of l. If $s = \frac{1}{3}$ when $w = 4$ and $l = 6$, find s when $w = 8$ and $l = 10$.

Solution: Recall that to solve a variation problem, first write the variation equation, evaluate the variation constant, then evaluate the unknown [see Examples 11-13, 11-14, 11-15, and 11-16].

(a) $y = kx$

$30 = k \cdot 2$

$15 = k$

$y = 15x$

$y = 15 \cdot 6$

$y = 90$

(b) $w = \dfrac{k}{x^3}$

$24 = \dfrac{k}{(\frac{1}{2})^3}$

$3 = k$

$w = \dfrac{3}{x^3}$

$w = \dfrac{3}{3^3}$

$w = \dfrac{1}{9}$

(c) $p = kqr^2$

$108 = k \cdot 2 \cdot 3^2$

$108 = k \cdot 18$

$\dfrac{108}{18} = k$

$6 = k$

$p = 6qr^2$

$p = 6 \cdot 4 \cdot 4^2$

$p = 384$

(d) $s = \dfrac{kw}{l^2}$

$\dfrac{1}{3} = \dfrac{k \cdot 4}{6^2}$

$\dfrac{1}{3} = \dfrac{k}{9}$

$3 = k$

$s = \dfrac{3w}{l^2}$

$s = \dfrac{3 \cdot 8}{10^2}$

$s = \dfrac{6}{25}$

PROBLEM 11-13 Solve the following applied variation problem:

Coulomb's Law: The force between two small electrical charges varies jointly as the charges and inversely as the square of the distance between the charges. If the distance between two particular charges is doubled and one of the charges is tripled, then what factor must multiply the other charge to maintain the original force between them?

Solution: Recall that to solve an applied variation problem, you:

1. *Read* the problem carefully, several times.
2. *Identify* each different quantity.
3. *Decide* how to represent the different quantities using different variables.
4. *Translate* to a general variation formula (if k is unknown).
5. *Find k* by evaluating the general variation formula.
6. *Replace k* with the constant found in step 5 to find a specific variation formula.
7. *Solve* the specific variation formula for the unknown variable.
8. *Interpret* the solution with respect to the unknown quantity to find the proposed solution of the original problem.
9. *Check* to see if the proposed solution satisfies all the conditions of the original problem [see Example 11-17].

Identify: The different quantities are $\begin{cases} \text{one small electrical charge} \\ \text{the other small electrical charge} \\ \text{the force between the electrical charges} \\ \text{the distance between the electrical charges} \\ \text{the unknown factor that multiplies one of} \\ \text{the charges} \end{cases}$.

Decide: Let $c_1 =$ one small electrical charge

and $c_2 =$ the other small electrical charge

and $f =$ the force between the electrical charges

and $d =$ the distance between the electrical charges

and $a =$ the factor that multiplies one of the charges.

Translate: f varies jointly as c_1 and c_2 and inversely as the square of d means:

$$f = \dfrac{kc_1c_2}{d^2} \quad \longleftarrow \quad \text{general variation formula}$$

Substitute: $\quad f = \dfrac{k(3c_1)(ac_2)}{(2d)^2}$ $\qquad$ Substitute: What factor a must multiply c_2 to maintain f when c_1 is tripled and d is doubled?

$$f = \frac{(3a)(kc_1c_2)}{4d^2}$$

$$f = \frac{3a}{4} \cdot \underbrace{\frac{kc_1c_2}{d^2}}$$

$$f = \frac{3a}{4} \cdot f \longleftarrow \text{ "to maintain } f\text{" means } f = \frac{kc_1c_2}{d^2}$$

Solve: $\quad 1 = \dfrac{3a}{4}$ $\qquad$ Divide both members by f.

$$\frac{4}{3} = a$$

Interpret: $\quad a = \frac{4}{3}$ means the factor that must multiply c_2 to maintain f when c_1 is tripled and d is doubled is $\frac{4}{3}$ or $1\frac{1}{3}$.

Check: $\quad$ Is f maintained as $\dfrac{kc_1c_2}{d^2}$ if c_1 is tripled, c_2 is multiplied by $\dfrac{4}{3}$, and d is doubled?

$$\text{Yes: } \quad \frac{k(3c_1)\left(\dfrac{4}{3}c_2\right)}{(2d)^2} = \frac{\left(3 \cdot \dfrac{4}{3}\right)(kc_1c_2)}{4d^2} = \frac{4(kc_1c_2)}{4d^2} = f.$$

Supplementary Exercises

PROBLEM 11-14 Find both the domain and range of each of the following relations which are specified by a set of ordered pairs:

(a) $\{(1, 5), (2, -5)\}$ $\qquad$ (b) $\{(0, 2), (3, 2)\}$ $\qquad$ (c) $\{(-4, -3), (-3, -2), (-2, -1)\}$

(d) $\{(2, 3), (3, 2)\}$ $\qquad$ (e) $\{(1, \frac{1}{2}), (2, \frac{1}{2}), (3, \frac{1}{2}), (4, \frac{1}{2})\}$ $\qquad$ (f) $\{(0.5, 1.5), (-0.25, 1.75), (-0.125, 1.875)\}$

PROBLEM 11-15 Find the domain and range of each of the following graphs of relations:

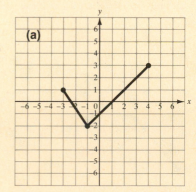 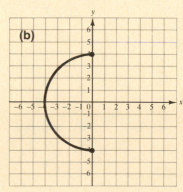

PROBLEM 11-16 Find the domain and range of each of the following equations of relations:

(a) $y = -3x - 1$ $\qquad$ (b) $x - y = 8$ $\qquad$ (c) $y = x^2$ $\qquad$ (d) $y = -\sqrt{x}$ $\qquad$ (e) $y = \sqrt{x + 1} - 3$

(f) $x - y = 2$ $\qquad$ (g) $y = |x| + 1$ $\qquad$ (h) $y = |x + 1|$ $\qquad$ (i) $y = x^3$ $\qquad$ (j) $y = \sqrt{16 - x^2}$

(k) $y = \dfrac{1}{x - 3}$

PROBLEM 11-17 Determine which of the following sets of ordered pairs specifies a function:

(a) $\{(1, 2), (2, 3), (3, 4)\}$ (b) $\{(0.5, 2), (0.75, 3)\}$ (c) $\{(-2, 5), (1, 6), (2, 5)\}$
(d) $\{(-4, 3), (-3, 3), (-2, 3), (-1, 3)\}$ (e) $\{(-7, 4), (-5, 2), (-3, 1), (-1, 0), (-7, 5)\}$
(f) $\{(\frac{1}{2}, \frac{1}{4}), (\frac{1}{2}, 1)\}$

PROBLEM 11-18 Determine which of the following graphs specify a function:

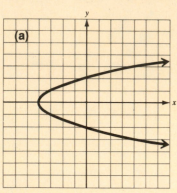
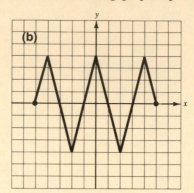
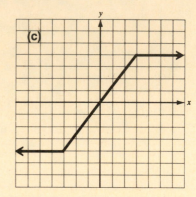

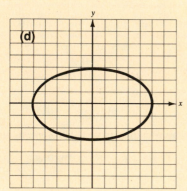
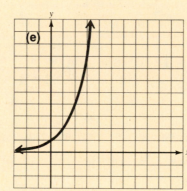
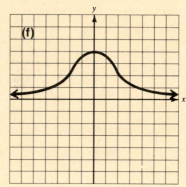

PROBLEM 11-19 Given that $f(x) = x^2 - 5$, $g(x) = 3x + 1$, and $h(x) = 5$; find each of the following:

(a) $f(2)$ (b) $g(2)$ (c) $h(2)$ (d) $f(3)$ (e) $g(3)$ (f) $h(3)$ (g) $f(a)$ (h) $g(a)$
(i) $h(a)$ (j) $f(5) - g(2)$ (k) $f(6) + h(1)$ (l) $f(4)g(4)$ (m) $2[f(1)] + g(1)$
(n) $f(a + 1)$ (o) $g(a + 1)$ (p) $h(a + 1)$

PROBLEM 11-20 Determine which of the following graphs specify a one-to-one function:

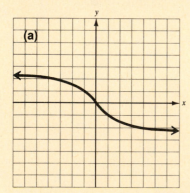
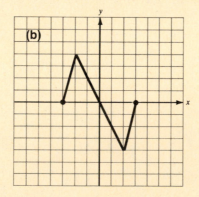
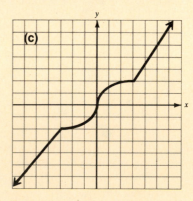

PROBLEM 11-21 Graph each of the following:

(a) $f(x) = 2x - 1$ (b) $g(x) = |x + 1|$ (c) $F(x) = -|x - 3|$ (d) $G(x) = \begin{cases} x & \text{if } x \leq 0 \\ 2 & \text{if } x > 0 \end{cases}$

PROBLEM 11-22 Find the inverse of the following functions specified by sets of ordered pairs:

(a) $\{(1, 4), (2, 5)\}$ (b) $\{(3, 5), (2, -1), (7, 11)\}$ (c) $\{(0, 2), (-1, 3), (5, 6)\}$
(d) $\{(-1, 2), (-2, 3), (-3, 4), (-4, 5)\}$ (e) $\{(6, 11), (12, 2), (-3, 1), (-2, 1), (7, 6)\}$

PROBLEM 11-23 Find the inverse of the following functions specified by equations:

(a) $f(x) = -3x + 6$ (b) $g(x) = 4x - 3$ (c) $y = 2x$ (d) $y = \dfrac{1}{x - 4}$ (e) $y = \dfrac{x + 1}{x}$

(f) $y = \dfrac{3x}{x - 1}$

PROBLEM 11-24 Graph the inverse of each function specified by the following graphs:

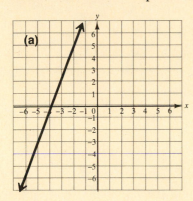

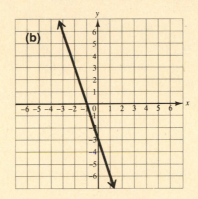

PROBLEM 11-25 Solve each of the following variation problems:

(a) If m varies directly as n, and $m = 24$ when $n = 3$, find m when $n = 10$.
(b) If u varies inversely as the square root of v, and $u = 10$ when $v = 9$, find u when $v = 3$.
(c) x varies jointly as y and the cube of z. If $x = 12$ when $y = 3$ and $z = 2$, find x when $y = 2$ and $z = 4$.
(d) r varies directly as the square of s and inversely as the cube of t. If $r = 1$ when $s = \frac{1}{2}$ and $t = 2$, find r when $s = 3$ and $t = 10$.

PROBLEM 11-26 Solve each applied variation problem:

(a) *Hooke's Law:* The distance that a vertical spring stretches varies directly with the weight hanging from it. A certain spring stretches 10 inches when a 25 pound weight is hung from it. How far will the spring stretch when a 40 pound weight is hung from it? How much weight is needed to make the spring stretch 40 inches?

(b) *Weight of an Astronaut:* The weight of an object varies inversely as the square of the distance from the object to the center of the earth. A certain astronaut weighs 200 pounds at sea level (4000 miles from the center of the earth). How much does the astronaut weigh when orbiting 16,000 miles above sea level? How far above the earth, to the nearest mile, will any astronaut weigh one-half of his or her sea level weight?

(c) *Electrical Resistance:* At a fixed temperature and chemical composition, the resistance of a wire varies directly as the length of the wire and inversely as the square of the diameter. At 20°C, the resistance of 20 feet of copper wire with a diameter of 20 mils (20 thousandths of an inch) is 2 ohms. What is the resistance of 90 feet of copper wire at 20°C if the diameter is 15 mils? At 20°C, what length of copper wire with a diameter of 25 mils has a resistance of 8 ohms?

(d) *Newton's Law of Universal Gravitation:* The force of attraction between any two masses in the universe varies jointly as the masses and inversely as the square of the distance between the masses. If the distance between particular masses is halved and one of the masses is doubled, then what factor must multiply the other mass (1) to maintain the original force between them, and (2) to produce 0.03125 of the original force?

Answers to Supplementary Exercises

(11-14) **(a)** domain $\{1, 2\}$, range $\{-5, 5\}$ **(b)** domain $\{0, 3\}$, range $\{2\}$

(c) domain $\{-4, -3, -2\}$, range $\{-3, -2, -1\}$ **(d)** domain $\{2, 3\}$, range $\{2, 3\}$

(e) domain $\{1, 2, 3, 4\}$, range $\{\frac{1}{2}\}$

(f) domain $\{-0.5, -0.25, -0.125\}$, range $\{1.5, 1.75, 1.875\}$

(11-15) **(a)** domain $\{x \mid -3 \le x \le 4\}$, range $\{y \mid -2 \le y \le 3\}$

(b) domain $\{x \mid -4 \le x \le 0\}$, range $\{y \mid -4 \le y \le 4\}$

(11-16) **(a)** domain $\{x \mid x \in R\}$, range $\{y \mid y \in R\}$ **(b)** domain $\{x \mid x \in R\}$, range $\{y \mid y \in R\}$

(c) domain $\{x \mid x \in R\}$, range $\{y \mid y \ge 0\}$ **(d)** domain $\{x \mid x \ge 0\}$, range $\{y \mid y \le 0\}$

(e) domain $\{x \mid x \ge -1\}$, range $\{y \mid y \ge -3\}$ **(f)** domain $\{x \mid x \in R\}$, range $\{y \mid y \in R\}$.

(g) domain $\{x \mid x \in R\}$, range $\{y \mid y \ge 1\}$ **(h)** domain $\{x \mid x \in R\}$, range $\{y \mid y \ge 0\}$

(i) domain $\{x \mid x \in R\}$, range $\{y \mid y \in R\}$

(j) domain $\{x \mid -4 \le x \le 4\}$, range $\{y \mid 0 \le y \le 4\}$

(k) domain $\{x \mid x \ne 3\}$, range $\{y \mid y \ne 0\}$

(11-17) **(a)** function **(b)** function **(c)** function **(d)** function **(e)** not a function

(f) not a function

(11-18) **(a)** not a function **(b)** function **(c)** function **(d)** not a function

(e) function **(f)** function

(11-19) **(a)** -1 **(b)** 7 **(c)** 5 **(d)** 4 **(e)** 10 **(f)** 5 **(g)** $a^2 - 5$ **(h)** $3a + 1$

(i) 5 **(j)** 13 **(k)** 36 **(l)** 143 **(m)** -4 **(n)** $a^2 + 2a - 4$ **(o)** $3a + 4$

(p) 5

(11-20) **(a)** a one-to-one function **(b)** not a one-to-one function **(c)** a one-to-one function

(11-21)

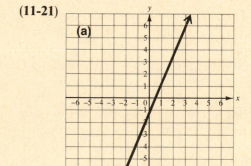

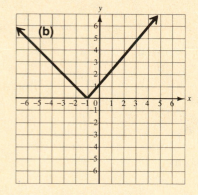

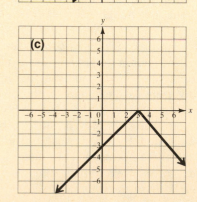

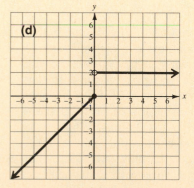

(11-22) **(a)** $\{(4, 1), (5, 2)\}$ **(b)** $\{(5, 3), (-1, 2), (11, 7)\}$ **(c)** $\{(2, 0), (3, -1), (6, 5)\}$

(d) $\{(2, -1), (3, -2), (4, -3), (5, -4)\}$ **(e)** $\{(11, 6), (2, 12), (1, -3), (1, -2), (6, 7)\}$

(11-23) (a) $f^{-1}(x) = -\frac{1}{3}x + 2$ (b) $g^{-1}(x) = \frac{1}{4}x + \frac{3}{4}$ (c) $y = \frac{1}{2}x$ (d) $y = \frac{1}{x} + 4$

(e) $y = \frac{1}{x-1}$ (f) $y = \frac{x}{x-3}$

(11-24)

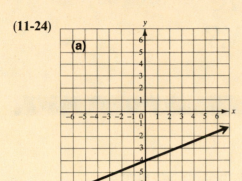 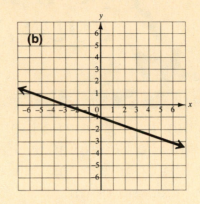

(11-25) (a) $m = 80$ (b) $u = 10\sqrt{3}$ (c) $x = 64$ (d) $r = \frac{36}{125}$ or 0.288

(11-26) (a) 16 in.; 100 lb   (b) 8 lb, 1657 mi   (c) 16 Ω, 125 ft   (d) $\frac{1}{8}$, 0.0039 or $\frac{1}{256}$

12 QUADRATIC EQUATIONS

THIS CHAPTER IS ABOUT

☑ Solving Incomplete Quadratic Equations
☑ Solving Quadratic Equations by Factoring
☑ Solving Quadratic Equations by Completing the Square
☑ Using the Quadratic Formula
☑ The Discriminant
☑ Writing Equations Given the Roots
☑ Solving Equations that are Quadratic in Form
☑ Solving Problems Using Quadratic Equations
☑ Solving Quadratic Inequalities

12-1. Solving Incomplete Quadratic Equations

A **quadratic equation** is a polynomial equation in which the highest degree of any term is two. An equation of the form $ax^2 + bx + c = 0$, where a, b, and c are real numbers and $a \neq 0$, is a **quadratic equation in one variable.** The term ax^2 is called the **quadratic term,** bx is called the **linear term,** and c is called the constant term. A **quadratic equation is in standard form** if

(1) one member is a polynomial in standard form,
(2) the quadratic term has a positive coefficient, and
(3) the other member is 0.

EXAMPLE 12-1: Write the following quadratic equations in standard form and identify the constants a, b, and c:

(a) $1 - 3x^2 = 2x$ and **(b)** $\dfrac{x^2}{4} + \dfrac{1}{3} = 3x$

Solution: **(a)**

$$1 - 3x^2 = 2x$$

$$-3x^2 - 2x + 1 = 0 \qquad \text{Write all non-zero terms in one member as a polynomial in standard form.}$$

$$3x^2 + 2x - 1 = 0 \qquad \text{Multiply by } -1 \text{ to get a positive coefficient in the quadratic term.}$$

$$a = 3, b = +2, c = -1 \qquad \text{Identify } a, b, \text{ and } c.$$

$3x^2 + 2x - 1 = 0$ is standard form; $a = 3$, $b = 2$, and $c = -1$.

(b) $$\frac{x^2}{4} + \frac{1}{3} = 3x$$

$$12\left(\frac{x^2}{4} + \frac{1}{3}\right) = 12(3x)$$ Clear fractions by multiplying by the LCD of 12.

$$3x^2 + 4 = 36x$$

$$3x^2 - 36x + 4 = 0$$ Write all non-zero terms in one member as a polynomial in standard form.

$$a = 3, b = -36, c = +4$$ Identify a, b, and c.

$$3x^2 - 36x + 4 = 0 \text{ is standard form; } a = 3, b = -36, \text{ and } c = 4.$$

A. Solving equations using the zero-product property

To solve **an equation that is the product of expressions,** we use the zero-product property to set each expression equal to 0, then solve each equation.

Recall: The zero-product property states that if $a \cdot b = 0$, then $a = 0$ or $b = 0$ (see Section 1-3).

EXAMPLE 12-2: Solve $(2x - 3)(3x + 4)(x - 5) = 0$ using the zero-product property.

Solution: $(2x - 3)(3x + 4)(x - 5) = 0$

$2x - 3 = 0$ or $3x + 4 = 0$ or $x - 5 = 0$ Set each expression equal to 0.

$2x = 3$ or $3x = -4$ or $x = 5$ Solve each equation.

$x = \frac{3}{2}$ or $x = -\frac{4}{3}$ or $x = 5$

$x = -\frac{4}{3}, \frac{3}{2}, \text{ or } 5$

Note: You may use the zero-product property ONLY if one member of the equation is 0 and the other member is the product of expressions.

B. Solving equations of the form $ax^2 + bx = 0$

The equation $ax^2 + bx = 0$ is called an **incomplete quadratic equation** because $c = 0$. To **solve an incomplete quadratic equation of the form $ax^2 + bx = 0$,** we factor the polynomial and use the zero-product property to solve the resulting equations.

EXAMPLE 12-3: Solve the following incomplete quadratic equations of the form $ax^2 + bx = 0$:

(a) $6x^2 - 9x = 0$ and **(b)** $\frac{2}{3}y^2 = \frac{3}{4}y$

Solution: **(a)** $6x^2 - 9x = 0$

$3x(2x - 3) = 0$ Factor $6x^2 - 9x$.

$3x = 0$ or $2x - 3 = 0$ Set each factor equal to 0.

$2x = 3$ Solve each equation.

$x = 0$ or $x = \frac{3}{2}$ ⟵ proposed solutions

Check:

$6x^2 - 9x = 0$ ⟵	original equation ⟶	$6x^2 - 9x = 0$
$6(0)^2 - 9(0) \mid 0$	Substitute 0 and $\frac{3}{2}$.	$6(\frac{3}{2})^2 - 9(\frac{3}{2}) \mid 0$
$0 - 0 \mid 0$		$6(\frac{9}{4}) - 9(\frac{3}{2}) \mid 0$
$0 = 0$ ⟵ 0 checks		$\frac{27}{2} - \frac{27}{2} \mid 0$
		$\frac{3}{2}$ checks ⟶ $0 = 0$

$x = 0$ or $\frac{3}{2}$

(b) $\frac{2}{3}y^2 = \frac{3}{4}y$

$\frac{2}{3}y^2 - \frac{3}{4}y = 0$ Write in standard form.

$8y^2 - 9y = 0$

$y(8y - 9) = 0$ Factor the polynomial.

$y = 0$ or $8y - 9 = 0$ Set each factor equal to zero.

$y = 0$ or $8y = 9$ Solve each equation.

$y = 0$ or $y = \dfrac{9}{8}$ Check as before.

Note: Do not solve equations by dividing by an expression containing a variable because you may lose a solution. For example, if you divide $6x^2 = 3x$ by $3x$, you get $2x = 1$, which has only one solution ($\frac{1}{2}$), whereas the solutions of $6x^2 = 3x$ are 0 and $\frac{1}{2}$.

C. Solving equations of the form $ax^2 + c = 0$

The equation $ax^2 + c = 0$ is an incomplete quadratic equation because $b = 0$. To **solve an incomplete quadratic equation of the form $ax^2 + c = 0$,** we solve for x^2 and use the **square root rule** to solve for x.

The Square Root Rule

If $p^2 = q$, then $p = \sqrt{q}$ or $p = -\sqrt{q}$

Note: $p = \sqrt{q}$ or $p = -\sqrt{q}$ may be written as $p = \pm\sqrt{q}$, and is read "p equals plus or minus the square root of q."

EXAMPLE 12-4: Solve the following incomplete quadratic equations of the form $ax^2 + c = 0$:

(a) $4x^2 - 36 = 0$ **(b)** $3x^2 - 5 = 0$ **(c)** $2x^2 = -3$

Solution: **(a)** $4x^2 - 36 = 0$

$4x^2 = 36$ Solve for x^2.

$x^2 = 9$

$x = \sqrt{9}$ or $-\sqrt{9}$ Solve for x using the square root rule.

$x = 3$ or -3

$x = \pm 3$ Check as before.

(b) $3x^2 - 5 = 0$

$3x^2 = 5$ Solve for x^2.

$x^2 = \dfrac{5}{3}$

$x = \sqrt{\dfrac{5}{3}}$ or $-\sqrt{\dfrac{5}{3}}$ Solve for x using the square root rule.

$x = \sqrt{\dfrac{15}{3}}$ or $-\sqrt{\dfrac{15}{3}}$ Rationalize the denominator.

$x = \pm\dfrac{\sqrt{15}}{3}$ Check as before.

(c) $2x^2 = -3$　　　　　　　　Solve for x^2.

$$x^2 = -\frac{3}{2}$$

$$x = \sqrt{-\frac{3}{2}} \text{ or } -\sqrt{-\frac{3}{2}}$$　　　　Solve for x using the square root rule.

$$x = \frac{\sqrt{-6}}{2} \text{ or } -\frac{\sqrt{-6}}{2}$$　　　　Rationalize the denominator.

$$x = \frac{i\sqrt{6}}{2} \text{ or } -\frac{i\sqrt{6}}{2}$$　　　　Write $\sqrt{-6}$ in terms of i.

Check: 　　　　$2x^2 + 3 = 0$ ⟵ original equation

$$2\left(\pm\frac{\sqrt{6}}{2}i\right)^2 + 3 \quad\Big|\quad 0 \qquad \text{Substitute } \pm\frac{\sqrt{6}}{2}i \text{ for } x.$$

$$2\left(-\frac{6}{2}\right) + 3 \quad\Big|\quad 0$$

$$-3 + 3 \quad\Big|\quad 0$$

$$0 = 0 \qquad \text{The expression } \pm\frac{\sqrt{6}}{2}i \text{ checks.}$$

$$x = \pm\frac{\sqrt{6}}{2}i$$

12-2. Solving Quadratic Equations by Factoring

To **solve a quadratic equation by factoring,** we write the quadratic equation so that one member is 0, then factor and solve it as before.

EXAMPLE 12-5: Solve the following quadratic equations by factoring:

(a) $3 - x^2 = 2x$ and (b) $7y = 12y^2 - 45$

Solution: (a) 　　　　$3 - x^2 = 2x$

$x^2 + 2x - 3 = 0$　　　　　　　Write in standard form.

$(x + 3)(x - 1) = 0$　　　　　　Factor the quadratic.

$x + 3 = 0$　or　$x - 1 = 0$　　　Set each factor equal to zero using the zero-product property.

$x = -3$ or 　　$x = 1$　　　Solve each equation.

$x = -3$ or 1　　　　　　　Check as before.

(b) 　　　　　　$7y = 12y^2 - 45$

$12y^2 - 7y - 45 = 0$　　　　　　Write in standard form.

$(4x - 9)(3x + 5) = 0$　　　　　Factor the quadratic.

$4x - 9 = 0$　　or $3x + 5 = 0$　　Set each factor equal to zero using the zero product property.

$4x = 9$　　　or　　$3x = -5$　　Solve each equation.

$x = \frac{9}{4}$　　　or　　$x = -\frac{5}{3}$

$x = -\frac{5}{3}$ or $\frac{9}{4}$　　　　　　Check as before.

12-3. Solve Equations by Completing the Square

A. Solving equations of the form $(x + d)^2 = e$

To **solve an equation of the form** $(x + d)^2 = e$, we use the square root rule to get $x + d = \pm\sqrt{e}$, then solve each equation.

EXAMPLE 12-6: Solve the following equations of the form $(x + d)^2 = e$:

(a) $(x - 3)^2 = 4$ and **(b)** $(y + 2)^2 = -3$

Solution: **(a)** $(x - 3)^2 = 4$

$x - 3 = \sqrt{4}$ or $x - 3 = -\sqrt{4}$	Solve for $x - 3$ using the square root rule.
$x - 3 = 2$ or $x - 3 = -2$	Solve each equation.
$x = 5$ or $x = 1$	Check as before.

(b) $(y + 2)^2 = -3$

$y + 2 = \pm\sqrt{3}$	Solve for $y + 2$ using the square root rule.
$y + 2 = \pm i\sqrt{3}$	Write $\sqrt{-3}$ in terms of i.
$y = -2 \pm i\sqrt{3}$	Check as before.

B. Solving quadratic equations by completing the square.

When it is inconvenient or impossible to solve a quadratic equation by factoring it, we can solve it by using a method called **completing the square.** To solve a quadratic equation in standard form $(ax^2 + bx + c = 0)$ using this method, we follow these steps:

1. Write in $ax^2 + bx = -c$ form.
2. Clear the coefficient of the quadratic term from the equation by dividing both members by a.
3. **Add the square of one-half of the coefficient of the linear term to both members;** this step "completes the square" by artificially creating a perfect square from the binomial in the left member.
4. Write the left member as a square; notice that this equation has the form $(x + d)^2 = e$.
5. Solve this equation using the square root rule (see previous section).

EXAMPLE 12-7: Solve the following equations by completing the square:

(a) $x^2 + 6x - 1 = 0$ and **(b)** $3y^2 - y + 1 = 0$

Solution:

(a)

$x^2 + 6x - 1 = 0$	
$x^2 + 6x = 1$	Write in $ax^2 + bx = -c$ form.
$x^2 + 6x + \left(\dfrac{6}{2}\right)^2 = 1 + \left(\dfrac{6}{2}\right)^2$	Add the square of $\frac{1}{2}$ of the coefficient of x to both members.
$(x + 3)^2 = 10$	Write as a perfect square.
$x + 3 = \pm\sqrt{10}$	Solve for $x + 3$ using the square root rule.
$x = -3 \pm \sqrt{10}$	Solve for x. Check as before.

(b)

$3y^2 - y + 1 = 0$	
$3y^2 - y = -1$	Write in $ax^2 + bx = -c$ form.
$y^2 - \dfrac{1}{3}y = -\dfrac{1}{3}$	Divide by the coefficient of y^2.

$$y^2 - \frac{1}{3}y + \left(-\frac{1}{6}\right)^2 = -\frac{1}{3} + \left(\frac{1}{6}\right)^2 \qquad \text{Add the square of } \tfrac{1}{2} \text{ of the coefficient of } y.$$

$$\left(y - \frac{1}{6}\right)^2 = -\frac{11}{36} \qquad \text{Write as a perfect square.}$$

$$y - \frac{1}{6} = \pm\sqrt{-\frac{11}{36}} \qquad \text{Solve for } y - \tfrac{1}{6} \text{ using the square root rule.}$$

$$y - \frac{1}{6} = \pm\frac{\sqrt{11}}{6}\,i \qquad \text{Write in terms of } i.$$

$$y = \frac{1}{6} \pm \frac{\sqrt{11}}{6}\,i \qquad \text{Solve for } y. \text{ Check as before.}$$

12-4. Using the Quadratic Formula

We can use the **quadratic formula** to solve any quadratic equation. The quadratic formula is derived from the standard quadratic equation $ax^2 + bx + c = 0$.

A. Deriving the quadratic formula

To derive the quadratic formula, we solve the standard quadratic equation for x by completing the square.

EXAMPLE 12-8: Solve $ax^2 + bx + c = 0$, where $a \neq 0$, by completing the square.

Solution: $\qquad ax^2 + bx + c = 0$

$$ax^2 + bx = -c \qquad \text{Write in } ax^2 + bx = -c \text{ form.}$$

$$x^2 + \frac{b}{a}x = -\frac{c}{a} \qquad \text{Divide by the coefficient of } x^2.$$

$$x^2 + \frac{b}{a}x + \left(\frac{b}{2a}\right)^2 = -\frac{c}{a} + \left(\frac{b}{2a}\right)^2 \qquad \begin{array}{l}\text{Complete the square by adding the square}\\ \text{of } \tfrac{1}{2} \text{ the coefficient of } x \text{ to both members.}\end{array}$$

$$\left(x + \frac{b}{2a}\right)^2 = -\frac{c}{a} + \left(\frac{b}{2a}\right)^2 \qquad \text{Write the left member as a square.}$$

$$\left(x + \frac{b}{2a}\right)^2 = -\frac{c}{a} + \frac{b^2}{4a^2} \qquad \text{Simplify the right member.}$$

$$\left(x + \frac{b}{2a}\right)^2 = \frac{-4ac}{4a^2} + \frac{b^2}{4a^2}$$

$$\left(x + \frac{b}{2a}\right)^2 = \frac{b^2 - 4ac}{4a^2}$$

$$x + \frac{b}{2a} = \pm\sqrt{\frac{b^2 - 4ac}{4a^2}} \qquad \text{Solve for } x + \frac{b}{2a} \text{ using the square root rule.}$$

$$x = -\frac{b}{2a} \pm \sqrt{\frac{b^2 - 4ac}{4a^2}} \qquad \text{Solve for } x.$$

$$x = -\frac{b}{2a} \pm \frac{\sqrt{b^2 - 4ac}}{2a}$$

$$x = \frac{-b \pm \sqrt{b^2 - 4ac}}{2a}$$

The Quadratic Formula

If $ax^2 + bx + c = 0$, where $a \neq 0$, then

$$x = \frac{-b \pm \sqrt{b^2 - 4ac}}{2a}$$

B. Solving quadratic equations using the quadratic formula

To **solve a quadratic equation using the quadratic formula,** we write the equation in standard form, identify the coefficients a, b, and c, substitute those values in the quadratic formula, and simplify.

EXAMPLE 12-9: Solve the following quadratic equations using the quadratic formula:

(a) $3x^2 - 2x - 1 = 0$ **(b)** $2x^2 + 3x - 1 = 0$ **(c)** $3z^2 - 5z + 4 = 0$

Solution: **(a)** $3x^2 - 2x - 1 = 0$

$a = 3, b = -2, c = -1$	Identify a, b, and c.
$x = \dfrac{-b \pm \sqrt{b^2 - 4ac}}{2a}$	Write the quadratic formula.
$x = \dfrac{-(-2) \pm \sqrt{(-2)^2 - 4(3)(-1)}}{2(3)}$	Substitute 3 for a, -2 for b, and -1 for c.
$x = \dfrac{2 \pm \sqrt{4 + 12}}{6}$	Simplify.
$x = \dfrac{2 \pm 4}{6}$	
$x = \dfrac{2 + 4}{6}$ or $x = \dfrac{2 - 4}{6}$	
$x = 1$ or $-\frac{1}{3}$	Check as before.

(b) $2x^2 + 3x - 1 = 0$

$a = 2, b = 3, c = -1$	Identify a, b, and c.
$x = \dfrac{-b \pm \sqrt{b^2 - 4ac}}{2a}$	Write the quadratic formula.
$x = \dfrac{-3 \pm \sqrt{3^2 - 4(2)(-1)}}{2(2)}$	Substitute 2 for a, 3 for b, and -1 for c.
$x = \dfrac{-3 \pm \sqrt{9 + 8}}{4}$	Simplify.
$x = \dfrac{-3 \pm \sqrt{17}}{4}$	Check as before.

(c) $3z^2 - 5z + 4 = 0$

$a = 3, b = -5, c = 4$	Identify a, b, and c.
$z = \dfrac{-(-5) \pm \sqrt{(-5)^2 - 4(3)(4)}}{2(3)}$	Substitute 3 for a, -5 for b, and 4 for c
$z = \dfrac{5 \pm \sqrt{25 - 48}}{6}$	Simplify.
$z = \dfrac{5 \pm \sqrt{-23}}{6}$	
$z = \dfrac{5 \pm i\sqrt{23}}{6}$	Write in terms of i.
$z = \dfrac{5}{6} \pm \dfrac{i\sqrt{23}}{6}$	Check as before.

C. Finding approximate solutions

To find the **approximate solutions of an equation,** we solve the equation using the quadratic formula and substitute the approximate value for the radical using the square root table (see Appendix 2, page 450).

EXAMPLE 12-10: Find the approximate solutions of the following quadratic equations:

(a) $3x^2 + 3x - 1 = 0$ and (b) $3y = 2y^2 - 1$

Solution: (a) $3x^2 + 3x - 1 = 0$

$a = 3, b = 3, c = -1$	Identify a, b, and c.
$x = \dfrac{-3 \pm \sqrt{9 + 12}}{6}$	Substitute 3 for a, 3 for b, and -1 for c.
$x = \dfrac{-3 \pm \sqrt{21}}{6}$	
$x \approx \dfrac{-3 \pm 4.583}{6}$	Substitute 4.583 for $\sqrt{21}$ using the square root table.
$x \approx \dfrac{1.583}{6}$ or $\dfrac{-7.583}{6}$	Simplify.
$x \approx 0.264$ or -1.264	

(b)

$3y = 2y^2 - 1$	
$2y^2 - 3y - 1 = 0$	Write in standard form.
$a = 2, b = -3, c = -1$	Identify a, b, and c.
$y = \dfrac{-(-3) \pm \sqrt{(-3)^2 - 4(2)(-1)}}{2(2)}$	Substitute 2 for a, -3 for b, -1 for c in the quadratic formula.
$y = \dfrac{3 \pm \sqrt{17}}{4}$	Simplify.
$y \approx \dfrac{3 \pm 4.123}{4}$	Substitute 4.123 for $\sqrt{17}$ using the square root table.
$y \approx \dfrac{7.123}{4}$ or $\dfrac{-1.123}{4}$	
$y \approx 1.781$ or -0.281	

Note: The answers obtained using the square root table are not exact, but approximate ($\approx$). Therefore these values will not check when substituted in the original equations.

12-5. The Discriminant

As you have seen, using the quadratic formula, you find the solutions of $ax^2 + bx + c = 0$ to be

$$x = \frac{-b \pm \sqrt{b^2 - 4ac}}{2a}$$

The expression under the radical, $b^2 - 4ac$, is called the **discriminant** of the quadratic formula. The discriminant determines the nature of the solutions, or **roots,** of an equation:

1. If $b^2 - 4ac = 0$, the equation has one real root.
2. If $b^2 - 4ac > 0$, the equation has two distinct real roots.
3. If $b^2 - 4ac < 0$, the equation has two distinct nonreal roots (complex conjugates).

A. Determining the nature of the solutions of quadratic equations using the discriminant

To determine whether an equation has one real solution, two distinct real solutions, or two distinct complex solutions, you evaluate the discriminant.

EXAMPLE 12-11: Determine whether the following quadratic equations have one real solution, two distinct real solutions, or two distinct complex solutions:

(a) $4x^2 + 4x + 1 = 0$ **(b)** $4y = 4y^2 - 1$ **(c)** $2z^2 + 6z + 5 = 0$

Solution: **(a)** $4x^2 + 4x + 1 = 0$

$$a = 4, b = 4, c = 1 \qquad \text{Identify } a, b, \text{ and } c.$$

$$D = 4^2 - 4(4)(1) \qquad \text{Substitute for } a, b, \text{ and } c \text{ in the discriminant } D.$$

$$= 0$$

The equation has one real solution because the discriminant is zero.

(b) $\qquad 4y = 4y^2 - 1$

$$4y^2 - 4y - 1 = 0 \qquad \text{Write in standard form.}$$

$$a = 4, b = -4, c = -1 \qquad \text{Identify } a, b, \text{ and } c.$$

$$D = (-4)^2 - 4(4)(-1) \qquad \text{Substitute for } a, b, \text{ and } c \text{ in the discriminant } D.$$

$$= 16 + 16$$

$$= 32$$

The equation has two distinct real solutions because the discriminant is positive.

(c) $2z^2 + 6z + 5 = 0$

$$a = 2, b = 6, c = 5 \qquad \text{Identify } a, b, \text{ and } c.$$

$$D = 6^2 - 4(2)(5) \qquad \text{Substitute for } a, b, \text{ and } c \text{ in the discriminant } D.$$

$$= 36 - 40$$

$$= -4$$

The equation has two complex roots because the discriminant is negative.

12-6. Writing Equations Given the Roots

To **write the equation of a quadratic equation given its roots,** we use the zero-product property in reverse.

EXAMPLE 12-12: Write a quadratic equation having the following numbers as solutions:

(a) $-3, 2$ **(b)** $-\frac{2}{3}, \frac{1}{2}$ **(c)** $2 \pm \sqrt{3}$ **(d)** $-1 \pm 3i$

Solution:

(a) $\qquad x = -3$ or $\qquad x = 2 \qquad$ Set the variable equal to the solutions.

$\qquad x + 3 = 0 \quad$ or $x - 2 = 0 \qquad$ Write each equation with one member 0.

$\qquad\qquad (x + 3)(x - 2) = 0 \qquad$ Write the product of the expressions.

$\qquad\qquad x^2 - 2x + 3x - 6 = 0 \qquad$ Multiply the binomials.

$\qquad\qquad x^2 + x - 6 = 0 \qquad$ Write in standard form.

(b) $\qquad y = -\frac{2}{3}$ or $\qquad y = \frac{1}{2} \qquad$ Set the variable equal to the solutions.

$\qquad y + \frac{2}{3} = 0 \quad$ or $\quad y - \frac{1}{2} = 0 \qquad$ Write each equation with one member 0.

$\qquad 3y + 2 = 0 \quad$ or $2y - 1 = 0 \qquad$ Clear fractions.

$\qquad\qquad (3y + 2)(2y - 1) = 0 \qquad$ Write the product of the expressions.

$\qquad\qquad 6y^2 - 3y + 4y - 2 = 0 \qquad$ Multiply the binomials.

$\qquad\qquad 6y^2 + y - 2 = 0 \qquad$ Write in standard form.

(c)
$$z = 2 - \sqrt{3} \text{ or} \qquad z = 2 + \sqrt{3}$$ Set the variable equal to the solutions.

$$z - 2 + \sqrt{3} = 0 \qquad \text{or } z - 2 - \sqrt{3} = 0$$ Write each equation with one member 0.

$$(z - 2 + \sqrt{3})(z - 2 - \sqrt{3}) = 0$$ Write the product of the expressions.

$$z^2 - 2z - z\sqrt{3} - 2z + 4 + 2\sqrt{3} + z\sqrt{3} - 2\sqrt{3} - 3 = 0$$ Multiply the polynomials.

$$z^2 - 2z - 2z + 4 - 3 = 0$$ Write in standard form.

$$z^2 - 4z + 1 = 0$$

(d)
$$x = -1 + 3i \text{ or} \qquad x = -1 - 3i$$ Set the variable equal to the solutions.

$$x + 1 - 3i = 0 \qquad \text{or } x + 1 + 3i = 0$$ Write each equation with one member 0.

$$(x + 1 - 3i)(x + 1 + 3i) = 0$$ Write the product of the expression.

$$x^2 + x + 3ix + x + 1 + 3i - 3ix - 3i - 9i^2 = 0$$ Multiply the polynomials.

$$x^2 + x + x + 1 + 9 = 0$$ Write in standard form.

$$x^2 + 2x + 10 = 0$$

Note: You can check your answer by solving the equation using the quadratic formula.

12-7. Solving Equations that Are Quadratic in Form

Some equations are not quadratic equations but can be reduced to quadratic equations using substitution or simplification. To solve an equation of this type, we rename it as a quadratic equation, solve the quadratic equation and check the solutions. It is necessary to check the solutions because renaming the original equation as a quadratic equation may introduce numbers (**extraneous solutions**) that are not solutions of the original equation.

A. Solving rational equations

To **solve a rational equation,** you clear the fractions and solve as before.

EXAMPLE 12-13: Solve the following rational equations:

(a) $\dfrac{3}{x + 1} = 1 - \dfrac{2}{3x - 1}$ and (b) $\dfrac{x + 1}{x} + \dfrac{x}{x - 1} = 3$

Solution: (a)
$$\frac{3}{x + 1} = 1 - \frac{2}{3x - 1}$$

$$\frac{3}{x + 1} + \frac{2}{3x - 1} = 1$$

$$(x + 1)(3x - 1)\left(\frac{3}{x + 1} + \frac{2}{3x - 1}\right) = (x + 1)(3x - 1)(1)$$ Clear fractions.

$$(3x - 1)3 + (x + 1)2 = 3x^2 + 2x - 1$$ Simplify.

$$9x - 3 + 2x + 2 = 3x^2 + 2x - 1$$

$$3x^2 - 9x = 0$$ Write the equation with one member as 0.

$$3x(x - 3) = 0$$ Solve the equation.

$$3x = 0 \text{ or } x - 3 = 0$$

$$x = 0 \text{ or} \qquad x = 3$$ Check as before.

(b)
$$\frac{x+1}{x} + \frac{x}{x-1} = 3$$

$$x(x-1)\left(\frac{x+1}{x} + \frac{x}{x-1}\right) = x(x-1)(3)$$ Clear fractions.

$$(x-1)(x+1) + x(x) = 3x^2 - 3x$$ Simplify.

$$x^2 - 1 + x^2 = 3x^2 - 3x$$

$$x^2 - 3x + 1 = 0$$ Write the equation with one member as 0.

$$x = \frac{3 \pm \sqrt{9-4}}{2}$$ Solve the equation using the quadratic formula.

$$x = \frac{3 \pm \sqrt{5}}{2}$$ Check as before.

B. Solving radical equations

To **solve a radical equation,** you use the power principle to clear radicals by squaring both members of the equation until all radicals are cleared, then solve the equation as before.

EXAMPLE 12-14: Solve the following radical equations:

(a) $\sqrt{x+5} = x - 1$ and **(b)** $1 = \sqrt{5y+21} - \sqrt{3y+16}$

Solution: **(a)** $\sqrt{x+5} = x - 1$

$$(\sqrt{x+5})^2 = (x-1)^2$$ Square both members using the power principle.

$$x + 5 = x^2 - 2x + 1$$ Simplify.

$$x^2 - 3x - 4 = 0$$ Write the equation with one member as 0.

$$(x-4)(x+1) = 0$$ Solve the equation.

$$x - 4 = 0 \text{ or } x + 1 = 0$$

$$x = 4 \text{ or } \quad x = -1$$ Checking will show that -1 is not a solution.

$$x = 4$$

(b) $1 = \sqrt{5y+21} - \sqrt{3y+16}$

$$\sqrt{5y+21} = 1 + \sqrt{3y+16}$$ Write the equation with one radical in each member.

$$(\sqrt{5y+21})^2 = (1 + \sqrt{3y+16})^2$$ Square each member using the power principle.

$$5y + 21 = 1 + 2\sqrt{3y+16} + 3y + 16$$ Simplify.

$$2y + 4 = 2\sqrt{3y+16}$$

$$y + 2 = \sqrt{3y+16}$$

$$(y+2)^2 = (\sqrt{3y+16})^2$$ Square both members.

$$y^2 + 4y + 4 = 3y + 16$$ Simplify.

$$y^2 + y - 12 = 0$$

$$(y+4)(y-3) = 0$$ Solve the equation.

$$y + 4 = 0 \quad \text{or } y - 3 = 0$$

$$y = -4 \text{ or } \quad y = 3$$ Checking will show that -4 is not a solution.

$$y = 3$$

C. Solving equations that are reducable to quadratic equations

To **solve an equation that is reducable to a quadratic equation,** we substitute a different variable in the equation and solve as before.

EXAMPLE 12-15: Solve the following equations by reducing them to quadratic equations:

(a) $x^4 + 4x^2 + 4 = 0$ **(b)** $3x - 5\sqrt{x} - 2 = 0$ **(c)** $y^{-2} + 3y^{-1} - 4 = 0$

Solution: **(a)** $x^4 + 4x^2 + 4 = 0$

$(x^2)^2 + 4x^2 + 4 = 0$

$u^2 + 4u + 4 = 0$ Substitute u for x^2.

$(u + 2)(u + 2) = 0$ Solve the equation.

$u + 2 = 0$

$u = -2$

$x^2 = -2$ Substitute x^2 for u.

$x = \pm\sqrt{-2}$

$x = \pm i\sqrt{2}$ Check as before.

(b) $3x - 5\sqrt{x} - 2 = 0$

$3u^2 - 5u - 2 = 0$ Substitute u^2 for x and u for $\sqrt{x}$.

$(3u + 1)(u - 2) = 0$ Solve for u.

$3u + 1 = 0$ or $u - 2 = 0$

$3u = -1$ or $u = 2$

$u = -\frac{1}{3}$ or $u = 2$

$u^2 = \frac{1}{9}$ or $u^2 = 4$ Find u^2.

$x = \frac{1}{9}$ or $x = 4$ Checking will show that $\frac{1}{9}$ is not a solution.

$x = 4$ is the solution.

(c) $y^{-2} + 3y^{-1} - 4 = 0$

$u^2 + 3u - 4 = 0$ Substitute u for y^{-1} and u^2 for $(y^{-1})^2$

$(u + 4)(u - 1) = 0$ Solve for u.

$u + 4 = 0$ or $u - 1 = 0$

$u = -4$ or $u = 1$

$y^{-1} = -4$ or $y^{-1} = 1$ Substitute y^{-1} for u.

$y = -\frac{1}{4}$ or $y = 1$ Use $y^{-1} = 1/y$ to solve for y.

$y = -\frac{1}{4}$ or 1   Check as before.

12-8. Solving Problems Using Quadratic Equations

To **solve a problem using a quadratic equation,** we:

1. *Read* the problem very carefully, several times.
2. *Draw a picture* (when appropriate) to help visualize the problem.
3. *Identify* the unknowns.
4. *Decide* how to represent the unknown using one variable.
5. *Make a table* (when appropriate) to help represent the unknowns.
6. *Translate* the problem to a quadratic equation.
7. *Simplify* the quadratic equation.

8. *Solve* the quadratic equation.
9. *Interpret* the solutions of the quadratic equation with respect to each represented unknown to find the proposed solutions of the original problem.
10. *Check* to see if the proposed solutions satisfy all the conditions of the original problem.

A. Solving number problems using quadratic equations

EXAMPLE 12-16: Solve the following number problem using a quadratic equation:

1. *Read:*   The sum of two whole numbers is 20. The positive difference of their squares is 16 less than their product. What are the two numbers?

2. *Identify:*   The unknown numbers are $\begin{cases} \text{the larger whole number} \\ \text{the smaller whole number} \end{cases}$.

3. *Decide:*   Let n = the larger whole number

   then $20 - n$ = the smaller whole number (see Notes 1 and 2).

4. *Translate:* The positive difference of their squares is 16 less than their product.

$$(n)^2 - (20 - n)^2 \qquad = \qquad n(20 - n) - 16$$

5. *Simplify:* $n^2 - (400 - 40n + n^2) = 20n - n^2 - 16$

$$n^2 - 400 + 40n - n^2 = 20n - n^2 - 16$$

$$-400 + 40n = 20n - n^2 - 16$$

$$n^2 + 20n - 384 = 0 \longleftarrow \text{simplest form}$$

6. *Solve:* $(n + 32)(n - 12) = 0$ Factor or use the quadratic formula.

$$n + 32 = 0 \quad \text{or } n - 12 = 0$$

$$n = -32 \text{ or } \qquad n = 12$$

7. *Interpret:* $n = -32$ is not a solution of the original problem because -32 is not a whole number.
   $n = 12$ means the larger whole number is 12.
   $20 - n = 20 - 12 = 8$ means the smaller whole number is 8.

8. *Check:*   Did you find two whole numbers? Yes: 8 and 12.
   Is the sum of the two whole numbers 20? Yes: $8 + 12 = 20$.
   Is the difference of their squares 16 less than their product?
   Yes: $12^2 - 8^2 = 144 - 64 = 80$, and $8(12) - 16 = 96 - 16 = 80$.

Note 1: If you let n = the larger whole number

   then $20 - n$ = the smaller whole number

   and $(n)^2 - (20 - n)^2$ = the positive difference of their squares,

because the larger number comes first when writing a positive difference.

Note 2: If you let n = the smaller whole number

   then $20 - n$ = the larger whole number

   and $(20 - n)^2 - (n)^2$ = the positive difference of their squares,

because the larger number comes first when writing a positive difference.

B. Solving geometry problems using quadratic equations

EXAMPLE 12-17: Solve the following geometry problem using a quadratic equation:

1. *Read:*   The hypotenuse of a right triangle is 1 ft longer than one leg and 2 ft longer than the other leg. How long is each side of the right triangle?

2. *Draw a picture:*

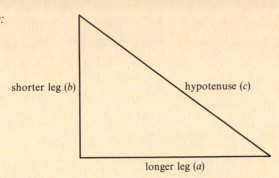

shorter leg (*b*)   hypotenuse (*c*)

longer leg (*a*)

3. *Identify:*   The unknown measures are $\left\{\begin{array}{l}\text{the length of the hypotenuse} \\ \text{the length of the longer leg} \\ \text{the length of the shorter leg}\end{array}\right\}$.

4. *Decide:*   Let c = the length of the hypotenuse (c)

then $c - 1$ = the length of the longer leg (a)

and $c - 2$ = the length of the shorter leg (b).

5. *Translate:* $c^2 = a^2 + b^2$ ⟵ Pythagorean theorem

$c^2 = \overbrace{(c - 1)^2} + \overbrace{(c - 2)^2}$ Substitute $a = c - 1$, and $b = c - 2$.

6. *Simplify:* $c^2 = c^2 - 2c + 1 + c^2 - 4c + 4$

$c^2 - 6c + 5 = 0$ ⟵ simplest form

7. *Solve:* $(c - 5)(c + 1) = 0$ Factor or use the quadratic formula.

$c - 5 = 0$ or $c - 1 = 0$

$c = 5$ or $c = 1$

8. *Interpret:* $c = 1$ is not a solution of the original problem because:
$c - 1 = 1 - 1 = 0$ [ft] and
$c - 2 = 1 - 2 = -1$ [ft] are not the required positive measures.

$c = 5$ means the length of the hypotenuse is 5 ft.
$c - 1 = 4$ means the length of the longer leg is 3 ft.
$c - 2 = 3$ means the length of the shorter leg is 3 ft.

9. *Check:*   Is a triangle with sides of 3, 4, and 5 a right triangle? Yes:

$$c^2 = a^2 + b^2$$

$(5)^2$	$(4)^2 + (3)^2$
25	16 + 9
25	25

Is the hypotenuse 1 ft longer than one leg? Yes: $5 = 1 + 4$.
Is the hypotenuse 2 ft longer than the other leg? Yes: $5 = 2 + 3$.

C. Solving work problems using quadratic equations

Recall 1: The unit rate for a person who completes a certain job (1 job) in n hours is $1/n$ job per hour [see Section 8-6B].

Recall 2: To find the amount of work completed in a given amount of time with respect to the whole job, you evaluate the work formula: work = (unit rate)(time period), or $w = rt$ [see Section 8-6B].

Recall 3: The total work is the whole completed job: total work = 1 (completed job).

EXAMPLE 12-18: Solve the following work problem using a quadratic equation:

1. *Read:* Bobbie can mow a certain lawn in 2 hours less time than it takes Warren. Working together, they can mow the same lawn in $2\frac{2}{5}$ hours. How long does it take each to mow the lawn alone?

2. *Identify:* The unknown time periods are $\begin{cases} \text{the time for Bobbie to mow the lawn alone} \\ \text{the time for Warren to mow the lawn alone} \end{cases}$.

3. *Decide:* Let x = the time for Bobbie to mow the lawn alone

   then $x + 2$ = the time for Warren to mow the lawn alone.

4. *Make a table:*

	unit rate (r) [per hour]	time period (t) [in hours]	work ($w = rt$) [portion of whole job]
Bobbie	$\dfrac{1}{x}$	$2\dfrac{2}{5}$ or $\dfrac{12}{5}$	$\dfrac{1}{x} \cdot \dfrac{12}{5}$ or $\dfrac{12}{5x}$
Warren	$\dfrac{1}{x + 2}$	$2\dfrac{2}{5}$ or $\dfrac{12}{5}$	$\dfrac{1}{x + 2} \cdot \dfrac{12}{5}$ or $\dfrac{12}{5(x + 2)}$

5. *Translate:* $\underbrace{\text{Bobbie's work}}$ plus $\underbrace{\text{Warren's work}}$ equals $\underbrace{\text{the total work}}$.

$$\frac{12}{5x} \qquad + \qquad \frac{12}{5(x + 2)} \qquad = \qquad 1$$

6. *Simplify:* The LCD of $\dfrac{12}{5x}$ and $\dfrac{12}{5(x + 2)}$ is $5x(x + 2)$.   Clear fractions.

$$5x(x + 2) \cdot \frac{12}{5x} + 5x(x + 2) \cdot \frac{12}{5(x + 2)} = 5x(x + 2)(1)$$

$$(x + 2)12 + x(12) = 5x(x + 2) \longleftarrow \text{fractions cleared}$$

$$12 + 24 + 12x = 5x^2 + 10x$$

$$5x^2 - 14x - 24 = 0 \longleftarrow \text{simplest form}$$

7. *Solve:* $(5x + 6)(x - 4) = 0$ Factor or use the quadratic formula.

$$5x + 6 = 0 \quad \text{or} \ x - 4 = 0$$

$$5x = -6 \ \text{or} \qquad x = 4$$

$$x = \frac{-6}{5} \ \text{or} \qquad x = 4$$

8. *Interpret:* $x = -\frac{6}{5}$ is not a solution of the original problem because $-\frac{6}{5}$ hours is a negative time period.

   $x = 4$ means the time for Bobbie to mow the lawn alone is 4 hours.
   $x + 2 = 4 + 2 = 6$ means the time for Warren to mow the lawn alone is 6 hours.

9. *Check:* Does Bobbie's unit rate (per hour) plus Warren's unit rate equal their unit rate together? Yes: $\frac{1}{4} + \frac{1}{6} = \frac{3}{12} + \frac{2}{12} = \frac{5}{12} = 1/(\frac{12}{5}) = 1/(2\frac{2}{5})$.

D. Solving uniform motion problems using quadratic equations ($d = rt$)

EXAMPLE 12-19: Solve the following uniform motion problem using a quadratic equation:

1. *Read:* An airplane took 1 hour more to fly 800 km against a headwind than it did on the return trip with a tailwind of the same velocity. If the constant airspeed of the plane was 180 km/h, how long did the round trip take?

2. *Identify:*   The unknown rates and times are $\left\{\begin{array}{l}\text{the constant velocity of the wind}\\\text{the groundspeed against the wind}\\\text{the groundspeed with the wind}\\\text{the time flying against the wind}\\\text{the time flying with the wind}\end{array}\right\}$.

3. *Decide:*   Let $x =$ the velocity of the wind

then $180 - x =$ the groundspeed against the wind

and $180 + x =$ the groundspeed with the wind.

4. *Make a table:*

	distance (d) [in kilometers]	rate (r) [groundspeed in km/h]	time (t = d/r) [in hours]
against the wind	800	$180 - x$	$\dfrac{800}{180 - x}$
wind the wind	800	$180 + x$	$\dfrac{800}{180 + x}$

5. *Translate:*   The flight against the wind took 1 hour more than the flight with the wind.

$$\frac{800}{180 - x} = 1 + \frac{800}{180 + x}$$

6. *Simplify:*   The LCD of $\dfrac{800}{180 - x}$ and $\dfrac{800}{180 + x}$ is $(180 - x)(180 + x)$.   Clear fractions.

$$(180 - x)(180 + x) \cdot \frac{800}{180 - x} = (180 - x)(180 + x)(1) + (180 - x)(180 + x) \cdot \frac{800}{180 + x}$$

$$(180 + x)800x = (180 - x)(180 + x) + (180 - x)800 \quad \longleftarrow \text{ fractions cleared}$$

$$144,000 + 800x = 32,400 - x^2 + 144,000 - 800x$$

$$800x = 32,400 - x^2 - 800x$$

$$x^2 + 1600x + 32,000 = 0 \quad \longleftarrow \text{ simplest form}$$

7. *Solve:* $(x - 20)(x + 1620) = 0$ Factor or use the quadratic formula.

$$x - 20 = 0 \quad \text{or} \quad x + 1620 = 0$$

$$x = 20 \quad \text{or} \qquad x = -1620$$

8. *Interpret:* $x = -1620$ (km/h) cannot be a solution of the original problem because rates are never negative.

$x = 20$ means the velocity of the wind is 20 km/h.

$180 - x = 180 - 20 = 160$ means the groundspeed against the wind is 160 km/h.

$180 + x = 180 + 20 = 200$ means the groundspeed with the wind is 200 km/h.

$\dfrac{800}{180 - x} = \dfrac{800}{160} = 5$ means the time flying against the wind is 5 hours.

$\dfrac{800}{180 + x} = \dfrac{800}{200} = 4$ means the time flying with the wind is 4 hours.

9. *Check:*   Is 5 hours one hour more than 4 hours? Yes.

Does it take 5 hours to fly 800 km against the wind if the groundspeed is 160 km/h?
Yes: $d = rt$; $800 = (160)(5)$.

Does it take 4 hours to fly 800 km with the wind if the groundspeed is 200 km/h?
Yes: $d = rt$; $800 = (200)(4)$.

12-9. Solving Quadratic Inequalities

A **quadratic inequality** is an inequality which can be written in the form $ax^2 + bx + c > 0$.

Note: Quadratic inequalities, like linear inequalities, may use any of the inequality symbols.

A. Solving quadratic inequalities using algebraic methods

The product principle of inequalities states that

· If $pq > 0$ then p and q are both positive, OR p and q are both negative.
· If $pq < 0$ then p is positive and q is negative, OR p is negative and q is positive.

To **solve a quadratic inequality that can be factored,** we factor and use the product principle of inequalities to find the solution set.

EXAMPLE 12-20: Solve the following quadratic inequalities:

(a) $x^2 - 3x - 4 > 0$ and **(b)** $y^2 - 4 \leq 0$.

Solution:

(a) $x^2 - 3x - 4 > 0$

$(x - 4)(x + 1) > 0$ 　　　　　　　　　　　　　　　　　Factor the quadratic.

$x - 4 > 0$ and $x + 1 > 0$　or $x - 4 < 0$ and $x + 1 < 0$　　　Use the product principle.

　　$x > 4$ and 　　$x > -1$ or 　　$x < 4$ and 　　$x < -1$　　Solve each inequality.

　　　　　　$x > 4$ 　　　　　　or 　　　　　　$x < -1$　　Solve each compound-*and* inequality.

$\{x \mid x > 4 \text{ or } x < -1\}$ 　　　　　　　　　　　Write the compound-*or* inequality in set notation.

(b) 　　　　$y^2 - 4 \leq 0$

$(y - 2)(y + 2) \leq 0$ 　　　　　　　　　　　　　　　Factor the quadratic.

$y - 2 \geq 0$ and $y + 2 \leq 0$　or $y - 2 \leq 0$ and $y + 2 \geq 0$　　Use the product principle.

　$y \geq 2$ and 　　$y \leq -2$ or 　　$y \leq 2$ and 　　$y \geq -2$　Solve the inequalities.

　　　　$\varnothing$ 　　　　　　or 　　　　$-2 \leq y \leq 2$　　Solve each compound-*and* inequality.

　　　　　　$-2 \leq y \leq 2$ 　　　　　　　　　　　Solve the compound-*or* inequality.

$\{y \mid -2 \leq y \leq 2\}$ 　　　　　　　　　　　　Write the solution in set notation.

B. Solving quadratic inequalities using the critical point method

To solve a quadratic inequality using the critical point method, we solve the inequality as an equation. The solutions to this equation (the **equality values** of the inequality) are called the **critical points.** We plot them on a real number line, which divides the line into intervals. We then test sample points from each interval in the original inequality; those points that are true indicate the intervals for which the inequality is true. The solution set, then, is the union of the sets of the intervals which satisfy the inequality. You may find the critical point method less cumbersome than the algebraic method shown in the previous section.

EXAMPLE 12-21: Solve the following quadratic inequalities using the critical point method:

(a) $x^2 - 2x - 3 > 0$ and **(b)** $6x^2 + x - 1 \leq 0$

Solution: **(a)** $x^2 - 2x - 3 > 0$

$\qquad\qquad x^2 - 2x - 3 = 0$ Find the equality values.

$\qquad\qquad (x - 3)(x + 1) = 0$

$\qquad\qquad x - 3 = 0 \text{ or } x + 1 = 0$

$\qquad\qquad x = 3 \text{ or } \qquad x = -1$

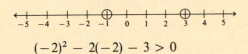

Plot the critical points (equality values).

$\qquad (-2)^2 - 2(-2) - 3 > 0$ Test the interval $x < -1$ using -2 as a test point.

$\qquad\qquad\qquad 5 > 0$ $x < -1$ satisfies the inequality because $5 > 0$.

$\qquad (0)^2 - 2(0) - 3 > 0$ Test the interval $-1 < x < 3$ using 0 as a test point.

$\qquad\qquad\qquad -3 > 0$ $-1 < x < 3$ does not satisfy the inequality because $-3 \not> 0$.

$\qquad 4^2 - 2(4) - 3 > 0$ Test the interval $x > 3$ using 4 as a test point.

$\qquad\qquad\qquad 5 > 0$ $x > 3$ satisfies the inequality because $5 > 0$.

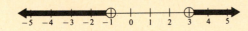

Graph the intervals that satisfy the inequality using open circles because -1 and 3 are not solutions.

$\qquad \{x \mid x < -1 \text{ or } x > 3\}$ Write the solutions in set notation.

(b) $\qquad 6x^2 + x - 1 \leq 0$

$\qquad\qquad 6x^2 + x - 1 = 0$ Find the equality values.

$\qquad (2x + 1)(3x - 1) = 0$

$\qquad 2x + 1 = 0 \quad \text{or } 3x - 1 = 0$

$\qquad x = -\tfrac{1}{2} \text{ or } \qquad x = \tfrac{1}{3}$

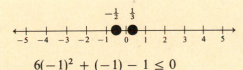

Plot the critical values.

$\qquad 6(-1)^2 + (-1) - 1 \leq 0$ Test the interval $x < -\tfrac{1}{2}$ using -1 as a test point.

$\qquad\qquad\qquad 4 \leq 0$ $x < -\tfrac{1}{2}$ does not satisfy the inequality because $4 \not\leq 0$.

$\qquad 6(0)^2 + 0 - 1 \leq 0$ Test the interval $-\tfrac{1}{2} < x < \tfrac{1}{3}$ using 0 as a test point.

$\qquad\qquad\qquad -1 \leq 0$ $-\tfrac{1}{2} \leq x \leq \tfrac{1}{3}$ satisfies the inequality because $-1 \leq 0$.

$\qquad 6(1)^2 + 1 - 1 \leq 0$ Test the interval $x > \tfrac{1}{3}$ using 1 as a test point.

$\qquad\qquad\qquad 6 \leq 0$ $x > \tfrac{1}{3}$ does not satisfy the inequality because $6 \not\leq 0$.

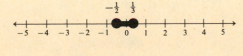

Graph the interval that satisfies the inequality using solid circles.

$\qquad \{x \mid -\tfrac{1}{2} \leq x \leq \tfrac{1}{3}\}$ Write the solutions in set notation.

Note: If the original inequality uses a > or < symbol, we plot the critical points using open circles, because the equality values are not solutions to the inequality. If the original inequality uses a ≥ or ≤ symbol, we plot the critical points using solid circles, because the equality values are solutions to the inequality.

C. Solving rational inequalities using the critical point method

To solve a rational inequality using the critical point method, you find the excluded values and equality values, plot the points and identify the intervals in which the inequality is true.

EXAMPLE 12-22: Solve the rational inequality $1 < \dfrac{x - 8}{x^2 - 3x - 4}$.

Solution:	
$x^2 - 3x - 4 = 0$	Find the excluded values.
$(x - 4)(x + 1) = 0$	
$x - 4 = 0$ or $x + 1 = 0$	
$x = 4$ or $x = -1$	
$1 = \dfrac{x - 8}{x^2 - 3x - 4}$	Find the equality values.
$x^2 - 3x - 4 = x - 8$	
$x^2 - 4x + 4 = 0$	
$(x - 2)^2 = 0$	
$x - 2 = 0$	
$x = 2$	

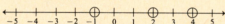

	Plot the critical points.
$1 < \dfrac{-2 - 8}{4 + 6 - 4}$	Test the interval $x < -1$ using -2 as a test point.
$1 < \dfrac{-5}{3}$	$x > -1$ does not satisfy the inequality because $1 \not< -\frac{5}{3}$.
$1 < \dfrac{0 - 8}{0 - 0 - 4}$	Test the interval $-1 < x < 2$ using 0 as a test point.
$1 < 2$	$-1 < x < 2$ satisfies the inequality because $1 < 2$.
$1 < \dfrac{3 - 8}{9 - 9 - 4}$	Test the interval $2 < x < 4$ using 3 as a test point.
$1 < \dfrac{5}{4}$	$2 < x < 4$ satisfies the inequality because $1 < \frac{5}{4}$.
$1 < \dfrac{5 - 8}{25 - 15 - 4}$	Test the interval $4 < x$ using 5 as a test point.
$1 < -\dfrac{1}{2}$	$4 < x$ does not satisfy the inequality because $1 \not< -\frac{1}{2}$.

Graph the intervals that satisfy the inequality.

$\{x | -1 < x < 2 \text{ or } 2 < x < 4\}$. Write in set notation.

Note: If an equality value is the same as an excluded value, it is not a solution and should be graphed with a hollow circle.

RAISE YOUR GRADES

Can you . . . ?

☑ write a quadratic equation in standard form and identify a, b, and c for the equation
☑ solve an equation using the zero-product property
☑ solve an incomplete quadratic equation of the form $ax^2 + bx = 0$
☑ solve an incomplete quadratic equation of the form $ax^2 + c = 0$
☑ solve a quadratic equation by factoring
☑ solve an equation of the form $(x + d)^2 = e$
☑ solve an equation by completing the square
☑ derive the quadratic formula
☑ solve a quadratic equation using the quadratic formula
☑ find approximate solutions
☑ determine the nature of the roots of a quadratic equation using the discriminant
☑ write a quadratic equation given its roots
☑ solve a rational equation
☑ solve a radical equation
☑ solve an equation that is reducable to a quadratic equation
☑ solve a number problem using a quadratic equation
☑ solve a geometry problem using a quadratic equation
☑ solve a work problem using a quadratic equation
☑ solve a uniform motion problem using a quadratic equation
☑ solve a quadratic inequality using the algebraic method
☑ solve a quadratic inequality using the critical point method
☑ solve a rational inequality using the critical point method

SOLVED PROBLEMS

PROBLEM 12-1 Write the following quadratic equations in standard form and identify a, b, and c:

(a) $1 - 3x = 2x^2$ (b) $1 = 3y - 5y^2$ (c) $1 - 3z^2 = 4z$
(d) $\frac{1}{2}x^2 - \frac{1}{3}x = 2$ (e) $\frac{3}{4}x - \frac{1}{2}x^2 = 4$ (f) $4 - \frac{2}{3}x^2 = \frac{1}{2}x$

Solution: Recall that to write a quadratic equation in standard form, you write one member as a polynomial in standard form, the coefficient of the first term as positive, and the other member as 0 [see Example 12-1].

(a) $\qquad 1 - 3x = 2x^2$ (b) $\qquad 1 = 3y - 5y^2$ (c) $\qquad 1 - 3z^2 = 4z$

$\qquad -2x^2 - 3x + 1 = 0$ $5y^2 - 3y + 1 = 0$ $-3z^2 - 4z + 1 = 0$

$\qquad 2x^2 + 3x - 1 = 0$ $a = 5, b = -3, c = 1$ $3z^2 + 4z - 1 = 0$

$\qquad a = 2, b = 3, c = -1$ $a = 3, b = 4, c = -1$

(d) $\qquad \frac{1}{2}x^2 - \frac{1}{3}x = 2$ (e) $\qquad \frac{3}{4}x - \frac{1}{2}x^2 = 4$ (f) $\qquad 4 - \frac{2}{3}x^2 = \frac{1}{2}x$

$\qquad \frac{1}{2}x^2 - \frac{1}{3}x - 2 = 0$ $-\frac{1}{2}x^2 + \frac{3}{4}x - 4 = 0$ $-\frac{2}{3}x^2 - \frac{1}{2}x + 4 = 0$

$\qquad 3x^2 - 2x - 12 = 0$ $2x^2 - 3x + 16 = 0$ $4x^2 + 3x - 24 = 0$

$\qquad a = 3, b = -2, c = -12$ $a = 2, b = -3, c = 16$ $a = 4, b = 3, c = -24$

PROBLEM 12-2 Solve the following equations using the zero-product principle:

(a) $(x + 4)(x - 3) = 0$ (b) $(2x - 3)(4x + 3) = 0$ (c) $(2 - 3y)(1 - 2y) = 0$
(d) $z(z - 2)(1 - z) = 0$ (e) $w(2w + 3)(2 - 5w) = 0$ (f) $p(2 - 3p)(1 + 4p) = 0$

Solution: Recall that to solve an equation that is the product of expressions, you use the zero-product property to set each expression equal to 0 and solve each equation [see Example 12-2].

(a) $(x + 4)(x - 3) = 0$

$x + 4 = 0$ or $x - 3 = 0$

$x = -4$ or $x = 3$

$x = -4$ or 3

(b) $(2x - 3)(4x + 3) = 0$

$2x - 3 = 0$ or $4x + 3 = 0$

$2x = 3$ or $4x = -3$

$x = \frac{3}{2}$ or $x = -\frac{3}{4}$

$x = -\frac{3}{4}$ or $\frac{3}{2}$

(c) $(2 - 3y)(1 - 2y) = 0$

$2 - 3y = 0$ or $1 - 2y = 0$

$-3y = -2$ or $-2y = -1$

$y = \frac{2}{3}$ or $y = \frac{1}{2}$

$y = \frac{1}{2}$ or $\frac{2}{3}$

(d) $z(z - 2)(1 - z) = 0$

$z = 0, z - 2 = 0$ or $1 - z = 0$

$z = 0,$ $z = 2,$ $1 = z$

$z = 0, 1$ or 2

(e) $w(2w + 3)(2 - 5w) = 0$

$w = 0, 2w + 3 = 0$ or $2 - 5w = 0$

$w = 0,$ $2w = -3,$ $2 = 5w$

$w = 0,$ $w = -\frac{3}{2},$ $\frac{2}{5} = w$

$w = -\frac{3}{2}, 0$ or $\frac{2}{5}$

(f) $p(2 - 3p)(1 + 4p) = 0$

$p = 0, 2 - 3p = 0$ or $1 + 4p = 0$

$p = 0,$ $2 = 3p,$ $1 = -4p$

$p = 0,$ $\frac{2}{3} = p,$ $-\frac{1}{4} = p$

$p = -\frac{1}{4}, 0$ or $\frac{2}{3}$

PROBLEM 12-3 Solve the following incomplete quadratic equations of the form $ax^2 + bx = 0$:

(a) $2x^2 - 6x = 0$ **(b)** $4y + 6y^2 = 0$ **(c)** $6z = 9z^2$
(d) $\frac{1}{4}w^2 - \frac{1}{2}w = 0$ **(e)** $\frac{2}{3}t - \frac{3}{4}t^2 = 0$ **(f)** $\frac{1}{2}p = \frac{3}{4}p^2$

Solution: Recall that to solve an incomplete quadratic equation of the form $ax^2 + bx = 0$, you write in standard form, factor the polynomial and solve using the zero-product property [see Example 12-3].

(a) $2x^2 - 6x = 0$

$2x(x - 3) = 0$

$2x = 0$ or $x - 3 = 0$

$2x = 0$ or $x = 3$

$x = 0$ or $x = 3$

$x = 0$ or 3

(b) $6y^2 + 4y = 0$

$2y(3y + 2) = 0$

$2y = 0$ or $3y + 2 = 0$

$2y = 0$ or $3y = -2$

$y = 0$ or $y = -\frac{2}{3}$

$y = -\frac{2}{3}$ or 0

(c) $9z^2 - 6z = 0$

$3z(3z - 2) = 0$

$3z = 0$ or $3z - 2 = 0$

$3z = 0$ or $3z = 2$

$z = 0$ or $z = \frac{2}{3}$

$z = 0$ or $\frac{2}{3}$

(d) $\frac{1}{4}w^2 - \frac{1}{2}w = 0$

$w^2 - 2w = 0$

$w(w - 2) = 0$

$w = 0$ or $w = 2$

$w = 0$ or 2

(e) $\frac{3}{4}t^2 - \frac{2}{3}t = 0$

$9t^2 - 8t = 0$

$t(9t - 8) = 0$

$t = 0$ or $9t - 8 = 0$

$t = 0$ or $9t = 8$

$t = 0$ or $t = \frac{8}{9}$

$t = 0$ or $\frac{8}{9}$

(f) $\frac{3}{4}p^2 - \frac{1}{2}p = 0$

$3p^2 - 2p = 0$

$p(3p - 2) = 0$

$p = 0$ or $3p - 2 = 0$

$p = 0$ or $3p = 2$

$p = 0$ or $p = \frac{2}{3}$

$p = 0$ or $\frac{2}{3}$

PROBLEM 12-4 Solve the following incomplete quadratic equations of the form $ax^2 + c = 0$:

(a) $x^2 - 4 = 0$ **(b)** $9y^2 - 16 = 0$ **(c)** $4z^2 - 3 = 0$
(d) $3w^2 - 1 = 0$ **(e)** $9t^2 + 4 = 0$ **(f)** $3p^2 + 2 = 0$

Solution: Recall that to solve an incomplete quadratic equation of the form $ax^2 + c = 0$, you solve for x^2 and use the square root rule to solve for x [see Example 12-4].

(a) $x^2 - 4 = 0$

$$x^2 = 4$$

$$x = \sqrt{4} \text{ or } -\sqrt{4}$$

$$x = 2 \quad \text{or } -2$$

(b) $9y^2 - 16 = 0$

$$9y^2 = 16$$

$$y^2 = \frac{16}{9}$$

$$y = \sqrt{\frac{16}{9}} \text{ or } -\sqrt{\frac{16}{9}}$$

$$y = \frac{4}{3} \quad \text{or } -\frac{4}{3}$$

(c) $4z^2 - 3 = 0$

$$4z^2 = 3$$

$$z^2 = \frac{3}{4}$$

$$z = \sqrt{\frac{3}{4}} \text{ or } -\sqrt{\frac{3}{4}}$$

$$z = \frac{\sqrt{3}}{2} \text{ or } -\frac{\sqrt{3}}{2}$$

(d) $3w^2 - 1 = 0$

$$3w^2 = 1$$

$$w^2 = \frac{1}{3}$$

$$w = \sqrt{\frac{1}{3}} \text{ or } -\sqrt{\frac{1}{3}}$$

$$w = \frac{\sqrt{3}}{3} \text{ or } -\frac{\sqrt{3}}{3}$$

(e) $9t^2 + 4 = 0$

$$9t^2 = -4$$

$$t^2 = -\frac{4}{9}$$

$$t = \sqrt{-\frac{4}{9}} \text{ or } -\sqrt{-\frac{4}{9}}$$

$$t = \frac{2}{3}i \quad \text{or } -\frac{2}{3}i$$

(f) $3p^2 + 2 = 0$

$$3p^2 = -2$$

$$p^2 = -\frac{2}{3}$$

$$p = \sqrt{-\frac{2}{3}} \text{ or } -\sqrt{-\frac{2}{3}}$$

$$p = \frac{\sqrt{6}}{3}i \quad \text{or } -\frac{\sqrt{6}}{3}i$$

PROBLEM 12-5 Solve the following quadratic equations by factoring:

(a) $x^2 + 2x - 3 = 0$ **(b)** $y^2 + 7y + 10 = 0$ **(c)** $z^2 - 2z - 24 = 0$
(d) $6x^2 + 7x + 2 = 0$ **(e)** $6x^2 + x = 12$ **(f)** $6x^2 = 4 - 23x$

Solution: Recall that to solve a quadratic equation by factoring, you write the quadratic equation so that one member is 0, factor the quadratic and solve as before [see Example 12-5].

(a) $x^2 + 2x - 3 = 0$

$$(x + 3)(x - 1) = 0$$

$$x + 3 = 0 \quad \text{or } x - 1 = 0$$

$$x = -3 \text{ or } \quad x = 1$$

$$x = -3 \text{ or } 1$$

(b) $y^2 + 7y + 10 = 0$

$$(y + 5)(y + 2) = 0$$

$$y + 5 = 0 \quad \text{or } y + 2 = 0$$

$$y = -5 \text{ or } \quad y = -2$$

$$y = -5 \text{ or } -2$$

(c) $z^2 - 2z - 24 = 0$

$$(z - 6)(z + 4) = 0$$

$$z - 6 = 0 \text{ or } z + 4 = 0$$

$$z = 6 \text{ or } \quad z = -4$$

$$z = -4 \text{ or } 6$$

(d) $6x^2 + 7x + 2 = 0$

$$(2x + 1)(3x + 2) = 0$$

$$2x + 1 = 0 \quad \text{or } 3x + 2 = 0$$

$$2x = -1 \text{ or } \quad 3x = -2$$

$$x = -\tfrac{1}{2} \text{ or } \quad x = -\tfrac{2}{3}$$

$$x = -\tfrac{2}{3} \text{ or } -\tfrac{1}{2}$$

(e) $6x^2 + x = 12$ **(f)** $6x^2 = 4 - 23x$

$6x^2 + x - 12 = 0$ $6x^2 + 23x - 4 = 0$

$(3x - 4)(2x + 3) = 0$ $(6x - 1)(x + 4) = 0$

$3x - 4 = 0$ or $2x + 3 = 0$ $6x - 1 = 0$ or $x + 4 = 0$

$3x = 4$ or $\quad 2x = -3$ $6x = 1$ or $\quad x = -4$

$x = \frac{4}{3}$ or $\quad x = -\frac{3}{2}$ $x = \frac{1}{6}$ or $\quad x = -4$

$x = -\frac{3}{2}$ or $\frac{4}{3}$ $x = -4$ or $\frac{1}{6}$

PROBLEM 12-6 Solve the following equations of the form $(x + d)^2 = e$:

(a) $(x - 3)^2 = 9$ **(b)** $(x + 4)^2 = 9$ **(c)** $(2x - 3)^2 = 1$
(d) $(x - 2)^2 = 3$ **(e)** $(2x - 3)^2 = 2$ **(f)** $(3x + 2)^2 = 3$

Solution: Recall that to solve an equation of the form $(x + d)^2 = e$, you use the square root rule to get $x + d = \pm\sqrt{e}$ and solve each equation [see Example 12-6].

(a) $(x - 3)^2 = 9$ **(b)** $(x + 4)^2 = 9$ **(c)** $(2x - 3)^2 = 1$

$x - 3 = \pm\sqrt{9}$ $x + 4 = \pm\sqrt{9}$ $2x - 3 = \pm\sqrt{1}$

$x - 3 = \pm 3$ $x + 4 = \pm 3$ $2x - 3 = \pm 1$

$x = 3 \pm 3$ $x = -4 \pm 3$ $2x = 3 \pm 1$

$x = 0$ or 6 $x = -7$ or -1 $2x = 2$ or 4

$x = 1$ or 2

(d) $(x - 2)^2 = 3$ **(e)** $(2x - 3)^2 = 2$ **(f)** $(3x + 2)^2 = 3$

$x - 2 = \pm\sqrt{3}$ $2x - 3 = \pm\sqrt{2}$ $3x + 2 = \pm\sqrt{3}$

$x = 2 \pm \sqrt{3}$ $2x = 3 \pm \sqrt{2}$ $3x = -2 \pm \sqrt{3}$

$x = 2 + \sqrt{3}$ or $2 - \sqrt{3}$ $x = \dfrac{3 \pm \sqrt{2}}{2}$ $x = \dfrac{-2 \pm \sqrt{3}}{3}$

$x = \dfrac{3 + \sqrt{2}}{2}$ or $\dfrac{3 - \sqrt{2}}{2}$ $x = \dfrac{-2 + \sqrt{3}}{3}$ or $\dfrac{-2 - \sqrt{3}}{3}$

PROBLEM 12-7 Solve the following equations by completing the square:

(a) $x^2 + 4x + 4 = 0$ **(b)** $2x^2 + x - 2 = 0$ **(c)** $x^2 + 4x - 3 = 0$
(d) $3x^2 - 6x - 1 = 0$ **(e)** $x^2 + x + 1 = 0$ **(f)** $2x^2 + 4x + 3 = 0$

Solution: Recall that to solve a quadratic equation by completing the square, you write the equation in $ax^2 + bx = -c$ form, divide by the coefficient of x^2, complete the square in one member to get an equation of the form $(x + d)^2 = e$ and solve as before [see Example 12-7].

(a) $x^2 + 4x + 4 = 0$ **(b)** $2x^2 + x - 2 = 0$

$x^2 + 4x = -4$ $x^2 + \frac{1}{2}x = 1$

$x^2 + 4x + \left(\dfrac{4}{2}\right)^2 = \left(\dfrac{4}{2}\right)^2 - 4$ $x^2 + \frac{1}{2}x + \left(\dfrac{1}{4}\right)^2 = \left(\dfrac{1}{4}\right)^2 + 1$

$(x + 2)^2 = 0$ $\left(x + \dfrac{1}{4}\right)^2 = \dfrac{17}{16}$

$x + 2 = 0$ $x + \dfrac{1}{4} = \pm\dfrac{\sqrt{17}}{4}$

$x = -2$

$x = -\dfrac{1}{4} \pm \dfrac{\sqrt{17}}{4}$

(c) $x^2 + 4x - 3 = 0$

$x^2 + 4x = 3$

$x^2 + 4x + \left(\dfrac{4}{2}\right)^2 = \left(\dfrac{4}{2}\right)^2 + 3$

$(x + 2)^2 = 7$

$x + 2 = \pm\sqrt{7}$

$x = -2 \pm \sqrt{7}$

(d) $3x^2 - 6x - 1 = 0$

$x^2 - 2x = \dfrac{1}{3}$

$x^2 - 2x + \left(\dfrac{-2}{2}\right)^2 = 1 + \dfrac{1}{3}$

$(x - 1)^2 = \dfrac{4}{3}$

$x - 1 = \pm\dfrac{2\sqrt{3}}{3}$

$x = 1 \pm \dfrac{2\sqrt{3}}{3}$

(e) $x^2 + x + 1 = 0$

$x^2 + x = -1$

$x^2 + x + \left(\dfrac{1}{2}\right)^2 = \dfrac{1}{4} - 1$

$\left(x + \dfrac{1}{2}\right)^2 = -\dfrac{3}{4}$

$x + \dfrac{1}{2} = \dfrac{i\sqrt{3}}{2}$

$x = -\dfrac{1}{2} \pm \dfrac{i\sqrt{3}}{2}$

(f) $2x^2 + 4x + 3 = 0$

$x^2 + 2x = -\dfrac{3}{2}$

$x^2 + 2x + \left(\dfrac{2}{2}\right)^2 = 1 - \dfrac{3}{2}$

$(x + 1)^2 = -\dfrac{1}{2}$

$x + 1 = \pm\dfrac{i\sqrt{2}}{2}$

$x = -1 \pm \dfrac{i\sqrt{2}}{2}$

PROBLEM 12-8 Solve the following equations using the quadratic formula:

(a) $x^2 + 3x - 4 = 0$ (b) $x^2 + 3x + 1 = 0$ (c) $x^2 + 3x + 3 = 0$
(d) $6x^2 - 5x - 6 = 0$ (e) $x^2 - x + 1 = 0$ (f) $3x^2 + 2x + 2 = 0$

Solution: Recall that to solve a quadratic equation using the quadratic formla, you identify a, b, and c, substitute the values in the quadratic formula and simplify [see Example 12-9].

(a) $x^2 + 3x - 4 = 0$

$a = 1, b = 3, c = -4$

$x = \dfrac{-b \pm \sqrt{b^2 - 4ac}}{2a}$

$x = \dfrac{-3 \pm \sqrt{9 - 4(1)(-4)}}{2(1)}$

$x = \dfrac{-3 \pm \sqrt{25}}{2}$

$x = \dfrac{-3 \pm 5}{2}$

$x = -4 \text{ or } 1$

(b) $x^2 + 3x + 1 = 0$

$a = 1, b = 3, c = 1$

$x = \dfrac{-b \pm \sqrt{b^2 - 4ac}}{2a}$

$x = \dfrac{-3 \pm \sqrt{9 - 4(1)(1)}}{2(1)}$

$x = \dfrac{-3 \pm \sqrt{5}}{2}$

(c) $x^2 + 3x + 3 = 0$

$a = 1, b = 3, c = 3$

$x = \dfrac{-b \pm \sqrt{b^2 - 4ac}}{2a}$

$x = \dfrac{-3 \pm \sqrt{9 - 4(1)(3)}}{2(1)}$

$x = \dfrac{-3 \pm \sqrt{-3}}{2}$

$x = \dfrac{-3 \pm i\sqrt{3}}{2}$

(d) $6x^2 - 5x - 6 = 0$

$a = 6, b = -5, c = -6$

$x = \dfrac{-(-5) \pm \sqrt{25 - 4(6)(-6)}}{12}$

$x = \dfrac{5 \pm \sqrt{25 + 144}}{12}$

$x = \dfrac{5 \pm 13}{12}$

$x = \dfrac{18}{12}$ or $-\dfrac{8}{12}$

$x = -\dfrac{2}{3}$ or $\dfrac{3}{2}$

(e) $x^2 - x + 1 = 0$

$a = 1, b = -1, c = 1$

$x = \dfrac{-(-1) \pm \sqrt{1 - 4(1)(1)}}{2(1)}$

$x = \dfrac{1 \pm \sqrt{-3}}{2}$

$x = \dfrac{1 \pm i\sqrt{3}}{2}$

(f) $3x^2 + 2x + 2 = 0$

$a = 3, b = 2, c = 2$

$x = \dfrac{-2 \pm \sqrt{4 - 4(3)(2)}}{2(3)}$

$x = \dfrac{-2 \pm \sqrt{4 - 24}}{6}$

$x = \dfrac{-2 \pm \sqrt{-20}}{6}$

$x = \dfrac{-2 \pm 2i\sqrt{5}}{6}$

$x = \dfrac{-1 \pm i\sqrt{5}}{3}$

PROBLEM 12-9 Find the approximate solutions of the following equations:

(a) $x^2 + x - 1 = 0$ (b) $x^2 + 4x + 2 = 0$ (c) $x^2 - 3x - 1 = 0$
(d) $2x^2 + 2x - 3 = 0$ (e) $3x^2 - 2x - 3 = 0$ (f) $4x^2 + 5x - 2 = 0$

Solution: Recall that to find the approximate solutions of an equation, you solve the equation using the quadratic formula and substitute the approximate value using a calculator or the square root table [see Example 12-10].

(a) $x^2 + x - 1 = 0$

$a = 1, b = 1, c = -1$

$x = \dfrac{-1 \pm \sqrt{1 + 4}}{2}$

$x = \dfrac{-1 \pm \sqrt{5}}{2}$

$x \approx \dfrac{-1 \pm 2.236}{2}$

$x \approx -1.618$ or 0.618

(b) $x^2 + 4x + 2 = 0$

$a = 1, b = 4, c = 2$

$x = \dfrac{-4 \pm \sqrt{16 - 8}}{2}$

$x = \dfrac{-4 \pm 2\sqrt{2}}{2}$

$x = -2 \pm \sqrt{2}$

$x \approx -2 \pm 1.414$

$x \approx -3.414$ or -0.586

(c) $x^2 - 3x - 1 = 0$

$a = 1, b = -3, c = -1$

$x = \dfrac{3 \pm \sqrt{9 + 4}}{2}$

$x = \dfrac{3 \pm \sqrt{13}}{2}$

$x \approx \dfrac{3 \pm 3.606}{2}$

$x \approx -0.303$ or 3.303

(d) $2x^2 + 2x - 3 = 0$

$a = 2, b = 2, c = -3$

$$x = \frac{-2 \pm \sqrt{4 + 24}}{4}$$

$$x = \frac{-2 \pm 2\sqrt{7}}{4}$$

$$x = \frac{-1 \pm \sqrt{7}}{2}$$

$$x \approx \frac{-1 \pm 2.646}{2}$$

$$x \approx -1.823 \text{ or } 0.823$$

(e) $3x^2 - 2x - 3 = 0$

$a = 3, b = -2, c = -3$

$$x = \frac{2 \pm \sqrt{4 + 36}}{6}$$

$$x = \frac{2 \pm 2\sqrt{10}}{6}$$

$$x = \frac{1 \pm \sqrt{10}}{3}$$

$$x \approx \frac{1 \pm 3.162}{3}$$

$$x \approx -1.081 \text{ or } 2.081$$

(f) $4x^2 + 5x - 2 = 0$

$a = 4, b = 5, c = -2$

$$x = \frac{-5 \pm \sqrt{25 + 32}}{8}$$

$$x = \frac{-5 \pm \sqrt{57}}{8}$$

$$x \approx \frac{-5 \pm 7.550}{8}$$

$$x \approx -1.569 \text{ or } 0.319$$

PROBLEM 12-10 Determine whether the following quadratic equations have one real solution, two real solutions, or two distinct complex or nonreal solutions:

(a) $x^2 + 2x + 1 = 0$ **(b)** $x^2 + 3x + 2 = 0$ **(c)** $x^2 + 1 = 0$
(d) $2x^2 + 2x + 1 = 0$ **(e)** $3x^2 + 2x - 3 = 0$ **(f)** $2x^2 + x + 1 = 0$

Solution: Recall that to determine whether a quadratic equation has one real solution, two real solutions, or two distinct complex or nonreal solutions, you evaluate the discriminant [see Example 12-11].

(a) $x^2 + 2x + 1 = 0$

$a = 1, b = 2, c = 1$

$b^2 - 4ac$

$4 - 4(1)(1) = 0$

One real solution.

(b) $x^2 + 3x + 2 = 0$

$a = 1, b = 3, c = 2$

$b^2 - 4ac$

$9 - 4(1)(2)$

$9 - 8 > 0$

Two real solutions.

(c) $x^2 + 1 = 0$

$a = 1, b = 0, c = 1$

$b^2 - 4ac$

$0 - 4(1)(1)$

$0 - 4 < 0$

Two nonreal solutions.

(d) $2x^2 + 2x + 1 = 0$

$a = 2, b = 2, c = 1$

$b^2 - 4ac$

$4 - 4(2)(1) < 0$

Two nonreal solutions.

(e) $3x^2 + 2x - 3 = 0$

$a = 3, b = 2, c = -3$

$b^2 - 4ac$

$4 - 4(3)(-3) > 0$

Two real solutions.

(f) $2x^2 + x + 1 = 0$

$a = 2, b = 1, c = 1$

$b^2 - 4ac$

$1 - 4(2)(1) < 0$

Two nonreal solutions.

PROBLEM 12-11 Write a quadratic equation having the following numbers as solutions:

(a) $-3, 1$ **(b)** $-\frac{1}{3}, \frac{2}{3}$ **(c)** $\pm 2\sqrt{5}$ **(d)** $3 \pm \sqrt{3}$ **(e)** $\pm 3i$ **(f)** $2 \pm i$

Solution: Recall that to write a quadratic equation given its roots, you use the reverse of the zero-product property [see Example 12-12].

(a) $x = -3$ or $x = 1$

$x + 3 = 0$ or $x - 1 = 0$

$(x + 3)(x - 1) = 0$

$x^2 + 2x - 3 = 0$

(b) $x = -\frac{1}{3}$ or $x = \frac{2}{3}$

$x + \frac{1}{3} = 0$ or $x - \frac{2}{3} = 0$

$3x + 1 = 0$ or $3x - 2 = 0$

$(3x + 1)(3x - 2) = 0$

$9x^2 - 3x - 2 = 0$

(c) $x = 2\sqrt{5}$ or $x = -2\sqrt{5}$

$x - 2\sqrt{5} = 0$ or $x + 2\sqrt{5} = 0$

$(x - 2\sqrt{5})(x + 2\sqrt{5}) = 0$

$x^2 - 20 = 0$

(d) $x = 3 + \sqrt{3}$ or $x = 3 - \sqrt{3}$

$x - 3 - \sqrt{3} = 0$ or $x - 3 + \sqrt{3} = 0$

$(x - 3 - \sqrt{3})(x - 3 + \sqrt{3}) = 0$

$x^2 - 6x + 6 = 0$

(e) $x = 3i$ or $x = -3i$

$x - 3i = 0$ or $x + 3i = 0$

$(x - 3i)(x + 3i) = 0$

$x^2 - 9i^2 = 0$

$x^2 + 9 = 0$

(f) $x = 2 + i$ or $x = 2 - i$

$x - 2 - i = 0$ or $x - 2 + i = 0$

$(x - 2 - i)(x - 2 + i) = 0$

$x^2 - 4x + 5 = 0$

PROBLEM 12-12 Solve the following rational equations:

(a) $\dfrac{1}{x} = 2x + 1$ **(b)** $\dfrac{x}{x - 1} = \dfrac{2x + 1}{x + 1}$ **(c)** $\dfrac{x + 5}{2x} = \dfrac{5}{x}$

(d) $\dfrac{x - 3}{2x + 3} = \dfrac{x - 2}{x + 2}$ **(e)** $\dfrac{x - 3}{x - 2} = \dfrac{3 - x}{x + 1}$ **(f)** $\dfrac{x - 3}{x^2 - 4} + 1 = \dfrac{1}{x - 2}$

Solution: Recall that to solve a rational equation, you clear the fractions and solve as before. Check your proposed solutions for extraneous solutions [see Example 12-13].

(a) $x \cdot \dfrac{1}{x} = x(2x + 1)$

$1 = 2x^2 + x$

$0 = 2x^2 + x - 1$

$0 = (2x - 1)(x + 1)$

$0 = 2x - 1$ or $x + 1 = 0$

$1 = 2x$ or $x = -1$

$\tfrac{1}{2} = x$

$x = -1$ or $\tfrac{1}{2}$

(b) $(x - 1)(x + 1)\dfrac{x}{x - 1} = (x - 1)(x + 1)\dfrac{2x + 1}{x + 1}$

$(x + 1)x = (x - 1)(2x + 1)$

$x^2 + x = 2x^2 - x - 1$

$0 = x^2 - 2x - 1$

$x = \dfrac{2 \pm \sqrt{4 + 4}}{2}$

$x = \dfrac{2 \pm 2\sqrt{2}}{2}$

$x = 1 \pm \sqrt{2}$

(c) $2x\dfrac{x + 5}{2x} = 2x\dfrac{5}{x}$

$x + 5 = 10$

$x = 5$

(d) $(2x + 3)(x + 2)\dfrac{x - 3}{2x + 3} = (2x + 3)(x + 2)\dfrac{x - 2}{x + 2}$

$(x + 2)(x - 3) = (2x + 3)(x - 2)$

$x^2 - x - 6 = 2x^2 - x - 6$

$0 = x^2$

$0 = x$

(e) $(x - 2)(x + 1)\dfrac{x - 3}{x - 2} = (x - 2)(x + 1)\dfrac{3 - x}{x + 1}$

$(x + 1)(x - 3) = (x - 2)(3 - x)$

$x^2 - 2x - 3 = -x^2 + 5x - 6$

$2x^2 - 7x + 3 = 0$

$(2x - 1)(x - 3) = 0$

$2x - 1 = 0$ or $x - 3 = 0$

$2x = 1$ or $x = 3$

$x = \tfrac{1}{2}$ or 3

(f) $(x^2 - 4)\dfrac{x - 3}{x^2 - 4} + 1 = (x^2 - 4)\dfrac{1}{x - 2}$

$x - 3 + x^2 - 4 = x + 2$

$x^2 - 9 = 0$

$x^2 = 9$

$x = \pm 3$

PROBLEM 12-13 Solve the following radical equations:

(a) $\sqrt{x + 4} = x + 2$ (b) $\sqrt{x + 2} = \sqrt{x} + 1$ (c) $\sqrt{x + 3} = \sqrt{x - 3}$

(d) $\sqrt{2 - 3x} + \sqrt{3x + 3} = 1$ (e) $\sqrt{3x + 1} - \sqrt{x + 4} = 1$ (f) $\sqrt{x + 3} - \sqrt{x - 2} = 1$

Solution: Recall that to solve a radical equation, you use the Power Principle to clear radicals by squaring both members of the equation until all radicals are cleared and solve the equation as before. Remember to separate radicals into both members if necessary [see Example 12-14].

(a) $\sqrt{x + 4} = x + 2$

$$x + 4 = x^2 + 4x + 4$$
$$0 = x^2 + 3x$$
$$0 = x(x + 3)$$
$$0 = x \text{ or } x + 3 = 0$$
$$x = 0 \text{ or } \quad x = -3$$

$x = 0$ since -3 does not check.

(b) $\sqrt{x + 2} = \sqrt{x} + 1$

$$x + 2 = x + 1 + 2\sqrt{x}$$
$$1 = 2\sqrt{x}$$
$$1 = 4x$$
$$\tfrac{1}{4} = x$$

(c) $\sqrt{x + 3} = \sqrt{x - 3}$

$$x + 3 = x - 3$$
$$6 = 0$$

No solution.

(d)

$$\sqrt{2 - 3x} = 1 - \sqrt{3x + 3}$$
$$2 - 3x = 1 + 3x + 3 - 2\sqrt{3x + 3}$$
$$-2 - 6x = -2\sqrt{3x + 3}$$
$$1 + 3x = \sqrt{3x + 3}$$
$$1 + 6x + 9x^2 = 3x + 3$$
$$(3x - 1)(3x + 2) = 0$$
$$3x - 1 = 0 \text{ or } 3x + 2 = 0$$
$$3x = 1 \text{ or } \quad 3x = -2$$
$$x = \tfrac{1}{3} \text{ or } \quad x = -\tfrac{2}{3}$$

No solution since neither check.

(e)

$$\sqrt{3x + 1} = 1 + \sqrt{x + 4}$$
$$3x + 1 = 1 + 2\sqrt{x + 4} + x + 4$$
$$2x - 4 = 2\sqrt{x + 4}$$
$$x - 2 = \sqrt{x + 4}$$
$$x^2 - 4x + 4 = x + 4$$
$$x^2 - 5x = 0$$
$$x(x - 5) = 0$$
$$x = 0 \text{ or } x - 5 = 0$$
$$x = 5$$

(f)

$$\sqrt{x + 3} = 1 + \sqrt{x - 2}$$
$$x + 3 = 1 + 2\sqrt{x - 2} + x - 2$$
$$4 = 2\sqrt{x - 2}$$
$$2 = \sqrt{x - 2}$$
$$4 = x - 2$$
$$6 = x$$

PROBLEM 12-14 Solve the following equations that are reducable to quadratic equations:

(a) $x^4 - 6x^2 + 9 = 0$ (b) $6x^4 + 5x^2 - 6 = 0$ (c) $6x + \sqrt{x} - 12 = 0$

(d) $6x - 11\sqrt{x} + 3 = 0$ (e) $x^{-2} + 8x^{-1} + 16 = 0$ (f) $3x^{-2} + 16x^{-1} - 12 = 0$

Solution: Recall that to solve an equation that is reducable to a quadratic equation, you substitute a different variable in the equation and solve as before (see Example 12-15].

(a) Substitute u for x^2

$$u^2 - 6u + 9 = 0$$
$$(u - 3)^2 = 0$$
$$u - 3 = 0$$
$$u = 3$$
$$x^2 = 3$$
$$x = \pm\sqrt{3}$$

(b) Substitute u for x^2

$$6u^2 + 5u - 6 = 0$$
$$(3u - 2)(2u + 3) = 0$$
$$3u - 2 = 0 \quad \text{or} \quad 2u + 3 = 0$$
$$u = \frac{2}{3} \quad \text{or} \quad u = -\frac{3}{2}$$
$$x^2 = \frac{2}{3} \quad \text{or} \quad -\frac{3}{2}$$
$$x = \pm\frac{\sqrt{6}}{3} \quad \text{or} \quad \pm\frac{\sqrt{6}}{2}i$$

(c) Substitute u for $\sqrt{x}$

$$6u^2 + u - 12 = 0$$
$$(3u - 4)(2u + 3) = 0$$
$$3u - 4 = 0 \quad \text{or} \quad 2u + 3 = 0$$
$$u = \tfrac{4}{3} \quad \text{or} \quad u = -\tfrac{3}{2}$$
$$\sqrt{x} = \tfrac{4}{3} \quad \text{or} \quad -\tfrac{3}{2}$$
$$x = \tfrac{16}{9} \quad \text{or} \quad \tfrac{9}{4}$$

$x = \tfrac{16}{9}$, since $\tfrac{9}{4}$ does not check.

(d) Substitute u for $\sqrt{x}$

$$6u^2 - 11u + 3 = 0$$
$$(3u - 1)(2u - 3) = 0$$
$$3u - 1 = 0 \quad \text{or} \quad 2u - 3 = 0$$
$$u = \tfrac{1}{3} \quad \text{or} \quad u = \tfrac{3}{2}$$
$$\sqrt{x} = \tfrac{1}{3} \quad \text{or} \quad \sqrt{x} = \tfrac{3}{2}$$
$$x = \tfrac{1}{9} \quad \text{or} \quad x = \tfrac{9}{4}$$

(e) Substitute u for x^{-1}

$$u^2 + 8u + 16 = 0$$
$$(u + 4)^2 = 0$$
$$u + 4 = 0$$
$$u = -4$$
$$x^{-1} = -4$$
$$x = -\tfrac{1}{4}$$

(f) Substitute u for x^{-1}

$$3u^2 + 16u - 12 = 0$$
$$(3u - 2)(u + 6) = 0$$
$$3u - 2 = 0 \quad \text{or} \quad u + 6 = 0$$
$$u = \tfrac{2}{3} \quad \text{or} \quad u = -6$$
$$x^{-1} = \tfrac{2}{3} \quad \text{or} \quad x^{-1} = -6$$
$$x = \tfrac{3}{2} \quad \text{or} \quad -\tfrac{1}{6}$$

PROBLEM 12-15 Solve each number problem using a quadratic equation:

(a) The sum of a positive number and its square is 1. Find the number.

(b) The difference between a negative number and its reciprocal is 1. What is the number?

Solution: Recall that to solve a number problem using a quadratic equation, you

1. *Read* the problem very carefully, several times.
2. *Identify* the unknown numbers.
3. *Decide* how to represent the unknown numbers using one variable.
4. *Translate* the problem to a quadratic equation.
5. *Simplify* the quadratic equation.
6. *Solve* the quadratic equation.
7. *Interpret* the solutions of the quadratic equation with respect to each represented unknown number to find the proposed solutions of the original problem.
8. *Check* to see if the proposed solutions satisfy all the conditions of the original problem [see Example 12-16].

(a) *Identify:*   The unknown numbers are $\begin{cases} \text{the positive number} \\ \text{the square of the positive number} \end{cases}$.

Decide: Let n = the positive number

then n^2 = the square of the positive number.

Translate: $\underbrace{\text{The sum of a positive number and its square}}$ is 1.

$$n + n^2 \qquad\qquad = 1$$

Simplify: $n^2 + n - 1 = 0$ ← simplest form

Solve: In $n^2 + n - 1 = 0$, $a = 1$, $b = 1$, $c = -1$.

$$n = \frac{-b \pm \sqrt{b^2 - 4ac}}{2a} \quad\longleftarrow \text{ quadratic formula}$$

$$n = \frac{-(1) \pm \sqrt{(1)^2 - 4(1)(-1)}}{2(1)} \qquad \text{Evaluate for } a = 1, b = 1, \text{ and } c = -1.$$

$$n = \frac{-1 \pm \sqrt{5}}{2}$$

Interpret: Since $n = \dfrac{-1 - \sqrt{5}}{2}$ is a negative number (≈ -1.618), it is not a solution of the original problem.

Since $n = \dfrac{-1 + \sqrt{5}}{2}$ is a positive number (≈ 0.618), it is the proposed solution of the original problem.

Check: Is the sum of the positive number $\dfrac{-1 + \sqrt{5}}{2}$ and its square equal to 1? Yes:

$$n + n^2 = \left(\frac{-1 + \sqrt{5}}{2}\right) + \left(\frac{-1 + \sqrt{5}}{2}\right)^2 = \frac{-1 + \sqrt{5}}{2} + \frac{1 - 2\sqrt{5} + 5}{4}$$

$$= \frac{-2 + 2\sqrt{5} + 6 - 2\sqrt{5}}{4} = 1$$

(b) *Identify:*   The unknown numbers are $\begin{cases} \text{the negative number} \\ \text{the reciprocal of the negative number} \end{cases}$.

Decide: Let n = the negative number

then $\dfrac{1}{n}$ = the reciprocal of the negative number.

Translate: $\underbrace{\text{The difference between a negative number and its reciprocal}}$ is 1.

$$n - \frac{1}{n} \qquad\qquad = 1$$

Simplify: The LCD of $\dfrac{n}{1}$, $\dfrac{1}{n}$, and $\dfrac{1}{1}$ is n.   Clear fractions.

$$n(n) - n\left(\frac{1}{n}\right) = n(1)$$

$$n^2 - 1 = n \quad\longleftarrow \text{ fraction cleared}$$

$$n^2 - n - 1 = 0 \quad\longleftarrow \text{ simplest form}$$

Solve: In $n^2 - n - 1 = 0$, $a = 1$, $b = -1$, $c = 1$.

$$n = \frac{-b \pm \sqrt{b^2 - 4ac}}{2a} \quad \longleftarrow \text{ quadratic formula}$$

$$n = \frac{-(-1) \pm \sqrt{(-1)^2 - 4(1)(-1)}}{2(1)} \qquad \text{Evaluate for } a = 1, b = -1, \text{ and } c = -1.$$

$$n = \frac{1 \pm \sqrt{5}}{2}$$

Interpret: Since $n = \dfrac{1 + \sqrt{5}}{2}$ is a positive number (≈ 1.618), it is not a solution of the original problem.

Since $n = \dfrac{1 - \sqrt{5}}{2}$ is a negative number (≈ -0.618), it is the proposed solution of the original problem.

Check: Is the difference between the negative number $\dfrac{1 - \sqrt{5}}{2}$ and its reciprocal 1? Yes:

$$n - \frac{1}{n} = \left(\frac{1 - \sqrt{5}}{2}\right) - \left(\frac{1}{\frac{1 - \sqrt{5}}{2}}\right) = \frac{1 - \sqrt{5}}{2} - \frac{2}{(1 - \sqrt{5})} = \frac{1 - \sqrt{5}}{2} - \frac{2 + 2\sqrt{5}}{-4} = \frac{2 - 2\sqrt{5} + 2 + 2\sqrt{5}}{4} = 1.$$

↑ rationalize the denominator

PROBLEM 12-16 Solve each geometry problem using a quadratic equation:

(a) The diagonal of a rectangle is 3 ft longer than its length and 5 ft longer than its width. Find the area of the rectangle.

(b) A rectangular swimming pool is surrounded by a concrete walkway of uniform width. The outside dimensions of the walkway are 13 ft by 18 ft. The water surface of the pool is 150 ft². What are the dimensions of the pool?

Solution: Recall that to solve a geometry problem using a quadratic equation, you:

1. *Read* the problem very carefully, several times.
2. *Draw a picture* to help visualize the problem.
3. *Identify* the unknown measures.
4. *Decide* how to represent the unknown measures using one variable.
5. *Translate* the problem to a quadratic equation using the correct geometry formula from Appendix Table 3.
6. *Simplify* the quadratic equation.
7. *Solve* the quadratic equation.
8. *Interpret* the solutions of the quadratic equation with respect to each represented unknown measure to find the proposed solutions of the original problem.
9. *Check* to see if the proposed solutions satisfy all the conditions of the original problem. [see Example 12-17].

(a) *Draw a picture:*

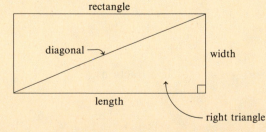

Identify: The unknown measures are $\begin{cases} \text{the length of the diagonal} \\ \text{the length of the rectangle} \\ \text{the width of the rectangle} \end{cases}$.

Decide: Let d = the length of the diagonal (hypotenuse)

then $d - 3$ = the length of the rectangle (longer leg)

and $d - 5$ = the width of the rectangle (shorter leg).

Translate:
$$c^2 = a^2 + b^2 \longleftarrow \text{Pythagorean theorem}$$
$$(d)^2 = (d - 3)^2 + (d - 5)^2$$

Simplify:
$$d^2 = d^2 - 6d + 9 + d^2 - 10d + 25$$
$$d^2 - 16d + 34 = 0 \longleftarrow \text{simplest form}$$

Solve:
$$d = \frac{-b \pm \sqrt{b^2 - 4ac}}{2a} \longleftarrow \text{quadratic formula}$$

$$d = \frac{-(-16) \pm \sqrt{(-16)^2 - 4(1)(34)}}{2(1)}$$

$$d = \frac{16 \pm \sqrt{120}}{2}$$

$$d = 8 \pm \sqrt{30}$$

Interpret: Since the diagonal (d) must be 3 ft longer than the length of the rectangle and 5 ft longer than the width, $d = (8 - \sqrt{30})$ ft (≈ 2.523 ft) cannot be the solution to the original problem.

If $d = 8 + \sqrt{30}$ (≈ 13.477), the length of the diagonal is $(8 + \sqrt{30})$ ft,

then $d - 3 = 5 + \sqrt{30}$ (≈ 10.477) means the length (l) of the rectangle is $(5 + \sqrt{30})$ ft,

and $d - 5 = 3 + \sqrt{30}$ (≈ 8.477) means the width (w) of the rectangle is $(3 + \sqrt{30})$ ft.

Therefore $A = lw = (5 + \sqrt{30})(3 + \sqrt{30}) = 15 + 8\sqrt{30} + 30 = 45 + 8\sqrt{30}$ means the area of the rectangle is $(45 + 8\sqrt{30})$ ft^2.

Check: Do sides of $3 + \sqrt{30}$, $5 + \sqrt{30}$, and $8\sqrt{30}$ form a right triangle? Yes:

$$c^2 = a^2 + b^2$$

hypotenuse [longest side] $\longrightarrow (8 + \sqrt{30})^2$	$(3 + \sqrt{30})^2 + (5 + \sqrt{30})^2 \longleftarrow$ legs [shorter sides]
$64 + 16\sqrt{30} + 30$	$(9 + 6\sqrt{30} + 30) + (25 + 10\sqrt{30} + 30)$
$94 + 16\sqrt{30}$	$94 + 16\sqrt{30} \longleftarrow$ a right triangle

Is the diagonal 3 ft longer than the length? Yes: $8 + \sqrt{30} = 3 + (5 + \sqrt{30})$.

Is the diagonal 5 ft longer than the width? Yes: $8 + \sqrt{30} = 5 + (3 + \sqrt{30})$.

(b) *Draw a picture:*

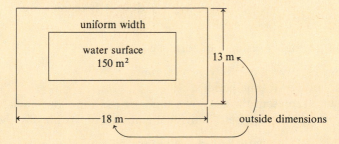

Identify: The unknown measures are $\left\{\begin{array}{l}\text{the uniform width of the walkway}\\ \text{length } (l) \text{ of the pool}\\ \text{the width } (w) \text{ of the pool}\end{array}\right\}$.

Decide: Let x = the uniform width of the walkway

and $18 - 2x$ = the length (l) of the pool (because $18 - x - x = 18 - 2x$)

and $13 - 2x$ = the width (w) of the pool (because $13 - x - x = 13 - 2x$).

Translate:

$$A = lw \longleftarrow \text{area formula for a rectangle}$$

$$150 = \overline{(18 - 2x)(13 - 2x)} \qquad \text{Substitute area and dimensions of pool.}$$

Simplify:

$$150 = 234 - 62x + 4x^2$$

$$4x^2 - 62x + 84 = 0$$

$$2x^2 - 31x + 42 = 0 \longleftarrow \text{simplest form}$$

Solve:

$$x = \frac{-b \pm \sqrt{b^2 - 4ac}}{2a}$$

$$x = \frac{-(-31) \pm \sqrt{(-31)^2 - 4(2)(42)}}{2(2)}$$

$$x = \frac{31 \pm \sqrt{625}}{4}$$

$$x = \frac{31 \pm 25}{4}$$

$$x = \frac{31 + 25}{4} \quad \text{or} \quad \frac{31 - 25}{4}$$

$$x = 14 \text{ or } 1.5$$

Interpret: $x = 14$ m is not the solution of the original problem because the outside width of the pool $[13 - 2x]$ is only 13 m.
$x = 1.5$ means the uniform width of the walkway is 1.5 m.
$18 - 2x = 18 - 2(1.5) = 15$ means the length (l) of the pool is 15 m.
$13 - 2x = 13 - 2(1.5) = 10$ means the width (w) of the pool is 10 m.

Check: Is the outside length of the walkway 18 m? Yes: 15 m + 1.5 m + 1.5 m = 18 m.
Is the outside width of the walkway 13 m? Yes: 10 m + 1.5 m + 1.5 m = 13 m.
Is the area of the pool 150 m²? Yes: (10 m)(15 m) = 150 m².

PROBLEM 12-17 Solve each work problem using a quadratic equation:

(a) Bert agreed to do a certain job for $216. Because it took him 3 hours longer than planned, he earned $6 less an hour than expected. How long did the job actually take?

(b) It takes one outlet pipe 10 days longer to empty a city water tank than it does a larger outlet pipe. If both outlet pipes are used, they can empty the tank in 12 days. How long does it take each outlet pipe to empty the tank alone?

Solution: To solve a work problem using a quadratic equation, you:

1. *Read* the problem very carefully, several times.
2. *Identify* the unknown time periods.
3. *Decide* how to represent the unknown time periods using one variable.
4. *Make a table* to help represent individual amounts of work using the formula

$$\text{work} = (\text{unit rate})(\text{time period}), \text{ or } w = rt.$$

5. *Translate* the problem to a quadratic equation.
6. *Simplify* the quadratic equation.
7. *Solve* the quadratic equation.
8. *Interpret* the solutions of the quadratic equations with respect to each represented unknown time period to find the proposed solutions of the original problem.
9. *Check* to see if the proposed solutions satisfy all the conditions of the original problem [see Example 12-18].

(a) *Identify:* The unknown time periods are $\left\{\begin{array}{l}\text{the planned time to complete the job}\\\text{the actual time to complete the job}\end{array}\right\}$.

Decide: Let x = the planned time to complete the job

then $x + 3$ = the actual time to complete the job.

Make a table:

	work(w) [in dollars]	time period (t) [in hours]	unit rate ($r = w/t$) [in dollars]
planned	216	x	$\dfrac{216}{x}$
actual	216	$x + 3$	$\dfrac{216}{x + 3}$

Translate: The planned unit rate was $6 an hour more than the actual unit rate.

$$\frac{216}{x} \qquad = \qquad 6 \qquad + \qquad \frac{216}{x + 3}$$

Simplify: The LCD of $\dfrac{216}{x}$ and $\dfrac{216}{x + 3}$ is $x(x + 3)$.   Clear fractions.

$$x(x + 3) \cdot \frac{216}{x} = x(x + 3)6 + x(x + 3) \cdot \frac{216}{x + 3}$$

$$(x + 3)(216) = 6x(x + 3) + x(216) \longleftarrow \text{fractions cleared}$$

$$216x + 648 = 6x^2 + 18x + 216x$$

$$6x^2 + 18x - 648 = 0 \longleftarrow \text{standard form}$$

$$x^2 + 3x - 108 = 0 \longleftarrow \text{simplest form}$$

$$(x - 9)(x + 12) = 0 \qquad\qquad \text{Factor or use the quadratic}$$
$$\qquad\qquad\qquad\qquad\qquad\qquad \text{formula.}$$

$$x - 9 = 0 \text{ or } x + 12 = 0$$

$$x = 9 \text{ or } \qquad x = -12$$

Interpret: $x = -12$ [hours] cannot be the planned time to complete the job.
$x = 9$ means the planned time to complete the job was 9 hours.
$x + 3 = 12$ means the actual time to complete the job was 12 hours.

Check: Is $216 for 9 hours $6 more an hour than $216 for 12 hours?
Yes: $\frac{216}{9} - \frac{216}{12} = 24 - 18 = 6$.

(b) *Identify:* The unknown time periods are $\left\{\begin{array}{l}\text{the time for the smaller outlet pipe}\\\text{the time for the larger outlet pipe}\end{array}\right\}$.

Decide: Let x = the time for the smaller outlet pipe

then $x - 10$ = the time for the larger outlet pipe.

Make a table:

	unit rate (r) [per day]	time period (t) [in days]	work ($w = rt$) [portion of the whole job]
smaller pipe	$\dfrac{1}{x}$	12	$\dfrac{1}{x} \cdot 12$ or $\dfrac{12}{x}$
larger pipe	$\dfrac{1}{x - 10}$	12	$\dfrac{1}{x - 10} \cdot 12$ $\dfrac{12}{x - 10}$

Translate: The smaller pipe's work plus the larger pipe's work equals the total work.

$$\frac{12}{x} \qquad + \qquad \frac{12}{x - 10} \qquad = \qquad 1$$

Simplify: The LCD of $\frac{12}{x}$ and $\frac{12}{x-10}$ is $x(x-10)$.   Clear fractions.

$$x(x-10) \cdot \frac{12}{x} + x(x-10) \cdot \frac{12}{x-10} = x(x-10)(1)$$

$$(x-10)12 + x(12) = x(x-10) \longleftarrow \text{fractions cleared}$$

$$12x - 120 + 12x = x^2 - 10x$$

$$x^2 - 34x + 120 = 0 \longleftarrow \text{simplest form}$$

Solve: $(x-4)(x-30) = 0$ Factor or use the quadratic formula.

$$x - 4 = 0 \text{ or } x - 30 = 0$$

$$x = 4 \text{ or } \qquad x = 30$$

Interpret: $x = 4$ [days] cannot be the time for the smaller outlet pipe because the larger outlet pipe takes 10 days fewer than the smaller outlet pipe:
$4 - 10 = -6$ [days].
$x = 30$ means the time for the smaller outlet pipe is 30 days.
$x - 10 = 30 - 10 = 20$ means the time for the larger outlet pipe is 20 days.

Check: Does the smaller pipe's unit rate [per day] plus the larger pipe's unit rate equal their unit rate together? Yes: $\frac{1}{20} + \frac{1}{30} = \frac{3}{60} + \frac{2}{60} = \frac{5}{60} = \frac{1}{12}$.

PROBLEM 12-18 Solve each uniform motion problem using a quadratic equation:

(a) If George had driven 5 mph faster than he actually did over a 660-mile trip, he would have saved one hour. How fast did George drive on the trip? How long did the trip take?

(b) Two cars start at the same time from the same place and travel at constant speeds that differ by 10 mph, on roads that are at right angles to each other. In two hours, the cars are 100 miles apart. What was the constant speed of the slower car? How far did the faster car travel?

Solution: To solve a uniform motion problem using a quadratic equation, you:

1. *Read* the problem very carefully, several times.
2. *Identify* the unknown distances, rates, and times.
3. *Decide* how to represent the unknown distances, rates, and times using one variable.
4. *Make* a table to help represent the unknown rates (or times) using $r = d/t$ (or $t = d/t$).
5. *Translate* the problem to a quadratic equation.
6. *Simplify* the quadratic equation.
7. *Solve* the quadratic equation.
8. *Interpret* the solutions of the quadratic equation with respect to each unknown distance, rate, and time to find the proposed solutions of the original problem.
9. *Check* to see if the proposed solutions satisfy all the conditions of the original problem. [see Example 12-19]:

(a) *Identify:*   The unknown rates and times are $\begin{cases} \text{the actual rate} \\ \text{the faster rate} \\ \text{the time for the actual rate} \\ \text{the time for the faster car} \end{cases}$.

Decide: Let r = the actual rate

then $r + 5$ = the faster rate.

Make a table:

	distance (d) [in miles]	rate (r) [in mph]	time ($t = d/r$) [in hours]
actual rate	660	r	$\frac{660}{r}$
faster rate	660	$r + 5$	$\frac{660}{r+5}$

Translate: The time for the actual rate is one hour longer than the time for the faster rate.

$$\frac{660}{t} = 1 + \frac{660}{r + 5}$$

Simplify: The LCD of $\dfrac{660}{r}$ and $\dfrac{660}{r + 5}$ is $r(r + 5)$. Clear fractions.

$$r(r + 5) \cdot \frac{660}{r} = r(r + 5)(1) + r(r + 5) \cdot \frac{660}{r + 5}$$

$$(r + 5)660 = r(r + 5) + r(660) \longleftarrow \text{fractions cleared}$$

$$660r + 3300 = r^2 + 5r + 660r$$

$$r^2 + 5r - 3300 = 0 \longleftarrow \text{simplest form}$$

Solve:
$$(r + 60)(r - 55) = 0 \qquad \text{Factor or use the quadratic equation.}$$

$$r + 60 = 0 \quad \text{or} \quad r - 55 = 0$$

$$r = -60 \text{ or} \qquad r = 55$$

Interpret: $r = -60$ [mph] cannot be the actual rate because car rates are not negative.
$r = 55$ means the actual rate is 55 mph.
$r + 5 = 55 + 5 = 60$ means the faster rate is 60 mph.

$$\frac{660}{r} = \frac{660}{55} = 12 \text{ means the time for the actual rate is 12 hours.}$$

$$\frac{660}{r + 5} = \frac{660}{60} = 11 \text{ means the time for the faster rate is 11 hours.}$$

Check: Is the time for the actual rate one hour more than the time for the faster rate?
Yes: 12 hours = 1 hour + 11 hours.

(b) *Draw a picture:*

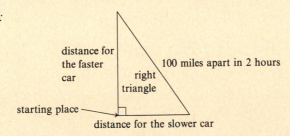

distance for the faster car

100 miles apart in 2 hours

right triangle

starting place

distance for the slower car

Identify: The unknown rates and distances are $\left\{\begin{array}{l}\text{the rate for the faster car} \\ \text{the rate for the slower car} \\ \text{the distance for the faster car} \\ \text{the distance for the slower car}\end{array}\right\}$.

Decide: Let r = the rate for the slower car.

then $r + 10$ = the rate for the faster car.

Make a table:

	rate (r) [in mph]	time (t) [in hours]	distance $(d = rt)$ [in miles]
slower car	r	2	$2r$
faster car	$r + 10$	2	$2(r + 10)$

Translate:
$$c^2 = a^2 + b^2 \longleftarrow \text{Pythagorean theorem}$$

hypotenuse $\longrightarrow$ $(100)^2 = (2r)^2 + [2(r + 10)]^2 \longleftarrow$ legs [see picture and table]

Simplify:
$$100^2 = 2^2 r^2 + 2^2 (r + 10)^2$$

$$10{,}000 = 4r^2 + 4(r^2 + 20r + 100)$$

$$10{,}000 = 4r^2 + 4r^2 + 80r + 400$$

$$8r^2 + 80r - 9600 = 0 \longleftarrow \text{standard form [the positive GCF is 8]}$$

$$r^2 + 10r - 1200 = 0 \longleftarrow \text{simplest form}$$

Solve: $(r + 40)(r - 30) = 0$ Factor or use the quadratic formula.

$$r + 40 = 0 \quad \text{or } r - 30 = 0$$

$$r = -40 \text{ or} \qquad r = 30$$

Interpret: $r = -40$ [mph] cannot be the rate of the slower car because car rates are never negative.

$r = 30$ means the rate for the slower car is 30 mph.

$r + 10 = 30 + 10 = 40$ means the rate for the faster car is 40 mph.

$2r = 2(30) = 60$ means the distance for the slower car is 60 miles.

$2(r + 10) = 2(40) = 80$ means the distance for the faster car is 80 miles.

Check: Is a triangle sides of 30, 40, and 50 a right triangle? Yes:

$$c^2 = a^2 + b^2$$

hypotenuse (longest side) $\longrightarrow$ $(50)^2$ | $(30)^2 + (40)^2$ $\longleftarrow$ legs (shorter sides)

2500 | 900 + 1600

2500 | 2500 $\longleftarrow$ a right triangle

Does it take 2 hours to travel 60 miles at 30 mph?
Yes: $t = d/r = 60/30 = 2$ (hours)
Does it take 2 hours to travel 80 miles at 40 mph?
Yes: $t = d/r = 80/40 = 2$ (hours)

PROBLEM 12-19 Solve the following quadratic inequalities using algebraic methods:

(a) $x^2 - 3x - 4 > 0$ **(b)** $x^2 + 4x + 4 \le 0$ **(c)** $x^2 - 9 \ge 0$
(d) $4x^2 + 4x + 1 > 0$ **(e)** $4x^2 > 4x + 3$ **(f)** $2x^2 \le 5x + 3$

Solution: Recall that to solve a quadratic inequality, you factor the quadratic and solve using the product property of inequalities [see Example 12-20].

(a) $(x - 4)(x + 1) > 0$

$x - 4 > 0$ and $x + 1 > 0$ or $x - 4 < 0$ and $x + 1 < 0$

$x > 4$ and $x > -1$ or $x < 4$ and $x < -1$

$x > 4$ or $x < -1$

$\{x \mid x < -1 \text{ or } x > 4\}$

(b) $(x + 2)(x + 2) \le 0$

$x + 2 \ge 0$ and $x + 2 \le 0$ or $x + 2 \le 0$ and $x + 2 \ge 0$

$x \ge -2$ and $x \le -2$ or $x \le -2$ and $x \ge -2$

$x = -2$ or $x = -2$

$\{x \mid x = -2\}$

(c) $(x - 3)(x + 3) \ge 0$

$x - 3 \ge 0$ and $x + 3 \ge 0$ or $x - 3 \le 0$ and $x + 3 \le 0$

$x \ge 3$ and $x \ge -3$ or $x \le 3$ and $x \le -3$

$x \ge 3$ or $x \le -3$

$\{x \mid x \le -3 \text{ or } x \ge 3\}$

(d) $(2x + 1)(2x + 1) > 0$

$2x + 1 > 0 \quad$ and $2x + 1 > 0 \quad$ or $2x + 1 < 0 \quad$ and $2x + 1 < 0$

$x > -\frac{1}{2}$ and $\qquad x > -\frac{1}{2}$ or $\qquad x < -\frac{1}{2}$ and $\qquad x < -\frac{1}{2}$

$x > -\frac{1}{2} \qquad\qquad$ or $\qquad\qquad x < -\frac{1}{2}$

$x \neq -\frac{1}{2}$

$\{x \mid x \neq -\frac{1}{2}\}$

(e) $\quad 4x^2 - 4x - 3 > 0$

$(2x - 3)(2x + 1) > 0$

$2x - 3 > 0$ and $2x + 1 > 0 \quad$ or $2x - 3 < 0$ and $2x + 1 < 0$

$x > \frac{3}{2}$ and $\qquad x > -\frac{1}{2}$ or $\qquad x < \frac{3}{2}$ and $\qquad x < -\frac{1}{2}$

$x > \frac{3}{2} \qquad\qquad$ or $\qquad\qquad x < -\frac{1}{2}$

$\{x \mid x < -\frac{1}{2} \text{ or } x > \frac{3}{2}\}$

(f) $\quad 2x^2 - 5x - 3 \leq 0$

$(2x + 1)(x - 3) \leq 0$

$2x + 1 \geq 0 \quad$ and $x - 3 \leq 0$ or $2x + 1 \leq 0 \quad$ and $x - 3 \geq 0$

$x \geq -\frac{1}{2}$ and $\qquad x \leq 3$ or $\qquad x \leq -\frac{1}{2}$ and $\qquad x \geq 3$

$-\frac{1}{2} \leq x \leq 3 \qquad\qquad$ or $\qquad\qquad \varnothing$

$\{x \mid -\frac{1}{2} \leq x \leq 3\}$

PROBLEM 12-20 Solve the following quadratic inequalities using the critical point method:

(a) $x^2 + 3x - 4 > 0$ **(b)** $x^2 + 2x - 8 \leq 0$ **(c)** $x^2 + 6x + 9 > 0$
(d) $2x^2 \leq 7x - 3$ **(e)** $6x^2 + 11x + 3 > 0$ **(f)** $3x^2 \leq 13x - 4$

Solution: Recall that to solve a quadratic inequality using the critical point method, you find the critical points by solving for the equality values and identify the intervals in which the inequality is true [see Example 12-21].

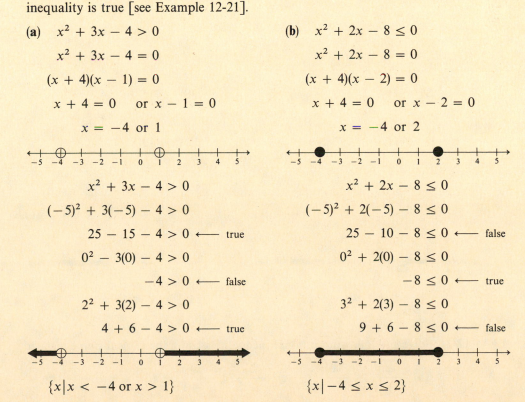

(a) $\quad x^2 + 3x - 4 > 0$

$x^2 + 3x - 4 = 0$

$(x + 4)(x - 1) = 0$

$x + 4 = 0 \quad$ or $x - 1 = 0$

$x = -4 \text{ or } 1$

$x^2 + 3x - 4 > 0$

$(-5)^2 + 3(-5) - 4 > 0$

$25 - 15 - 4 > 0 \longleftarrow$ true

$0^2 - 3(0) - 4 > 0$

$-4 > 0 \longleftarrow$ false

$2^2 + 3(2) - 4 > 0$

$4 + 6 - 4 > 0 \longleftarrow$ true

$\{x \mid x < -4 \text{ or } x > 1\}$

(b) $\quad x^2 + 2x - 8 \leq 0$

$x^2 + 2x - 8 = 0$

$(x + 4)(x - 2) = 0$

$x + 4 = 0 \quad$ or $x - 2 = 0$

$x = -4 \text{ or } 2$

$x^2 + 2x - 8 \leq 0$

$(-5)^2 + 2(-5) - 8 \leq 0$

$25 - 10 - 8 \leq 0 \longleftarrow$ false

$0^2 + 2(0) - 8 \leq 0$

$-8 \leq 0 \longleftarrow$ true

$3^2 + 2(3) - 8 \leq 0$

$9 + 6 - 8 \leq 0 \longleftarrow$ false

$\{x \mid -4 \leq x \leq 2\}$

(c) $x^2 + 6x + 9 > 0$

$(x + 3)^2 = 0$

$x = -3$

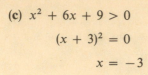

$x^2 + 6x + 9 > 0$

$(-5)^2 + 6(-5) + 9 > 0$

$4 > 0 \longleftarrow$ true

$5^2 + 6(5) + 9 > 0$

$64 > 0 \longleftarrow$ true

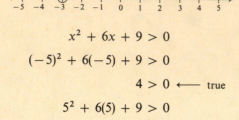

$\{x \mid x \neq -3\}$

(d) $2x^2 - 7x + 3 \leq 0$

$(2x - 1)(x - 3) = 0$

$2x - 1 = 0$ or $x - 3 = 0$

$x = \frac{1}{2}$ or $\qquad x = 3$

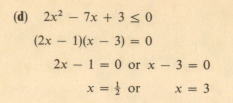

$2x^2 - 7x + 3 \leq 0$

$2(0)^2 - 7(0) + 3 \leq 0$

$3 \leq 0 \longleftarrow$ false

$2(1)^2 - 7(1) + 3 \leq 0$

$-2 \leq 0 \longleftarrow$ true

$2(4)^2 - 7(4) + 3 \leq 0$

$7 \leq 0 \longleftarrow$ false

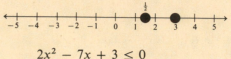

$\{x \mid \frac{1}{2} \leq x \leq 3\}$

(e) $6x^2 + 11x + 3 > 0$

$6x^2 + 11x + 3 = 0$

$(2x + 3)(3x + 1) = 0$

$2x + 3 = 0 \quad$ or $3x + 1 = 0$

$x = -\frac{3}{2}$ or $\qquad x = -\frac{1}{3}$

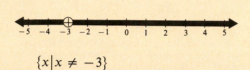

$6x^2 + 11x + 3 > 0$

$6(-2)^2 + 11(-2) + 3 > 0$

$5 > 0 \longleftarrow$ true

$6(-1)^2 + 11(-1) + 3 > 0$

$-2 > 0 \longleftarrow$ false

$6(0)^2 + 11(0) + 3 > 0$

$3 > 0 \longleftarrow$ true

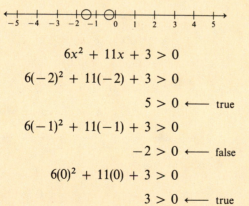

$\{x \mid x < -\frac{3}{2} \text{ or } x > -\frac{1}{3}\}$

(f) $3x^2 - 13x + 4 \leq 0$

$3x^2 - 13x + 4 = 0$

$(3x - 1)(x - 4) = 0$

$3x - 1 = 0$ or $x - 4 = 0$

$x = \frac{1}{3}$ or $\qquad x = 4$

$3x^2 - 13x + 4 \leq 0$

$3(0)^2 - 13(0) + 4 \leq 0$

$4 \leq 0 \longleftarrow$ false

$3(1)^2 - 13(1) + 4 \leq 0$

$-6 \leq 0 \longleftarrow$ true

$3(5)^2 - 13(5) + 4 \leq 0$

$14 \leq 0 \longleftarrow$ false

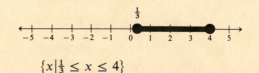

$\{x \mid \frac{1}{3} \leq x \leq 4\}$

PROBLEM 12-21 Solve the following rational inequalities using the critical point method:

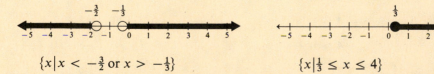

(a) $\dfrac{1}{x} > 2x + 1$ **(b)** $\dfrac{x}{x + 1} \geq \dfrac{1}{x + 1}$ **(c)** $\dfrac{2}{x + 2} < \dfrac{1}{x - 1}$ **(d)** $\dfrac{2x}{x^2 + 2x + 1} \geq \dfrac{1}{x + 1}$

Solution: Recall that to solve a rational inequality using the critical point method, you find the critical values (the excluded values and the equality values), and identify the intervals in which the inequality is true [see Example 12-22].

(a) $\dfrac{1}{x} > 2x + 1$

$x = 0 \longleftarrow$ excluded value

$\dfrac{1}{x} = 2x + 1$

$1 = 2x^2 + x$

$0 = 2x^2 + x - 1$

$0 = (2x - 1)(x + 1)$

$0 = 2x - 1$ or $x + 1 = 0$

$x = \tfrac{1}{2}$ or $-1 \longleftarrow$ equality values

$\dfrac{1}{x} > 2x + 1$

$\dfrac{1}{-2} > 2(-2) + 1$

$-\tfrac{1}{2} > -3 \longleftarrow$ true

$\dfrac{1}{-\frac{1}{2}} > 2(-\tfrac{1}{2}) + 1$

$-2 > 0 \longleftarrow$ false

$\dfrac{1}{\frac{1}{4}} > 2(\tfrac{1}{4}) + 1$

$4 > \tfrac{3}{2} \longleftarrow$ true

$\dfrac{1}{1} > 2(1) + 1$

$1 > 3 \longleftarrow$ false

$\{x \mid x < -1 \text{ or } 0 < x < \tfrac{1}{2}\}$

(b) $\dfrac{x}{x + 1} \geq \dfrac{1}{x + 1}$

$x + 1 = 0$

$x = -1 \longleftarrow$ excluded value

$\dfrac{x}{x + 1} = \dfrac{1}{x + 1}$

$x^2 + x = x + 1$

$x^2 = 1$

$x = \pm 1 \longleftarrow$ equality values

$\dfrac{x}{x + 1} \geq \dfrac{1}{x + 1}$

$\dfrac{-2}{-2 + 1} \geq \dfrac{1}{-2 + 1}$

$2 \geq -1 \longleftarrow$ true

$\dfrac{0}{0 + 1} \geq \dfrac{1}{0 + 1}$

$0 \geq 1 \longleftarrow$ false

$\dfrac{2}{2 + 1} \geq \dfrac{1}{2 + 1}$

$\tfrac{2}{3} \geq \tfrac{1}{3} \longleftarrow$ true

$\{x \mid x < -1 \text{ or } x \geq 1\}$

(c) $\dfrac{2}{x + 2} < \dfrac{1}{x - 1}$

$x + 2 = 0 \quad$ or $\quad x - 1 = 0$

$x = -2$ or $1 \longleftarrow$ excluded values

$\dfrac{2}{x + 2} = \dfrac{1}{x - 1}$

$2x - 2 = x + 2$

$x = 4 \longleftarrow$ equality values

(d) $\dfrac{2x}{x^2 + 2x + 1} \geq \dfrac{1}{x + 1}$

$x^2 + 2x + 1 = 0$ or $x + 1 = 0$

$x = -1 \longleftarrow$ excluded

$\dfrac{2x}{x^2 + 2x + 1} = \dfrac{1}{x + 1}$

$2x^2 + 2x = x^2 + 2x + 1$

$x^2 = 1$

$x = \pm 1$

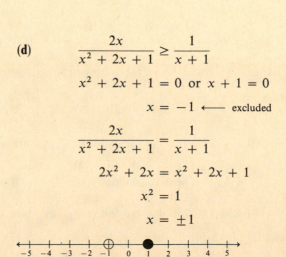

$$\frac{2}{x+2} < \frac{1}{x-1}$$

$$\frac{2}{-3+2} < \frac{1}{-3-1}$$

$$-2 < -\tfrac{1}{4} \longleftarrow \text{true}$$

$$\frac{2}{-1+2} < \frac{1}{-1-1}$$

$$2 < -\tfrac{1}{2} \longleftarrow \text{false}$$

$$\frac{2}{2+2} < \frac{1}{2-1}$$

$$\tfrac{1}{2} < 1 \longleftarrow \text{true}$$

$$\frac{2}{5+2} < \frac{1}{5-1}$$

$$\tfrac{2}{7} < \tfrac{1}{4} \longleftarrow \text{false}$$

$$\frac{2x}{x^2+2x+1} \geq \frac{1}{x+1}$$

$$\frac{2(-2)}{(-2)^2+2(-2)+1} \geq \frac{1}{-2+1}$$

$$-4 \geq -1 \longleftarrow \text{false}$$

$$\frac{2(0)}{0^2+2(0)+1} \geq \frac{1}{0+1}$$

$$0 \geq 1 \longleftarrow \text{false}$$

$$\frac{2(2)}{2^2+2(2)+1} \geq \frac{1}{2+1}$$

$$\tfrac{4}{9} \geq \tfrac{1}{3} \longleftarrow \text{true}$$

$$\{x \mid x \geq 1\}$$

$$\{x \mid x < -2 \text{ or } 1 < x < 4\}$$

Supplementary Exercises

PROBLEM 12-22 Write the following quadratic equations in standard form and identify a, b, and c:

(a) $2 - 3x = 5x^2$ (b) $3 = y - 3y^2$ (c) $3 - z^2 = 2z$
(d) $\tfrac{2}{3}x^2 - 1 = \tfrac{1}{2}x$ (e) $\tfrac{1}{2} - \tfrac{1}{4}y^2 = \tfrac{2}{3}y$ (f) $\tfrac{2}{3}z = \tfrac{1}{4} - \tfrac{5}{6}z^2$

PROBLEM 12-23 Solve the following equations using the zero-product property:

(a) $(x - 3)(x + 2) = 0$ (b) $(3x - 2)(3 - x) = 0$ (c) $(4y - 3)(3y + 4) = 0$
(d) $z(z - 3)(z + 2) = 0$ (e) $w(2w - 3)(3 - 2w) = 0$ (f) $t(2 - 3t)(1 - t) = 0$

PROBLEM 12-24 Solve the following incomplete quadratic equations of the form $ax^2 + bx = 0$:

(a) $3x^2 - 6x = 0$ (b) $6y + 4y^2 = 0$ (c) $8z^2 = 12z$
(d) $\tfrac{3}{4}w^2 - \tfrac{1}{2}w = 0$ (e) $\tfrac{1}{2}t^2 + \tfrac{1}{3}t = 0$ (f) $\tfrac{1}{6}q^2 = \tfrac{2}{3}q$

PROBLEM 12-25 Solve the following incomplete quadratic equations of the form $ax^2 + c = 0$:

(a) $x^2 - 9 = 0$ (b) $4y^2 - 25 = 0$ (c) $9z^2 - 5 = 0$
(d) $2w^2 - 3 = 0$ (e) $4t^2 + 9 = 0$ (f) $3p^2 + 5 = 0$

PROBLEM 12-26 Solve the following quadratic equations by factoring:

(a) $x^2 - x - 2 = 0$ (b) $y^2 - 7y + 12 = 0$ (c) $z^2 + 2z - 15 = 0$
(d) $4t^2 + 4t + 1 = 0$ (e) $6p^2 - 5p - 6 = 0$ (f) $6w^2 - 5w - 4 = 0$

PROBLEM 12-27 Solve the following equations of the form $(x + d)^2 = e$:

(a) $(x - 1)^2 = 1$ (b) $(x + 1)^2 = 4$ (c) $(3x - 2)^2 = 4$
(d) $(x - 5)^2 = 2$ (e) $(3x - 1)^2 = 3$ (f) $(2x + 3)^2 = -6$

PROBLEM 12-28 Solve the following equations by completing the square:

(a) $x^2 - 6x + 9 = 0$ (b) $3x^2 - 5x - 2 = 0$ (c) $x^2 - 4x - 2 = 0$
(d) $6x^2 + x - 2 = 0$ (e) $x^2 + 2x + 2 = 0$ (f) $3x^2 - x + 1 = 0$

PROBLEM 12-29 Solve the following quadratic equation using the quadratic formulas:

(a) $x^2 - x - 6 = 0$ (b) $x^2 - 3x - 4 = 0$ (c) $x^2 - x + 6 = 0$
(d) $2x^2 - x - 6 = 0$ (e) $2x^2 - x - 2 = 0$ (f) $2x^2 - 3x + 2 = 0$

PROBLEM 12-30 Find the approximate solutions of the following quadratic equations:

(a) $x^2 + 3x + 1 = 0$ (b) $x^2 + 2x - 2 = 0$ (c) $x^2 + 3x - 2 = 0$
(d) $2x^2 + 5x + 1 = 0$ (e) $3x^2 - 5x + 1 = 0$ (f) $4x^2 - 6x + 1 = 0$

PROBLEM 12-31 Determine whether the following quadratic equations have one real solution, two distinct real solutions, or two distinct complex solutions:

(a) $x^2 - 6x + 9 = 0$ (b) $x^2 + 5x + 2 = 0$ (c) $x^2 + 2x + 2 = 0$
(d) $2x^2 - 3x + 1 = 0$ (e) $3x^2 + 2x - 1 = 0$ (f) $4x^2 - 5x + 2 = 0$

PROBLEM 12-32 Write a quadratic equation having the following numbers as solutions:

(a) ± 2 (b) $-\frac{2}{3}, \frac{1}{4}$ (c) $\pm\sqrt{3}$ (d) $2 \pm \sqrt{5}$ (e) $\pm 2i$ (f) $3 \pm 2i$

PROBLEM 12-33 Solve the following rational equations:

(a) $\dfrac{1}{x + 2} = 3x + 2$ (b) $\dfrac{1}{x - 3} = \dfrac{x}{2x - 1}$ (c) $\dfrac{x + 3}{2x - 3} = \dfrac{x}{x - 3}$

(d) $\dfrac{x}{x + 1} = 1 - \dfrac{x}{x - 1}$ (e) $\dfrac{x + 1}{2x} - \dfrac{1}{x} = \dfrac{x - 1}{x + 1}$ (f) $\dfrac{x}{x^2 - 1} = 1 - \dfrac{x}{x - 1}$

PROBLEM 12-34 Solve the following radical equations:

(a) $\sqrt{x + 1} = x + 1$ (b) $\sqrt{x - 2} = \sqrt{x} - 1$ (c) $\sqrt{2x + 1} = \sqrt{2x - 1}$
(d) $\sqrt{3x - 2} - 2 = \sqrt{2 - x}$ (e) $1 + \sqrt{x + 1} = \sqrt{2x + 3}$ (f) $\sqrt{3x + 1} - \sqrt{2x - 1} = \sqrt{x - 4}$

PROBLEM 12-35 Solve the following equations that are reducable to quadratic equations:

(a) $x^4 - 8x + 16 = 0$ (b) $6x^4 + x^2 - 12 = 0$ (c) $x - 6\sqrt{x} + 9 = 0$
(d) $3x + 16\sqrt{x} - 12 = 0$ (e) $6x^{-2} - 11x^{-1} + 3 = 0$ (f) $6x^{-2} + 5x^{-1} - 6 = 0$

PROBLEM 12-36 Solve each number problem using a quadratic equation:

(a) The square of a positive number is 12 more than the number itself. What is the number?

(b) One negative number is three times another number. The sum of their squares is 360. Find the numbers.

(c) One negative number is 7 more than 3 times another number. The product of the numbers is 48. What are the numbers?

(d) The product of two consecutive positive integers is 72. Find the integers.

(e) One positive number is 2 more than another number. The sum of their reciprocals is 3. What are the numbers?

(f) The sum of a number and its reciprocal is 4. Find the two possible solutions.

PROBLEM 12-37 Solve each geometry problem using a quadratic equation:

(a) The length of a rectangle is 3 ft longer than the width. The area of the rectangle is 154 ft^2. Find the perimeter of the rectangle.

(b) The perimeter of a rectangle is 32 m. The area of the rectangle is 48 m^2. Find the dimensions of the rectangle.

(c) The side of one square is 3 in. longer than twice the length of the side of another square. The difference between the area of the two squares is 105 in.^{2}. Find the perimeter of each square.

(d) A guy wire reaches from the top of a vertical 84-ft TV antenna to a ground anchor. The base of the TV antenna is 63 ft from the anchor. How long is the guy wire?

(e) A square has a perimeter of 12 cm. Find the length of the diagonal.

(f) A square has a 12-cm diagonal. Find the perimeter.

(g) The hypotenuse of a right triangle is 5 m longer than one leg and 2 cm longer than the other leg. Find the perimeter and area of the right triangle.

(h) A strip of uniform width is mowed around the outside edge of a rectangular lawn that is 80 m by 120 m. What is the uniform width if the lawn is one-fourth mowed?

PROBLEM 12-38 Solve each work problem using a quadratic equation:

(a) Working alone, John can complete a certain job in 3 hours less time than Agnes. Working together, they can complete the same job in 2 hours. How long does it take each to do the job alone?

(b) It takes one outlet pipe 6 hours longer to empty a swimming pool than it does an inlet pipe to fill the pool. If both pipes are left open, the pool can be filled in 20 hours. How long does it take the inlet pipe to fill the pool if the outlet pipe is closed?

(c) Elizabeth took a job for $192. It took her 4 hours longer than expected and so she earned $2.40 less than planned. How long did Elizabeth expect the job to take?

(d) Serena can paint a car in 3 hours less time than her competitor. If they work together, they can paint the same car in 4 hours. How long does it take Serena to paint the car?

PROBLEM 12-39 Solve each uniform motion problem using a quadratic equation:

(a) A car made a 400-mile trip at 10 mph faster than on the return trip over the same 400 miles which took 2 hours longer. Find the rates on both legs of the trip.

(b) Brett can row 16 miles downstream and then make the return trip upstream in a total time of 6 hours. The rate of the current is 2 mph. Find the landspeed downstream. How long did it take to row upstream?

(c) Byron wants to fly 300 miles due north. He takes the wrong course and flys in a straight line to end up 50 miles due west of his planned destination. How far did Byron fly?

(d) An airplane flys between two cities that are 3200 km apart. It takes 20 minutes longer to make the trip against a head-wind of 40 km/h than it does normally in still air. What is the normal airspeed of the plane? How long does the trip take with a 40 km/h tailwind?

PROBLEM 12-40 Solve the following quadratic inequalities using the algebraic method:

(a) $x^2 - x - 2 > 0$　　**(b)** $x^2 + 7x + 12 < 0$　　**(c)** $x^2 - 1 \geq 0$
(d) $4x^2 - 4x + 1 \leq 0$　　**(e)** $4x^2 + 4x - 3 > 0$　　**(f)** $2x^2 + 5x - 3 < 0$

PROBLEM 12-41 Solve the following quadratic inequalities using the critical point method:

(a) $x^2 - 2x + 1 > 0$　　**(b)** $x^2 - x - 12 < 0$　　**(c)** $x^2 - 4x + 4 < 0$
(d) $2x^2 + x - 3 \geq 0$　　**(e)** $3x^2 + 11x + 6 \leq 0$　　**(f)** $3x^2 - 7x + 4 \geq 0$

PROBLEM 12-42 Solve the following rational inequalities using the critical point method:

(a) $\dfrac{1}{x + 2} > x + 2$　　**(b)** $\dfrac{1}{x - 1} < \dfrac{x}{x - 1}$　　**(c)** $\dfrac{1}{x - 3} < \dfrac{x}{x + 5}$

(d) $\dfrac{3}{x^2 - 9} \geq \dfrac{1}{x - 3}$　　**(e)** $\dfrac{x}{x^2 - x - 2} \geq \dfrac{1}{2 - x}$　　**(f)** $\dfrac{x + 1}{2x} \leq \dfrac{x + 1}{x - 3}$

Answers to Supplementary Exercises

(12-22) **(a)** $5x^2 + 3x - 2 = 0$, $a = 5$, $b = 3$, $c = -2$ **(b)** $3y^2 - y + 3 = 0$, $a = 3$, $b = -1$, $c = 3$
(c) $z^2 + 2z - 3 = 0$, $a = 1$, $b = 2$, $c = -3$ **(d)** $4x^2 - 3x - 6 = 0$, $a = 4$, $b = -3$, $c = -6$
(e) $3y^2 + 8y - 6 = 0$, $a = 3$, $b = 8$, $c = -6$ **(f)** $10z^2 + 8z - 3 = 0$, $a = 10$, $b = 8$, $c = -3$

(12-23) **(a)** $-2, 3$ **(b)** $\frac{2}{3}, 3$ **(c)** $-\frac{4}{3}, \frac{3}{4}$ **(d)** $-2, 0, 3$ **(e)** $0, \frac{3}{2}$ **(f)** $0, \frac{2}{3}, 1$

(12-24) **(a)** $0, 2$ **(b)** $-\frac{3}{2}, 0$ **(c)** $0, \frac{3}{2}$ **(d)** $0, \frac{2}{3}$ **(e)** $-\frac{2}{3}, 0$ **(f)** $0, 4$

(12-25) **(a)** ± 3 **(b)** $\pm\frac{5}{2}$ **(c)** $\pm\frac{\sqrt{5}}{3}$ **(d)** $\pm\frac{\sqrt{6}}{2}$ **(e)** $\pm\frac{3}{2}i$ **(f)** $\pm\frac{\sqrt{15}}{3}i$

(12-26) **(a)** $-1, 2$ **(b)** $-3, 1$ **(c)** $-5, 3$ **(d)** $-\frac{1}{2}$ **(e)** $-\frac{2}{3}, \frac{3}{2}$ **(f)** $-\frac{1}{2}, \frac{4}{3}$

(12-27) **(a)** $0, 2$ **(b)** $3, 4$ **(c)** $0, \frac{4}{3}$ **(d)** $5 \pm \sqrt{2}$ **(e)** $\frac{1 \pm \sqrt{3}}{3}$ **(f)** $\frac{-3 \pm i\sqrt{6}}{2}$

(12-28) **(a)** 3 **(b)** $-\frac{1}{3}, 2$ **(c)** $2 \pm \sqrt{6}$ **(d)** $-\frac{2}{3}, \frac{1}{2}$ **(e)** $-1 \pm i$ **(f)** $\frac{1 \pm i\sqrt{11}}{6}$

(12-29) **(a)** $-2, 3$ **(b)** $-1, 4$ **(c)** $\frac{1 \pm i\sqrt{23}}{2}$ **(d)** $-\frac{3}{2}, 2$ **(e)** $\frac{1 \pm \sqrt{17}}{4}$ **(f)** $\frac{3 \pm i\sqrt{7}}{4}$

(12-30) **(a)** $-0.382, -2.618$ **(b)** $-2.732, 0.732$ **(c)** $-3.562, 0.562$ **(d)** $-2.281, -0.219$
(e) $0.232, 1.434$ **(f)** $0.191, 1.309$

(12-31) **(a)** one real solution   **(b)** two real solutions   **(c)** two nonreal solutions
(d) two real solutions   **(e)** two real solutions   **(f)** two nonreal solutions

(12-32) **(a)** $x^2 - 4 = 0$ **(b)** $12x^2 + 5x - 2 = 0$ **(c)** $x^2 - 3 = 0$ **(d)** $x^2 - 4x - 1 = 0$
(e) $x^2 + 4 = 0$ **(f)** $x^2 - 6x + 13 = 0$

(12-33) **(a)** $\frac{-4 \pm \sqrt{7}}{3}$ **(b)** $\frac{5 \pm \sqrt{21}}{2}$ **(c)** $\frac{3 \pm 3i\sqrt{3}}{2}$ **(d)** $\pm i$ **(e)** 1 **(f)** $-\frac{1}{2}$

(12-34) **(a)** $-1, 0$ **(b)** $\frac{9}{4}$ **(c)** $\varnothing$ **(d)** 2 **(e)** $-1, 3$ **(f)** 5

(12-35) **(a)** ± 2 **(b)** $\pm\frac{2}{3}\sqrt{3}, \pm\frac{i\sqrt{6}}{2}$ **(c)** 9 **(d)** $\frac{4}{9}$ **(e)** $\frac{2}{3}, 3$ **(f)** $\frac{3}{2}, -\frac{2}{3}$

(12-36) **(a)** 4 **(b)** $-6, -18$ **(c)** $-\frac{16}{3}, -9$ **(d)** $8, 9$ **(e)** $\frac{-2 + \sqrt{10}}{3}, \frac{4 + \sqrt{10}}{3}$
(f) $2 \pm \sqrt{3}$

(12-37) **(a)** 50 ft   **(b)** 4 m by 12 m   **(c)** 16 in., 44 in.   **(d)** 105 ft   **(e)** $3\sqrt{2}$ cm
(f) $24\sqrt{2}$ cm   **(g)** $(14 + 6\sqrt{5})$ m, $(30 + 14\sqrt{5})$ m²   **(h)** $(50 - 10\sqrt{19})$ m

(12-38) **(a)** John [3 hr], Agnes [6 hr]   **(b)** $(-6 + \sqrt{129})$ hr   **(c)** 16 hr   **(d)** $\left(\frac{5 + \sqrt{73}}{2}\right)$ hr

(12-39) **(a)** 50 mph going, 40 mph returning   **(b)** 8 mph, 4 hr   **(c)** $(50\sqrt{37})$ mi
(d) 640 km/h, $4\frac{12}{17}$ hr

(12-40) **(a)** $\{x \mid x < -1 \text{ or } x > 2\}$ **(b)** $\{x \mid -4 < x < -3\}$ **(c)** $\{x \mid x \le -1 \text{ or } x \ge 1\}$
(d) $\{\frac{1}{2}\}$ **(e)** $\{x \mid x < -\frac{3}{2} \text{ or } x > \frac{1}{2}\}$ **(f)** $\{x \mid -3 < x < \frac{1}{2}\}$

(12-41) (a) $\{x|x \neq 1\}$ (b) $\{x|-3 < x < 4\}$ (c) $\varnothing$ (d) $\{x|x \leq -\frac{3}{2} \text{ or } x \geq 1\}$
(e) $\{x|-3 \leq x \leq -\frac{2}{3}\}$ (f) $\{x|x \leq 1 \text{ or } x \geq \frac{4}{3}\}$

(12-42) (a) $\{x|x < -3 \text{ or } -2 < x < -1\}$ (b) $\{x|x \neq 1\}$
(c) $\{x|x < -5 \text{ or } -1 < x < 3 \text{ or } x > 5\}$ (d) $\{x|x < -3 \text{ or } 0 \leq x < 3\}$
(e) $\{x|-1 < x \leq -\frac{1}{2} \text{ or } x > 2\}$ (f) $\{x|x \leq -3 \text{ or } -1 \leq x < 0 \text{ or } x > 3\}$

13 THE CONIC SECTIONS

The **conic sections** are the graphs associated with four special types of equations. These graphs are called **parabolas, circles, ellipses,** and **hyperbolas.** The equations associated with these graphs are all second-degree equations in two variables.

We call these graphs the conic sections because they can be found by slicing a cone with a plane, as shown in Figure 13-1.

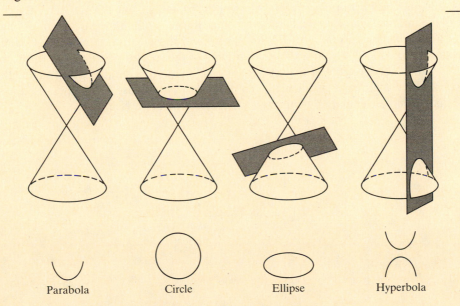

Parabola   Circle   Ellipse   Hyperbola

Figure 13-1. The conic sections.

13-1. The Distance Formula

A. Using the Pythagorean theorem to derive the distance formula

The **Pythagorean theorem** states that in a right triangle, if c is the length of the hypotenuse and a and b are the length of the two legs, then $a^2 + b^2 = c^2$.

To **find the length of the hypotenuse** of a right triangle, we substitute the given numbers for a and b in the equation $a^2 + b^2 = c^2$, then solve for c.

EXAMPLE 13-1: Find the length of the hypotenuse of the right triangle having the following pairs of legs: **(a)** 6, 8 **(b)** 4, 8

Solution: **(a)** $c^2 = a^2 + b^2$ Write the Pythagorean theorem.

$c^2 = 6^2 + 8^2$ Substitute 6 for a and 8 for b.

$c^2 = 36 + 64$ Simplify.

$c^2 = 100$

$c = 10$ Solve for c.

(b) $c^2 = a^2 + b^2$ Write the Pythagorean theorem.

$c^2 = 4^2 + 8^2$ Substitute 4 for a and 8 for b.

$c^2 = 16 + 64$ Simplify.

$c^2 = 80$

$c = \pm \sqrt{80}$

$c = 4\sqrt{5}$ Solve for c.

Note: Because c represents a length, we only use the nonnegative value.

B. Finding the distance between two points

On a number line, the **distance** between two points is the absolute value of the difference of the coordinates. If x_1 is the coordinate of A and x_2 is the coordinate of B, then the **distance between A and B** is $|x_1 - x_2|$ or $|x_2 - x_1|$. To find the distance between two points (P_1 and P_2) on Cartesian coordinates, we must find a third point (R) that will determine a right triangle with the others. This third point is the intersection of vertical and horizontal lines containing the first two points, and may be found by taking the x-coordinate of the first point and the y-coordinate of the second point (or the y-coordinate of the first point and the x-coordinate of the second point). Once this third point of the triangle is plotted, the lengths of the triangle's "legs" may be calculated using the method shown above for points on a number line. Then the length of its "hypotenuse," or the distance between the two points, may then be calculated using the Pythagorean theorem (see Figure 13-2).

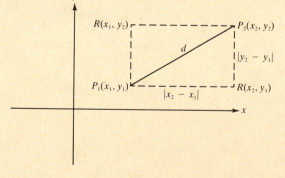

Figure 13-2. Preparing to find the distance between two points.

EXAMPLE 13-2: Find the distance between the points $P_1(x_1, y_1)$ and $P_2(x_2, y_2)$.

Solution: Form the right triangle: The vertical and horizontal lines containing P_1 and P_2 intersect at $R(x_2, y_1)$.

Find the length of one leg: The distance between (x_1, y_1) and (x_2, y_1) is $|x_2 - x_1|$.

Find the length of the other leg: The distance between (x_2, y_2) and (x_2, y_1) is $|y_2 - y_1|$.

Write the Pythagorean theorem: $c^2 = a^2 + b^2$
Substitute $|x_2 - x_1|$ for a,
and $|y_2 - y_1|$ for b: $\qquad c^2 = |x_2 - x_1|^2 + |y_2 - y_1|^2$
Write the squares without
absolute value signs: $\qquad c^2 = (x_2 - x_1)^2 + (y_2 - y_1)^2$
Solve for c: $\qquad c = \sqrt{(x_2 - x_1)^2 + (y_2 - y_1)^2}$

The distance between P_1 and P_2 (the hypotenuse of the right triangle) is equal to c.

The distance d between any two points (x_1, y_1) and (x_2, y_2) is given by the formula derived above.

The Distance Formula

$$d = \sqrt{(x_2 - x_1)^2 + (y_2 - y_1)^2}$$

To find the distance between two points, we substitute their coordinates for (x_1, y_1) and (x_2, y_2) in the distance formula, then simplify the expression (the order of substitution makes no difference).

EXAMPLE 13-3: Find the distance between the following pairs of points:
(a) $(2, 0)$ and $(5, 4)$ (b) $(-2, 3)$ and $(0, -3)$

Solution: (a) $\quad d = \sqrt{(x_2 - x_1)^2 + (y_2 - y_1)^2}$ — Write the distance formula.

$\qquad d = \sqrt{(5 - 2)^2 + (4 - 0)^2}$ — Substitute $(2, 0)$ for (x_1, y_1) and $(5, 4)$ for (x_2, y_2).

$\qquad d = \sqrt{3^2 + 4^2}$ — Simplify.

$\qquad d = \sqrt{9 + 16}$

$\qquad d = \sqrt{25}$

$\qquad d = 5$

(b) $\quad d = \sqrt{(x_2 - x_1)^2 + (y_2 - y_1)^2}$ — Write the distance formula.

$\qquad d = \sqrt{(-2 - 0)^2 + (3 - (-3))^2}$ — Substitute $(-2, 3)$ for (x_2, y_2) and $(0, -3)$ for (x_1, y_1).

$\qquad d = \sqrt{(-2)^2 + 6^2}$ — Simplify.

$\qquad d = \sqrt{4 + 36}$

$\qquad d = \sqrt{40}$

$\qquad d = 2\sqrt{10}$

13-2. Solving Problems Using the Pythagorean Theorem

To **solve a problem using the Pythagorean theorem** we:

1. *Read* the problem carefully, several times.
2. *Draw a picture* to help visualize the right triangle.
3. *Identify* each unknown measure in the right triangle.
4. *Decide* how to represent the unknown measures using one variable.
5. *Translate* the problem to a quadratic equation using the Pythagorean Theorem.
6. *Simplify* the quadratic equation.
7. *Solve* the quadratic equation.
8. *Interpret* the solutions of the quadratic equation with respect to each represented unknown measure to find the proposed solutions of the original problem.
9. *Check* to see if the proposed solutions satisfy all the conditions of the original problem.

EXAMPLE 13-4: Solve this problem using the Pythagorean theorem:

1. *Read:* $\qquad$ A 15-ft ladder is placed on level ground so that it exactly reaches a window that is 12 ft above the ground. How far is the base of the ladder from the based of the vertical building wall?

2. *Draw a picture:*

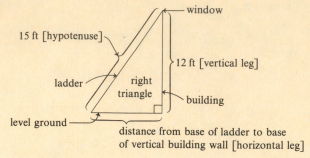

3. *Identify:*   The unknown measure is the length of the horizontal leg.

4. *Decide:*   Let x = the length of the horizontal leg.

5. *Translate:*

$$a^2 \; + \; b^2 \; = \; c^2 \longleftarrow \text{Pythagorean theorem}$$

$$(12)^2 + (x)^2 = (15)^2 \qquad \text{Substitute: } a = 12, b = x, \text{ and } c = 15 \text{ (see picture).}$$

6. *Simplify:*

$$144 + x^2 = 225$$

$$x^2 = 81 \longleftarrow x^2 = d \text{ form}$$

7. *Solve:*

$$x = \pm\sqrt{81} \qquad \text{Use the square root rule.}$$

$$x = \pm 9$$

8. *Interpret:* $x = -9$ (ft) cannot be the solution of the original problem because the required measure is not negative.

$x = 9$ means the distance from the base of the ladder to the base of the vertical building wall is 9 ft.

9. *Check:*   Does a 9-ft measure form a right triangle? Yes.

$$a^2 + b^2 = c^2 \longleftarrow \text{Pythagorean theorem}$$

$(12)^2 + (9)^2$	$(15)^2$	
$144 + 81$	225	Substitute: $a = 12, b = 9$, and $c = 15$.
225	$225 \longleftarrow 9 \text{ ft checks}$	Compute.

13-3. Parabolas

A. Graphing quadratic equations by plotting points

A quadratic equation in two variables of the form $y = ax^2 + bx + c$, where $a \neq 0$ is **a quadratic function** because it satisfyies the definition of function introduced in Chapter 11. The equation $y = ax^2 + bx + c$ may be written in functional notation $f(x) = ax^2 + bx + c$; recall that y and $f(x)$ are interchangable. The graph of a quadratic function $f(x) = ax^2 + bx + c$ is **a parabola.** To **graph a quadratic function,** we can make a table of ordered pairs that satisfy the equation, plot the points, and draw a smooth curve through them.

EXAMPLE 13-5: Graph the following quadratic functions:

(a) $y = x^2$ **(b)** $f(x) = (x + 2)^2 - 3$

Solution: **(a)** $y = x^2$

x	$x^2 = y$	
-2	$(-2)^2 = 4$	Make a table of ordered pairs.
-1	$(-1)^2 = 1$	
0	$0^2 = 0$	
1	$1^2 = 1$	
2	$2^2 = 4$	

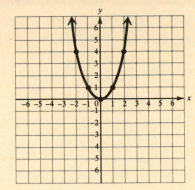

Plot the ordered pairs.

Draw a smooth curve through the points.

(b) $f(x) = (x + 2)^2 - 3$

x	$(x + 2)^2 - 3 = f(x)$
-5	$(-5 + 2)^2 - 3 = 6$
-4	$(-4 + 2)^2 - 3 = 1$
-3	$(-3 + 2)^2 - 3 = -2$
-2	$(-2 + 2)^2 - 3 = -3$
-1	$(-1 + 2)^2 - 3 = -2$
0	$(0 + 2)^2 - 3 = 1$
1	$(1 + 2)^2 - 3 = 6$

Make a table of ordered pairs.

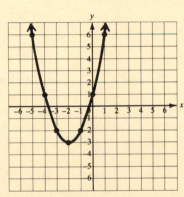

Plot the ordered pairs.

Draw a smooth curve through the points.

In the previous example you should note that the graph of $f(x) = (x + 2)^2 - 3$ is the same as $y = x^2$, except that it is shifted two units to the left and three units down. The equation of the quadratic function $y = ax^2 + bx + c$ may be written in the form $y = a(x - h)^2 + k$, where $a \neq 0$. The constant k represents a vertical shift from the origin, h represents a horizontal shift from the origin, and a determines the "fatness" of the curve.

B. Finding the intercepts of parabolas

To **find the y-intercept** of the graph of $y = ax^2 + bx + c$, we substitute 0 for x in the equation and solve for y.

To **find the x-intercept(s)** of the graph of $y = ax^2 + bx + c$, we substitute 0 for y in the equation and solve for x.

EXAMPLE 13-6: Find the y-intercept of the graph of $y = 3x^2 + 4x - 3$.

Solution: $y = 3x^2 + 4x - 3$

$\qquad y = 3(0)^2 + 4(0) - 3$ Substitute 0 for x.

$\qquad y = -3$

$(0, -3)$ is the y-intercept.

EXAMPLE 13-7: Find the x-intercept(s) of the graphs of:

(a) $y = x^2 + 4x + 4$ **(b)** $y = x^2 - 4$ **(c)** $y = x^2 - 4x + 6$

Solution: **(a)** $y = x^2 + 4x + 4$

$$0 = y^2 + 4x + 4 \qquad \text{Substitute 0 for } y.$$

$$0 = (x + 2)^2 \qquad \text{Solve for } x.$$

$$0 = x + 2$$

$$x = -2$$

$(-2, 0)$ is the x-intercept.

(b) $y = x^2 - 4$

$$0 = x^2 - 4 \qquad \text{Substitute 0 for } y.$$

$$4 = x^2 \qquad \text{Solve for } x.$$

$$\pm 2 = x$$

$(-2, 0)$ and $(2, 0)$ are the x-intercepts.

(c) $y = x^2 - 4x + 6$

$$0 = x^2 - 4x + 6 \qquad \text{Substitute 0 for } y.$$

$$x = \frac{4 \pm \sqrt{16 - 24}}{2} \qquad \text{Solve for } x.$$

$$x = 2 \pm i\sqrt{2}$$

$y = x^2 - 4x + 6$ has no x-intercept.

C. Finding the vertices and axes symmetry of parabolas

The **vertex of a parabola** is the lowest or the highest point of the parabola. The **axis of symmetry of a parabola** is the line about which the parabola is symmetric. The y-value of the vertex is the point at which the parabola and the axis of symmetry intersect. The highest point of a parabola that opens downward is called the **maximum** value of the parabola, and the y-value of the lowest point of a parabola that opens upward is called the **minimum** value of the parabola.

To develop a formula for the vertex of $y = ax^2 + bx + c$, where $a > 0$, we complete the square of this quadratic equation. To determine the equation of the axis of symmetry, we set x equal to the x-coordinate of the vertex.

EXAMPLE 13-8: Find the vertex and axis of symmetry of the graph of $y = ax^2 + bx + c$ where $a > 0$.

Solution: $y = ax^2 + bx + c$

$$y = (ax^2 + bx) + c$$

$$y = a\left(x^2 + \frac{b}{a}x\right) + c$$

$$y = a\left(x^2 + \frac{b}{a}x + \left(\frac{b}{2a}\right)^2\right) + c - a\left(\frac{b}{2a}\right)^2 \qquad \text{Add } \left(\frac{b}{2a}\right)^2 \text{ inside the parentheses and subtract the same quantity outside.}$$

$$y = a\left(x + \frac{b}{2a}\right)^2 + c - \frac{b^2}{4a}$$

$$y = a\left(x + \frac{b}{2a}\right)^2 + \frac{4ac - b^2}{4a}$$

$$x + \frac{b}{2a} = 0$$

Find the minimum value of y by setting $x + \frac{b}{2a} = 0$.

$$x = -\frac{b}{2a}$$

Solve for x.

$$y = a\left(-\frac{b}{2a} + \frac{b}{2a}\right)^2 + \frac{4ac - b^2}{4a}$$

Substitute $-\frac{b}{2a}$ for x and solve for y.

$$y = a(0)^2 + \frac{4ac - b^2}{4a}$$

$$y = \frac{4ac - b^2}{4a}$$

$$\left(-\frac{b}{2a}, \frac{4ac - b^2}{4a}\right) \text{ is the vertex.}$$

Substituting for (x, y).

$$x = -\frac{b}{2a} \text{ is the equation of the axis of symmetry.}$$

Set x equal to the x-coordinate.

The graph of $y = ax^2 + bx + c$ has its **vertex** at

$$\left(-\frac{b}{2a}, \frac{4ac - b^2}{4a}\right)$$

and its **axis of symmetry** is determined by

$$x = -\frac{b}{2a}$$

To find the vertex and axis of symmetry of the graph of a quadratic equation, we identify a, b, and c, substitute in the above ordered pair of the vertex and the equation of the axis of symmetry, then evaluate.

EXAMPLE 13-9: Find the vertex and the axis of symmetry of:

(a) $y = x^2 - 3x - 4$ and **(b)** $f(x) = -4x^2 + 4x - 1$

Solution: **(a)** $y = x^2 - 3x - 4$

$$a = 1, b = -3, c = -4$$

Identify a, b, and c.

$$\left(-\frac{b}{2a}, \frac{4ac - b^2}{4a}\right)$$

Write the ordered pair of the vertex.

$$\left(-\frac{-3}{2 \cdot 1}, \frac{4(1)(-4) - (-3)^2}{4(1)}\right)$$

Substitute for a, b, and c.

$$\left(\frac{3}{2}, -\frac{25}{4}\right) \text{ is the vertex.}$$

Evaluate.

$$x = \frac{3}{2} \text{ or } 2x - 3 = 0 \text{ is}$$
the axis of symmetry.

Write x equal to the first coordinate since $x = -\frac{b}{2a}$ is the axis of symmetry.

(b) $f(x) = -4x^2 + 4x - 1$

$$a = -4, b = 4, c = -1$$

Identify a, b, and c.

$$\left(-\frac{b}{2a}, \frac{4ac - b^2}{4a}\right)$$

Write the ordered pair of the vertex.

$$\left(-\frac{4}{2(-4)}, \frac{4(-4)(-1) - 4^2}{4(-4)}\right)$$

Substitute for a, b, and c.

$\left(\dfrac{1}{2}, 0\right)$ is the vertex.　　　　　　　　Evaluate.

$x = \dfrac{1}{2}$ or $2x - 1 = 0$ is　　　　　Write x equal to the first coordinate since

the axis of symmetry.　　　　　　$x = -\dfrac{b}{2a}$ is the axis of symmetry.

D. Graphing equations of the form y or $f(x) = ax^2 + bx + c$

To **graph an equation of the form** y or $f(x) = ax^2 + bx + c$, we find the vertex, x-intercept(s), y-intercept (and other solutions if necessary), plot the ordered pairs, and draw a smooth curve through the points as before.

Note: You may find it easier to sketch the graph of a quadratic equation by finding the vertex and intercepts than by making a table of solutions. It is difficult to identify the vertex of some parabolas by making a table.

EXAMPLE 13-10: Graph the following quadratic equation: $y = x^2 + 2x - 3$.

Solution: $y = x^2 + 2x - 3$

$a = 1, b = 2, c = -3$　　　　　　Identify a, b, and c.

$\left(-\dfrac{2}{2 \cdot 1}, \dfrac{4(1)(-3) - 2^2}{4(1)}\right)$　　　Find the vertex.

$(-1, -4)$

$0 = x^2 + 2x - 3$　　　　　　　Find the x-intercepts.

$0 = (x + 3)(x - 1)$

$0 = x + 3$ or $x - 1 = 0$

$x = 1$ or -3

$(1, 0), (-3, 0)$ are the x-intercepts.

$y = x^2 + 2x - 3$　　　　　　　Find the y-intercept.

$y = 0^2 + 2(0) - 3$

$y = -3$

$(0, -3)$ is the y-intercept.

x	$x^2 + 2x - 3 = y$	Find additional solutions.
2	$2^2 + 2 \cdot 2 - 3 = 5$	
-4	$16 - 8 - 3 = 5$	
-2	$4 - 4 - 3 = -3$	

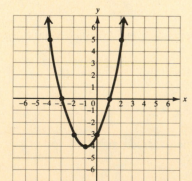

Plot the solutions.

Draw a smooth curve through the points.

E. Graph equations of the form $x = ay^2 + by + c$

When x and y are interchanged in the equation $y = ax^2 + by + c$ to get the equation $x = ay^2 + by + c$, the latter is the **inverse** of the former. Its graph will still be a parabola, but one whose axis of symmetry will be horizontal. Equations of the form $x = ay^2 + by + c$ are not functions (see Section 11-2).

EXAMPLE 13-11: Graph the quadratic equation: $x = y^2 - 4$.

Solution: $x = y^2 - 4$

$a = 1, b = 0, c = -4$

$\left(\dfrac{4ac - b^2}{4a}, -\dfrac{b}{2a} \right)$ Find the vertex.

$(-4, 0)$

$0 = y^2 - 4$ Find the y-intercepts.

$4 = y^2$

$y = \pm 2$

$(0, 2)$ and $(0, -2)$ are the y-intercepts.

$x = 0^2 - 4$ Find the x-intercept.

$x = -4$

$(-4, 0)$ is the x-intercept.

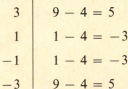

y	$y^2 - 4 = x$
3	$9 - 4 = 5$
1	$1 - 4 = -3$
-1	$1 - 4 = -3$
-3	$9 - 4 = 5$

Find other solutions.

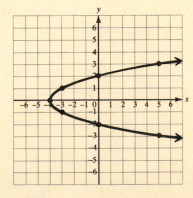

Plot the ordered pairs.

Draw a smooth curve through the points.

13-4. Solving Problems by Finding Maximum and Minimum Values

If $a < 0$, then the vertex is the highest point on the graph of $y = ax^2 + bx + c$ and the y-coordinate of the vertex is called the **maximum value.** If $a > 0$, the vertex is the lowest point on the graph of $y = ax^2 + bx + c$ and the y-coordinate of the vertex is called the **minimum value.**

To **solve a problem by finding the maximum or minimum value,** we:

1. *Read* the problem carefully, several times.
2. *Draw a picture* (when appropriate) to help visualize the problem.
3. *Identify* the unknown measures.
4. *Decide* how to represent the unknown measures using different variables.

5. *Translate* to a quadratic equation using geometry formulas (see Appendix Table 5).

6. *Find* the vertex of the quadratic equation using $\left(-\dfrac{b}{2a}, \dfrac{4ac - b^2}{4a}\right)$.

7. *Interpret* the vertex with respect to the unknown measures to find the proposed maximum or minimum value.

8. *Check* to see if the proposed maximum or minimum value satisfies all the conditions of the original problem.

EXAMPLE 13-12: Solve this problem by finding the maximum or minimum value.

1. *Read:*   What is the largest rectangular field that can be enclosed with 8 km of fencing?

2. *Draw a picture:*

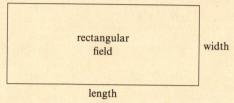

rectangular
field   width

length

3. *Identify:*   The unknown measures are $\begin{Bmatrix} \text{the length of the rectangular field} \\ \text{the width of the rectangular field} \end{Bmatrix}$.

4. *Decide:*   Let l = the length of the rectangular field

and w = the width of the rectangular field.

5. *Translate:*
$$P = 2l + 2w \longleftarrow \text{perimeter formula for a rectangle}$$

$$8 = 2l + 2w \qquad \text{Substitute 8 (km) for } P.$$

$$8 - 2w = 2l \qquad \text{Solve for one of the remaining variables.}$$

$$4 - w = l \longleftarrow \text{solved for } l$$

$$A = lw \longleftarrow \text{area formula for a rectangle}$$

$$A = (4 - w)w \qquad \text{Substitute } 4 - w \text{ for } l.$$

$$A = 4w - w^2 \qquad \text{Write in standard form.}$$

$$A = -1w^2 + 4w + 0 \longleftarrow \text{standard form } (A = aw^2 + bw + c)$$

6. *Find the vertex:* $\text{vertex} = \left(-\dfrac{b}{2a}, \dfrac{4ac - b^2}{4a}\right)$

$$= \left(-\dfrac{4}{2(-1)}, \dfrac{4(-1)(0) - (4)^2}{4(-1)}\right) \qquad \begin{array}{l} \text{Substitute: } a = -1, b = 4, \\ \text{and } c = 0. \end{array}$$

$$= (2, 4) \longleftarrow (w, A)$$

7. *Interpret:*   In $A = -lw^2 + 4w + 0$, $a = -1 < 0$ means the vertex $(2, 4)$ is a maximum value.
$(w, A) = (2, 4)$ means a width (w) of 2 km will maximize the area (A) at 4 km².
$l = 4 - w = 4 - 2 = 2$ means the length (l) of the maximum area is also 2 km [see note].

8. *Check:*   Is the perimeter of the maximum rectangular area 8 km?
    Yes: $P = 2l + 2w = 2(2) + 2(2) = 8$.
Does the maximum area equal the length times the width?
    Yes: $A = lw = 2(2) = 4$.

Solution: The largest rectangular field that can be enclosed with 8 km of fencing is a square field measuring 2 km on each side with an area of 4 km².

13-5. Circles

A **circle** is the set of all points in a plane equidistant from a fixed point. The fixed point is called the **center** of the circle and the distance is called the **radius.**

A. Finding equations of circles given the center points and radii

To **find an equation of a circle** with center (h, k), radius r, and (x, y) as any point on the circle, we use the distance formula to get $r = \sqrt{(x - h)^2 + (y - k)^2}$, then square both members to get the equation of the circle.

The Equation of a Circle

A circle with center (h, k) and radius r has an equation

$$(x - h)^2 + (y - k)^2 = r^2$$

EXAMPLE 13-13: Write an equation of a circle with the following center point and radius:

(a) $(0, 0), r = 3$ (b) $(2, 3), r = 1$ (c) $(-2, 1), r = 2$

Solution: (a) $(x - h)^2 + (y - k)^2 = r^2$ Write the general equation of a circle.

$(x - 0)^2 + (y - 0)^2 = 3^2$ Substitute for (h, k) and r.

$x^2 + y^2 = 9$

(b) $(x - h)^2 + (y - k)^2 = r^2$

$(x - 2)^2 + (y - 3)^2 = 1^2$ Substitute for (h, k) and r.

$(x - 2)^2 + (y - 3)^2 = 1$

(c) $(x - h)^2 + (y - k)^2 = r^2$

$[x - (-2)]^2 + (y - 1)^2 = 2^2$ Substitute for (h, k) and r.

$(x + 2)^2 + (y - 1)^2 = 4$

B. Writing equations of circles in standard form

To write an equation of the form $x^2 + y^2 + Cx + Dx + E = 0$ as an equation of a circle in standard form, we complete the square for each variable, then write the resulting equation in standard form.

EXAMPLE 13-14: Find the center and radius of each of the following circles:

(a) $x^2 + (y + 3)^2 = 9$ and (b) $x^2 + y^2 + 6x - 8y = 0$.

Solution: (a) $x^2 + (y + 3)^2 = 9$

$(x - 0)^2 + [y - (-3)]^2 = 3^2$ Write in standard form.

$(0, -3)$ is the center.   Identify (h, k) as the center point.

$r = 3$ Identify the radius.

(b) $x^2 + y^2 + 6x - 8y = 0$

$[x^2 + 6x] + [y^2 - 8y] = 0$ Group the variables.

$[x^2 + 6x + 9] + [y^2 - 8y + 16] = 9 + 16$ Complete each square and add the same amount to each member.

$[x + 3]^2 + [y - 4]^2 = 25$ Factor each perfect square.

$[x - (-3)]^2 + (y - 4)^2 = 5^2$ Write equation in standard form.

$(-3, 4)$ is the center   Identify (h, k) as the center point.

$r = 5$ Identify the radius.

C. Graphing circles

To **graph an equation of a circle,** you identify the center and radius, plot the center and draw the set of all points at distance r from the center.

EXAMPLE 13-15: Graph the circle of (a) $x^2 + (y + 2)^2 = 9$ and (b) $x^2 + y^2 - 6x + 2y + 6 = 0$.

Solution: (a) $x^2 + (y + 2)^2 = 9$

$(x - 0)^2 + [y - (-2)]^2 = 3^2$ Write equation in standard form.

$(0, -2)$ is the center   Identify (h, k) as the center point.

$r = 3$ Identify the radius.

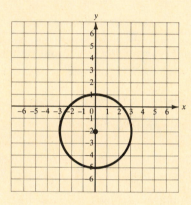

Plot the center.

Draw the circle with radius 3.

(b) $x^2 + y^2 - 6x + 2y + 6 = 0$

$(x^2 - 6x) + (y^2 + 2y) = -6$ Write equation in standard form by completing the square for each variable.

$[x^2 - 6x + 9] + [y^2 + 2y + 1] = -6 + 9 + 1$

$[x - 3]^2 + [y + 1]^2 = 4$

$[x - 3]^2 + [y - (-1)]^2 = 2^2$

$(3, -1)$ is the center   Identify (h, k) as the center.

$r = 2$ Identify the radius.

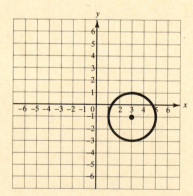

Plot the center.

Draw the circle with radius 2.

13-6. Ellipses

An **ellipse** is the set of all points in a plane such that the sum of the distances of each point from two fixed points is constant. Each of the fixed points is a **focus** of the ellipse. A **chord** of an ellipse is any line segment with end points on the ellipse. The **major axis** is the longest chord and the **minor axis** is the shortest chord. The points of intersection of the axes and the ellipse are the **intercepts.** The **center** of the ellipse is the point of intersection of the axes. The **vertices** of the ellipse are the end-points of the major axis.

A. Writing equations of ellipses

The Equation of an Ellipse

The standard equation of an ellipse with its center at the origin and intercepts $(-a, 0)$, $(a, 0)$, $(0, -b)$, and $(0, b)$ is

$$\frac{x^2}{a^2} + \frac{y^2}{b^2} = 1$$

To write the equation of a given ellipse, substitute for a and b in the standard equation.

EXAMPLE 13-16: Write an equation of an ellipse with center $(0, 0)$ and intercepts $(-3, 0)$, $(3, 0)$, $(0, -5)$, and $(0, 5)$.

Solution: $a = 3, b = 5$ Identify a and b.

$$\frac{x^2}{a^2} + \frac{y^2}{b^2} = 1$$ Write the standard equation of an ellipse.

$$\frac{x^2}{3^2} + \frac{y^2}{5^2} = 1$$ Substitute 3 for a and 5 for b.

B. Graphing ellipses

To **graph an ellipse** with its center at the origin, we identify and plot the intercepts, then draw a smooth curve through the points.

EXAMPLE 13-17: Graph the following equations:

(a) $\dfrac{x^2}{4^2} + \dfrac{y^2}{2^2} = 1$ **(b)** $9x^2 + 4y^2 = 36$

Solution: **(a)** $\dfrac{x^2}{4^2} + \dfrac{y^2}{2^2} = 1$

     $a = 4, b = 2$ Identify a and b.

     $(-4, 0), (4, 0), (0, -2), (0, 2)$ are the intercepts.

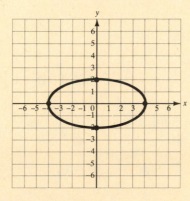

Plot the intercepts.

Draw a smooth curve through the points.

(b) $9x^2 + 4y^2 = 36$

$$\frac{9x^2}{36} + \frac{4y^2}{36} = 1$$ Write the equation in standard form.

$$\frac{x^2}{4} + \frac{y^2}{9} = 1$$

$$\frac{x^2}{2^2} + \frac{y^2}{3^2} = 1$$

$(-2, 0), (2, 0), (0, -3), (0, 3)$ are the intercepts.

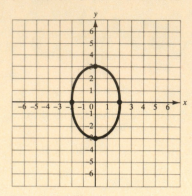

Plot the intercepts.

Draw a smooth curve through the points.

Note: If $a > b$, the major axis of the ellipse is horizontal and if $a < b$, the major axis is vertical.

13-7. Hyperbolas

A **hyperbola** is the set of all points in a plane such that the difference of the distances of each point from two fixed points is a constant. Each of the fixed points is a **focus** of the hyperbola. The line containing the foci contains the **major** or **transverse axis.** The end-points of the major axis are the **vertices** of the hyperbola.

Hyperbolas with centers at the origin have standard equations as follows:

$$\frac{x^2}{a^2} - \frac{y^2}{b^2} = 1 \qquad \text{if the hyperbola has a horizontal major axis}$$

$$-\frac{x^2}{a^2} + \frac{y^2}{b^2} = 1 \qquad \text{if the hyperbola has a vertical major axis}$$

Asymptotes are lines associated with hyperbolas. As a hyperbola gets farther from its center it gets closer and closer to, but NEVER TOUCHES, its asymptotes. For hyperbolas with the above equations, the equations of the asymptotes are

$$y = -\frac{b}{a}x \quad \text{and} \quad y = \frac{b}{a}x$$

A. Graphing hyperbolas

To **graph a hyperbola,** identify a and b, graph the asymptotes and vertices, and sketch the graph.

EXAMPLE 13-18: Graph the hyperbolas with centers at the origin and the equations:

(a) $\dfrac{x^2}{3^2} - \dfrac{y^2}{2^2} = 1$ **(b)** $4y^2 - 9x^2 = 36$

Solution: **(a)** $\dfrac{x^2}{3^2} - \dfrac{y^2}{2^2} = 1$

$a = 3, b = 2$

$(-3, 0)$ and $(3, 0)$ are the vertices.   Find the vertices using the equation with horizontal major axis.

$y = -\dfrac{2}{3}x$ and $y = \dfrac{2}{3}x$ Substitute for a and b in $y = \pm(b/a)x$ to find the asymptotes.

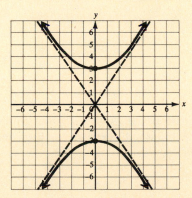

(b) $4y^2 - 9x^2 = 36$

$$\frac{4y^2}{36} - \frac{9x^2}{36} = 1$$ Write the equation in standard form.

$$-\frac{x^2}{2^2} + \frac{y^2}{3^2} = 1$$

$a = 2, b = 3$ Identify a and b.

$(0, 3)$ and $(0, -3)$ are the vertices.   Find the vertices on the vertical major axis.

$$y = -\frac{3}{2}x \text{ and } y = \frac{3}{2}x$$ Substitute for a and b in $y = \pm(b/a)x$ to find the asymptotes.

Graph the vertices.

Graph the asymptotes.

Draw a smooth curve through the vertices and approaching the asymptotes.

B. Graphing hyperbolas of the form $xy = k$

To **graph a hyperbola of the form $xy = k$,** we make a table of solutions, plot the ordered pairs, and draw a smooth curve through the points.

EXAMPLE 13-19: Graph the hyperbolas with equations **(a)** $xy = 6$ and **(b)** $xy = -4$.

Solution: **(a)** $xy = 6$

$$y = \frac{6}{x}$$ Solve for one variable.

x	-6	-3	-2	-1	1	2	3	6
$\frac{6}{x} = y$	-1	-2	-3	-6	6	3	2	1

Make a table of solutions.

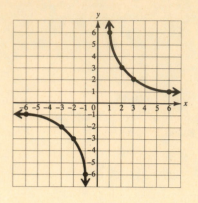

Plot the solutions.
Draw a smooth curve through the points.

(b) $xy = -4$

x	-4	-2	-1	1	2	4
$-\dfrac{4}{x} = y$	1	2	4	-4	-2	-1

Make a table of solutions.

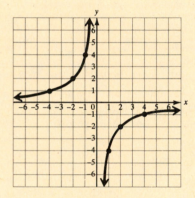

Plot the solutions.
Draw a smooth curve through the points.

13-8. Graphing Quadratic Inequalities

To **graph a quadratic inequality,** we graph the associated equation and use a test point to determine which region satisfies the inequality. (Use a solid curve or a dashed curve depending on the inequality symbol used, as before.)

EXAMPLE 13-20: Graph the inequalities **(a)** $x^2 + y^2 + 6y < 0$ and **(b)** $y \le x^2 + 2x$.

Solution: **(a)** $x^2 + y^2 + 6y < 0$

$\qquad x^2 + (y + 3)^2 = 3^2$ Write the equation in standard form.

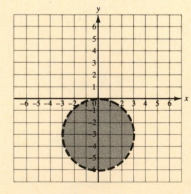

Graph $x^2 + (y + 3)^2 = 3^2$ using a dashed curve.

Use $(2, 0)$ as a test point and find that $2^2 + 0^2 + 6(0) \not< 0$.

Shade the region that does not contain $(2, 0)$.

(b) $y \le x^2 + 2x$

$\qquad y = (x + 1)^2 - 1$ Write the equation in standard form.

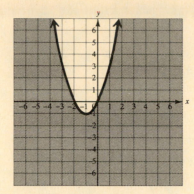

Graph $y = (x + 1)^2 - 1$ using a solid curve.
Use $(2, 0)$ as a test point and find that $0 \leq 2^2 + 2(2)$.
Shade the region that contains $(2, 0)$.

RAISE YOUR GRADES

Can you . . . ?

☑ find the length of the hypotenuse of a right triangle given the length of the legs
☑ find the distance between two points
☑ solve a problem using the Pythagorean theorem
☑ graph quadratic functions by plotting points
☑ find the y-intercept of a quadratic equation
☑ find the x-intercept(s) of a quadratic equation
☑ find the vertex and axis of symmetry of the graph of a quadratic equation
☑ graph a quadratic equation of the form $y = ax^2 + bx + c$ by finding the vertex
☑ graph a quadratic equation of the form $x = ay^2 + by + c$
☑ solve problems by finding maximum or minimum values
☑ write the equation of a circle given the center and the radius
☑ find the center and radius of a circle with equation $x^2 + y^2 + Cx + Dy + E = 0$
☑ graph an equation of a circle
☑ write an equation of an ellipse with its center at the origin and given intercepts
☑ graph a hyperbola with center at the origin
☑ graph a hyperbola with equation of the form $xy = k$
☑ graph a quadratic inequality

SOLVED PROBLEMS

PROBLEM 13-1 Find the length of the hypotenuse of a right triangle having the following pairs of legs: **(a)** 3, 4   **(b)** 5, 12   **(c)** 1, 1   **(d)** 1, 2   **(e)** 3, 7   **(f)** 4, 6

Solution: Recall that to find the length of the hypotenuse of a right triangle, you substitute the given numbers for a and b in the equation $a^2 + b^2 = c^2$ and solve for c [see Example 13-1].

(a) $a^2 + b^2 = c^2$ **(b)** $c^2 = a^2 + b^2$ **(c)** $c^2 = a^2 + b^2$

$\quad 3^2 + 4^2 = c^2$ $c^2 = 5^2 + 12^2$ $c^2 = 1^2 + 1^2$

$\quad 9 + 16 = c^2$ $c^2 = 25 + 144$ $c^2 = 1 + 1$

$\quad\quad\quad 25 = c^2$ $c^2 = 169$ $c^2 = 2$

$\quad\quad\quad\quad 5 = c$ $c = 13$ $c = \sqrt{2}$

(d) $c^2 = a^2 + b^2$

$c^2 = 1^2 + 2^2$

$c^2 = 1 + 4$

$c^2 = 5$

$c = \sqrt{5}$

(e) $c^2 = a^2 + b^2$

$c^2 = 3^2 + 7^2$

$c^2 = 9 + 49$

$c^2 = 58$

$c = \sqrt{58}$

(f) $c^2 = a^2 + b^2$

$c^2 = 4^2 + 6^2$

$c^2 = 16 + 36$

$c^2 = 52$

$c = \sqrt{52}$

$c = 2\sqrt{13}$

PROBLEM 13-2 Find the distance between the following pairs of points:

(a) $(2, 3), (6, 6)$ **(b)** $(3, 2), (-1, 5)$ **(c)** $(-1, 3), (2, -1)$

(d) $(-2, -2), (-2, 3)$ **(e)** $(2, 5), (-3, 5)$ **(f)** $(2, -3), (-1, -2)$

Solution: Recall that to find the distance between two points, you substitute for (x_1, y_1) and (x_2, y_2) in the distance formula and simplify [see Example 13-3].

(a) $d = \sqrt{(x_2 - x_1)^2 + (y_2 - y_1)^2}$

$d = \sqrt{(6 - 2)^2 + (6 - 3)^2}$

$d = \sqrt{4^2 + 3^2}$

$d = \sqrt{16 + 9}$

$d = \sqrt{25}$

$d = 5$

(b) $d = \sqrt{(x_2 - x_1)^2 + (y_2 - y_1)^2}$

$d = \sqrt{(-1 - 3)^2 + (5 - 2)^2}$

$d = \sqrt{(-4)^2 + 3^2}$

$d = \sqrt{16 + 9}$

$d = \sqrt{25}$

$d = 5$

(c) $d = \sqrt{(x_2 - x_1)^2 + (y_2 - y_1)^2}$

$d = \sqrt{(-1 - 2)^2 + (3 - (-1))^2}$

$d = \sqrt{(-3)^2 + 4^2}$

$d = \sqrt{9 + 16}$

$d = \sqrt{25}$

$d = 5$

(d) $d = \sqrt{(x_2 - x_1)^2 + (y_2 - y_1)^2}$

$d = \sqrt{(-2 - (-2))^2 + (-2 - 3)^2}$

$d = \sqrt{0^2 + (-5)^2}$

$d = \sqrt{25}$

$d = 5$

(e) $d = \sqrt{(x_2 - x_1)^2 + (y_2 - y_1)^2}$

$d = \sqrt{(-3 - 2)^2 + (5 - 5)^2}$

$d = \sqrt{(-5)^2 + 0^2}$

$d = \sqrt{25}$

$d = 5$

(f) $d = \sqrt{(x_2 - x_1)^2 + (y_2 - y_1)^2}$

$d = \sqrt{(-1 - 2)^2 + (-2 - (-3))^2}$

$d = \sqrt{(-3)^2 + 1^2}$

$d = \sqrt{9 + 1}$

$d = \sqrt{10}$

PROBLEM 13-3 Solve each problem using the Pythagorean theorem:

(a) The hypotenuse of a right triangle is 1 ft longer than one leg and 2 ft longer than the other leg. How long is each side of the right triangle?

(b) Two cars start at the same time from the same place and travel at constant speeds that differ by 10 mph on roads that are at right angles to each other. In 2 hours, the cars are 100 miles apart. What was the speed of the faster car? How far did the slower car travel?

Solution: Recall that to solve a problem using the Pythagorean theorem, you:

1. *Read* the problem carefully, several times.
2. *Draw a picture* to help visualize the right triangle.
3. *Identify* each unknown measure in the right triangle.
4. *Decide* how to represent the unknown measures using one variable.

5. *Translate* the problem to a quadratic equation using the Pythagorean theorem.
6. *Simplify* the quadratic equation.
7. *Solve* the quadratic equation.
8. *Interpret* the solutions of the quadratic equation with respect to each represented unknown measure to find the proposed solutions.
9. *Check* to see if the proposed solutions satisfy all the conditions of the original problem [see Examples 13-4].

(a) *Draw a picture:*

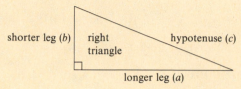

Identify: The unknown measures are $\begin{cases} \text{the length of the hypotenuse} \\ \text{the length of the shorter leg} \\ \text{the length of the longer leg} \end{cases}$.

Decide: Let $c = $ the length of the hypotenuse (c),

               then $c - 1 = $ the length of the shorter leg (b),

               and $c - 2 = $ the length of the longer leg (a).

Translate: $c^2 = a^2 + b^2$ ⟵ Pythagorean theorem

               $c^2 = (c - 2)^2 + (c - 1)^2$ Substitute $c - 2$ for a, and $c - 1$ for b.

Simplify: $c^2 = c^2 - 2c + 1 + c^2 - 4c + 4$

         $c^2 - 6c + 5 = 0$ ⟵ simplest form

Solve: $(c - 5)(c - 1) = 0$ Factor or use the quadratic formula.

         $c - 5 = 0$ or $c - 1 = 0$

            $c = 5$ or $c = 1$

Interpret: $c = 1$ cannot be a solution of the original problem because $c - 1 = 1 - 1 = 0$ (ft), and $c - 2 = 1 - 2 = -1$(ft), and the lengths of the sides of a triangle cannot be negative.

          $c = 5$ means the length of the hypotenuse is 5 ft.

          $c - 1 = 5 - 1 = 4$ means the length of the longer leg is 4 ft. ⟵ solutions

          $c - 2 = 5 - 2 = 3$ means the length of the shorter leg is 3 ft. [check as before]

Draw a picture:

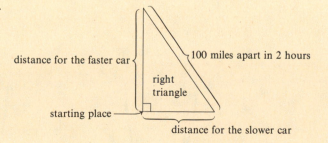

(b) *Identify:*   The unknown measures are $\begin{cases} \text{the rate for the slower car} \\ \text{the rate for the faster car} \\ \text{the distance for the slower car} \\ \text{the distance for the faster car} \end{cases}$.

Decide: Let $r = $ the rate for the slower car

               then $r + 10 = $ the rate for the faster car.

Make a table:

	rate (r) [in mph]	time (t) [in hours]	distance ($d = rt$) [in miles]
slower car	r	2	$2r$
faster car	$r + 10$	2	$2(r + 10)$

Translate:

$$c^2 = a^2 + b^2 \longleftarrow \text{Pythagorean theorem}$$

$$\text{hypotenuse} \longrightarrow (100)^2 = (2r)^2 + [2(r + 10)]^2 \longleftarrow \text{legs}$$

Simplify:

$$100^2 = 2^2 r^2 + 2^2(r + 10)^2$$

$$10{,}000 = 4r^2 + 4r^2 + 80r + 400$$

$$8r^2 + 80r - 9600 = 0 \longleftarrow \text{standard form [the GCF is 8]}$$

$$r^2 + 10r - 1200 = 0 \longleftarrow \text{simplest form}$$

$$(r + 40)(r - 30) = 0 \qquad \text{Factor or use the quadratic equation.}$$

$$r + 40 = 0 \qquad \text{or} \ \ r - 30 = 0$$

$$r = \cancel{-40} \ \text{or} \qquad r = 30 \ [\text{mph}] \longleftarrow \text{rate for the slower car}$$

$$\text{rate for the faster car} \longrightarrow r + 10 = 30 + 10 = 40 \ (\text{mph}) \searrow \text{solutions}$$

$$\text{distance for the slower car} \longrightarrow 2r = 2(30) = 60 \ (\text{mi}) \nwarrow$$

PROBLEM 13-4 Graph the following quadratic functions:

(a) $y = -x^2$ **(b)** $f(x) = (x - 2)^2 - 3$ **(c)** $f(x) = 2x^2$

Solution: Recall that to graph a quadratic function, you make a table of ordered pairs that satisfy the equation, plot the ordered pairs and draw a smooth curve through the points [see Example 13-5].

(a) $y = -x^2$

x	$-x^2 = y$
-2	$-(-2)^2 = -4$
-1	$-(-1)^2 = -1$
0	$-0^2 = 0$
1	$-1^2 = -1$
2	$-2^2 = -4$

(b) $f(x) = (x - 2)^2 - 3$

x	$(x - 2)^2 - 3 = f(x)$
0	$(0 - 2)^2 - 3 = 1$
1	$(1 - 2)^2 - 3 = -2$
2	$(2 - 2)^2 - 3 = -3$
3	$(3 - 2)^2 - 3 = -2$
4	$(4 - 2)^2 - 3 = 1$
5	$(5 - 2)^2 - 3 = 6$

(c)

x	$2x^2 = f(x)$
-2	$2(-2)^2 = 8$
-1	$2(-1)^2 = 2$
0	$2(0)^2 = 0$
1	$2(1)^2 = 2$
2	$2(2)^2 = 8$

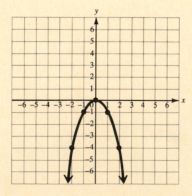
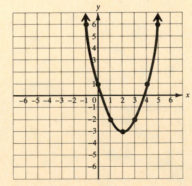
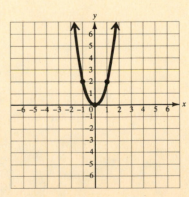

PROBLEM 13-5 Find the y-intercept of the graph of the following equations:

(a) $y = x^2 + 2x - 3$ (b) $y = -x^2 + 4x$ (c) $y = 2x^2 + 4$

Solution: Recall that to find the y-intercept of the graph of an equation, you substitute 0 for x in the equation and solve for y [see Example 13-6].

(a) $y = x^2 + 2x - 3$ (b) $y = -x^2 + 4x$ (c) $y = 2x^2 + 4$

$y = 0^2 + 2(0) - 3$ $y = -0^2 + 4(0)$ $y = 2(0)^2 + 4$

$y = -3$ $y = 0$ $y = 4$

$(0, -3)$ is the y-intercept. $(0, 0)$ is the y-intercept. $(0, 4)$ is the y-intercept.

PROBLEM 13-6 Find the x-intercept(s) of the graph of the following equations:

(a) $y = x^2 + 4$ (b) $y = 4x^2 - 12x + 9$ (c) $y = 2x^2 - 3x - 2$

Solution: Recall that to find the x-intercepts of the graph of an equation, you substitute 0 for y and solve for x [see Example 13-7].

(a) $y = x^2 + 4$ (b) $y = 4x^2 - 12x + 9$ (c) $y = 2x^2 - 3x - 2$

$0 = x^2 + 4$ $0 = 4x^2 - 12x + 9$ $0 = 2x^2 - 3x - 2$

$x^2 = -4$ $0 = (2x - 3)^2$ $0 = (2x + 1)(x - 2)$

$x = \pm\sqrt{-4}$ $0 = 2x - 3$ $0 = 2x + 1$ or $x - 2 = 0$

Since x has no real solution, $3 = 2x$ $-\frac{1}{2} = x$ or $x = 2$
the graph has no x-intercept.
$\frac{3}{2} = x$ $(-\frac{1}{2}, 0)$ and $(2, 0)$ are the
$(\frac{3}{2}, 0)$ is the x-intercept. x-intercepts.

PROBLEM 13-7 Find the vertex and axis of symmetry of the graphs of each of the following equations:

(a) $y = x^2 - 5$ (b) $y = x^2 + 6x$ (c) $y = 2x^2 + 5x - 3$

Solution: Recall that to find the vertex and axis of symmetry of the graph of a quadratic equation, you identify a, b, and c, substitute each value in $\left(-\dfrac{b}{2a}, \dfrac{4ac - b^2}{4a}\right)$ to get the vertex and in $x = -\dfrac{b}{2a}$ to get the axis of symmetry and evaluate [see Example 13-8].

(a) $y = x^2 - 5$ (b) $y = x^2 + 6x$ (c) $y = 2x^2 + 5x - 3$

$a = 1, b = 0, c = -5$ $a = 1, b = 6, c = 0$ $a = 2, b = 5, c = -3$

$\left(-\dfrac{b}{2a}, \dfrac{4ac - b^2}{4a}\right)$ $\left(-\dfrac{b}{2a}, \dfrac{4ac - b^2}{4a}\right)$ $\left(-\dfrac{b}{2a}, \dfrac{4ac - b^2}{4a}\right)$

$\left(-\dfrac{0}{2}, \dfrac{4(1)(-5) - 0^2}{4(1)}\right)$ $\left(-\dfrac{6}{2(1)}, \dfrac{4(1)(0) - 6^2}{4(1)}\right)$ $\left(-\dfrac{5}{4}, \dfrac{4(2)(-3) - 5^2}{4(2)}\right)$

$(0, -5)$ is the vertex. $(-3, -9)$ is the vertex. $\left(-\dfrac{5}{4}, -\dfrac{49}{8}\right)$ is the vertex.

$x = 0$ is the axis of $x = -3$ is the axis
symmetry.   of symmetry. $x = -\frac{5}{4}$, or $4x + 5 = 0$, is
the axis of symmetry.

PROBLEM 13-8 Graph the following quadratic equations:

(a) $y = x^2 - 2$ (b) $y = x^2 + 2x$ (c) $y = -2x^2 + 2x$
(d) $f(x) = 2x^2 + 3x - 2$ (e) $f(x) = 4x^2 + 12x + 9$ (f) $y = 2x^2 + 2x - 3$

Solution: Recall that to graph an equation of the form $y = f(x) = ax^2 + bx + c$, you find the vertex, x-intercept(s), y-intercept, and other solutions if necessary, plot the ordered pairs and draw a smooth curve through the points as before [see Example 13-9].

(a) $y = x^2 - 2$

$a = 1, b = 0, c = -2$

$(0, -2)$ is the vertex.

$0 = x^2 - 2$

$x = \pm\sqrt{2}$ or ± 1.414

$(-1.414, 0)$ and $(1.414, 0)$

$y = 0^2 - 2$

$(0, -2)$

x	$x^2 - 2 = y$
2	$2^2 - 2 = 2$
1	$1^2 - 2 = -1$
-1	$1 - 2 = -1$
-2	$4 - 2 = 2$

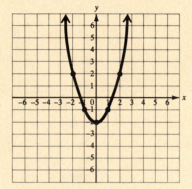

(b) $y = x^2 + 2x$

$a = 1, b = 2, c = 0$

$(-1, -1)$ is the vertex.

$0 = x^2 + 2x$

$0 = x(x + 2)$

$x = 0$ or -2

$(0, 0)$ and $(-2, 0)$

$y = 0^2 + 2(0)$

$y = 0$

$(0, 0)$

x	$x^2 + 2x = y$
1	$1 + 2 = 3$
-3	$9 - 6 = 3$

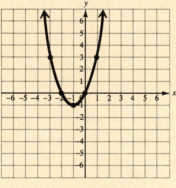

(c) $y = -2x^2 + 2x$

$a = -2, b = 2, c = 0$

$(\frac{1}{2}, \frac{1}{2})$ is the vertex.

$0 = -2x^2 + 2x$

$0 = -2x(x - 1)$

$x = 0$ or 1

$(0, 0)$ and $(1, 0)$

$y = -2(0)^2 + 2(0)$

$y = 0$

$(0, 0)$

x	$-2x^2 + 2x = y$
2	$-8 + 4 = -4$
-1	$-2 - 2 = -4$

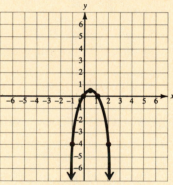

(d) $f(x) = 2x^2 + 3x - 2$

$a = 2, b = 3, c = -2$

$(-\frac{3}{4}, -\frac{25}{8})$ is the vertex.

$0 = 2x^2 + 3x - 2$

$0 = (2x - 1)(x + 2)$

$x = \frac{1}{2}$ or -2

$(-2, 0)$ and $(\frac{1}{2}, 0)$

$f(x) = 2(0)^2 + 3(0) - 2$

$f(x) = -2$

$(0, -2)$

x	$2x^2 + 3x - 2 = f(x)$
1	$1 + 3 - 2 = 2$
-1	$2 - 3 - 2 = -3$
-3	$18 - 9 - 2 = 7$

(e) $f(x) = 4x^2 + 12x + 9$

$a = 4, b = 12, c = 9$

$(-\frac{3}{2}, 0)$ is the vertex.

$0 = 4x^2 + 12x + 9$

$0 = (2x + 3)^2$

$x = -\frac{3}{2}$

$(-\frac{3}{2}, 0)$

$f(x) = 4(0)^2 + 12(0) + 9$

$f(x) = 9$

$(0, 9)$

x	$4x^2 + 12x + 9 = f(x)$
-1	$4 - 12 + 9 = 1$
-2	$16 - 24 + 9 = 1$

(f) $y = 2x^2 + 2x - 3$

$a = 2, b = 2, c = -3$

$(-\frac{1}{2}, -\frac{7}{2})$ is the vertex.

$0 = 2x^2 + 2x - 3$

$x = \dfrac{-2 \pm \sqrt{4 + 24}}{4}$

≈ -1.823 or 0.823

$\approx (-1.823, 0)$ and $(0.823, 0)$

$y = 2(0)^2 + 2(0) - 3$

$y = -3$

$(0, -3)$

x	$2x^2 + 2x - 3 = y$
1	$2 + 2 - 3 = 1$
-1	$2 - 2 - 3 = -3$
-2	$8 - 4 - 3 = 1$

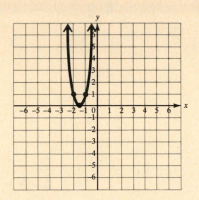
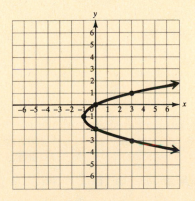
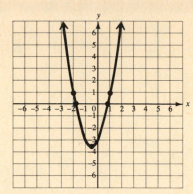

PROBLEM 13-9 Graph the following equations of the form $x = ax^2 + bx + c$:

(a) $x = y^2$ **(b)** $x = y^2 + 2y$ **(c)** $x = 2y^2 + 3y - 2$

(d) $x = y^2 - 4y + 4$ **(e)** $x = 2y^2 + y - 1$ **(f)** $x = 2y^2 - 5y + 2$

Solution: Recall that to graph an equation of the form $x = ay^2 + by + c$, you graph by the same method as before [see Example 13-11].

(a) $x = y^2$

   $a = 1, b = 0, c = 0$

$$\left(\frac{4ac - b^2}{4a}, -\frac{b}{2a}\right)$$

   $(0, 0)$ is the vertex.

   $(0, 0)$ is the x-intercept.

   $(0, 0)$ is the y-intercept.

y	$y^2 = x$
2	$2^2 = 4$
1	$1^2 = 1$
-1	$1 = 1$
-2	$4 = 4$

(b) $x = y^2 + 2y$

   $a = 1, b = 2, c = 0$

$$\left(\frac{4ac - b^2}{4a}, -\frac{b}{2a}\right)$$

   $(-1, -1)$ is the vertex.

   $(0, 0)$ is the x-intercept.

   $(0, 0)$ and $(0, -2)$ are the y-intercepts.

y	$y^2 + 2y = x$
1	$1 + 2 = 3$
-3	$9 - 6 = 3$

(c) $x = 2y^2 - 3y - 2$

   $a = 2, b = -3, c = -2$

$$\left(\frac{4ac - b^2}{4a}, -\frac{b}{2a}\right)$$

   $\left(-\frac{25}{8}, \frac{3}{4}\right)$ is the vertex.

   $(-2, 0)$ is the x-intercept.

   $(0, -\frac{1}{2})$ and $(0, 2)$ are the y-intercepts.

y	$2y^2 - 3y - 2 = x$
3	$18 - 9 - 2 = 7$
1	$2 - 3 - 2 = -3$
-1	$2 + 3 - 2 = 3$

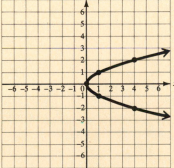
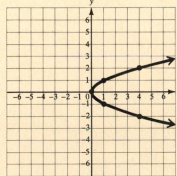
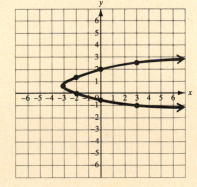

(d) $x = y^2 - 4y + 4$

   $a = 1, b = -4, c = 4$

   $(0, 2)$ is the vertex.

   $(4, 0)$ is the x-intercept.

   $(0, 2)$ is the y-intercept.

(e) $x = 2y^2 + y - 1$

   $a = 2, b = 1, c = -1$

   $\left(-\frac{9}{8}, -\frac{1}{4}\right)$ is the vertex.

   $(-1, 0)$ is the x-intercept.

   $(0, -1)$ and $(0, \frac{1}{2})$ are the y-intercepts.

(f) $x = 2y^2 - 5y + 2$

   $a = 2, b = -5, c = 2$

   $\left(-\frac{9}{8}, \frac{5}{4}\right)$ is the vertex.

   $(2, 0)$ is the x-intercept.

   $(0, \frac{1}{2})$ and $(0, 2)$ are the y-intercepts.

y	$y^2 - 4y + 4 = x$
1	$1 - 4 + 4 = 1$
3	$9 - 12 + 4 = 1$
4	$16 - 16 + 4 = 4$

y	$2y^2 + y - 1 = x$
-2	$8 - 2 - 1 = 5$
1	$2 + 1 - 1 = 2$
2	$8 + 2 - 1 = 9$

y	$2y^2 - 5y + 2 = x$
3	$18 - 15 + 2 = 5$
1	$2 - 5 + 2 = -1$
-1	$2 + 5 + 2 = 9$

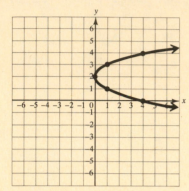
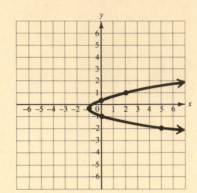
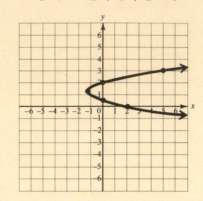

PROBLEM 13-10 Solve the following problem by finding the maximum or minimum value:

Peter has 16 m of fencing to enclose a garden. What is the area of the garden if it is rectangular and each diagonal is as short as possible?

Solution: Recall that to solve a problem by finding the maximum or minimum value, you:

1. *Read* the problem carefully, several times.
2. *Draw a picture* (when appropriate) to help visualize the problem.
3. *Identify* the unknown measures.
4. *Decide* how to represent the unknown measures using different variables.
5. *Translate* to a quadratic equation using geometry formulas [see Appendix Table 3].
6. *Find the vertex* of the quadratic equation using $\left(-\dfrac{b}{2a}, \dfrac{4ac - b^2}{4a}\right)$.
7. *Interpret* the vertex with respect to the unknown measures to find the proposed maximum or minimum value.
8. *Check* to see if the proposed maximum or minimum value satisfies all the conditions of the original problem [see Example 13-12].

Draw a picture:

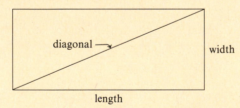

Identify:

The unknown measures are $\left\{\begin{array}{l}\text{the length of the rectangular garden}\\ \text{the width of the rectangular garden}\\ \text{the diagonal length of the rectangular garden}\end{array}\right\}$.

Decide:

Let $l =$ the length of the rectangular garden,

and $w =$ the width of the rectangular garden,

and $d =$ the diagonal length of the rectangular garden.

Translate:

$P = 2l + 2w$ ⟵ perimeter formula for a rectangle

$16 = 2l + 2w$ Substitute 16 (m) for P.

$16 - 2w = 2l$ Solve for one of the remaining variables.

$8 - w = l$ ⟵ solved for l

$$c^2 = a^2 + b^2 \longleftarrow \text{Pythagorean theorem}$$
$$\downarrow \quad \downarrow \quad \downarrow$$
$$d^2 = l^2 + w^2 \qquad \text{Substitute: } a = l, b = w, \text{ and } c = d \text{ (see picture).}$$
$$d^2 = (8 - w)^2 + w^2 \qquad \text{Substitute } 8 - w \text{ for } l.$$
$$d^2 = (64 - 16w + w^2) + w^2 \qquad \text{Write standard form.}$$
$$d^2 = 2w^2 - 16w + 64 \longleftarrow \text{standard form } (d^2 = aw^2 + bw + c)$$

Find the vertex:
$$\text{vertex} = \left(-\frac{b}{2a}, \frac{4ac - b^2}{4a}\right)$$
$$= \left(-\frac{-16}{2(2)}, \frac{4(2)(64) - (-16)^2}{4(2)}\right) \qquad \text{Substitute: } a = 2, b = -16, \text{ and } c = 64.$$
$$= (4, 32) \longleftarrow (w, d^2)$$

Interpret: In $d^2 = 2w^2 - 16w + 64$, $a = 2 > 0$ means the vertex $(4, 32)$ is a minimum value.
$(w, d^2) = (4, 32)$ means a width (w) of 4 m will minimize the diagonal length (d) at $\sqrt{32}$ m $[d^2 = 32$ means $d = \pm\sqrt{32}]$.
$l = 8 - w = 8 - 4 = 4$ means the length (l) of the rectangle with the minimum diagonal length is also 4 m.

Check: Is the perimeter of the rectangle with a minimum diagonal length 16 m? Yes:
$P = 2l + 2w = 2(4) + 2(4) = 16$.
Does the Pythagorean theorem hold for the proposed length, width, and minimum diagonal length? Yes:

$$c^2 = a^2 + b^2$$

$(\sqrt{32})^2$	$(4)^2 + (4)^2$
32	$16 + 16$
32	$32 \longleftarrow$ checks

PROBLEM 13-11 Write an equation of a circle with the following center and radius:

(a) $(0, 1), r = 1$ (b) $(-1, 0), r = 2$ (c) $(-2, -3), r = 3$
(d) $(1, -4), r = 4$ (e) $(-3, 1), r = \sqrt{2}$ (f) $(2, -1), r = \sqrt{3}$

Solution: Recall that to write an equation of a circle, you substitute for (h, k) and r in the equation $(x - h)^2 + (y - k)^2 = r^2$ [see Example 13-13].

(a) $(0, 1), r = 1$
$$(x - h)^2 + (y - k)^2 = r^2$$
$$(x - 0)^2 + (y - 1)^2 = 1^2$$
$$x^2 + (y - 1)^2 = 1$$

(b) $(-1, 0), r = 2$
$$(x - h)^2 + (y - k)^2 = r^2$$
$$[x - (-1)]^2 + (y - 0)^2 = 2^2$$
$$(x + 1)^2 + y^2 = 4$$

(c) $(-2, -3), r = 3$
$$(x - h)^2 + (y - k)^2 = r^2$$
$$[x - (-2)]^2 + [y - (-3)]^2 = 3^2$$
$$(x + 2)^2 + (y + 3)^2 = 9$$

(d) $(1, -4), r = 4$
$$(x - h)^2 + (y - k)^2 = r^2$$
$$(x - 1)^2 + [y - (-4)]^2 = 4^2$$
$$(x - 1)^2 + (y + 4)^2 = 16$$

(e) $(-3, 1), r = \sqrt{2}$
$$(x - h)^2 + (y - k)^2 = (\sqrt{2})^2$$
$$[x - (-3)]^2 + (y - 1)^2 = 2$$
$$(x + 3)^2 + (y - 1)^2 = 2$$

(f) $(2, -1), r = \sqrt{3}$
$$(x - h)^2 + (y - k)^2 = (\sqrt{3})^2$$
$$(x - 2)^2 + [y - (-1)]^2 = 3$$
$$(x - 2)^2 + (y + 1)^2 = 3$$

PROBLEM 13-12 Find the center and radius of each of the following circles:

(a) $(x - 3)^2 + y^2 = 9$ (b) $(x + 2)^2 + (y - 1)^2 = 4$ (c) $(x + 1)^2 + y^2 = 8$
(d) $x^2 + y^2 - 2x + 2y = 2$ (e) $x^2 + y^2 + 4x - 6y = 12$ (f) $x^2 + y^2 - 6x + 4y = -12$

Solution: Recall that to find the center and radius of a circle, you write the equation in standard form and identify (h, k) and r [see Example 13-14].

   (a) $(x - 3)^2 + y^2 = 9$ (b) $(x + 2)^2 + (y - 1)^2 = 4$

       $(x - 3)^2 + (y - 0)^2 = 3^2$ $[x - (-2)]^2 + (y - 1)^2 = 2^2$

       $(3, 0), r = 3$ $(-2, 1), r = 2$

   (c) $(x + 1)^2 + y^2 = 8$ (d) $x^2 - 2x + y^2 + 2y = 2$

       $[x - (-1)]^2 + (y - 0)^2 = 8^2$ $(x^2 - 2x + 1) + (y^2 + 2y + 1) = 4$

       $(-1, 0), r = 2\sqrt{2}$ $(x - 1)^2 + (y + 1)^2 = 2^2$

                                     $(x - 1)^2 + [y - (-1)]^2 = 2^2$

                                     $(1, -1), r = 2$

   (e) $x^2 + 4x + y^2 - 6y = 12$ (f) $x^2 - 6x + y^2 + 4y = -12$

       $(x + 2)^2 + (y - 3)^2 = 25$ $(x - 3)^2 + (y + 2)^2 = 1$

       $[x - (-2)]^2 + (y - 3)^2 = 5^2$ $(3, -2), r = 1$

       $(-2, 3), r = 5$

PROBLEM 13-13 Graph the following equations of circles:

(a) $x^2 + y^2 = 9$ (b) $(x - 3)^2 + (y + 2)^2 = 4$ (c) $x^2 + (y + 3)^2 = 9$
(d) $x^2 + y^2 + 6y = 0$ (e) $x^2 + y^2 - 4x + 2y - 4 = 0$ (f) $x^2 + y^2 + 6x - 2y + 6 = 0$

Solution: Recall that to graph a circle, you identify the center and radius, plot the center and draw the set of all points at distance r from the center [see Example 13-15].

(a) $x^2 + y^2 = 9$ (b) $(x - 3)^2 + (y + 2)^2 = 4$ (c) $x^2 + (y + 3)^2 = 9$

   $(x - 0)^2 + (y - 0)^2 = 3^2$ $(x - 3)^2 + [y - (-2)]^2 = 2^2$ $(x - 0)^2 + [y - (-3)]^2 = 3^2$

   $(0, 0), r = 3$ $(3, -2), r = 2$ $(0, -3), r = 3$

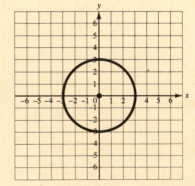   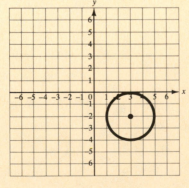   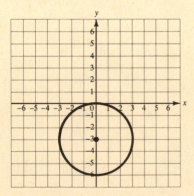

(d) $x^2 + y^2 + 6y = 0$ (e) $x^2 - 4x + y^2 + 2y = 4$ (f) $x^2 + 6x + y^2 - 2y = -6$

   $(x + 0)^2 + (y + 3)^2 = 9$ $(x - 2)^2 + (y + 1)^2 = 9$ $(x + 3)^2 + (y - 1)^2 = 4$

   $(0, -3), r = 3$ $(2, -1), r = 3$ $(-3, 1), r = 2$

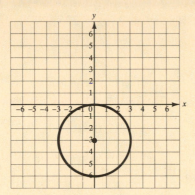 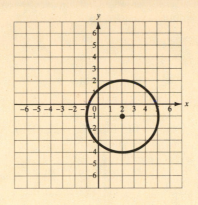 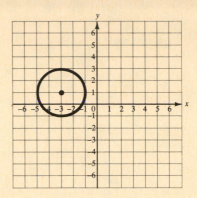

PROBLEM 13-14 Write an equation for each of the following ellipses with center at $(0,0)$ and intercepts at:

(a) $(-2,0),(2,0),(0,-3),(0,3)$ **(b)** $(-4,0),(4,0),(0,-1),(0,1)$ **(c)** $(-3,0),(3,0),(0,-2),(0,2)$

Solution: Recall that to write an equation for an ellipse, you identify a and b and substitute in the general equation of an ellipse with center at $(0,0)$ [see Example 13-16].

(a) $\dfrac{x^2}{a^2}+\dfrac{y^2}{b^2}=1$ **(b)** $\dfrac{x^2}{a^2}+\dfrac{y^2}{b^2}=1$ **(c)** $\dfrac{x^2}{a^2}+\dfrac{y^2}{b^2}=1$

$a=2, b=3$ $a=4, b=1$ $a=3, b=2$

$\dfrac{x^2}{2^2}+\dfrac{y^2}{3^2}=1$ $\dfrac{x^2}{4^2}+\dfrac{y^2}{1^2}=1$ $\dfrac{x^2}{3^2}+\dfrac{y^2}{2^2}=1$

PROBLEM 13-15 Graph the following equations:

(a) $\dfrac{x^2}{2^2}+\dfrac{y^2}{3^2}=1$ **(b)** $\dfrac{x^2}{4}+\dfrac{y^2}{1}=1$ **(c)** $\dfrac{x^2}{9}+\dfrac{y^2}{16}=1$

(d) $x^2+9y^2=9$ **(e)** $4x^2+y^2=4$ **(f)** $9x^2+25y^2=225$

Solution: Recall that to graph an equation of an ellipse, you identify and plot the intercepts and draw a smooth curve through the points [see Example 13-17].

(a) $\dfrac{x^2}{2^2}+\dfrac{y^2}{3^2}=1$ **(b)** $\dfrac{x^2}{2^2}+\dfrac{y^2}{1^2}=1$ **(c)** $\dfrac{x^2}{3^2}+\dfrac{y^2}{4^2}=1$

$a=2, b=3$ $a=2, b=1$ $a=3, b=4$

$(-2,0),(2,0),(0,-3),(0,3)$ $(-2,0),(2,0),(0,-1),(0,1)$ $(-3,0),(3,0),(0,-4),(0,4)$

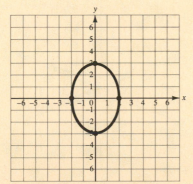 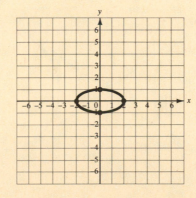 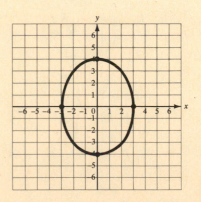

(d) $x^2 + 9y^2 = 9$

$$\frac{x^2}{9} + \frac{y^2}{1} = 1$$

$$\frac{x^2}{3^2} + \frac{y^2}{1^2} = 1$$

$(-3, 0), (3, 0), (0, -1), (0, 1)$

(e) $4x^2 + y^2 = 4$

$$\frac{x^2}{1} + \frac{y^2}{4} = 1$$

$$\frac{x^2}{1^2} + \frac{y^2}{2^2} = 1$$

$(-1, 0), (1, 0), (0, -2), (0, 2)$

(f) $9x^2 + 25y^2 = 225$

$$\frac{9x^2}{225} + \frac{25y^2}{225} = 1$$

$$\frac{x^2}{5^2} + \frac{y^2}{3^2} = 1$$

$(-5, 0), (5, 0), (0, -3), (0, 3)$

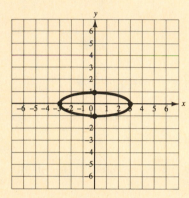
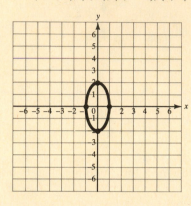
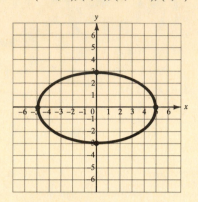

PROBLEM 13-16 Graph the following equations of hyperbolas:

(a) $\dfrac{x^2}{4} - \dfrac{y^2}{25} = 1$ **(b)** $\dfrac{x^2}{16} - \dfrac{y^2}{9} = 1$ **(c)** $\dfrac{y^2}{1} - \dfrac{x^2}{9} = 1$

(d) $\dfrac{y^2}{9} - \dfrac{x^2}{16} = 1$ **(e)** $9x^2 - 16y^2 = 144$ **(f)** $y^2 - x^2 = 16$

Solution: Recall that to graph a hyperbola, you identify a and b, graph the vertices and asymptotes and sketch the curve [see Example 13-18].

(a) $\dfrac{x^2}{2^2} - \dfrac{y^2}{5^2} = 1$

$a = 2, b = 5$

$y = -\dfrac{b}{a}x$ or $y = \dfrac{b}{a}x$

$y = -\frac{5}{2}x$ or $y = \frac{5}{2}x$

$(-2, 0), (2, 0)$ are vertices.

(b) $\dfrac{x^2}{4^2} - \dfrac{y^2}{3^2} = 1$

$a = 4, b = 3$

$y = -\dfrac{b}{a}x$ or $y = \dfrac{b}{a}x$

$y = -\frac{3}{4}x$ or $y = \frac{3}{4}x$

$(-4, 0), (4, 0)$ are vertices.

(c) $\dfrac{y^2}{1^2} - \dfrac{x^2}{3^2} = 1$

$a = 3, b = 1$

$y = -\dfrac{b}{a}x$ or $y = \dfrac{b}{a}x$

$y = -\frac{1}{3}x$ or $y = \frac{1}{3}x$

$(0, -1), (0, 1)$ are vertices.

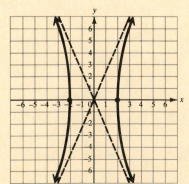
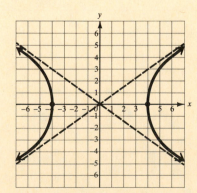
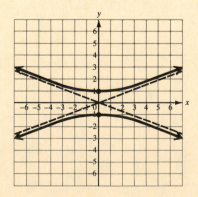

(d) $\dfrac{y^2}{3^2} - \dfrac{x^2}{4^2} = 1$

$a = 4, b = 3$

$y = -\dfrac{b}{a}x$ or $y = \dfrac{b}{a}x$

$y = -\tfrac{3}{4}x$ or $y = \tfrac{3}{4}x$

$(0, -3), (0, 3)$ are vertices.

(e) $9x^2 - 16y^2 = 144$

$\dfrac{x^2}{16} - \dfrac{y^2}{9} = 1$

$a = 4, b = 3$

$y = -\dfrac{b}{a}x$ or $y = \dfrac{b}{a}x$

$y = -\tfrac{3}{4}x$ or $y = \tfrac{3}{4}x$

$(-4, 0), (4, 0)$ are vertices.

(f) $-x^2 + y^2 = 16$

$-\dfrac{x^2}{16} + \dfrac{y^2}{16} = 1$

$a = 4, b = 4$

$y = -\dfrac{b}{a}x$ or $y = \dfrac{b}{a}x$

$y = -x$ or $y = x$

$(0, -4), (0, 4)$ are vertices.

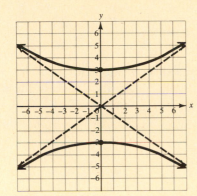

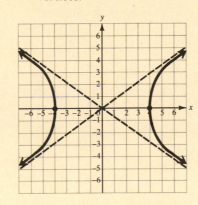

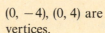
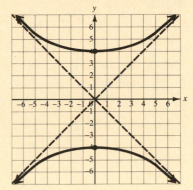

PROBLEM 13-17 Graph the following equations of the form $xy = k$:

(a) $xy = 8$ **(b)** $xy = -6$ **(c)** $xy = -2$ **(d)** $xy + x = 6$

Solution: Recall that to graph an equation of the form $xy = k$, you make a table of solutions, plot the ordered pairs and draw a smooth curve through the points [see Example 13-19].

(a) $xy = 8$

x	8	4	2	1	-1	-2	-4	-8
y	1	2	4	8	-8	-4	-2	-1

(b) $xy = -6$

x	-6	-3	-2	-1	1	2	3	6
y	1	2	3	6	-6	-3	-2	-1

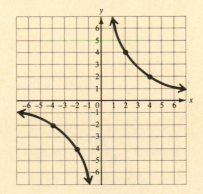

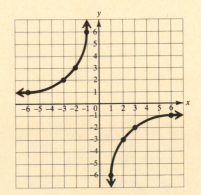

(c) $xy = -2$

x	$\tfrac{1}{2}$	1	2	4	-4	-2	-1	$-\tfrac{1}{2}$
y	-4	-2	-1	$-\tfrac{1}{2}$	$\tfrac{1}{2}$	1	2	4

(d) $x(y + 1) = 6$

x	6	3	2	1	-1	-2	-3	-6
y	0	1	2	5	-7	-4	-3	-2

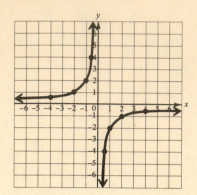 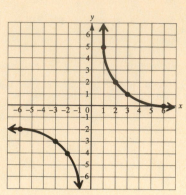

PROBLEM 13-18 Graph the following quadratic inequalities:

(a) $y \leq 2x^2 - 3x - 2$ **(b)** $x^2 + y^2 > 9$ **(c)** $9x^2 + 4y^2 \leq 36$

(d) $x^2 > y^2 + 1$ **(e)** $9x^2 + 36 \geq 4y^2$ **(f)** $xy < 6$

Solution: Recall that to graph a quadratic inequality, you graph the associated equation and use a test point or test points to determine which region or regions satisfy the inequality and shade the solution sets. You use a solid curve or broken curve as before [see Example 13-20].

(a) $y \leq 2x^2 - 3x - 2$

$(-\frac{1}{2}, 0), (2, 0)$ x-intercepts

$(\frac{3}{4}, -\frac{25}{8})$ vertex

Test point $(0, 0)$

(b) $x^2 + y^2 > 9$

$(0, 0), r = 3$

Test point $(0, 0)$

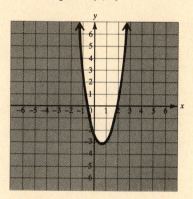 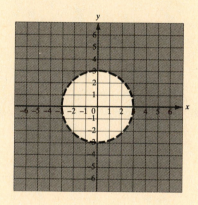

(c) $\dfrac{x^2}{4} + \dfrac{y^2}{9} \leq 1$

$(-2, 0), (2, 0), (0, -3), (0, 3)$ are vertices.

Test point $(0, 0)$

(d) $x^2 > y^2 + 1$

$\dfrac{x^2}{1^2} - \dfrac{y^2}{1^2} = 1$

$(-1, 0), (1, 0), (0, -1), (0, 1)$ are vertices.

Test point $(0, 0)$

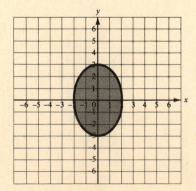 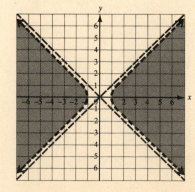

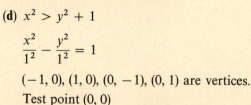

(e) $-\dfrac{x^2}{4} + \dfrac{y^2}{9} \le 1$

(0, −3), (0, 3) are vertices.

$y = -\frac{3}{2}x$ and $y = \frac{3}{2}x$

Test point (0, 0)

(f) $xy < 6$

x	6	3	2	1	−1	−2	−3	−6
y	1	2	3	6	−6	−3	−2	−1

Test point (0, 0)

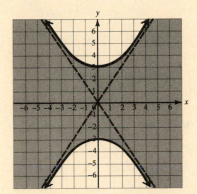

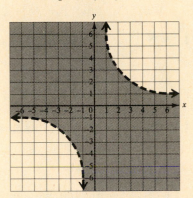

Supplementary Exercises

PROBLEM 13-19 Find the length of the hypotenuse of a right triangle having the following pair of legs: **(a)** 9, 12   **(b)** 10, 24   **(c)** 2, 3   **(d)** 2, 2   **(e)** 3, 5   **(f)** 4, 10

PROBLEM 13-20 Find the distance between the following pairs of points:

(a) (1, 2), (5, 5)   **(b)** (3, −2), (−1, 1)   **(c)** (−2, 1), (2, −2)
(d) (−1, 0), (1, −3)   **(e)** (−3, 2), (2, 2)   **(f)** (−2, 4), (−2, −1)

PROBLEM 13-21 Solve each problem using the Pythagorean theorem:

(a) A pilot takes off to fly 100 miles due east. The pilot takes a wrong course and flys in a straight line to land at an airfield that is 20 miles due north of the planned destination. How far was the actual flight, to the nearest whole mile?

(b) Two planes take off at the same time from the same airport and travel at constant speeds that differ by 100 km/h. The faster plane travels due east and the slower plane travels due south. In 1 hour, the two planes are 500 km apart. What was the speed of the slower plane? How far east did the faster plane travel?

(c) A square has a perimeter of 12 cm. Find the length of the diagonal.

(d) A square has a 12-cm diagonal. Find its perimeter.

(e) The hypotenuse of a right triangle is 5 m longer than one leg and 2 m longer than the other leg. Find the perimeter of the right triangle.

(f) The diagonal of a rectangle is 3 ft longer than its length and 5 ft longer than its width. What is the area of the rectangle?

PROBLEM 13-22 Graph the following quadratic functions:

(a) $y = 3x^2$ **(b)** $y = x^2 + 1$ **(c)** $f(x) = (x - 2)^2 + 1$
(d) $f(x) = -x^2 + 3$ **(e)** $y = -\frac{1}{3}x^2$ **(f)** $f(x) = (x + 3)^2 - 3$

PROBLEM 13-23 Find the *y*-intercept of the graph of each of the following equations:

(a) $y = x^2 - 2$ (b) $y = 2x^2 + 3x$ (c) $y = 4x^2$
(d) $y = 3x^2 - 4x + 3$ (e) $y = 4x^2 + 3x - 5$ (f) $y = -3x^2 + 2x + 4$

PROBLEM 13-24 Find the *x*-intercept(s) of the graph of each of the following equations:

(a) $y = x^2$ (b) $y = x^2 - 2x + 1$ (c) $y = 3x^2 + x - 2$
(d) $y = 4x^2 - 13x + 3$ (e) $y = x^2 + 2$ (f) $y = -x^2 + 2x + 3$

PROBLEM 13-25 Find the vertex and axis of symmetry of the graphs of each of the following equations:

(a) $y = x^2 + 6x + 9$ (b) $y = -x^2 + 2x - 1$ (c) $y = 2x^2 - 3x + 2$
(d) $y = 3x^2 - 5x - 2$ (e) $y = -2x^2 + 3x - 1$ (f) $y = x^2 + x + 1$

PROBLEM 13-26 Graph the following quadratic equations:

(a) $y = x^2 + 3$ (b) $y = x^2 - 4x$ (c) $y = x^2 - 3x - 4$
(d) $y = -2x^2 - x + 6$ (e) $y = x^2 - x - 6$ (f) $y = -2x^2 + 9x - 9$

PROBLEM 13-27 Graph the following quadratic equations:

(a) $x = y^2 + 2$ (b) $x = y^2 - 2y$ (c) $x = y^2 - 2y + 1$
(d) $x = 2y^2 - 5y + 2$ (e) $x = -y^2 - y + 2$ (f) $x = -2y^2 + 7y - 6$

PROBLEM 13-28 Solve each problem by finding the maximum or minimum value:

(a) What is the largest rectangular field that can be enclosed with 8 km of fencing if a straight river is used as one side instead of fencing.

(b) What is the largest rectangular field that can be enclosed with 8 km of fencing if a straight river is used as one side instead of fencing and the diagonal length is minimized?

(c) The height (*h*) of a given object after a given time (*t*) thrown vertically upward is given by $h = -16t^2 + v_0 t + h_0$ where v_0 is the *initial velocity* and h_0 is the *initial height*. Find the maximum height of a ball that is thrown upward with an initial velocity of 64 fps (feet per second) from 4 feet above the ground. How long does it take the ball to reach its maximum height? How long does it take the ball to hit the ground?

(d) How should a piece of wire that is 2 yards long be cut into two pieces so that when each piece is bent into the shape of a square the sum of the areas will be minimized?

PROBLEM 13-29 Write an equation of a circle with the following center and radius:

(a) $(0, 2), r = 2$ (b) $(-2, 0), r = 2$ (c) $(1, -2), r = 3$
(d) $(-3, 2), r = 3$ (e) $(-1, -3), r = \sqrt{5}$ (f) $(4, -2), r = \sqrt{6}$

PROBLEM 13-30 Find the center and radius of each of the following circles:

(a) $x^2 + y^2 = 9$ (b) $(x - 2)^2 + (y - 1)^2 = 4$ (c) $(x + 3)^2 + (y + 1)^2 = 1$
(d) $(x - 1)^2 + (y + 3)^2 = 9$ (e) $x^2 + y^2 + 4x = 0$ (f) $x^2 + y^2 + 6x - 2y = -9$

PROBLEM 13-31 Graph the following equations of circles:

(a) $(x - 2)^2 + (y - 1)^2 = 9$ (b) $(x + 2)^2 + (y + 3)^2 = 4$ (c) $(x + 3)^2 + y^2 = 4$
(d) $x^2 + (y + 2)^2 = 9$ (e) $x^2 + y^2 - 6x = 0$ (f) $x^2 + y^2 + 2x - 4y = 4$

PROBLEM 13-32 Write an equation for each of the following ellipses with center at $(0, 0)$ and intercepts at:

(a) $(-1, 0), (1, 0), (0, -2), (0, 2)$ (b) $(-2, 0), (2, 0), (0, -1), (0, 1)$ (c) $(-3, 0), (3, 0), (0, -4), (0, 4)$
(d) $(-4, 0), (4, 0), (0, -2), (0, 2)$ (e) $(-2, 0), (2, 0), (0, -4), (0, 4)$ (f) $(-1, 0), (1, 0), (0, -3), (0, 3)$

PROBLEM 13-33 Graph the following equations of ellipses:

(a) $\dfrac{x^2}{3^2} + \dfrac{y^2}{1^2} = 1$ (b) $\dfrac{x^2}{4} + \dfrac{y^2}{16} = 1$ (c) $\dfrac{x^2}{9} + \dfrac{y^2}{4} = 1$

(d) $x^2 + 9y^2 = 9$ (e) $16x^2 + 9y^2 = 144$ (f) $4x^2 + 16y^2 = 64$

PROBLEM 13-34 Graph the following equations of hyperbolas:

(a) $\dfrac{x^2}{2^2} - \dfrac{y^2}{4^2} = 1$ (b) $\dfrac{x^2}{25} - \dfrac{y^2}{16} = 1$ (c) $-\dfrac{x^2}{4} + \dfrac{y^2}{9} = 1$

(d) $-\dfrac{x^2}{25} + \dfrac{y^2}{9} = 1$ (e) $x^2 - y^2 = 9$ (f) $-9x^2 + 4y^2 = 36$

PROBLEM 13-35 Graph the following equations of the form $xy = k$:

(a) $xy = 3$ (b) $xy = -3$ (c) $xy = 12$
(d) $xy = -12$ (e) $xy = -8$ (f) $xy = 2$

PROBLEM 13-36 Graph the following quadratic inequalities:

(a) $y \geq x^2$ (b) $x \leq 2y^2 - 3y - 2$ (c) $x^2 + y^2 < 4$
(d) $x^2 + 9y^2 \geq 36$ (e) $4x^2 - 16y^2 \leq 64$ (f) $xy \geq 4$

Answers to Supplementary Exercises

(13-19) (a) 15 (b) 26 (c) $\sqrt{13}$ (d) $2\sqrt{2}$ (e) $\sqrt{34}$ (f) $2\sqrt{29}$

(13-20) (a) 5 (b) 5 (c) 5 (d) 13 (e) 5 (f) 5

(13-21) (a) 102 mi (b) 300 km/h, 400 km (c) $(3\sqrt{2})$ cm (d) $(24\sqrt{2})$
(e) $(14 + 6\sqrt{5})$ m (f) $(45 + 8\sqrt{30})$ ft^2

(13-22)

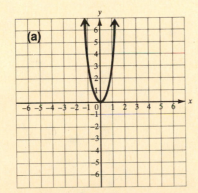

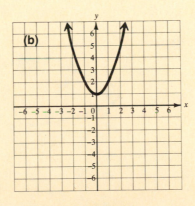

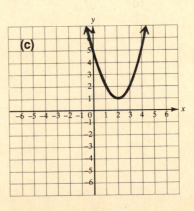

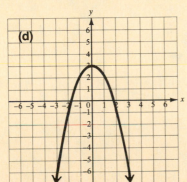

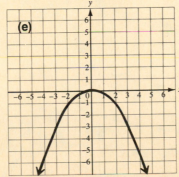

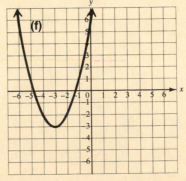

(13-23) **(a)** $(0, -2)$ **(b)** $(0, 0)$ **(c)** $(0, 0)$ **(d)** $(0, 3)$ **(e)** $(0, -5)$ **(f)** $(0, 4)$

(13-24) **(a)** $(0, 0)$ **(b)** $(1, 0)$ **(c)** $(-1, 0), (\frac{2}{3}, 0)$ **(d)** $(\frac{1}{4}, 0), (3, 0)$ **(e)** none
(f) $(-1, 0), (3, 0)$

(13-25) **(a)** $(-3, 0), x = -3$ **(b)** $(1, 0), x = 1$ **(c)** $(\frac{3}{4}, \frac{7}{8}), x = \frac{3}{4}$ **(d)** $(\frac{5}{6}, -\frac{49}{12}), x = \frac{5}{6}$
(e) $(\frac{3}{4}, \frac{1}{8}), x = \frac{3}{4}$ **(f)** $(-\frac{1}{2}, \frac{3}{4}), x = -\frac{1}{2}$

(13-26)

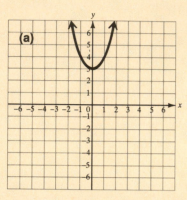

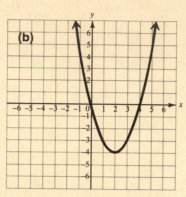

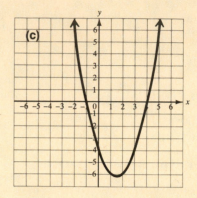

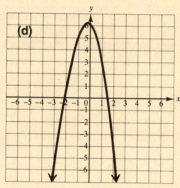

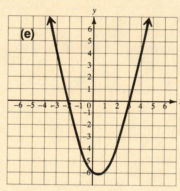

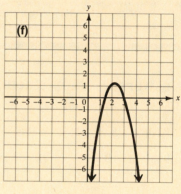

(13-27)

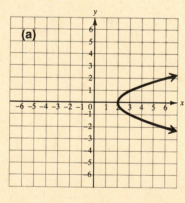

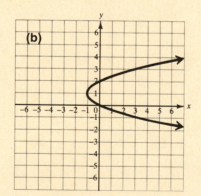

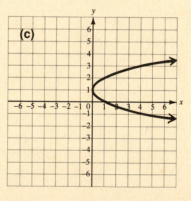

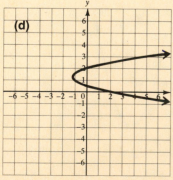

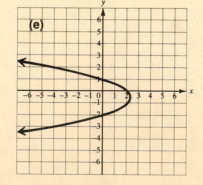

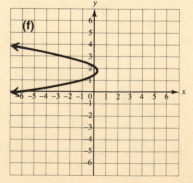

(13-28) **(a)** $l = 4$ km, $w = 2$ km, $A = 8$ km^2 **(b)** $l = 1.6$ km, $w = 3.2$ km, $d = \sqrt{12.8}$ km, $A = 5.12$ km^2

(c) 68 ft, 2 sec, $\left(\dfrac{4 + \sqrt{17}}{2} \text{ sec } [\approx 4.1 \text{ sec}]\right)$

(d) The wire should be cut in half so that each piece is 1 yd long and the area of each is $\frac{1}{16}$ yd^2.

(13-29) **(a)** $x^2 + (y - 2)^2 = 4$ **(b)** $(x + 2)^2 + y^2 = 4$ **(c)** $(x - 1)^2 + (y + 2)^2 = 9$
(d) $(x + 3)^2 + (y - 2)^2 = 9$ **(e)** $(x + 1)^2 + (y + 3)^2 = 5$
(f) $(x - 4)^2 + (y + 2)^2 = 6$

(13-30) **(a)** $(0, 0)$, $r = 3$ **(b)** $(2, 1)$, $r = 2$ **(c)** $(-3, -1)$, $r = 1$ **(d)** $(1, -3)$, $r = 3$
(e) $(-2, 0)$, $r = 2$ **(f)** $(-3, 1)$, $r = 1$

(13-31)

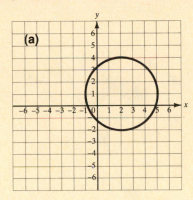

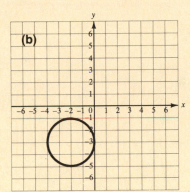

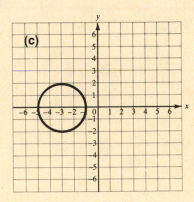

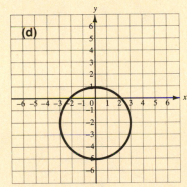

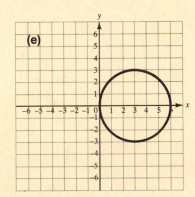

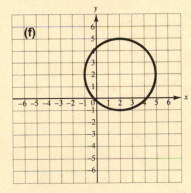

(13-32) **(a)** $\dfrac{x^2}{1} + \dfrac{y^2}{4} = 1$ **(b)** $\dfrac{x^2}{4} + \dfrac{y^2}{1} = 1$ **(c)** $\dfrac{x^2}{9} + \dfrac{y^2}{16} = 1$ **(d)** $\dfrac{x^2}{16} + \dfrac{y^2}{4} = 1$

(e) $\dfrac{x^2}{4} + \dfrac{y^2}{16} = 1$ **(f)** $\dfrac{x^2}{1} + \dfrac{y^2}{9} = 1$

(13-33)

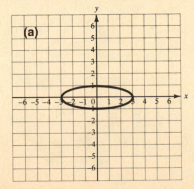

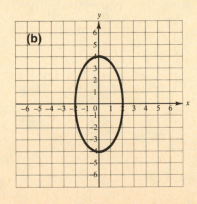

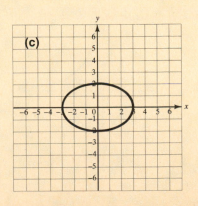

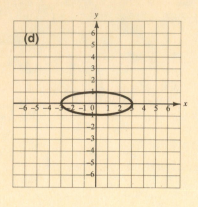

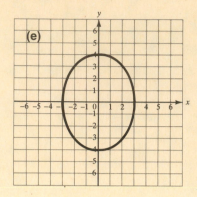

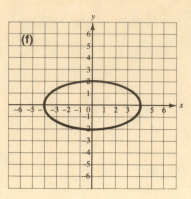

(13-34)

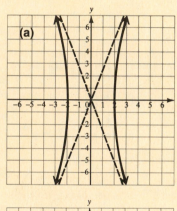

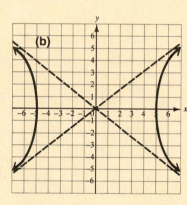

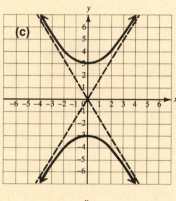

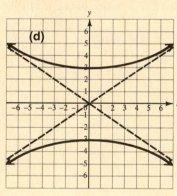

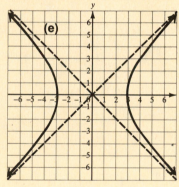

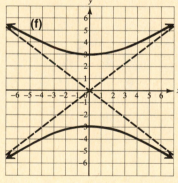

(13-35)

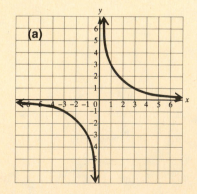

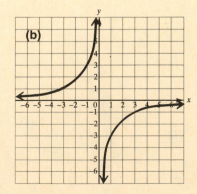

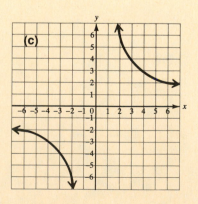

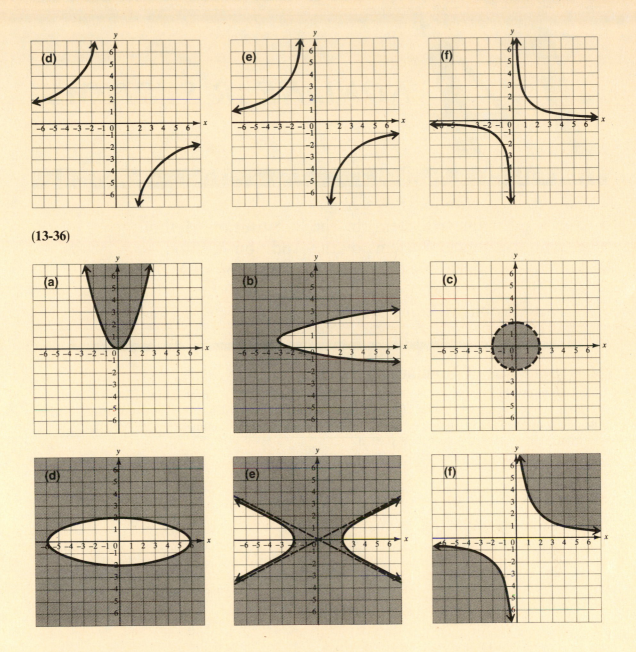

(13-36)

14 NONLINEAR SYSTEMS

THIS CHAPTER IS ABOUT

☑ Solutions of Nonlinear Systems
☑ Solving Nonlinear Systems by the Substitution Method
☑ Solving Nonlinear Systems by the Addition Method
☑ Solving Nonlinear Systems by a Combination of Methods
☑ Graphing Second-degree Systems of Inequalities
☑ Solving Problems Using Nonlinear Systems

14-1. Solutions of Nonlinear Systems

An equation in which some terms have more than one variable, or a variable of degree two or higher, is a nonlinear equation. A system of equations that contains at least one nonlinear equation is called **a nonlinear system of equations.** An ordered pair is a solution of a nonlinear system of equations if it is a solution of each equation of the system.

EXAMPLE 14-1: Determine whether $(2, -1)$ is a solution of:

(a) $\begin{cases} x^2 + y^2 + 4x = 13 \\ x^2 - 9y^2 = 5 \end{cases}$ and **(b)** $\begin{cases} x^2 + y^2 - 2x + 4y + 3 = 0 \\ x^2 - x - y - 3 = 0 \end{cases}$

Solution:

(a)

$$x^2 + y^2 + 4x = 13 \longleftarrow \text{original equations} \longrightarrow x^2 - 9y^2 = 5$$

$$\frac{2^2 + (-1)^2 + 4(2) \mid 13}{4 + 1 + 8 \mid 13} \quad \text{Substitute } (2, -1). \quad \frac{2^2 - 9(-1)^2 \mid 5}{4 - 9 \mid 5}$$

$$13 = 13 \qquad\qquad -5 \neq 5$$

Since $(2, -1)$ does not check in $x^2 - 9y^2 = 5$, it is not a solution of this system.

(b)

$$x^2 + y^2 - 2x + 4y + 3 = 0 \longleftarrow \text{original equations} \longrightarrow x^2 - x - y - 3 = 0$$

$$\frac{2^2 + (-1)^2 - 2(2) + 4(-1) + 3 \mid 0}{4 + 1 - 4 - 4 + 3 \mid 0} \quad \text{Substitute } (2, -1). \quad \frac{2^2 - 2 - (-1) - 3 \mid 0}{4 - 2 + 1 - 3 \mid 0}$$

$$0 = 0 \qquad\qquad 0 = 0$$

Since $(2, -1)$ checks in both equations, it is a solution of this system.

To **determine the number of real solutions** of some nonlinear systems, we graph both equations on the same coordinate system and count the number of points at which the graphs intersect.

EXAMPLE 14-2: Determine the number of real solutions of the systems:

(a) $\begin{cases} x^2 + y^2 = 9 \\ x^2 + y^2 - 6x = 0 \end{cases}$ and **(b)** $\begin{cases} x^2 - 4y^2 = 4 \\ x^2 + y^2 = 9 \end{cases}$

Solution: (a)

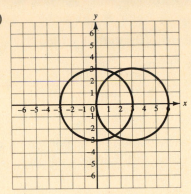

Graph $x^2 + y^2 = 9$.

Graph $x^2 + y^2 - 6x = 0$.

Identify the points of intersection of the graphs.

This system has two real solutions.

(b)

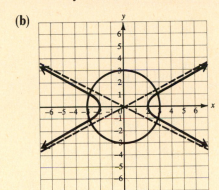

Graph $x^2 - 4y^2 = 4$.

Graph $x^2 + y^2 = 9$.

Identify the points of intersection of the graphs.

This system has four real solutions.

Note: Graphing does not assist you in determining the number of complex solutions of a system.

14-2. Solving Nonlinear Systems by the Substitution Method

To **solve a nonlinear system** by the substitution method, we

1. Solve one of the equations for one of the variables.
2. Substitute the expression from step 1 in the other equation.
3. Solve the resulting equation for one variable.
4. Solve for the other variable by substituting the value obtained in step 3 in the first equation.
5. Check the solution(s) by substituting the ordered pair in both the equations of the system.

A. Solving a system of one linear and one nonlinear equation by the substitution method

It is generally easier to solve the linear equation first for one of the variables, then substitute that expression in the quadratic equation.

EXAMPLE 14-3: Solve the system $\begin{Bmatrix} x^2 + y^2 = 17 \\ x - 4y = 0 \end{Bmatrix}$.

Solution: $x - 4y = 0$ ⟵ linear equation

$x = 4y$ Solve for x.

$x^2 + y^2 = 17$ ⟵ quadratic equation

$(4y)^2 + y^2 = 17$ Substitute $4y$ for x.

$16y^2 + y^2 = 17$ Solve for y.

$17y^2 = 17$

$y^2 = 1$

$$y = \pm 1$$

$$x = 4y$$ Solve for x by substituting ± 1 for y.

$$x = 4(-1) \text{ or } 4(1)$$ Check as before.

The solutions of this system are $(4, 1)$ and $(-4, -1)$.

B. Solving a system of two nonlinear equations

To solve a **system of two nonlinear equations,** we solve one equation for one of the variables, or for the square of one of the variables, substitute that expression in the second equation, and solve as before.

EXAMPLE 14-4: Solve the system $\begin{cases} x^2 + y^2 - 6y = 0 \\ x^2 - y = 0 \end{cases}$.

Solution: $x^2 - y = 0$ Solve for x^2.

$$x^2 = y$$

$$x^2 + y^2 - 6y = 0$$ Substitute y for x^2.

$$y + y^2 - 6y = 0$$ Solve for y.

$$y^2 - 5y = 0$$

$$y(y - 5) = 0$$

$$y = 0 \text{ or } y = 5$$

$$x^2 = 0 \text{ or } 5$$

$$x = 0 \text{ or } \pm\sqrt{5}$$

The solutions are $(0, 0), (-\sqrt{5}, 5), (\sqrt{5}, 5)$. Check as before.

Note: Solve the first equation for a variable, the square of a variable, or a term containing either, depending on which term you wish to substitute for in the second equation.

14-3. Solving Nonlinear Systems by the Addition Method

To **solve a nonlinear system by the addition method,** we

1. Write each equation in standard form.
2. Multiply each equation by the appropriate constants so that the coefficients of one variable are additive inverses of each other.
3. Add the equations to eliminate one variable.
4. Solve for the remaining variable.
5. Substitute that value in one of the original equations to solve for the other variable.
6. Check the solutions by substituting them in the original equations.

A. Solving systems of one linear and one nonlinear equation by the addition method

To solve systems of one linear and one nonlinear equation by the addition method, we align like terms in columns, leaving space where appropriate for nonexisting terms.

EXAMPLE 14-5: Solve the system $\begin{cases} y = x^2 - 2x + 1 \\ x + y = 0 \end{cases}$.

Solution: $\quad x^2 - 2x - y + 1 = 0$ Write equations in standard form, leaving space for nonexisting terms.

$$x + y = 0$$

$$x^2 - 2x - y + 1 = 0$$

$$\underline{\qquad x + y \qquad = 0}$$

$$x^2 - x \qquad + 1 = 0$$ Add the equations to eliminate y.

$$x^2 - x + 1 = 0 \qquad \text{Solve for } x.$$

$$x = \frac{1 \pm \sqrt{1-4}}{2}$$

$$x = \frac{1 \pm i\sqrt{3}}{2}$$

$$x + y = 0$$

$$y = -x \qquad \text{Solve for } y.$$

$$y = -\frac{1 \pm i\sqrt{3}}{2}$$

The solutions are $\left(\dfrac{1+i\sqrt{3}}{2}, -\dfrac{1+i\sqrt{3}}{2}\right)$ Check as before.

and $\left(\dfrac{1-i\sqrt{3}}{2}, -\dfrac{1-i\sqrt{3}}{2}\right)$.

B. Solving systems with two nonlinear equations by the addition method

To solve a system of two nonlinear equations, we follow the steps as before. It is not necessary to set the nonlinear equations equal to zero before adding them.

EXAMPLE 14-6: Solve the system $\begin{Bmatrix} 4x^2 + 9y^2 = 36 \\ 16x^2 - 9y^2 = 144 \end{Bmatrix}$.

Solution:
$$4x^2 + 9y^2 = 36$$
$$\underline{16x^2 - 9y^2 = 144}$$
$$20x^2 \qquad = 180 \qquad \text{Add the equations to eliminate } y^2.$$

$$x^2 = 9 \qquad \text{Solve for } x.$$

$$x = \pm 3$$

$$4(\pm 3)^2 + 9y^2 = 36 \qquad \text{Solve for } y.$$

$$36 + 9y^2 = 36$$

$$9y^2 = 0$$

$$y^2 = 0$$

$$y = 0$$

The solutions are $(-3, 0)$ and $(3, 0)$.   Check as before.

14-4. Solving Nonlinear Systems by a Combination of Methods

To solve some systems, you may need to use both the addition and substitution methods.

EXAMPLE 14-7: Solve the systems (a) $\begin{Bmatrix} x^2 + y^2 - 10y = 0 \\ x^2 + y^2 - 10x = 0 \end{Bmatrix}$ and (b) $\begin{Bmatrix} x^2 - 4xy + y^2 = -2 \\ x^2 + y^2 = 2 \end{Bmatrix}$.

Solution: (a)
$$x^2 + y^2 - 10y = 0$$
$$\underline{-x^2 - y^2 + 10x = 0} \qquad \text{Multiply by } -1 \text{ to get additive inverses.}$$
$$10x - 10y = 0 \qquad \text{Add the equations to eliminate } x^2 \text{ and } y^2.$$

$$10x = 10y$$

$$x = y$$

$$x^2 + y^2 - 10x = 0 \qquad \text{Substitute } x \text{ for } y.$$

$$x^2 + x^2 - 10x = 0 \qquad \text{Solve for } x.$$

$$2x^2 - 10x = 0$$

$$2x(x - 5) = 0$$

$$x = 0 \text{ or } x = 5$$

The solutions are $(0, 0)$ and $(5, 5)$.   Check as before.

(b)
$$x^2 - 4xy + y^2 = -2$$
$$\underline{-x^2 \qquad\quad - y^2 = -2} \qquad \text{Multiply by } -1 \text{ to get additive inverses.}$$
$$- 4xy \qquad\quad = -4 \qquad \text{Add the equations to eliminate } x^2 \text{ and } y^2.$$

$$xy = 1 \qquad \text{Solve for } y.$$

$$y = \frac{1}{x}$$

$$x^2 + y^2 = 2 \qquad \text{Substitute } \frac{1}{x} \text{ for } y.$$

$$x^2 + \left(\frac{1}{x}\right)^2 = 2 \qquad \text{Solve for } x.$$

$$x^2 + \frac{1}{x^2} = 2$$

$$x^4 + 1 = 2x^2$$

$$x^4 - 2x^2 + 1 = 0$$

$$(x^2 - 1)^2 = 0$$

$$x^2 - 1 = 0$$

$$x = \pm 1$$

$$y = \frac{1}{x} \qquad \text{Solve for } y.$$

$$y = \frac{1}{\pm 1}$$

$$y = \pm 1$$

$(-1, -1)$ and $(1, 1)$ are solutions.   Check as before.

Note: Although we have used the addition method first in this example, it may be easier to solve some systems by using the substitution method first. Deciding which method to use first is a matter of experience.

14-5. Graphing Second-degree Systems of Inequalities

If the graphs of the inequalities in a system are drawn on the same coordinate system, the **solution set of the system** is the set of ordered pairs represented by the points in the INTERSECTION of the graphs of the inequalities.

EXAMPLE 14-8: Graph the following second-degree systems of inequalities:

(a) $\begin{cases} x^2 - 4y^2 \geq 4 \\ x^2 + y^2 < 9 \end{cases}$ and **(b)** $\begin{cases} x^2 + y^2 \leq 16 \\ x^2 - y \leq 0 \\ 9x^2 + 16y^2 \geq 144 \end{cases}$.

Solution:

(a) Graph $x^2 - 4y^2 \geq 4$.
    Graph $x^2 + y^2 < 9$.

(b) Graph $x^2 + y^2 \leq 16$.
    Graph $x^2 - y \leq 0$.
    Graph $9x^2 + 16y^2 \geq 144$.

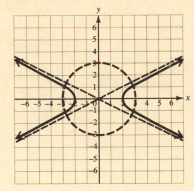

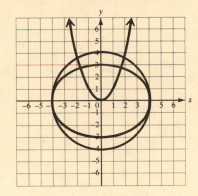

Shade the intersection.

Shade the intersection.

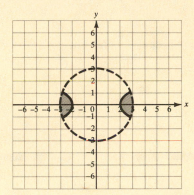

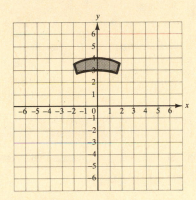

14-6. Solving Problems Using Nonlinear Systems

To **solve a problem using a nonlinear system,** we:

1. *Read* the problem carefully, several times.
2. *Identify* the unknown quantities.
3. *Decide* how to represent the unknown quantities using two variables.
4. *Make a table* (when necessary) to help represent the unknown quantities.
5. *Translate* the problem to a nonlinear system.
6. *Simplify* the nonlinear system when necessary.
7. *Solve* the nonlinear system.
8. *Interpret* the solutions of the nonlinear system with respect to each represented unknown quantity to find the proposed solutions of the original problem.
9. *Check* to see if the proposed solutions satisfy all the conditions of the original problem.

EXAMPLE 14-9: Solve the following problem using a nonlinear system:

Read: The sum of two numbers is 27. The product of the numbers is 180. What are the numbers?

Identify: The unknown quantities are $\begin{cases} \text{the larger number} \\ \text{the smaller number} \end{cases}$.

Decide: Let x = the larger number

and y = the smaller number.

Translate:

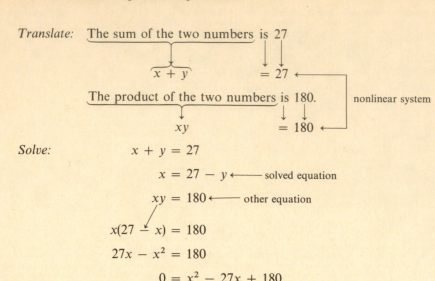

Solve:
$$x + y = 27$$
$$x = 27 - y \longleftarrow \text{solved equation}$$
$$xy = 180 \longleftarrow \text{other equation}$$
$$x(27 - x) = 180$$
$$27x - x^2 = 180$$
$$0 = x^2 - 27x + 180$$
$$0 = (x - 12)(x - 15)$$
$$x - 12 = 0 \quad \text{or} \quad x - 15 = 0$$
$$x = 12 \quad \text{or} \qquad x = 15$$

Interpret: $x = 12$ cannot be the solution of the original problem because x was chosen to be the larger number, but $x = 12$ and $x + y = 27$ means $y = 15$ is the larger number.
$x = 15$ means the larger number is 15.
$y = 27 - x = 27 - 15 = 12$ means the smaller number is 12.

Check: Is the sum of the two numbers 27? Yes: $15 + 12 = 27$.
Is the product of the two numbers 180? Yes: $15(12) = 180$.

RAISE YOUR GRADES

Can you . . . ?

☑ determine whether an ordered pair is a solution of a nonlinear system
☑ determine the number of real solutions of a nonlinear system
☑ solve a system of one linear and one nonlinear equation by the substitution method
☑ solve a system of two nonlinear equations by the substitution method
☑ solve a system of one linear and one nonlinear equation by the addition method
☑ solve a system of two nonlinear equations by the addition method
☑ solve a nonlinear system by a combination of methods
☑ graph a second-degree system of inequalities
☑ solve a problem using a nonlinear system

SOLVED PROBLEMS

PROBLEM 14-1 Determine whether the given ordered pairs are solutions of the following nonlinear systems:

(a) $(0, 1)$, $\begin{cases} y = x^2 + 4x + 1 \\ y = 4x + 1 \end{cases}$　　**(b)** $(3, -4)$, $\begin{cases} x^2 + y = 5 \\ x^2 + y^2 = 25 \end{cases}$　　**(c)** $(0, 3)$, $\begin{cases} x^2 + y^2 = 9 \\ x^2 - y = 9 \end{cases}$

(d) $(3, 2\sqrt{2})$, $\begin{cases} x^2 - y^2 = 1 \\ x^2 + y^2 = 17 \end{cases}$　　**(e)** $(2, 3i)$, $\begin{cases} x^2 - y^2 = 13 \\ 2x + y^2 = -5 \end{cases}$　　**(f)** $(i, \sqrt{2})$, $\begin{cases} x^2 + y^2 = 3 \\ x^2 - y = 3 \end{cases}$

Solution: Recall that to determine whether an ordered pair is a solution of a nonlinear system, you check to see if it is a solution of each equation in the system [see Example 14-11].

(a) $y = x^2 + 4x + 1$

$$\frac{1 \mid 0^2 + 4(0) + 1}{}$$

$1 = 1$

$y = 4x + 1$

$$\frac{1 \mid 4(0) + 1}{}$$

$1 = 1$

(0, 1) is a solution.

(b) $x^2 + y = 5$

$$\frac{3^2 + (-4) \mid 5}{9 - 4 \mid 5}$$

$5 = 5$

$x^2 + y^2 = 25$

$$\frac{3^2 + (-4)^2 \mid 25}{9 + 16 \mid 25}$$

$25 = 25$

(3, −4) is a solution.

(c) $x^2 + y^2 = 9$

$$\frac{0^2 + 3^2 \mid 9}{0 + 9 \mid 9}$$

$9 = 9$

$x^2 - y = 9$

$$\frac{0^2 - 3 \mid 9}{}$$

$-3 \neq 9$

(0, 3) is not a solution.

(d) $x^2 - y^2 = 1$

$$\frac{3^2 - (2\quad2)^2 \mid 1}{9 - 8 \mid 1}$$

$1 = 1$

$x^2 + y^2 = 17$

$$\frac{3^2 + (2\quad2)^2 \mid 17}{9 + 8 \mid 17}$$

$17 = 17$

$(3, 2\sqrt{2})$ is a solution

(e) $x^2 - y^2 = 13$

$$\frac{2^2 - (3i)^2 \mid 13}{4 - (-9) \mid 13}$$
$$4 + 9 \mid 13$$

$13 = 13$

$2x + y^2 = -5$

$$\frac{2(2) + (3i)^2 \mid -5}{4 + (-9) \mid -5}$$
$$4 - 9 \mid -5$$

$-5 = -5$

(2, 3i) is a solution.

(f) $x^2 + y^2 = 3$

$$\frac{i^2 + (\sqrt{2})^2 \mid 3}{-1 + 2 \mid 3}$$

$1 \neq 3$

$x^2 - y = 3$

$$\frac{i^2 - 2 \mid 3}{}$$

$-1 - 2 \neq 3$

(i, 2) is not a solution.

PROBLEM 14-2 Determine the number of real solutions of the following nonlinear systems:

(a) $\begin{cases} x^2 + y^2 = 25 \\ x - 7y = -25 \end{cases}$

(b) $\begin{cases} x^2 + y^2 = 9 \\ x^2 - y = 9 \end{cases}$

(c) $\begin{cases} y = x^2 - 4 \\ x^2 + 4y^2 = 4 \end{cases}$

(d) $\begin{cases} x^2 + y^2 - 6x = 0 \\ x - y^2 = 0 \end{cases}$

(e) $\begin{cases} x^2 + y^2 - 6x = 0 \\ 4x^2 + 9y^2 = 36 \end{cases}$

(f) $\begin{cases} 16x^2 + 9y^2 = 144 \\ x^2 - y^2 = 4 \end{cases}$

Solution: Recall that to determine the number of real solutions of a nonlinear system, you graph each equation and count the number of points at which the graphs intersect [see Example 14-2].

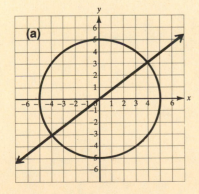

Two real solutions.

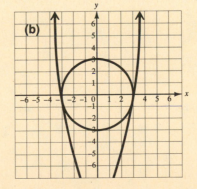

Two real solutions.

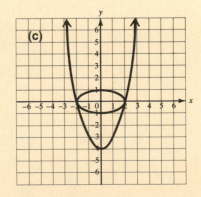

Two real solutions.

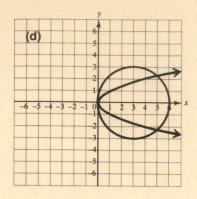

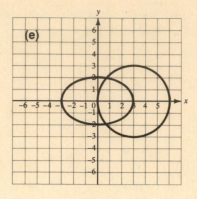

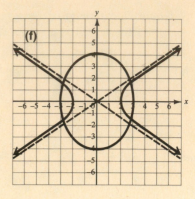

Three real solutions.   Two real solutions.   Four real solutions.

PROBLEM 14-3 Solve the following systems, which consist of one linear and one nonlinear equation, by the substitution method:

(a) $\begin{cases} 9x^2 + 16y^2 = 144 \\ 3x - 4y = 12 \end{cases}$ **(b)** $\begin{cases} x^2 + y^2 = 25 \\ 4x - 3y = 0 \end{cases}$ **(c)** $\begin{cases} x^2 + y^2 - 6x + 4y + 4 = 0 \\ x - y = 2 \end{cases}$

(d) $\begin{cases} x^2 + y^2 - 6x + 8y = 0 \\ 4x - 3y = 0 \end{cases}$ **(e)** $\begin{cases} x^2 + 6x - y = -5 \\ x - y = 0 \end{cases}$ **(f)** $\begin{cases} 2x^2 + 3y^2 = 6 \\ 2x + 3y = 0 \end{cases}$

Solution: Recall that to solve a system by the substitution method, you solve one of the equations for one of the variables, substitute that expression in the other equation, solve the resulting equation for one variable, substitute that value in the original equation to solve for the other variable, and check the solution [see Example 14-3].

(a) $3x - 4y = 12$

$3x = 12 + 4y$

$(3x)^2 = (12 + 4y)^2$

$9x^2 + 16y^2 = 144$

$(12 + 4y)^2 + 16y^2 = 144$

$144 + 96y + 16y^2 + 16y^2 = 144$

$32y^2 + 96y = 0$

$32y(y + 3) = 0$

$y = 0 \text{ or } -3$

$3x - 4y = 12$

$3x - 4(0) = 12$

$x = 4$

$3x - 4y = 12$

$3x - 4(-3) = 12$

$3x = 0$

$x = 0$

The solutions are $(0, -3)$ and $(4, 0)$.

(b) $4x - 3y = 0$

$4x = 3y$

$x = \frac{3}{4}y$

$x^2 + y^2 = 25$

$(\frac{3}{4}y)^2 + y^2 = 25$

$\frac{9}{16}y^2 + y^2 = 25$

$9y^2 + 16y^2 = 400$

$25y^2 = 400$

$y^2 = 16$

$y = \pm 4$

$4x - 3y = 0$

$4x - 3(4) = 0$

$4x = 12$

$x = 3$

$4x - 3y = 0$

$4x - 3(-4) = 0$

$4x + 12 = 0$

$4x = -12$

$x = -3$

The solutions are $(3, 4)$ and $(-3, -4)$.

(c)

$$x - y = 2$$
$$x = y + 2$$
$$x^2 + y^2 - 6x + 4y + 4 = 0$$
$$(y + 2)^2 + y^2 - 6(y + 2) + 4y + 4 = 0$$
$$y^2 + 4y + 4 + y^2 - 6y - 12 + 4y + 4 = 0$$
$$2y^2 + 2y - 4 = 0$$
$$2(y^2 + y - 2) = 0$$
$$(y + 2)(y - 1) = 0$$
$$y = -2 \text{ or } 1$$
$$x = y + 2$$
$$-2 + 2 \text{ or } 1 + 2$$
$$x = 0 \text{ or } 3$$

The solutions are $(0, -2)$ and $(3, 1)$.

(d)

$$4x - 3y = 0$$
$$4x = 3y$$
$$x = \frac{3}{4}y$$
$$x^2 + y^2 - 6x + 8y = 0$$
$$\left(\frac{3}{4}y\right)^2 + y^2 - 6\left(\frac{3}{4}y\right) + 8y = 0$$
$$\frac{9}{16}y^2 + y^2 - \frac{9}{2}y + 8y = 0$$
$$9y^2 + 16y^2 - 72y + 128y = 0$$
$$25y^2 + 56y = 0$$
$$y(25y + 56) = 0$$
$$y = 0 \text{ or } -\frac{56}{25}$$

$$x = \frac{3}{4}y$$
$$x = \frac{3}{4}(0) \text{ or } \frac{3}{4}\left(-\frac{56}{25}\right)$$
$$x = 0 \text{ or } -\frac{42}{25}$$

The solutions are $(0, 0)$ and $\left(-\frac{42}{25}, -\frac{56}{25}\right)$.

(e)

$$x - y = 0$$
$$x = y$$
$$x^2 + 6x - y = -5$$
$$x^2 + 6x - x = -5$$
$$x^2 + 5x + 5 = 0$$
$$x = \frac{-5 \pm \sqrt{25 - 20}}{2}$$
$$x = \frac{-5 \pm \sqrt{5}}{2}$$
$$y = x$$
$$y = \frac{-5 \pm \sqrt{5}}{2}$$

The solutions are $\left(\frac{-5 + \sqrt{5}}{2}, \frac{-5 + \sqrt{5}}{2}\right)$ and $\left(\frac{-5 - \sqrt{5}}{2}, \frac{-5 - \sqrt{5}}{2}\right)$.

(f)

$$2x + 3y = 0$$
$$2x = -3y$$
$$x = -\frac{3}{2}y$$
$$2x^2 + 3y^2 = 6$$
$$2\left(-\frac{3}{2}y\right)^2 + 3y^2 = 6$$
$$\frac{9}{2}y^2 + 3y^2 = 6$$
$$9y^2 + 6y^2 = 12$$
$$15y^2 = 12$$
$$y^2 = \frac{4}{5}$$
$$y = \pm\frac{2\sqrt{5}}{5}$$
$$x = -\frac{3}{2}\left(\pm\frac{2\sqrt{5}}{5}\right)$$
$$= +\frac{-3\sqrt{5}}{5}$$

The solutions are $\left(-\dfrac{3\sqrt{5}}{5}, \dfrac{2\sqrt{5}}{5}\right)$

and $\left(\dfrac{3\sqrt{5}}{5}, -\dfrac{2\sqrt{5}}{5}\right)$.

PROBLEM 14-4 Solve the following systems of two nonlinear equations by the substitution method:

(a) $\begin{cases} 16x^2 - 25y^2 = 400 \\ 16x^2 + 25y^2 = 400 \end{cases}$
(b) $\begin{cases} 9x^2 + 4y^2 = 9 \\ x^2 - y = 0 \end{cases}$
(c) $\begin{cases} y + x^2 - 4 = 0 \\ y - x^2 + 4 = 0 \end{cases}$

(d) $\begin{cases} x^2 + y^2 = 4 \\ x^2 - 2y = 4 \end{cases}$
(e) $\begin{cases} x^2 + y^2 = 25 \\ x^2 + y^2 - 6x = 7 \end{cases}$
(f) $\begin{cases} 4x^2 + y^2 = 4 \\ x^2 + 4y^2 = 4 \end{cases}$

Solution: Recall that to solve a system consisting of two nonlinear equations by the substitution method, you solve for one variable, or the square of one variable, substitute the expression in the second equation, and solve the system as before [see Example 14-4].

(a) $16x^2 - 25y^2 = 400$

$16x^2 = 25y^2 + 400$

$16x^2 + 25y^2 = 400$

$25^2 + 400 + 25y^2 = 400$

$50y^2 = 0$

$y = 0$

$16x^2 = 25(0)^2 + 400$

$16x^2 = 400$

$x^2 = 25$

$x = \pm 5$

The solutions are $(-5, 0)$ and $(5, 0)$.

(b) $x^2 - y = 2$

$x^2 = y$

$9x^2 + 4y^2 = 9$

$9y + 4y^2 = 9$

$4y^2 + 9y - 9 = 0$

$(4y - 3)(y + 3) = 0$

$y = \dfrac{3}{4}$ or -3

$x^2 = y$

$x^2 = \dfrac{3}{4}$ or -3

$x = \pm\dfrac{\sqrt{3}}{2}$ or $\pm i\sqrt{3}$

The solutions are $\left(-\dfrac{\sqrt{3}}{2}, \dfrac{3}{4}\right), \left(\dfrac{\sqrt{3}}{2}, \dfrac{3}{4}\right)$ and $(-i\sqrt{3}, -3), (i\sqrt{3}, -3)$.

(c) $y - x^2 + 4 = 0$

$y = x^2 - 4$

$y + x^2 - 4 = 0$

$x^2 - 4 + x^2 - 4 = 0$

$2x^2 = 8$

$x^2 = 4$

$x = \pm 2$

$y = x^2 - 4$

$y = (\pm 2)^2 - 4$

$y = 0$

The solutions are $(-2, 0)$ and $(2, 0)$.

(d) $x^2 - 2y = 4$

$x^2 = 2y + 4$

$x^2 + y^2 = 4$

$2y + 4 + y^2 = 4$

$y^2 + 2y = 0$

$y(y + 2) = 0$

$y = 0$ or -2

$x^2 = 2y + 4$

$x^2 = 2(0) + 4$

$x^2 = 4$

$x = \pm 2$

$x^2 = 2(-2) + 4$

$x^2 = 0$

$x = 0$

The solutions are $(0, -2)$, $(-2, 0)$ and $(2, 0)$.

(e) $x^2 + y^2 = 25$

$$y^2 = 25 - x^2$$

$$x^2 + y^2 - 6x = 7$$

$$x^2 + 25 - x^2 - 6x = 7$$

$$-6x = -18$$

$$x = 3$$

$$y^2 = 25 - 3^2$$

$$y^2 = 16$$

$$y = \pm 4$$

The solutions are $(3, 4)$ and $(3, -4)$.

(f) $4x^2 + y^2 = 4$

$$y^2 = 4 - 4x^2$$

$$x^2 + 4y^2 = 4$$

$$x^2 + 4(4 - 4x^2) = 4$$

$$x^2 + 16 - 16x^2 = 4$$

$$-15x^2 = -12$$

$$x^2 = \frac{12}{15}$$

$$x = \pm\frac{2\sqrt{5}}{5}$$

$$y^2 = 4 - 4\left(\frac{4}{5}\right)$$

$$y^2 = \frac{4}{5}$$

$$y = \pm\frac{2\sqrt{5}}{5}$$

The solutions are $\left(\dfrac{2\sqrt{5}}{5}, \dfrac{2\sqrt{5}}{5}\right)$,

$\left(-\dfrac{2\sqrt{5}}{5}, \dfrac{2\sqrt{5}}{5}\right)$, $\left(-\dfrac{2\sqrt{5}}{5}, -\dfrac{^2\sqrt{5}}{5}\right)$

and $\left(\dfrac{2\sqrt{5}}{5}, -\dfrac{2\sqrt{5}}{5}\right)$.

PROBLEM 14-5 Solve the following systems consisting of one linear and one nonlinear equation by the addition method:

(a) $\begin{cases} x^2 + y = 0 \\ x + y = 0 \end{cases}$
(b) $\begin{cases} x^2 + 4x - y + 4 = 0 \\ x + y = 0 \end{cases}$
(c) $\begin{cases} x^2 - 6x + y + 9 = 0 \\ 4x + y = 0 \end{cases}$

(d) $\begin{cases} x - y^2 + 6y - 9 = 0 \\ x - y + 1 = 0 \end{cases}$
(e) $\begin{cases} x + y^2 = 0 \\ x - y + 2 = 0 \end{cases}$
(f) $\begin{cases} x^2 + 6x - y + 6 = 0 \\ x + y = 0 \end{cases}$

Solution: Recall that to solve a nonlinear system, you write the equations in standard form, add the equations to eliminate one variable, solve for the remaining variable, substitute to find the remaining variable, and check the solutions [see Example 14-5].

(a)
$$\begin{array}{r} x^2 + y = 0 \\ -x - y = 0 \\ \hline x^2 - x = 0 \end{array}$$

$$x(x - 1) = 0$$

$$x = 0 \text{ or } x = 1$$

$$x + y = 0$$

$$0 + y = 0$$

$$1 + y = 0$$

$$y = 0 \text{ or } -1$$

The solutions are $(0, 0)$ and $(1, -1)$.

(b)
$$\begin{array}{r} x^2 + 4x - y + 4 = 0 \\ x + y \quad\quad = 0 \\ \hline x^2 + 5x \quad\quad + 4 = 0 \end{array}$$

$$(x + 4)(x + 1) = 0$$

$$x = -4 \text{ or } -1$$

$$x + y = 0$$

$$-4 + y = 0$$

$$-1 + y = 0$$

$$y = 4 \text{ or } 1$$

The solutions are $(-4, 4)$ and $(-1, 1)$.

(c)
$$\begin{array}{r} x^2 - 6x + y + 9 = 0 \\ - 4x - y \quad\quad = 0 \\ \hline x^2 - 10x \quad\quad + 9 = 0 \end{array}$$

$$(x - 9)(x - 1) = 0$$

$$x = 9 \text{ or } 1$$

$$4x + y = 0$$

$$36 + y = 0$$

$$4 + y = 0$$

$$y = -36 \text{ or } -4$$

The solutions are $(1, -4)$ and $(9, -36)$.

(d)
$$x - y^2 + 6y - 9 = 0$$
$$\underline{-x \qquad + y - 1 = 0}$$
$$-y^2 + 7y - 10 = 0$$

$$y^2 - 7y + 10 = 0$$

$$(y - 5)(y - 2) = 0$$

$$y = 5 \text{ or } y = 2$$

$$x - y + 1 = 0$$

$$x - 5 + 1 = 0$$

$$x - 2 + 1 = 0$$

$$x = 4 \text{ or } 1$$

The solutions are (4, 5) and (1, 2).

(e)
$$x + y^2 \qquad = 0$$
$$\underline{-x \qquad + y - 2 = 0}$$
$$y^2 + y - 2 = 0$$

$$(y + 2)(y - 1) = 0$$

$$y = -2 \text{ or } 1$$

$$x - y + 2 = 0$$

$$x - (-2) + 2 = 0$$

$$x - 1 + 2 = 0$$

$$x = -4 \text{ or } -1$$

The solutions are $(-4, -2)$ and $(-1, 1)$.

(f)
$$x^2 + 6x - y + 6 = 0$$
$$\underline{x + y \qquad = 0}$$
$$x^2 + 7x \qquad + 6 = 0$$

$$(x + 6)(x + 1) = 0$$

$$x = -6 \text{ or } x = -1$$

$$x + y = 0$$

$$-6 + y = 0$$

$$-1 + y = 0$$

$$y = 6 \text{ or } 1$$

The solutions are $(-1, 1)$ and $(-6, 6)$.

PROBLEM 14-6 Solve the following systems of two nonlinear equations by the addition method:

(a) $\begin{cases} 4x^2 - y^2 = 4 \\ x^2 + y^2 = 1 \end{cases}$
(b) $\begin{cases} x^2 - y^2 = -1 \\ 2x^2 + y^2 = 4 \end{cases}$
(c) $\begin{cases} 9x^2 + 16y^2 = 144 \\ x^2 + y^2 = 16 \end{cases}$

(d) $\begin{cases} 9x^2 + 4y^2 = 36 \\ x^2 + y = 3 \end{cases}$
(e) $\begin{cases} xy - 3x = 6 \\ xy = -6 \end{cases}$
(f) $\begin{cases} -9x^2 + 4y^2 = 36 \\ 4x^2 - 9y^2 = 36 \end{cases}$

Solution: Recall that to solve a nonlinear system by the addition method, you write the equations in standard form, multiply if necessary to get opposite coefficients of one variable, add the equations to eliminate one variable, solve for the remaining variable, substitute that value in the original equation to solve for the other variable, and check the solutions [see Example 14-6].

(a)
$$4x^2 - y^2 = 4$$
$$\underline{x^2 + y^2 = 1}$$
$$5x^2 \qquad = 5$$

$$x^2 = 1$$

$$x = \pm 1$$

$$x^2 + y^2 = 1$$

$$(\pm 1)^2 + y^2 = 1$$

$$y^2 = 0$$

$$y = 0$$

The solutions are $(-1, 0)$ and $(1, 0)$.

(b)
$$x^2 - y^2 = -1$$
$$\underline{2x^2 + y^2 = \quad 4}$$
$$3x^2 \qquad = \quad 3$$

$$x^2 = 1$$

$$x = \pm 1$$

$$x^2 - y^2 = -1$$

$$(\pm 1)^2 - y^2 = -1$$

$$-y^2 = -2$$

$$y^2 = 2$$

$$y = \pm\sqrt{2}$$

The solutions are $(-1, -\sqrt{2})$ $(-1, \sqrt{2})$, $(1, -\sqrt{2}$ and $(1, \sqrt{2})$.

(c)
$$9x^2 + 16y^2 = 144$$
$$x^2 + \quad y^2 = \quad 16$$

$$9x^2 + 16y^2 = \quad 144$$
$$\underline{-9x^2 - \quad 9y^2 = -144}$$
$$7y^2 = 0$$

$$y^2 = 0$$

$$y = 0$$

$$x^2 + y^2 = 16$$

$$x^2 + 0 = 16$$

$$x^2 = 16$$

$$x = \pm 4$$

The solutions are $(-4, 0)$ and $(4, 0)$.

(d)
$$9x^2 + 4y^2 = 36$$
$$-9x^2 - 9y = -27$$
$$\overline{\quad 4y^2 - 9y = 9}$$

$$4y^2 - 9y - 9 = 0$$

$$y = \frac{9 \pm \sqrt{81 + 144}}{8}$$

$$y = \frac{9 \pm 15}{8}$$

$$y = 3 \text{ or } -\frac{3}{4}$$

$$x^2 + y = 3$$
$$x^2 + 3 = 3$$
$$x^2 = 0$$
$$x = 0$$

$$x^2 + y = 3$$
$$x^2 + \left(-\frac{3}{4}\right) = 3$$
$$x^2 = 3 + \frac{3}{4}$$
$$x^2 = \frac{15}{4}$$
$$x = \pm\frac{\sqrt{15}}{2}$$

The solutions are (0, 3), $\left(\frac{\sqrt{15}}{2}, -\frac{3}{4}\right)$ and $\left(-\frac{\sqrt{15}}{2}, -\frac{3}{4}\right)$.

(e)
$$xy - 3x = 6$$
$$-xy \quad\quad = 6$$
$$\overline{\quad -3x = 12}$$

$$x = -4$$

$$xy = -6$$
$$-4y = -6$$
$$y = \frac{3}{2}$$

The solution is $\left(-4, \frac{3}{2}\right)$.

(f)
$$-9x^2 + 4y^2 = 36$$
$$4x^2 - 9y^2 = 36$$

$$-81x^2 + 36y^2 = 324$$
$$16x^2 - 36y^2 = 144$$
$$\overline{\quad -65x^2 \quad\quad = 468}$$

$$x^2 = -\frac{36}{5}$$

$$x = \pm\frac{6i\sqrt{5}}{5}$$

$$4x^2 - 9y^2 = 36$$
$$4\left(-\frac{468}{65}\right) - 9y^2 = 36$$
$$9y^2 = -\frac{1872}{65} - 36$$
$$9y^2 = -\frac{4212}{65}$$
$$y^2 = -\frac{468}{65}$$
$$y = \pm\frac{6i\sqrt{5}}{5}$$

The solutions are $\left(\frac{6i\sqrt{5}}{5}, \frac{6i\sqrt{5}}{5}\right)$, $\left(-\frac{6i\sqrt{5}}{5}, \frac{6i\sqrt{5}}{5}\right)$, $\left(\frac{6i\sqrt{5}}{5}, -\frac{6i\sqrt{5}}{5}\right)$ and $\left(-\frac{6i\sqrt{5}}{5}, -\frac{6i\sqrt{5}}{5}\right)$.

PROBLEM 14-7 Solve the following systems using a combination of methods:

(a) $\begin{cases} x^2 + y^2 + 6x = 0 \\ x^2 + y^2 - 6y = 0 \end{cases}$

(b) $\begin{cases} -x^2 + 2xy + y^2 = 7 \\ -x^2 + y^2 = 3 \end{cases}$

(c) $\begin{cases} x^2 - 4xy - y^2 = -4 \\ x^2 - y^2 = 8 \end{cases}$

(d) $\begin{cases} x^2 + y^2 - 4x + 6y = 0 \\ x^2 + y^2 - y = 0 \end{cases}$

(e) $\begin{cases} x^2 + y^2 + 3x - 5y = -3 \\ x^2 + y^2 + 2x - 4y = -2 \end{cases}$

(f) $\begin{cases} x^2 + y^2 - 2x + 3y = -2 \\ x^2 + y^2 - 3x + 4y = -1 \end{cases}$

Solution: Recall that to solve a system of equations, you may need to use the addition method and then the substitution method [see Example 14-7].

(a)
$$x^2 + y^2 + 6x = 0$$
$$-x^2 - y^2 + 6y = 0$$
$$\overline{\quad 6x + 6y = 0}$$
$$x + y = 0$$
$$x = -y$$

(b)
$$-x^2 + 2xy + y^2 = 7$$
$$x^2 \quad\quad - y^2 = -3$$
$$\overline{\quad 2xy \quad = 4}$$
$$xy = 2$$
$$y = \frac{2}{x}$$

$$x^2 + y^2 - 6y = 0$$

$$(-y)^2 + y^2 - 6y = 0$$

$$y^2 + y^2 - 6y = 0$$

$$2y^2 - 6y = 0$$

$$2y(y - 3) = 0$$

$$y = 0 \text{ or } 3$$

$$x = -y$$

$$x = -0 \text{ or } -3$$

The solutions are $(0, 0)$ and $(-3, 3)$.

$$-x^2 + y^2 = 3$$

$$-x^2 + \left(\frac{2}{x}\right)^2 = 3$$

$$-x^2 + \frac{4}{x^2} = 3$$

$$-x^4 + 4 = 3x^2$$

$$x^4 + 3x^2 + 4 = 0$$

$$(x^2 + 4)(x^2 - 1) = 0$$

$$x^2 = -4 \text{ or } 1$$

$$x = \pm 2i \text{ or } \pm 1$$

$$y = \frac{2}{x}$$

$$y = \frac{2}{1} \text{ or } \frac{2}{-1} \text{ or } \frac{2}{2i} \text{ or } \frac{2}{-2i}$$

$$y = 2 \text{ or } -2 \text{ or } -i \text{ or } i$$

The solutions are $(1, 2)$, $(-1, -2)$, $(2i, -i)$ and $(-2i, i)$.

(c)
$$x^2 - 4xy - y^2 = -4$$
$$\underline{-x^2 \qquad + y^2 = -8}$$
$$- 4xy \qquad = -12$$

$$xy = 3$$

$$y = \frac{3}{x}$$

$$x^2 - y^2 = 8$$

$$x^2 - \left(\frac{3}{x}\right)^2 = 8$$

$$x^2 - \frac{9}{x^2} = 8$$

$$x^4 - 9 = 8x^2$$

$$x^4 - 8x^2 - 9 = 0$$

$$(x^2 - 9)(x^2 + 1) = 0$$

$$x^2 = 9 \text{ or } -1$$

$$x = \pm 3 \text{ or } \pm i$$

$$y = \frac{3}{x}$$

$$y = \frac{3}{3} \text{ or } \frac{3}{-3} \text{ or } \frac{3}{i} \text{ or } \frac{3}{-i}$$

$$y = 1 \text{ or } -1 \text{ or } -3i \text{ or } 3i$$

The solutions are $(1, 3)$, $(-3, -1)$, $(i, -3i)$ and $(-1, 3i)$.

(d)
$$x^2 + y^2 - 4x + 6y = 0$$
$$\underline{-x^2 - y^2 \qquad + y = 0}$$
$$-4x + 7y = 0$$

$$x = \frac{7}{4}y$$

$$x^2 + y^2 - y = 0$$

$$\left(\frac{7}{4}y\right)^2 + y^2 - y = 0$$

$$\frac{49}{16}y^2 + y^2 - y = 0$$

$$49y^2 + 16y^2 - 16y = 0$$

$$65y^2 - 16y = 0$$

$$y(65y - 16) = 0$$

$$y = 0 \text{ or } +\frac{16}{65}$$

$$x = \frac{7}{4}y$$

$$x = \frac{7}{4}(0) \text{ or } \frac{7}{4}\left(+\frac{16}{65}\right)$$

$$x = 0 \text{ or } +\frac{28}{65}$$

The solutions are $(0, 0)$ and $\left(\frac{28}{65}, \frac{16}{65}\right)$.

(e)

$$x^2 + y^2 + 3x - 5y = -3$$
$$-x^2 - y^2 - 2x + 4y = 2$$
$$\overline{x - y = -1}$$

$$x = y - 1$$

$$x^2 + y^2 + 2x - 4y = -2$$

$$(y - 1)^2 + y^2 + 2(y - 1) - 4y = -2$$

$$y^2 - 2y + 1 + y^2 + 2y - 2 - 4y = -2$$

$$2y^2 - 4y + 1 = 0$$

$$y = \frac{4 \pm \sqrt{16 - 8}}{4}$$

$$y = \frac{2 \pm \sqrt{2}}{2}$$

$$x = y - 1$$

$$x = \frac{2 \pm \sqrt{2}}{2} - 1$$

$$x = \pm \frac{\sqrt{2}}{2}$$

The solutions are $\left(\frac{\sqrt{2}}{2}, \frac{2 + \sqrt{2}}{2}\right)$,

$\left(-\frac{\sqrt{2}}{2}, \frac{2 - \sqrt{2}}{2}\right)$.

(f)

$$x^2 + y^2 - 2x + 3y = -2$$
$$-x^2 - y^2 + 3x - 4y = 1$$
$$\overline{x - y = -1}$$

$$x = y - 1$$

$$x^2 + y^2 - 3x + 4y = -1$$

$$(y - 1)^2 + y^2 - 3(y - 1) + 4y = -1$$

$$y^2 - 2y + 1 + y^2 - 3y + 3 + 4y = -1$$

$$2y^2 - y + 5 = 0$$

$$y = \frac{1 \pm \sqrt{1 - 40}}{4}$$

$$y = \frac{1 \pm i\sqrt{39}}{4}$$

$$x = y - 1$$

$$x = \frac{1 \pm i\sqrt{39}}{4} - 1$$

$$x = \frac{-3 \pm i\sqrt{39}}{4}$$

The solutions are $\left(\frac{-3 + i\sqrt{39}}{4}, \frac{1 + i\sqrt{39}}{4}\right)$

and $\left(\frac{-3 - i\sqrt{39}}{4}, \frac{1 - i\sqrt{39}}{4}\right)$.

PROBLEM 14-8 Graph the following second-degree systems:

(a) $\begin{cases} x^2 + y^2 = 9 \\ x - y \geq -2 \end{cases}$
 (b) $\begin{cases} x^2 + y^2 \leq 9 \\ x^2 + y^2 - 6x \leq 0 \end{cases}$
 (c) $\begin{cases} 4x^2 + 9y^2 \leq 36 \\ x^2 + y^2 \geq 4 \end{cases}$

(d) $\begin{cases} x^2 + y^2 \leq 9 \\ xy \geq 2 \end{cases}$
 (e) $\begin{cases} x^2 - y^2 \geq 4 \\ x^2 + y^2 \leq 16 \\ x + y \geq 0 \end{cases}$
 (f) $\begin{cases} x - y^2 \geq 0 \\ 4x^2 - y^2 \geq 4 \\ x^2 + y^2 \leq 16 \end{cases}$

Solution: Recall that to graph a second-degree system of inequalities, you graph each inequality on the same coordinate system, then shade the intersection of the individual graphs [see Example 14-8].

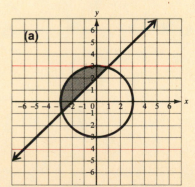
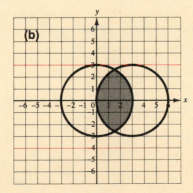
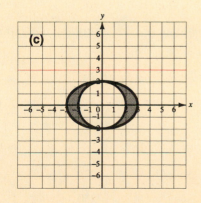

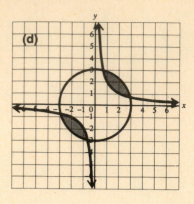

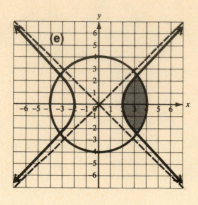

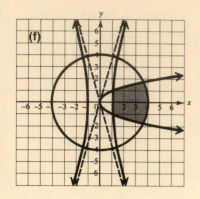

PROBLEM 14-9 Solve the following problem using a nonlinear system:

A plane completed a 2250-mile trip with a 25 mph tailwind. The trip would have taken one hour longer in still air. How long did the trip take with the wind? What is the constant airspeed of the plane?

Solution: Recall that to solve a problem using a nonlinear system, you:

1. *Read* the problem carefully, several times.
2. *Identify* the unknown quantities.
3. *Decide* how to represent the unknown quantities using two variables.
4. *Make a table* (when necessary) to help represent the unknown quantities.
5. *Translate* the problem to a nonlinear system.
6. *Simplify* the nonlinear system when necessary.
7. *Solve* the nonlinear system.
8. *Interpret* the solutions of the nonlinear system with respect to each represented unknown quantity to find the proposed solutions of the original problem.
9. *Check* to see if the proposed solutions satisfy all the conditions of the original problem [see Example 14-9].

Identify: The unknown quantities are $\begin{cases} \text{the constant airspeed of the plane} \\ \text{the time for the trip with the wind} \end{cases}$.

Decide: Let x = the constant airspeed of the plane

and y = the time for the trip with the wind.

Make a table:

	distance (d) [in miles]	rate (r) [in mph]	time (t) [in hours]
trip with the wind	2250	$x + 25$	y
trip in still air	2250	x	$y + 1$

Translate: With the wind, the distance equals the rate times the time

$$2250 = (x + 25) \cdot y$$

In still air, the distance equals the rate times the time

$$2250 = x \cdot (y + 1)$$

Simplify: $2250 = (x + 25)y \longrightarrow xy + 25y = 2250 \longleftarrow$

 $2250 = x(y + 1) \longrightarrow xy + x = 2250 \longleftarrow$ nonlinear system

Solve:

$$xy + 25y = 2250 \xrightarrow{\text{multiply by } -1} -xy - 25y = -2250$$

$$xy + x = 2250 \xrightarrow[\text{same}]{} \underline{xy + x = 2250}$$

$$x - 25y = 0$$

$$x = 25y \longleftarrow \text{solved equation}$$

$$xy + 25y = 2250 \longleftarrow \text{system equation}$$

$$\downarrow$$

$$(25y)y + 25y = 2250 \qquad \text{Substitute } 25y \text{ for } x.$$

$$25y^2 + 25y - 2250 = 0 \longleftarrow \text{standard form (the GCF is 25)}$$

$$y^2 + y - 90 = 0 \longleftarrow \text{simplest form}$$

$$(y + 10)(y - 9) = 0$$

$$y + 10 = 0 \qquad \text{or } y - 9 = 0$$

$$y = -10 \text{ or } \qquad y = 9$$

Interpret: $y = -10$ [hours] cannot be the solution to the original problem because the required time measure is positive.

$y = 9$ means the time for the trip with the wind is 9 hours.

$x = 25y = 25(9) = 225$ means the constant airspeed of the plane is 225 mph.

Check: Does it take 9 hours to travel 2250 miles at 225 mph with a tailwind of 25 mph?
Yes: $d = rt = (225 + 25)(9) = 250(9) = 2250$.
Does it take 1 hour longer than 9 hours to travel 2250 miles at 225 mph in still air?
Yes: $d = rt = (225)(9 + 1) = 225(10) = 2250$.

Supplementary Exercises

PROBLEM 14-10 Determine whether the given ordered pairs are solutions of the following non-linear systems:

(a) $(-3, -4)$, $\begin{Bmatrix} x^2 + y = 5 \\ x^2 + y^2 = 25 \end{Bmatrix}$ (b) $(3, 0)$, $\begin{Bmatrix} x^2 + y^2 = 9 \\ x^2 - y = 9 \end{Bmatrix}$ (c) $(2, i)$, $\begin{Bmatrix} x^2 + y^2 = 3 \\ x^2 - y^2 = 7 \end{Bmatrix}$

(d) $(2, i\sqrt{3})$, $\begin{Bmatrix} x^2 + y^2 = 1 \\ x^2 - y^2 = 7 \end{Bmatrix}$ (e) $(\sqrt{2}, 3i)$, $\begin{Bmatrix} x^2 - y^2 = 11 \\ x^2 + y^2 = -6 \end{Bmatrix}$ (f) $(i\sqrt{2}, i)$, $\begin{Bmatrix} x^2 - y^2 = -1 \\ x^2 + y^2 = 1 \end{Bmatrix}$

PROBLEM 14-11 Determine the number of real solutions of the following systems:

(a) $\begin{Bmatrix} x^2 + y = 5 \\ x^2 + y^2 = 25 \end{Bmatrix}$ (b) $\begin{Bmatrix} x^2 - y^2 = 9 \\ x^2 + y^2 = 9 \end{Bmatrix}$ (c) $\begin{Bmatrix} y = 2x^2 - 3x - 2 \\ 9x^2 + 16y^2 = 144 \end{Bmatrix}$

(d) $\begin{Bmatrix} x = y^2 \\ x^2 + y^2 - 4x = 0 \end{Bmatrix}$ (e) $\begin{Bmatrix} 4x^2 + 9y^2 = 36 \\ 4x^2 - 9y^2 = 36 \end{Bmatrix}$ (f) $\begin{Bmatrix} xy = 4 \\ x^2 + y^2 = 9 \end{Bmatrix}$

PROBLEM 14-12 Solve the following systems of one linear and one nonlinear equation by the substitution method:

(a) $\begin{Bmatrix} x^2 + y^2 = 25 \\ 3x + 4y = 0 \end{Bmatrix}$ (b) $\begin{Bmatrix} x^2 + y^2 + 8x + 4y = -12 \\ x - y = 2 \end{Bmatrix}$ (c) $\begin{Bmatrix} 9x^2 + 4y^2 = 144 \\ 2x - y = 0 \end{Bmatrix}$

(d) $\begin{Bmatrix} x - y = 0 \\ xy = -9 \end{Bmatrix}$ (e) $\begin{Bmatrix} 9x^2 + 16y^2 = 144 \\ x + y = 0 \end{Bmatrix}$ (f) $\begin{Bmatrix} x^2 - 8x - y + 19 = 0 \\ 4x + 4y = 31 \end{Bmatrix}$

PROBLEM 14-13 Solve the following systems of two nonlinear equations by the substitution method:

(a) $\begin{cases} 9x^2 + 16y^2 = 144 \\ 9x^2 - 16y^2 = 144 \end{cases}$
(b) $\begin{cases} x^2 - 6x - y + 9 = 0 \\ x^2 - y = 0 \end{cases}$
(c) $\begin{cases} xy + 4x = 12 \\ xy = 4 \end{cases}$

(d) $\begin{cases} x^2 + y^2 = 9 \\ x^2 + y = 9 \end{cases}$
(e) $\begin{cases} x^2 + 2y^2 = 4 \\ x^2 + y^2 = 3 \end{cases}$
(f) $\begin{cases} 4x^2 + 9y^2 = 36 \\ 9x^2 + 4y^2 = 36 \end{cases}$

PROBLEM 14-14 Solve the following systems of one linear and one nonlinear equation by the addition method:

(a) $\begin{cases} x^2 - y = 0 \\ x - y = 0 \end{cases}$
(b) $\begin{cases} x^2 + 4x - y + 4 = 0 \\ 3x + y - 4 = 0 \end{cases}$
(c) $\begin{cases} x^2 - 6x + y + 9 = 0 \\ 2x + y - 3 = 0 \end{cases}$

(d) $\begin{cases} x - y^2 + 6y - 9 = 0 \\ x - 3y + 9 = 0 \end{cases}$
(e) $\begin{cases} x + y^2 + 2 = 0 \\ x + 3y = 0 \end{cases}$
(f) $\begin{cases} x - y^2 - 3 = 0 \\ x - 2y = 0 \end{cases}$

PROBLEM 14-15 Solve the following systems of two nonlinear equations by the addition method:

(a) $\begin{cases} 9x^2 - y^2 = 9 \\ 9x^2 + y^2 = 9 \end{cases}$
(b) $\begin{cases} 4x^2 + 9y^2 = -36 \\ x^2 - y = 2 \end{cases}$
(c) $\begin{cases} x^2 + y^2 - 6y = 0 \\ x^2 + y^2 = 9 \end{cases}$

(d) $\begin{cases} 9x^2 + 16y^2 = 144 \\ x^2 + y^2 = 16 \end{cases}$
(e) $\begin{cases} x^2 + 4y^2 = 25 \\ x^2 + y^2 = 16 \end{cases}$
(f) $\begin{cases} 9x^2 - 16y^2 = 144 \\ x^2 + y^2 = 9 \end{cases}$

PROBLEM 14-16 Solve the following systems by a combination of methods:

(a) $\begin{cases} x^2 + y^2 - 6x = 0 \\ x^2 + y^2 - 8y = 0 \end{cases}$
(b) $\begin{cases} x^2 + 6xy + y^2 = 24 \\ x^2 + y^2 = 6 \end{cases}$
(c) $\begin{cases} x^2 + 5xy + y^2 = 28 \\ x^2 + y^2 = 8 \end{cases}$

(d) $\begin{cases} x^2 + y^2 - 6x + 8y = 11 \\ x^2 + y^2 + 6x - 8y = 11 \end{cases}$
(e) $\begin{cases} 4x^2 + 9y^2 = 36 \\ 4x^2 + 9y^2 - 8x + 36y = -4 \end{cases}$
(f) $\begin{cases} x^2 + y^2 + 8x = 0 \\ x^2 + y^2 + y = 0 \end{cases}$

PROBLEM 14-17 Graph the following second-degree systems of inequalities:

(a) $\begin{cases} x^2 + y^2 \leq 25 \\ x^2 \geq y \end{cases}$
(b) $\begin{cases} x^2 - 4y^2 \geq 4 \\ x^2 - y \leq 9 \end{cases}$
(c) $\begin{cases} 4x^2 + 9y^2 \leq 36 \\ 9x^2 + 4y^2 \leq 36 \end{cases}$

(d) $\begin{cases} x^2 + y^2 \geq 9 \\ x - y \leq -2 \\ x + y \geq 2 \end{cases}$
(e) $\begin{cases} x^2 + y^2 + 6x + 6y \geq -9 \\ x^2 + y^2 + 6x \leq 0 \\ x^2 + y^2 + 6y \leq 0 \end{cases}$
(f) $\begin{cases} x^2 + y^2 \leq 16 \\ 9x^2 + 4y^2 \geq 36 \\ -9x^2 + 4y^2 \geq 36 \end{cases}$

PROBLEM 14-18 Solve each problem using a nonlinear system:

(a) The sum of two numbers is 26. The sum of their squares is 340. Find the numbers.

(b) The perimeter of a rectangle is 46 m. The area of the rectangle is 102 m². Find the dimensions of the rectangle.

(c) The annual income from an investment at simple interest is $180 per year. If the simple interest rate is increased by 1%, the same investment would earn $210. How much is being invested? What is the original simple interest rate?

(d) A car travels 200 km at a constant rate. On the return trip, the car travels 10 km/h faster and takes one hour less time. What was the constant rate of the car going? How long did the return trip take?

(e) At a constant rowing rate, it takes 6 hours to row 16 miles downstream and then back again. The landspeed upstream is twice the rate of the current. What is the waterspeed of the boat? What is the rate of the current?

(f) It takes 8 hours to fill a pool using an inlet pipe and then empty it using an outlet pipe. With both pipes open, it takes $7\frac{1}{2}$ hours to fill the pool. How long does it take to fill the pool with the outlet pipe closed?

Answers to Supplementary Exercises

(14-10) (a) yes   (b) yes   (c) no   (d) yes   (e) no   (f) no

(14-11) (a) 3   (b) 2   (c) 4   (d) 3   (e) 2   (f) 4

(14-12) (a) $(-4, 3), (4, -3)$ (b) $(-2, -4)$ (c) $(\frac{12}{5}, \frac{24}{5}), (-\frac{12}{5}, -\frac{24}{5})$ (d) $(-3i, -3i), (3i, 3i)$
     (e) $(-\frac{12}{5}, \frac{12}{5}), (\frac{12}{5}, -\frac{12}{5})$ (f) $(\frac{9}{2}, \frac{13}{4}), (\frac{5}{2}, \frac{21}{4})$

(14-13) (a) $(-4, 0), (4, 0)$ (b) $(\frac{3}{2}, \frac{9}{4})$ (c) $(2, 2)$ (d) $(-3, 0), (3, 0), (-2\sqrt{2}, 1), (2\sqrt{2}, 1)$
     (e) $(-\sqrt{2}, -1), (-\sqrt{2}, 1), (\sqrt{2}, -1), (\sqrt{2}, 1)$
     (f) $\left(-\dfrac{6\sqrt{13}}{13}, -\dfrac{6\sqrt{13}}{13}\right), \left(-\dfrac{6\sqrt{13}}{13}, \dfrac{6\sqrt{13}}{13}\right), \left(\dfrac{6\sqrt{13}}{13}, -\dfrac{6\sqrt{13}}{13}\right), \left(\dfrac{6\sqrt{13}}{13}, \dfrac{6\sqrt{13}}{13}\right)$

(14-14) (a) $(0, 0), (1, 1)$ (b) $(0, 4), (-7, 25)$ (c) $(2, -1), (6, -9)$ (d) $(0, 3), (9, 6)$
     (e) $(-6, 2), (-3, 1)$ (f) $(2 - 2i\sqrt{2}, 1 - i\sqrt{2}), (2 + 2i\sqrt{2}, 1 + i\sqrt{2}$

(14-15) (a) $(-1, 0), (1, 0)$ (b) $(0, -2), \left(-\dfrac{4\sqrt{2}}{3}, \dfrac{14}{9}\right), \left(\dfrac{4\sqrt{2}}{3}, \dfrac{14}{9}\right)$

     (c) $\left(-\dfrac{3\sqrt{3}}{2}, \dfrac{3}{2}\right), \left(\dfrac{3\sqrt{3}}{2}, \dfrac{3}{2}\right)$ (d) $(-4, 0), (4, 0)$

     (e) $(-\sqrt{13}, -\sqrt{3}), (-\sqrt{13}, \sqrt{3}), (\sqrt{13}, -\sqrt{3}), (\sqrt{13}, \sqrt{3})$

     (f) $\left(-\dfrac{12\sqrt{2}}{5}, -\dfrac{3i\sqrt{7}}{5}\right), \left(-\dfrac{12\sqrt{2}}{5}, \dfrac{3i\sqrt{7}}{5}\right), \left(\dfrac{12\sqrt{2}}{5}, -\dfrac{3i\sqrt{7}}{5}\right), \left(\dfrac{12\sqrt{2}}{5}, \dfrac{3i\sqrt{7}}{5}\right)$

(14-16) (a) $(0, 0), \left(\dfrac{96}{25}, \dfrac{72}{25}\right)$ (b) $(-\sqrt{3}, -\sqrt{3}), (\sqrt{3}, \sqrt{3})$ (c) $(2, 2), (-2, -2)$

     (d) $\left(-\dfrac{4}{5}\sqrt{11}, -\dfrac{3}{5}\sqrt{11}\right), \left(\dfrac{4}{5}\sqrt{11}, \dfrac{3}{5}\sqrt{11}\right)$

     (e) $\left(\dfrac{5 + 3\sqrt{65}}{10}, \dfrac{-15 + \sqrt{65}}{15}\right), \left(\dfrac{5 - 3\sqrt{65}}{10}, \dfrac{-15 - \sqrt{65}}{15}\right)$ (f) $(0, 0), \left(-\dfrac{8}{65}, -\dfrac{64}{65}\right)$

(14-17)

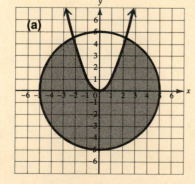
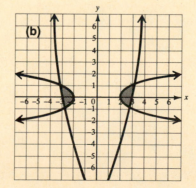
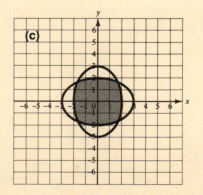

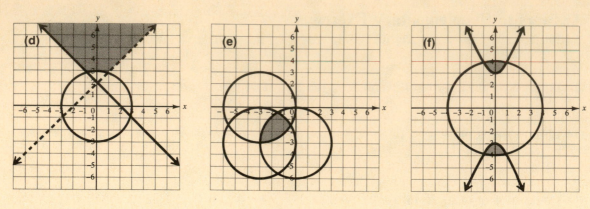

(**14-18**) (**a**) 12, 14   (**b**) 17 m by 6 m   (**c**) $3000, 6%   (**d**) 40 km/h, 4 hr

(**e**) 6 mph (water-speed), 2 mph (current)   (**f**) 3 hr

15 EXPONENTIAL AND LOGARITHMIC FUNCTIONS

15-1. Exponential Functions

The function $f(x) = b^x$, where b is a positive constant not equal to 1, is called **an exponential function.** The constant b is called the **base.** To **graph an exponential function,** we find several ordered pairs that are solutions of the function.

A. Evaluating expressions of the form b^x

To **evaluate an expression** of the form b^x using a calculator, enter the base, press the exponential key, enter the exponent and press the equal ($=$) key.

EXAMPLE 15-1: Evaluate the expressions (a) $3^{1.73}$ and (b) $3^{1.732}$. Round each answer to the nearest thousandth.

Solution: (a) $3^{1.73}$

3	calculator display	Enter base 3 in calculator.
3		Press $\boxed{y^x}$ key.
1.73		Enter exponent 1.73.
6.68990		Press $\boxed{=}$ key.
6.690		Write number to the nearest thousandth.

(b) $3^{1.732}$

3	calculator display	Enter base 3 in calculator.
3		Press $\boxed{y^x}$ key.
1.732		Enter exponent 1.732.
6.70461		Press $\boxed{=}$ key.
6.705		Write number to the nearest thousandth.

B. Graphing exponential functions

To **graph an exponential function,** we make a table of solutions of the exponential function, plot the ordered pairs and draw a smooth curve through the points.

EXAMPLE 15-2: Graph the exponential functions (**a**) $f(x) = 2^x$ and (**b**) $f(x) = (\frac{1}{2})^x$.

Solution: (**a**)

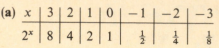

x	3	2	1	0	−1	−2	−3
2^x	8	4	2	1	$\frac{1}{2}$	$\frac{1}{4}$	$\frac{1}{8}$

Make a table.

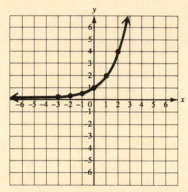

Plot the ordered pairs.

Draw a smooth curve through the points.

(**b**)

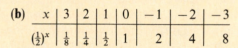

x	3	2	1	0	−1	−2	−3
$(\frac{1}{2})^x$	$\frac{1}{8}$	$\frac{1}{4}$	$\frac{1}{2}$	1	2	4	8

Make a table.

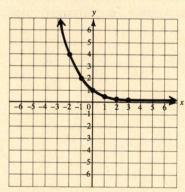

Plot the ordered pairs.

Draw a smooth curve through the points.

C. Graphing inverses of exponential functions

The **inverse** of an exponential function $y = b^x$ is $x = b^y$. To graph the inverse of an exponential function, we make a table of solutions of the function, plot the ordered pairs, and draw a smooth curve through the points.

EXAMPLE 15-3: Graph the exponential functions (**a**) $x = 2^y$ and (**b**) $x = (\frac{1}{2})^y$.

Solution: (**a**)

y	3	2	1	0	−1	−2	−3
$2^y = x$	8	4	2	1	$\frac{1}{2}$	$\frac{1}{4}$	$\frac{1}{8}$

Make a table of solutions.

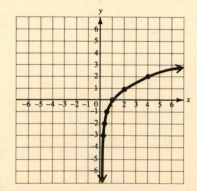

Plot the ordered pairs (x, y).

Draw a smooth curve through the points.

(b) $x = (\frac{1}{2})^y$

y	3	2	1	0	-1	-2	-3
$x = (\frac{1}{2})^y$	$\frac{1}{8}$	$\frac{1}{4}$	$\frac{1}{2}$	1	2	4	8

Make a table of solutions.

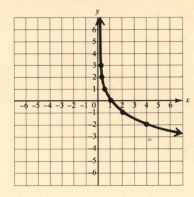

Plot the ordered pairs (x, y).

Draw a smooth curve through the points.

Note: The graph of $x = 2^y$ looks like the graph of $y = 2^x$ except that it is reflected across the graph of the line $y = x$.

D. The number e and the natural exponential function

A number that we encounter frequently when working with exponential (and logarithmic) functions is the irrational number e. Like π, e has a nonrepeating, nonterminating decimal representation. (For the purposes of this book we will use the value 2.7182818. Many hand-held calculators are programmed to use such an approximation of e.) We see e as a base in all the exponential and logarithmic formulas used by scientists to describe natural phenomena, such as growth or radioactive decay. For this reason the function $f(x) = e^x$ is called the **natural exponential function.** Calculus also uses e to simplify many of its formulas.

15-2. Logarithmic Functions

The inverse of an exponential function is called a **logarithmic function.** We define $y = \log_b x$ to mean $x = b^y$, where $x > 0$ and b is a positive constant and is not equal to 1. The number y is the **logarithm,** b the **base,** and x the **argument.** We read $y = \log_b x$ as "y is the logarithm, base b, of x."

Note: Logarithms are exponents!

A. Writing logarithmic equations in exponential form

To **write a logarithmic equation in exponential form,** we identify the parts of the logarithmic equation as given above, then rearrange them using the definition $y = \log_b x$ means $x = by$.

EXAMPLE 15-4: Write these logarithmic equations in exponential form:

(a) $3 = \log_5 125$ **(b)** $\log_{32} 2 = \frac{1}{5}$ **(c)** $\log_2 \frac{1}{8} = -3$

Solution: **(a)** $3 = \log_5 125$

           $b = 5$ Identify the base (b).

           $x = 125$ Identify the argument (x).

           $y = 3$ Identify the logarithm or exponent (y).

        $125 = 5^3$ Write in exponential form.

     (b) $\log_{32} 2 = \frac{1}{5}$

           $b = 32$ Identify the base.

           $x = 2$ Identify the argument.

           $y = \frac{1}{5}$ Identify the logarithm or exponent.

        $2 = 32^{\frac{1}{5}}$ Write in exponential form.

(c) $\log_2 \frac{1}{8} = -3$

$b = 2$	Identify the base.
$x = \frac{1}{8}$	Identify the argument.
$y = -3$	Identify the logarithm or exponent.
$\frac{1}{8} = 2^{-3}$	Write in exponential form.

B. Writing exponential equations in logarithmic form

To **write an exponential equation in logarithmic form,** we identify the parts of the exponential equation, then rearrange them using the definition of a logarithm given above.

EXAMPLE 15-5: Write these exponential equations in logarithmic form:

(a) $2^5 = 32$ **(b)** $81^{\frac{1}{4}} = 3$ **(c)** $3^{-2} = \frac{1}{9}$

Solution: **(a)** $2^5 = 32$

$b = 2$	Identify the base.
$y = 5$	Identify the exponent or logarithm.
$x = 32$	Identify the argument.
$\log_2 32 = 5$	Write in logarithmic form.

(b) $81^{\frac{1}{4}} = 3$

$b = 81$	Identify the base.
$y = \frac{1}{4}$	Identify the exponent or logarithm.
$x = 3$	Identify the argument.
$\log_{81} 3 = \frac{1}{4}$	Write in logarithmic form.

(c) $3^{-2} = \frac{1}{9}$

$b = 3$	Identify the base.
$y = -2$	Identify the exponent or logarithm.
$x = \frac{1}{9}$	Identify the argument.
$\log_3 \frac{1}{9} = -2$	Write in logarithmic form.

C. Solving logarithmic equations

To **solve a logarithmic equation,** we write the equation in exponential form and solve.

EXAMPLE 15-6: Solve the following equations:

(a) $\log_2 x = 4$ **(b)** $\log_4 16 = x$ **(c)** $\log_x 128 = 7$

Solution: **(a)** $\log_2 x = 4$

$x = 2^4$	Write in exponential form.
$x = 16$	Solve.

(b) $\log_4 16 = x$

$16 = 4^x$	Write in exponential form.
$4^2 = 4^x$	Write both members with the same base.
$2 = x$	Equate the exponents of the same base.

(c) $\log_x 128 = 7$

$\quad\quad 128 = x^7$ Write in exponential form.

$\quad\quad 2^7 = x^7$ Write both members with the same base.

$\quad\quad 2 = x$ Take the seventh root of both members.

D. Graphing logarithmic functions

To **graph a logarithmic equation,** we write it in exponential form, write a table of solutions, and graph as before.

EXAMPLE 15-7: Graph the logarithmic equations (**a**) $y = \log_2 x$ and (**b**) $\log_{\frac{1}{3}} x = y$.

Solution: (**a**) $y = \log_2 x$

$\quad\quad x = 2^y$ Write in exponential form.

Make a table of solutions.

y	3	2	1	0	-1	-2	-3
$x = 2^y$	8	4	2	1	$\frac{1}{2}$	$\frac{1}{4}$	$\frac{1}{8}$

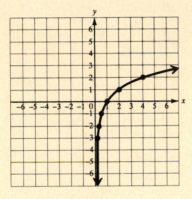

Plot the ordered pairs.

Draw a smooth curve through the points.

(**b**) $y = \log_{\frac{1}{3}} x$

$\quad\quad x = \left(\frac{1}{3}\right)^y$ Write the equation in exponential form.

Make a table of solutions.

y	3	2	1	0	-1	-2
$\left(\frac{1}{3}\right)^y = x$	$\frac{1}{27}$	$\frac{1}{9}$	$\frac{1}{3}$	1	3	9

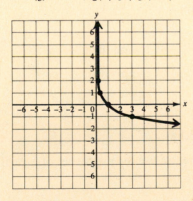

Plot the ordered pairs.

Draw a smooth curve through the points.

15-3. Using Properties of Logarithmic Functions

Because logarithms are exponents, the **properties of exponents** apply to logarithms. The specific application of these properties to logarithms are stated as follows:

- Product Property: $\log_b (MN) = \log_b M + \log_b N$

- Quotient Property: $\log_b \left(\dfrac{M}{N}\right) = \log_b M - \log_b N$

- Power Property: $\log_b (M^p) = p \log_b M$

- Change-of-Base Property: $\log_N M = \dfrac{\log_b M}{\log_b N}$

A. Proving logarithmic properties

To **prove the properties of logarithms,** we use the definition of a logarithm to rewrite the logarithmic expression.

EXAMPLE 15-8: Prove $\log_b (MN) = \log_b M + \log_b N$ (the product property of logarithms).

Solution: Let $\log_b M = x$ and $\log_b N = y$

$b^x = M$ and $b^y = N$	Write in exponential form.
$MN = b^x b^y$	
$MN = b^{x+y}$	Use the product property for exponents.
$\log_b (MN) = x + y$	Use the definition $y = \log_b x$.
$\log_b (MN) = \log_b M + \log_b N$	Substitute $\log_b M$ for x and $\log_b N$ for y.

Note: See Problem 15-8 for a proof of the other logarithmic properties.

B. Expressing logarithms in terms of simpler logarithms

To **express a logarithm in terms of simpler logarithms,** we use the properties of logarithms to rewrite and expand the original logarithms.

EXAMPLE 15-9: Express these logarithms in terms of simpler logarithms:

(a) $\log_b 3x^3 y^2 z$ **(b)** $\log_b \dfrac{4\sqrt{x}}{7y^2}$

Solution: **(a)** $\log_b 3x^3 y^2 z$

$= \log_b 3 + \log_b x^3 + \log_b y^2 + \log_b z$	Use the product property.
$= \log_b 3 + 3\log_b x + 2\log_b y + \log_b z$	Use the power property.

(b) $\log_b \dfrac{4\sqrt{x}}{7y^2}$

$= \log_b 4\sqrt{x} - \log_b 7y^2$	Use the quotient property.
$= \log_b 4 + \log_b x^{\frac{1}{2}} - (\log_b 7 + \log_b y^2)$	Use the product property.
$= \log_b 4 + \frac{1}{2}\log_b x - \log_b 7 - 2\log_b y$	Use the power property.

C. Expressing logarithmic expressions as a single logarithm

To **express simple logarithmic expressions as a single logarithm,** we use the properties of logarithms to condense the simple logarithms as a single compound one.

EXAMPLE 15-10: Express these logarithmic expressions as single logarithms:

(a) $2\log_b x + 3\log_b y$ **(b)** $\frac{1}{2}\log_b x - \frac{1}{3}\log_b y$

Solution: **(a)** $2\log_b x + 3\log_b y$

$= \log_b x^2 + \log_b y^3$	Use the power property.
$= \log_b x^2 y^3$	Use the product property.

(b) $\frac{1}{2} \log_b x - \frac{1}{3} \log_b y$

$$= \log_b x^{\frac{1}{2}} - \log_b y^{\frac{1}{3}} \qquad \text{Use the power property.}$$

$$= \log_b \frac{x^{\frac{1}{2}}}{y^{\frac{1}{3}}} \qquad \text{Use the quotient property.}$$

$$= \log_b \frac{\sqrt{x}}{\sqrt[3]{y}}$$

D. Approximating logarithms using given values

To **approximate** a logarithm using given logarithmic values, we express the logarithm in terms of simpler logarithms, substitute the given value, then evaluate the expression.

EXAMPLE 15-11: Approximate the following logarithms if $\log_{10} 2 \approx 0.3010$, $\log_{10} 3 \approx 0.4771$ and $\log_{10} 5 \approx 0.6990$:

(a) $\log_{10} 72$ **(b)** $\log_{10} \sqrt{5}$

Solution: **(a)** $\log_{10} 72$

$$= \log_{10} 8 \cdot 9 \qquad \text{Factor argument.}$$

$$= \log_{10} 8 + \log_{10} 9 \qquad \text{Use the product property.}$$

$$= \log_{10} 2^3 + \log_{10} 3^2$$

$$= 3 \log_{10} 2 + 2 \log_{10} 3 \qquad \text{Use the power property.}$$

$$\approx 3(0.3010) + 2(0.4771) \qquad \text{Substitute for } \log_{10} 2 \text{ and } \log_{10} 3.$$

$$\approx 0.9030 + 0.9542 \qquad \text{Evaluate.}$$

$$\approx 1.8572$$

(b) $\log_{10} \sqrt{5}$

$$= \log_{10} 5^{\frac{1}{2}}$$

$$= \frac{1}{2} \log_{10} 5 \qquad \text{Use power property.}$$

$$\approx \frac{1}{2}(0.6990) \qquad \text{Substitute for } \log_{10} 5.$$

$$\approx 0.3495 \qquad \text{Evaluate.}$$

15-4. Finding Logarithms and Antilogarithms

Logarithms with base 10 are called **common logarithms.** When writing common logarithms the 10 is usually omitted, so that $\log_{10} M$ is written simply as $\log M$. To **find a common logarithm,** we use a common logarithmic table (see Appendix 2) or a calculator that has a $\boxed{\log x}$ key.

A. Finding common logarithms using a table

To **find a common logarithm using a table,** we must first write the number in scientific notation. We then use the product property of logarithms to determine the **characteristic,** or integer part of the logarithm, and read the **mantissa,** or decimal part of the logarithm, in the table.

EXAMPLE 15-12: Find the common logarithm of: **(a)** 1.57   **(b)** 342   **(c)** 0.0417

Solution: **(a)** $\log 1.57 = \log (1.57 \times 10^0)$ Write in scientific notation.

$$= \log 1.57 + \log 10^0 \qquad \text{Use the product property.}$$

$$= \log 1.57 + 0 \qquad \text{Substitute 0 for } \log 10^0$$

$$\approx 0.1959 + 0 \qquad \begin{array}{l}\text{Read the intersection of 1.5 in the left} \\ \text{column and 7 in the top row of the table.}\end{array}$$

$$\approx 0.1959$$

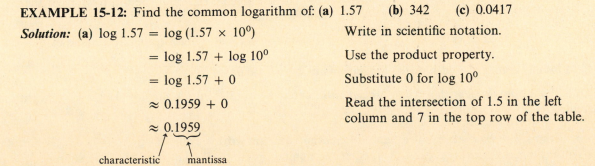

characteristic   mantissa

(b) $\log 342 = \log (3.42 \times 10^2)$ Write in scientific notation.

$= \log 3.42 + \log 10^2$ Use the product property.

$= \log 3.42 + 2$ Substitute 2 for $\log 10^2$.

$\approx 0.5340 + 2$ Read the intersection of 3.4 in the left column and 2 in the top row of the table.

≈ 2.5340

characteristic ⸻ mantissa

(c) $\log 0.0417 = \log (4.17 \times 10^{-2})$ Write in scientific notation.

$= \log 4.17 + \log 10^{-2}$ Use the product property.

$= \log 4.17 + (-2)$ Substitute -2 for $\log 10^{-2}$.

$\approx 0.6201 + (-2)$ Read the intersection of 4.1 in the left column and 7 in the top row of the table.

$\approx 0.6201 - 2$ or

≈ -1.3799 or ⎫

$\approx 8.6201 - 10$ ⎭ ⟵ other forms of the same logarithm

Note 1: In example **(c)**, -2 is the characteristic and 0.6201 is the mantissa.

Note 2: The common logarithm of a number greater than zero but less than one may be written in any of these three ways.

B. Finding common antilogarithms using a table

The **antilogarithm** of $\log x = y$ is defined as $x = $ antilog y. To **find the common antilogarithm,** we locate the mantissa in the table, read the corresponding numbers in the left column and the top row of the table, and use the characteristic to position the decimal point.

EXAMPLE 15-13: Find the common antilogarithm of: **(a)** 0.5966   **(b)** 3.6618   **(c)** -2.5969

Solution: **(a)** Antilog 0.5966

≈ 3.95 Read the left column (3.9) and the top row (5) for the mantissa 0.5966.

(b) Antilog 3.6618

$\approx 4.59 \times 10^3$ Read the left column (4.5) and the top row (9) for the mantissa 0.6618.

≈ 4590 Position the decimal point for a characteristic of 3.

(c) Antilog -2.5969

$= $ Antilog $0.4031 - 3$ Write the argument -2.5969 as a number between 0 and 1 by adding 3 to it, then subtracting 3 from the new expression.

$\approx 2.53 \times 10^{-3}$ Read the left column (2.5) and the top row (3) for the mantissa 0.4031.

≈ 0.00253 Position the decimal point for a characteristic of -3.

C. Finding common and natural logarithms using a calculator

To **find a common logarithm** using a calculator, enter the number, press the $\boxed{\log}$ key, and read the solution.

Logarithms with base e are called **natural logarithms** and are written as $\ln x$ ($\log_e M = \ln M$). To **find a natural logarithm** using a calculator, enter the number, press the $\boxed{\ln}$ key, and read the solution.

EXAMPLE 15-14: Find the common logarithm and natural logarithm of (a) 3450 and (b) 0.0321.

Solution: (a) log 3450 ≈ 3.5378   Enter 3450 and press $\boxed{\log}$ key.

In 3450 ≈ 8.1461   Enter 3450 and press $\boxed{\ln}$ key.

(b) log 0.0321 ≈ −1.4935   Enter 0.0321 and press $\boxed{\log}$ key.

In 0.0321 ≈ −3.4389   Enter 0.0321 and press $\boxed{\ln}$ key.

D. Finding the common and natural antilogarithms using a calculator

To find a **common antilogarithm,** recall that antilog $x = 10^x$, so enter the number and press the $\boxed{10^x}$ key. To **find a natural antilogarithm,** recall that antiln $x = e^x$, so enter the number in the calculator and press the $\boxed{e^x}$ key.

EXAMPLE 15-15: Find the common and natural antilogarithms (to 5 significant digits) of (a) 2.4 and (b) 0.24.

Solution: (a) Antilog 2.4 ≈ 251.19   Enter 2.4 and press $\boxed{10^x}$ key.

antiln 2.4 ≈ 11.023   Enter 2.4 and press $\boxed{e^x}$ key.

(b) Antilog 0.24 ≈ 1.7378   Enter 0.24 and press $\boxed{10^x}$ key.

Antiln 0.24 ≈ 1.2712   Enter 0.24 and press $\boxed{e^x}$ key.

E. Finding logarithms using the change-of-base formula

Since logarithm tables and calculators are for use with common logarithms, it is often convenient to be able to change logarithms of different bases into common logarithms. To do this we use the following **change-of-base formula:**

$$\log_N M = \frac{\log_a M}{\log_a N}, \qquad \text{where } a > 0 \text{ and } a \neq 1$$

Although we almost always use $a = 10$ or $a = e$ in the change-of-base formula, we could use any convenient number greater than zero and not equal to one, if we had suitable tables or calculators.

EXAMPLE 15-16: Find the logarithms using a table or calculator and the change-of-base formula: (a) $\log_4 64$ (b) ln 2.7183

Solution: (a) $\log_4 64 = \dfrac{\log 64}{\log 4}$ Express $\log_N M$ as $\dfrac{\log_a M}{\log_a N}$.

$\approx \dfrac{1.8062}{0.6021}$ Substitute using a table or calculator.

≈ 2.9998 Evaluate.

≈ 3

(b) $\ln 2.7183 = \dfrac{\log 2.7183}{\log e}$ Express $\ln M$ as $\dfrac{\log M}{\log e}$.

$\approx \dfrac{0.43429}{0.4343}$ Substitute using a table or calculator.

≈ 1 Evaluate.

15-5. Computing with Logarithms

To **compute with logarithms,** we use the properties of logarithms. You must preserve both mantissas and characteristics, so you may need to add integers to the characteristics.

EXAMPLE 15-17: Compute $\dfrac{0.432 \cdot \sqrt{2}}{84}$.

Solution: $\log \dfrac{0.432 \cdot \sqrt{2}}{84}$

$= \log 0.432 + \log 2^{\frac{1}{2}} - \log 84$	Express as simpler logarithms.
$= \log 0.432 + \frac{1}{2} \log 2 - \log 84$	Use the power property.
$\approx (0.6355 - 1) + \frac{1}{2}(0.3010) - 1.9243$	Substitute using table or calculator.
$\approx 0.6355 - 1 + 0.1505 - 1.9243$	
≈ -2.1383	
$\approx 0.8617 - 3$	
$\text{antilog}\,(0.8617 - 3) \approx 0.007273$	Find the antilogarithm.

$$\frac{0.432 \cdot \sqrt{2}}{84} \approx 0.007273$$

15-6. Solving Exponential and Logarithmic Equations

A. Solving exponential equations by equating exponents of like bases

Equations with variables in the exponents are called **exponential equations.** We can solve exponential equations by writing both members as powers of a common base, then equating the exponents and solving that equation in the usual manner.

EXAMPLE 15-18: Solve the equations **(a)** $3^{2x+3} = 81$ and **(b)** $2^{x^2} = 128$.

Solution: **(a)** $3^{2x+3} = 81$

$3^{2x+3} = 3^4$	Write 81 as a power of 3.
$2x + 3 = 4$	Equate the exponents.
$2x = 1$	Solve for x.
$x = \frac{1}{2}$	Check as before.

(b) $2^{x^2} = 128$

$2^{x^2} = 2^7$	Write 128 as a power of 2.
$x^2 = 7$	Equate the exponents.
$x = \pm\sqrt{7}$	Solve for x.

B. Solving equations using the logarithms of both members

Logarithms may be used as an aid to simplify calculations. To **solve an equation using logarithms,** we write the logarithms of both members, then solve the resulting equation.

EXAMPLE 15-19: Solve the equations **(a)** $3^{2x+3} = 81$ and **(b)** $e^{2x-1} = 20$.

Solution: **(a)** $3^{2x+3} = 81$

$\log 3^{2x+3} = \log 81$	Write the log of each member.
$(2x + 3) \log 3 = \log 81$	Solve for $2x + 3$.
$2x + 3 = \dfrac{\log 81}{\log 3}$	Use the power property.
$2x = \dfrac{\log 81}{\log 3} - 3$	
$x = \dfrac{\dfrac{\log 81}{\log 3} - 3}{2}$	

(b)
$$e^{2x-1} = 20$$

$$\ln e^{2x-1} = \ln 20 \qquad \text{Write the natural log of both members.}$$

$$(2x - 1)\ln e = \ln 20 \qquad \text{Use the power property.}$$

$$2x - 1 = \frac{\ln 20}{\ln e} \qquad \text{Solve for } x.$$

$$2x = \frac{\ln 20}{\ln e} + 1$$

$$x = \frac{\ln 20 + 1}{2} \qquad \text{Substitute 1 for } \ln e.$$

C. Solving logarithmic equations

Equations containing logarithmic expressions are called **logarithmic equations.** To solve a logarithmic equation, we obtain a single logarithmic expression in each member, then equate their arguments.

EXAMPLE 15-20: Solve the following equations:

(a) $\log x + \log (x - 3) = 1$ **(b)** $\log x = \log (x + 3) - \log (x - 1)$

Solution: **(a)** $\log x + \log (x - 3) = 1$

$$\log [x(x - 3)] = 1 \qquad \text{Use the product property.}$$

$$\log (x^2 - 3x) = \log 10 \qquad \text{Substitute log 10 for 1.}$$

$$x^2 - 3x = 10 \qquad \text{Equate the arguments.}$$

$$x^2 - 3x - 10 = 0 \qquad \text{Solve for } x.$$

$$(x - 5)(x + 2) = 0$$

$$x - 5 = 0 \text{ or } x + 2 = 0$$
$$x = 5 \text{ or } \qquad x = -2 \qquad -2 \text{ is not a solution because the argument cannot be negative.}$$

(b)
$$\log x = \log (x + 3) - \log (x - 1)$$

$$\log x = \log \frac{x + 3}{x - 1} \qquad \text{Use the quotient property.}$$

$$x = \frac{x + 3}{x - 1} \qquad \text{Equate antilogarithms.}$$

$$x(x - 1) = x + 3 \qquad \text{Solve for } x.$$

$$x^2 - x = x + 3$$

$$x^2 - 2x - 3 = 0$$

$$(x - 3)(x + 1) = 0$$

$$x - 3 = 0 \text{ or } x + 1 = 0$$

$$x = 3 \text{ or } \qquad x = -1 \qquad -1 \text{ is not a solution because the argument cannot be negative.}$$

15-7 Solving Problems Using Exponential Formulas

To **solve a problem using an exponential formula,** we:

1. *Read* the problem carefully, several times.
2. *Identify* the **dependent quantity** and **independent quantity.**
3. *Decide* how to represent the dependent and independent quantities using two different variables.

4. *Translate* to a **general exponential formula** of the form $y = y_0 e^{kx}$ or $y = y_0 e^{-kx}$.
5. *Replace* y_0 by the **initial value** of the dependent variable y when $x = 0$.
6. *Find k* by evaluating the exponential formula from step 5.
7. *Replace k* with the constant found in step 6 to find a **specific exponential formula**.
8. *Solve* the specific exponential formula for the unknown variable.
9. *Interpret* the solution from step 8 with respect to the unknown quantity to find the proposed solution of the original problem.
10. *Check* to see if the proposed solution satisfies all the conditions of the original problem.

A. Solving problems using exponential growth formulas

If x and y are any two different variables, then $y = y_0 e^{kx}$ is called a general **exponential growth formula**, where y_0 is the initial value at time 0 ($x = 0$) and k is a positive **exponential growth constant**.

EXAMPLE 15-21: Solve this problem using an exponential growth formula:

Read: The number of bacteria present in a culture at time zero (beginning of experiment) is 100. At the end of 2 hours, there are 300 bacteria in the culture. How many bacteria will be in the culture after one full day?

Identify: The dependent quantity is the number of bacteria in the culture at the end of a given time period.
The independent quantity is the amount of elapsed time from time zero.

Decide: Let b = the number of bacteria at the end of a given time period

and t = the amount of elapsed time from time zero.

Translate: $b = b_0 e^{kt}$ ⟵ general exponential growth formula

Replace b_0: $b = (100)e^{kt}$ $b = 100$ (bacteria) when $t = 0$ (time zero) means $b_0 = 100$.

Find k: $300 = 100e^{k(2)}$ $b = 300$ (bacteria) when $t = 2$ (hours).

$$\ln 300 = \ln \left[100e^{2k}\right]$$

$$\ln 300 = \ln 100 + \ln e^{2k}$$

$$\ln 300 = \ln 100 + 2k$$

$$\ln 300 - \ln 100 = 2k$$

$$\ln \left(\frac{300}{100}\right) = 2k$$

$$\ln 3 = 2k$$

$$\frac{\ln 3}{2} = k \quad ⟵ \text{ exponential growth constant } [\approx 0.5493]$$

Replace k: $b = 100e^{\left(\frac{\ln 3}{2}\right)t}$ ⟵ specific exponential growth formula

Solve: $b = 100e^{\left(\frac{\ln 3}{2}\right)24}$ Find b (number of bacteria) when $t = 24$ (hours).

$$b \approx 100e^{\left(\frac{1.09861}{2}\right)24}$$ Use a calculator with a $\ln x$ key.

$$b \approx 100e^{13.18332}$$

$$b = 100(531,426)$$ Use a calculator with an e^x key.

$$b \approx 53,142,600$$

Interpret: $b \approx 53,142,600$ means after one full day the culture contains approximately 53,142,600 bacteria.

Check: Does $t \approx 24$ (hours) when $b = 100e^{\left(\frac{\ln 3}{2}\right)t}$ is evaluated for $b = 53,142,600$ (bacteria)?
Yes: $53,142,600 = 100e^{\left(\frac{\ln 3}{2}\right)t}$ Find t (the number of hours) when $b = 53,142,600$ (bacteria).

$$\ln(53,142,600) = \ln\left[100e^{\left(\frac{\ln 3}{2}\right)t}\right]$$

$$\ln(53,142,600) = \ln 100 + \ln e^{\left(\frac{\ln 3}{2}\right)t}$$

$$\ln(53,142,600) - \ln 100 = \left(\frac{\ln 3}{2}\right)t$$

$$\ln\left(\frac{53,142,600}{100}\right) = \left(\frac{\ln 3}{2}\right)t$$

$$\frac{2\ln(531,426)}{\ln 3} = t$$

$$23.9999 \approx t$$

$$24 = t \longleftarrow b \approx 53,142,600 \text{ checks}$$

B. Solving problems using exponential decay formulas

If x and y are any two different variables, then $y = y_0 e^{-kx}$ is called a general **exponential decay formula** where y_0 is the initial value at time zero ($x = 0$) and k is a positive **exponential decay constant.**

EXAMPLE 15-22: Solve this problem using an exponential decay formula:

Read: The amount of carbon 14 in a given organism at the time of that organism's death (time zero) is 100%. The **half-life** of carbon 14 is 5750 years, which means only 50% of the original carbon 14 will remain in the given object 5750 years later. How old is an object that is found to contain only 20% of its original carbon 14, to the nearest ten years?

Identify: The dependent quantity is the percent of carbon 14 remaining at the end of a given time period.
The independent quantity is the amount of elapsed time from time zero.

Decide: Let c = the percent of carbon 14 remaining at the end of a given time period.

and t = the amount of elapsed time from time zero.

Translate: $c = c_0 e^{-kt} \longleftarrow$ general exponential decay formula
Replace c_0: $c = (1)e^{-kt}$ $c = 100\%$ (of the carbon 14) when $t = 0$ (at time zero) means $c_0 = 100\% = 1$.

Find k: $0.5 = e^{-k(5750)}$ $c = 50\%$ (of the carbon 14) when $t = 5750$ [years].

$\ln 0.5 = \ln e^{-5750k}$

$\ln 0.5 = -5750k$

$\dfrac{\ln 0.5}{-5750} = k \longleftarrow$ exponential decay constant (≈ 0.0001)

Replace k: $c = e^{\left(\frac{\ln 0.5}{-5750}\right)t} \longrightarrow$ specific exponential decay formula

$c = e^{\left(\frac{\ln 0.5}{5750}\right)t}$ Simplify.

Solve: $\qquad 0.2 = e^{\left(\frac{\ln 0.5}{5750}\right)t}$ $\qquad$ Find t (elapsed time in years) when $c = 20\%$ (of the carbon 14).

$$\ln 0.2 = \ln e^{\left(\frac{\ln 0.5}{5750}\right)t}$$

$$\ln 0.2 = \left(\frac{\ln 0.5}{5750}\right)t$$

$$t = \frac{5750(\ln 0.2)}{\ln 0.5}$$

$$t \approx \frac{5750(-1.609437912)}{-0.6931471806} \qquad \text{Use a calculator with a } \ln x \text{ key.}$$

$$t \approx 13{,}351.08655$$

$$t \approx 13{,}350 \qquad \text{Round to the nearest ten.}$$

Interpret: $\quad t \approx 13{,}350$ means the age of an object that contains only 20% of its original carbon 14 is about 13,350 years old.

Check: $\quad$ Does $c \approx 20\%$ (of the carbon 14) when $c = e^{\left(\frac{\ln 0.5}{5750}\right)t}$ is evaluated for $t = 13{,}350$ (years)?

$\qquad$ Yes: $c = e^{\left(\frac{\ln 0.5}{5750}\right)(13{,}350)}$ $\qquad$ Find c (the percent of carbon 14 that remains) when $t = 13{,}350$ (years).

$$c \approx e^{(-0.0001205473)(13{,}350)}$$

$$c = e^{-16.09306932}$$

$$c \approx 0.2000261978$$

$$c \approx 0.2$$

$$c = 20\% \longleftarrow t = 13{,}350 \text{ checks}$$

15-9. Solving Problems Using Logarithmic Formulas

To **solve a problem using a logarithmic formula,** we:

1. *Read* the problem very carefully, several times.
2. *Identify* the dependent quantity and the independent quantity.
3. *Decide* how to represent the dependent and independent quantities using two different variables.
4. *Translate* to a **general logarithmic formula** of the form $y = k \log\left(\dfrac{x}{x_0}\right)$.
5. *Replace* x_0 by the initial value of the independent variable x when $y = 0$.
6. *Find* k by evaluating the logarithmic formula from step 5.
7. *Replace* k with the constant found in step 6 to find a **specific logarithmic formula.**
8. *Solve* the specific logarithmic formula for the unknown variable.
9. *Interpret* the solution from step 8 with respect to the unknown quantity to find the proposed solution of the original problem.
10. *Check* to see if the proposed solution satisfies all the conditions of the original problem.

EXAMPLE 15-23: Solve this problem using a logarithmic formula:

Read: $\qquad$ The amplitude of an earthquake wave is measured on a seismograph in millimeters (mm). The magnitude of an earthquake is measured on the Richter scale. By international agreement, the measure of a zero-level earthquake is 10^{-3} mm of amplitude at 100 km from the epicenter (middle) of the earthquake. The magnitude of the San Francisco earthquake of 1906 was 8.3 on the Richter scale, and had an amplitude of 199,526 mm at 100 km from the epicenter. The Japan earthquake of 1933 had an amplitude of 794,328 mm at 100 km from the epicenter. What was the magnitude of the Japan earthquake of 1933 on the Richter scale?

Identify: The dependent quantity is the magnitude of the earthquake.
The independent quantity is the amplitude of the earthquake.

Decide: Let m = the magnitude of the earthquake

and a = the amplitude of the earthquake.

Translate: $m = k \log\left(\dfrac{a}{a_0}\right)$ ⟵ general logarithmic formula

Replace a_0: $m = k \log\left(\dfrac{a}{10^{-3}}\right)$ By international agreement, $a_0 = 10^{-3}$.

$m = k \log(10^3 a)$ Simplify.

Find k: $8.3 = k \log[10^3(199{,}526)]$ $m = 8.3$ (on the Richter scale)
when $a = 199{,}526$ (mm).

$8.3 = k \log(199{,}526{,}000)$

$8.3 \approx k(8.29999)$ Use a calculator with log x key.

$8.3 \approx k(8.3)$ Round to the nearest tenth.

$1 = k$ ⟵ logarithmic constant

Replace k: $m = 1 \log(10^3 a)$ ⟵ specific logarithmic formula

$m = \log(10^3 a)$ Simplify.

$m = \log 10^3 + \log a$

$m = \log a + 3$ ⟵ simplest form

Solve: $m = \log(794{,}328) + 3$ Find m (on the Richter scale) when $a = 794{,}328$ (mm).

$m \approx 5.89999 + 3$ Use a calculator with a log x key.

$m = 8.89999$ Add.

$m \approx 8.9$ Round to the nearest tenth.

Interpret: $m \approx 8.9$ means the magnitude of the Japan earthquake of 1933 was 8.9 on the Richter scale.

Check: Does $a \approx 794{,}328$ (mm) when $m = \log(10^3 a)$ is evaluated for $m = 8.9$ (on the Richter scale)? Yes:

$8.9 = \log a + 3$ Find a (mm) when $m = 8.9$ (on the Richter scale).

$5.9 = \log a$

$10^{5.9} = a$

$794{,}328.2347 = a$ Use a calculator with a y^x key or 10^x key.

$794{,}328 = a$ ⟵ $m \approx 8.9$ checks

RAISE YOUR GRADES
Can you . . . ?

☑ evaluate expressions of the form p^x to the nearest thousandth using a calculator
☑ graph exponential functions
☑ graph the inverses of exponential functions
☑ write logarithmic equations in exponential form
☑ write exponential equations in logarithmic form
☑ solve logarithmic equations

☑ graph logarithmic functions
☑ prove logarithmic properties
☑ express complex logarithms in terms of simpler logarithms
☑ express logarithmic expressions as a single logarithm
☑ find logarithms using given logarithmic values
☑ find common logarithms using a table
☑ find common antilogarithms using a table
☑ find common and natural logarithms using a calculator
☑ find common and natural antilogarithms using a calculator
☑ find logarithms using the change-of-base formula
☑ compute using logarithms
☑ find logarithms when four digits occur
☑ find antilogarithms when five digits occur
☑ solve exponential equations by equating exponents of like bases
☑ solve exponential equations using the logarithms of both members
☑ solve logarithmic equations
☑ solve a problem using an exponential growth formula
☑ solve a problem using an exponential decay formula
☑ solve a problem using a logarithmic formula

SOLVED PROBLEMS

PROBLEM 15-1 Evaluate the following expressions. Round each answer to the nearest thousandth:

(a) $2^{1.5}$ **(b)** $3^{1.75}$ **(c)** $9^{0.5}$ **(d)** $9^{0.25}$ **(e)** e^1 **(f)** $e^{1.4}$

Solution: Recall that to evaluate an expression of the form b^x, you enter the base, press the exponential key, enter the exponent and press the equal ($=$) key [see Example 15-1].

(a) $2^{1.5}$ **(b)** $3^{1.75}$ **(c)** $9^{0.5}$

(a)	(b)	(c)	
2	3	9	Enter the base.
2	3	9	Press the $\boxed{y^x}$ key.
1.5	1.75	0.5	Enter the exponent.
2.8284271	6.8385212	3	Press the $\boxed{=}$ key.
2.828	6.839	3.000	Round to the nearest thousandth.

(d) $9^{0.25}$ **(e)** e^1 **(f)** $e^{1.4}$

(d)	(e)	(f)	
9	2.7182818	2.7182818	Enter the base.
9	2.7182818	2.7182818	Press the $\boxed{y^x}$ key.
0.25	1	1.4	Enter the exponent.
1.7320508	2.7182818	4.0551999	Press the $\boxed{=}$ key.
1.732	2.718	4.0552	Round to the nearest thousandth.

PROBLEM 15-2 Graph the following exponential functions:

(a) $y = 3^x$ **(b)** $y = (\frac{1}{3})^x$ **(c)** $y = 2^{2x}$ **(d)** $y = 2^{x-3}$

Solution: Recall that to graph an exponential function, you make a table of solutions of the function, plot the ordered pairs and draw a smooth curve through the points [see Example 15-2].

(a) $y = 3^x$

x	-2	-1	0	1	2
3^x	$\frac{1}{9}$	$\frac{1}{3}$	1	3	9

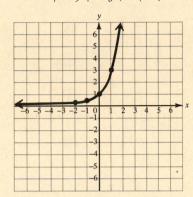

(b) $y = \left(\frac{1}{3}\right)^x$

x	-2	-1	0	1	2
y	9	3	1	$\frac{1}{3}$	$\frac{1}{9}$

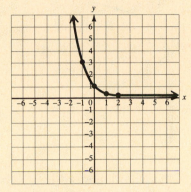

(c) $y = 2^{2x}$

x	-2	-1	0	1	2
2^{2x}	$\frac{1}{16}$	$\frac{1}{4}$	1	4	16

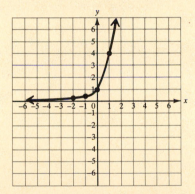

(d) $y = 2^{x-3}$

x	0	1	2	3	4	5	6
y	$\frac{1}{8}$	$\frac{1}{4}$	$\frac{1}{2}$	1	2	4	8

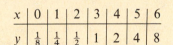
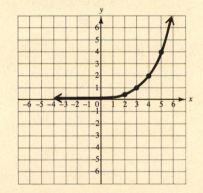

PROBLEM 15-3 Graph the following exponential functions:

(a) $x = 4^y$ **(b)** $x = \left(\frac{3}{4}\right)^y$ **(c)** $x = 2^{y-2}$ **(d)** $x = 2^{y+1}$

Solution: Recall that to graph an exponential function, you make a table of solutions, plot the ordered pairs and draw a smooth curve through the points [see Example 15-3].

(a) $x = 4^y$

y	-2	-1	0	1	2
4^y	$\frac{1}{16}$	$\frac{1}{4}$	1	4	16

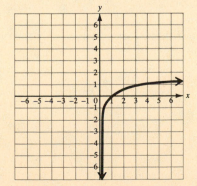

(b) $x = \left(\frac{3}{4}\right)^y$

y	-3	-5	-2	-1	0	1	2
$\left(\frac{3}{4}\right)^y$	2.4	4.2	$\frac{16}{9}$	$\frac{4}{3}$	1	$\frac{3}{4}$	$\frac{9}{16}$

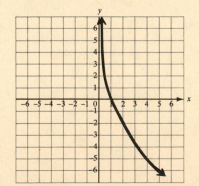

(c) $x = 2^{y-2}$

y	-2	-1	0	1	2	4
2^{y-2}	$\frac{1}{16}$	$\frac{1}{8}$	$\frac{1}{4}$	$\frac{1}{2}$	1	4

(d) $x = 2^{y+1}$

y	-4	-3	-2	-1	0	1	2
2^{y+1}	$\frac{1}{8}$	$\frac{1}{4}$	$\frac{1}{2}$	1	2	4	8

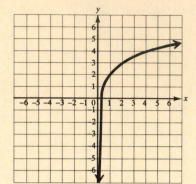

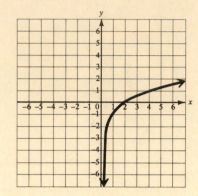

PROBLEM 15-4 Write the following logarithmic equations in exponential form:

(a) $\log_{10} 100 = 2$ **(b)** $\log_6 6 = 1$ **(c)** $\log_{\frac{1}{2}} 16 = -4$
(d) $\log_{\frac{1}{2}} 4 = -2$ **(e)** $\log_{\sqrt{3}} = 2$ **(f)** $\log_e 1 = 0$

Solution: Recall that to write a logarithmic equation in exponential form, you identify the parts and use the definition of a logarithm [see Example 15-4].

(a) $\log_{10} 100 = 2$ **(b)** $\log_6 6 = 1$ **(c)** $\log_{\frac{1}{2}} 16 = 4$

$\qquad b = 10 \qquad\qquad b = 6 \qquad\qquad b = \frac{1}{2}$

$\qquad y = 2 \qquad\qquad y = 1 \qquad\qquad y = -4$

$\qquad x = 100 \qquad\qquad x = 6 \qquad\qquad x = 16$

$\qquad 100 = 10^2 \qquad\qquad 6 = 6^1 \qquad\qquad 16 = \left(\frac{1}{2}\right)^{-4}$

(d) $\log_{\frac{1}{2}} 4 = 2$ **(e)** $\log_{\sqrt{3}} 3 = 2$ **(f)** $\log_e 1 = 0$

$\qquad b = \frac{1}{2} \qquad\qquad b = \sqrt{3} \qquad\qquad b = e$

$\qquad y = -2 \qquad\qquad y = 2 \qquad\qquad y = 0$

$\qquad x = 4 \qquad\qquad x = 3 \qquad\qquad x = 1$

$\qquad 4 = \left(\frac{1}{2}\right)^{-2} \qquad\qquad 3 = (\sqrt{3})^2 \qquad\qquad 1 = e^0$

PROBLEM 15-5 Write the following exponential equations in logarithmic form:

(a) $10^3 = 1000$ **(b)** $10^{-2} = 0.01$ **(c)** $10^0 = 1$
(d) $2^7 = 128$ **(e)** $5^4 = 625$ **(f)** $9^{\frac{1}{2}} = 3$

Solution: Recall that to write an exponential equation in logarithmic form, you identify the parts and use the definition of a logarithm

(a) $\qquad 10^3 = 1000$ **(b)** $\qquad 10^{-2} = 0.01$ **(c)** $\qquad 10^0 = 1$

$\qquad b = 10 \qquad\qquad b = 10 \qquad\qquad b = 10$

$\qquad y = 3 \qquad\qquad y = -2 \qquad\qquad y = 0$

$\qquad x = 1000 \qquad\qquad x = 0.01 \qquad\qquad x = 1$

$\quad \log_{10} 1000 = 3 \qquad \log_{10} 0.01 = -2 \qquad \log_{10} 1 = 0$

(d)
$$2^7 = 128$$
$$b = 2$$
$$y = 7$$
$$x = 128$$
$$\log_2 128 = 7$$

(e)
$$5^4 = 625$$
$$b = 5$$
$$y = 4$$
$$x = 625$$
$$\log_5 625 = 4$$

(f)
$$9^{\frac{1}{2}} = 3$$
$$b = 9$$
$$y = \tfrac{1}{2}$$
$$x = 3$$
$$\log_9 3 = \tfrac{1}{2}$$

PROBLEM 15-6 Solve the following equations:

(a) $\log_4 x = 3$ **(b)** $\log_2 x = -1$ **(c)** $\log_x 64 = 3$
(d) $\log_x 64 = 6$ **(e)** $\log_{36} 6 = x$ **(f)** $\log_{27} 81 = x$

Solution: Recall that to solve a logarithmic equation you write the equation in exponential form and both members on like bases, equate the exponents and solve [see Example 15-6].

(a) $\log_4 x = 3$
$$x = 4^3$$
$$x = 64$$

(b) $\log_2 x = -1$
$$x = 2^{-1}$$
$$x = \tfrac{1}{2}$$

(c) $\log_x 64 = 3$
$$64 = x^3$$
$$4^3 = x^3$$
$$4 = x$$

(d) $\log_x 64 = 6$
$$64 = x^6$$
$$2^6 = x^6$$
$$2 = x$$

(e) $\log_{36} 6 = x$
$$6 = 36^x$$
$$6^1 = (6^2)^x$$
$$6^1 = 6^{2x}$$
$$1 = 2x$$
$$\tfrac{1}{2} = x$$

(f) $\log_{27} 81 = x$
$$81 = 27^x$$
$$3^4 = (3^3)^x$$
$$3^3 = 3^{3x}$$
$$4 = 3x$$
$$\tfrac{4}{3} = x$$

PROBLEM 15-7 Graph the following logarithmic functions:

(a) $f(x) = \log_3 x$ **(b)** $f(x) = -\log_{\frac{1}{2}} x$

Solution: Recall that to graph a logarithmic equation, you make a table of solutions, plot the ordered pairs and draw a smooth curve through the points [see Example 15-7].

(a) $y = \log_3 x$
$$x = 3^y$$

y	-2	-1	0	1	2
x	$\frac{1}{9}$	$\frac{1}{3}$	1	3	9

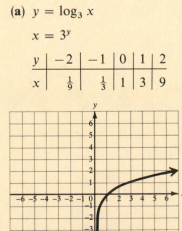

(b) $f(x) = -\log_{\frac{1}{2}} x$
$$x = \left(\tfrac{1}{2}\right)^{-y}$$

y	-3	-2	-1	0	1	2	3
x	$\frac{1}{8}$	$\frac{1}{4}$	$\frac{1}{2}$	1	2	4	8

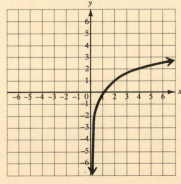
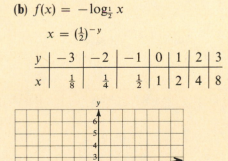

Fig. No. 15-68

PROBLEM 15-8 Prove the following properties of logarithms:

(a) $\log_b \dfrac{M}{N} = \log_b M - \log_b N$ (b) $\log_b (M^p) = p \log_b M$ (c) $\log_N M = \dfrac{\log_a M}{\log_a N}$

Solution: Recall that to prove a property of logarithm, you use the definition of logarithms [see Example 15-8].

(a) $\log_b \dfrac{M}{N} = \log_b M - \log_b N$

Let $\log_b M = x$ and $\log_b N = y$

Then $M = b^x$ and $N = b^y$

$\dfrac{M}{N} = \dfrac{b^x}{b^y}$

$\dfrac{M}{N} = b^{x-y}$

$\log_b \dfrac{M}{N} = x - y$

$\log_b \dfrac{M}{N} = \log_b M - \log_b N$

(b) $\log_b (M^p) = p \log_b M$

Let $\log_b M = x$

Then $M = b^x$

$M^p = (b^x)^p$

$\quad\ = b^{xp}$

$\log_b (M^p) = px$

$\log_b (M^p) = p \log_b M$

(c) $\log_N M = \dfrac{\log_a M}{\log_a N}$

Let $\log_N M = x$

$M = N^x$

$\log_a M = \log_a (N^x)$

$\log_a M = x \log_a N$

$\dfrac{\log_a M}{\log_a N} = x$

$\dfrac{\log_a M}{\log_a N} = \log_N M$

PROBLEM 15-9 Express the following logarithms in terms of simpler logarithms:

(a) $\log_b w^2 x^4 y^3$ (b) $\log_b \dfrac{u^5}{v^3 w^2}$ (c) $\log_b \dfrac{\sqrt[3]{x^2}}{\sqrt{x}}$ (d) $\log_b \sqrt[5]{\dfrac{x^2 y^3}{z^4}}$

Solution: Recall that to express a logarithm in terms of simpler logarithms, you use the properties of logarithms [see Example 15-9].

(a) $\log_b w^2 x^4 y^3$

$= \log_b w^2 + \log_b x^4 + \log_b y^3$

$= 2 \log_b w + 4 \log_b x + 3 \log_b y$

(b) $\log_b \dfrac{u^5}{v^3 w^2}$

$= \log_b u^5 - \log_b v^3 w^2$

$= \log_b u^5 - (\log_b v^3 + \log_b w^2)$

$= \log_b u^5 - \log_b v^3 - \log_b w^2$

$= 5 \log_b u - 3 \log_b v - 2 \log_b w$

(c) $\log_b \dfrac{\sqrt[3]{x^2}}{\sqrt{x}}$

$= \log_b \sqrt[3]{x^2} - \log_b \sqrt{x}$

$= \log_b x^{\frac{2}{3}} - \log_b x^{\frac{1}{2}}$

$= \dfrac{2}{3} \log_b x - \dfrac{1}{2} \log_b x$

$= \left(\dfrac{2}{3} - \dfrac{1}{2} \right) \log_b x$

$= \dfrac{1}{6} \log_b x$

(d) $\log_b \sqrt[5]{\dfrac{x^2 y^3}{z^4}}$

$= \log_b \left[\dfrac{x^2 y^3}{z^4} \right]^{\frac{1}{5}}$

$= \log_b \dfrac{x^{\frac{2}{5}} y^{\frac{3}{5}}}{z^{\frac{4}{5}}}$

$= \log_b x^{\frac{2}{5}} y^{\frac{3}{5}} - \log_b z^{\frac{4}{5}}$

$= \log_b x^{\frac{2}{5}} + \log_b y^{\frac{3}{5}} - \log_b z^{\frac{4}{5}}$

$= \dfrac{2}{5} \log_b x + \dfrac{3}{5} \log_b y - \dfrac{4}{5} \log_b z$

PROBLEM 15-10 Express the following logarithmic expressions as a single logarithm:

(a) $3 \log_b p + 4 \log_b q$ (b) $\frac{1}{2} \log_b r - 2 \log_b q$ (c) $\frac{1}{2} \log_b x + \frac{1}{4} \log_b y - \frac{3}{4} \log_b z$
(d) $\frac{1}{3} \log_b x - \frac{2}{3} \log_b y - \frac{1}{3} \log_b z$ (e) $\frac{2}{3} \log_b x - \frac{1}{3} \log_b y + \log_b z - \log_b w$

Solution: Recall that to express a logarithmic expression as a single logarithm, you use the properties of logarithms as before but in reverse [see Example 15-10].

(a) $3 \log_b p + 4 \log_b q$

$= \log_b p^3 + \log_b q^4$

$= \log_b p^3 q^4$

(b) $\frac{1}{2} \log_b r - 2 \log_b q$

$= \log_b r^{\frac{1}{2}} - \log_b q^2$

$= \log_b \dfrac{\sqrt{r}}{q^2}$

(c) $\frac{1}{2} \log_b x + \frac{1}{4} \log_b y - \frac{3}{4} \log_b z$

$= \log_b x^{\frac{1}{2}} + \log_b y^{\frac{1}{4}} - \log_b z^{\frac{3}{4}}$

$= \log_b x^{\frac{2}{4}} y^{\frac{1}{4}} - \log_b z^{\frac{3}{4}}$

$= \log_b \dfrac{x^{\frac{2}{4}} y^{\frac{1}{4}}}{z^{\frac{3}{4}}}$

$= \log_b \sqrt[4]{\dfrac{x^2 y}{z^3}}$

(d) $\frac{1}{3} \log_b x - \frac{2}{3} \log_b y - \frac{1}{3} \log_b z$

$= \log_b x^{\frac{1}{3}} - \log_b y^{\frac{2}{3}} - \log_b z^{\frac{1}{3}}$

$= \log_b x^{\frac{1}{3}} - (\log_b y^{\frac{2}{3}} + \log_b z^{\frac{1}{3}})$

$= \log_b x^{\frac{1}{3}} - \log_b y^{\frac{2}{3}} z^{\frac{1}{3}}$

$= \log_b \dfrac{x^{\frac{1}{3}}}{y^{\frac{2}{3}} z^{\frac{1}{3}}}$

$= \log_b \sqrt[3]{\dfrac{x}{y^2 z}}$

(e) $\frac{2}{3} \log_b x - \frac{1}{3} \log_b y + \log_b z - \log_b w$

$= \log_b x^{\frac{2}{3}} + \log_b z - (\log_b y^{\frac{1}{3}} + \log_b w)$

$= \log_b x^{\frac{2}{3}} z^{\frac{3}{3}} - \log_b y^{\frac{1}{3}} w^{\frac{3}{3}}$

$= \log_b \dfrac{x^{\frac{2}{3}} z^{\frac{3}{3}}}{y^{\frac{1}{3}} w^{\frac{3}{3}}}$

$= \log_b \sqrt[3]{\dfrac{x^2 z^3}{y w^3}}$

PROBLEM 15-11 Approximate the following logarithms in terms of $\log_{10} 2 \approx 0.3010$, $\log_{10} 3 \approx 0.4771$, and $\log_{10} 5 \approx 0.6990$:

(a) $\log_{10} 36$ **(b)** $\log_{10} \sqrt{3}$ **(c)** $\log_{10} \sqrt[4]{6}$ **(d)** $\log_{10} 2.5$

Solution: Recall that to approximate a logarithm using the given logarithmic values, you express the logarithm in terms of the simpler logarithms, substitute the given values and evaluate the expression [see Example 15-11].

(a) $\log_{10} 36$

$= \log_{10} (4 \cdot 9)$

$= \log_{10} 4 + \log_{10} 9$

$= \log_{10} 2^2 + \log_{10} 3^2$

$= 2 \log_{10} 2 + 2 \log_{10} 3$

$\approx 2(0.3010) + 2(0.4771)$

≈ 1.5562

(b) $\log_{10} \sqrt{3}$

$= \log_{10} 3^{\frac{1}{2}}$

$= \frac{1}{2} \log_{10} 3$

$\approx \frac{1}{2}(0.4771)$

≈ 0.2386

(c) $\log_{10} \sqrt[4]{6}$

$= \log_{10} 6^{\frac{1}{4}}$

$= \frac{1}{4} \log_{10} (2 \cdot 3)$

$= \frac{1}{4}(\log_{10} 2 + \log_{10} 3)$

$\approx \frac{1}{4}(0.3010 + 0.4771)$

$\approx \frac{1}{4}(0.7781)$

≈ 0.1945

(d) $\log_{10} 2.5$

$= \log_{10} \frac{5}{2}$

$= \log_{10} 5 - \log_{10} 2$

$\approx 0.6990 - 0.3010$

≈ 0.3980

406 Intermediate Algebra College Outline

PROBLEM 15-12 Find the common logarithms of the following numbers:

(a) 1.43   (b) 8.17   (c) 0.432   (d) 18.3   (e) 18500   (f) 0.00657

Solution: Recall that to find the logarithm of a number, you write the number in scientific notation, use the product property of logarithms and read the logarithm in the table [see Example 15-12].

(a) log 1.43

$= \log 1.43 \times 10^0$

$= \log 1.43 + \log 10^0$

$= \log 1.43 + 0$

≈ 0.1553

(b) log 8.17

$= \log 8.17 \times 10^0$

$= \log 8.17 + \log 10^0$

$= \log 8.17 + 0$

≈ 0.9122

(c) log 0.432

$= \log 4.32 \times 10^{-1}$

$= \log 4.32 + \log 10^{-1}$

$= \log 4.32 + (-1)$

$\approx 0.6355 - 1$

(d) log 18.3

$= \log 1.83 \times 10^1$

$= \log 1.83 + \log 10^1$

$= \log 1.83 + 1$

$\approx 0.2625 + 1$

≈ 1.2625

(e) log 18500

$= \log 1.85 \times 10^4$

$= \log 1.85 + \log 10^4$

$= \log 1.85 + 4$

$\approx 0.2672 + 4$

≈ 4.2672

(f) log 0.00657

$= \log 6.57 \times 10^{-3}$

$= \log 6.57 + \log 10^{-3}$

$= \log 6.57 + (-3)$

$\approx 0.8176 - 3$

PROBLEM 15-13 Find the antilogarithms of the following numbers:

(a) 0.6075   (b) 2.4871   (c) 3.8142   (d) 1.9289   (e) 0.3160 − 1   (f) 0.8785 − 3

Solution: Recall that to find the antilogarithm of a number, you locate the mantissa in the table, read the corresponding numbers in the left column and the top row of the table and use the characteristic to place the decimal point [see Example 15-13].

(a) antilog 0.6075
4.05

(b) antilog 2.4871
3.07×10^2
307.0
307

(c) antilog 3.8142
6.52×10^3
6520.0
6520

(d) antilog 1.9289
8.49×10^1
84.9

(e) antilog (0.3160 − 1)
2.07×10^{-1}
0.207

(f) antilog (0.8785 − 3)
7.56×10^{-3}
0.00756

PROBLEM 15-14 Find the common logarithms and the natural logarithms of the following numbers using a calculator: (a) 358   (b) 74.7   (c) 2.59   (d) 0.341   (e) 0.0901   (f) 3840

Solution: Recall that to find the common logarithm of a number, you enter the number, press the $\boxed{\log}$ key and read the solution. To find the natural logarithm of a number, you enter the number, press the $\boxed{\ln}$ key and read the solution [see Example 15-14].

(a) log 358
2.553883
ln 358
5.880533

(b) log 74.7
1.8733206
ln 74.7
4.3134801

(c) log 2.59
0.4132998
ln 2.59
0.9516579

(d) log 0.341
−0.4672456
ln 0.341
−1.0758728

(e) log 0.0901
−1.0452752
ln 0.0901
−2.4068351

(f) log 3840
3.5843312
ln 3840
8.2532276

PROBLEM 15-15 Find the common and natural antilogarithm of each of the following:

(a) 2.4   **(b)** 5.72   **(c)** 2.433   **(d)** 1.643   **(e)** 0.514   **(f)** 0.0416

Solution: Recall that to find the common antilogarithm of a number, you enter the number, press the $\boxed{10^x}$ key and read the solution. To find the natural antilogarithm of a number, you enter the number, press the $\boxed{e^x}$ key and read the solution [see Example 15-15].

(a) antilog 2.4   **(b)** antilog 5.72   **(c)** antilog 2.433

≈ 251.18864 ≈ 524807.46 ≈ 271.01916

antiln 2.4   antiln 5.72   antiln 2.433

≈ 11.023176 ≈ 304.90492 ≈ 11.39301

(d) antilog 1.643   **(e)** antilog 0.514   **(f)** antilog 0.0416

≈ 43.954162 ≈ 3.2658783 ≈ 1.1005252

antiln 1.643   antiln 0.514   antiln 0.0416

≈ 5.1706582 ≈ 1.6719657 ≈ 1.0424774

PROBLEM 15-16 Find the logarithms of the following numbers using the change-of-base formula:

(a) $\log_8 72$ **(b)** $\log_6 27$ **(c)** $\log_3 42$ **(d)** $\log_2 50$ **(e)** $\ln 41$ **(f)** $\ln 81$

Solution: Recall that to find the logarithm of a number using the change-of-base formula, you write

$$\log_N M = \frac{\log_a M}{\log_a N} \text{ [see Example 15-61].}$$

(a) $\log_8 72 = \dfrac{\log 72}{\log 8}$ **(b)** $\log_6 27 = \dfrac{\log 27}{\log 6}$ **(c)** $\log_3 42 = \dfrac{\log 42}{\log 3}$

$\approx \dfrac{1.8573}{0.9031}$ $\approx \dfrac{1.4314}{0.7782}$ $\approx \dfrac{1.6232}{0.4771}$

≈ 2.0566 ≈ 1.8394 ≈ 3.4022

(d) $\log_2 50 = \dfrac{\log 50}{\log 2}$ **(e)** $\ln 41 = \dfrac{\log 41}{\log e}$ **(f)** $\ln 81 = \dfrac{\log 81}{\log e}$

$\approx \dfrac{1.6990}{0.3010}$ $\approx \dfrac{1.6128}{0.4343}$ $\approx \dfrac{1.9085}{0.4343}$

≈ 5.6439 ≈ 3.7136 ≈ 4.3944

PROBLEM 15-17 Compute the following expressions:

(a) 327^2 **(b)** 432×2.34 **(c)** $5830 \div 2.34$ **(d)** $\sqrt[4]{0.0687}$ **(e)** $6^{3.1}$ **(f)** $\dfrac{427\sqrt[3]{2.11}}{11.3^2}$

Solution: Recall that to compute with logarithms, you use the properties of logarithms. You must preserve both mantissas and characteristics [see Example 15-17].

(a) $\log 327^2 = 2 \log 327$ **(b)** $\log (432 \times 2.34)$ **(c)** $\log (5839 \div 2.35)$

$\approx 2(2.5145)$ $= \log 432 + \log 2.34$ $= \log 5830 - \log 2.35$

≈ 5.0291 $\approx 2.6355 + 0.3692$ $\approx 3.7657 - 0.3711$

$327^2 = 106929$ ≈ 3.0047 ≈ 3.3946

$432 \times 2.34 \approx 1010.88$ $5830 \div 2.35 \approx 2480.85$

(d) $\log \sqrt[4]{0.0687}$

$= \frac{1}{4} \log 0.0687$

$\approx \frac{1}{4}(0.8370 - 2)$

$\approx \frac{1}{4}(2.8370 - 4)$

$\approx 0.7093 - 1$

$\sqrt[4]{0.0687} \approx 0.5120$

(e) $\log 6^{3.1}$

$= 3.1 \log 6$

$\approx 3.1(0.7782)$

≈ 2.4123

$6^{3.1} \approx 258.39$

(f) $\log \dfrac{427 \sqrt[3]{2.11}}{11.3^2}$

$= \log 427 + \frac{1}{3} \log 2.11 - 2 \log 11.3$

$\approx 2.6304 + \frac{1}{3}(0.3243) - 2(1.0531)$

$\approx 2.6304 + 0.1081 - 2.1062$

≈ 0.6323

$\dfrac{427 \sqrt[3]{2.11}}{11.3^2} \approx 4.2884$

PROBLEM 15-18 Solve the following exponential equations:

(a) $3^x = 27$ (b) $4^{x-3} = 16$ (c) $4^x = 8$ (d) $3^{2-x} = 81$ (e) $2^{x^2-2} = 64$
(f) $3^{x^2+2} = 81$

Solution: Recall that to solve an exponential equation, you may write both members as powers of a common base, equate the exponents and solve [see Example 15-18].

(a) $3^x = 27$

$3^x = 3^3$

$x = 3$

(b) $4^{x-3} = 16$

$4^{x-3} = 4^2$

$x - 3 = 2$

$x = 5$

(c) $4^x = 8$

$2^{2x} = 2^3$

$2x = 3$

$x = \frac{3}{2}$

(d) $3^{2-x} = 81$

$3^{2-x} = 3^4$

$2 - x = 4$

$-x = 2$

$x = -2$

(e) $2^{x^2-2} = 64$

$2^{x^2-2} = 2^6$

$x^2 - 2 = 6$

$x^2 = 8$

$x = \pm 2\sqrt{2}$

(f) $3^{x^2+2} = 81$

$3^{x^2+2} = 3^4$

$x^2 + 2 = 4$

$x^2 = 2$

$x = \pm\sqrt{2}$

PROBLEM 15-19 Solve the following exponential equations using the logarithm of both members:

(a) $2^x = 27$ (b) $3^{x-1} = 16$ (c) $3^{x+2} = 1$ (d) $2^{3x} = 3$ (e) $4^{x^2} = 3$
(f) $e^{x+2} = 4$

Solution: Recall that to solve an exponential equation using the logarithm of both members, you write the logarithm of both members, use the properties of logarithms and solve [see Example 15-19].

(a) $2^x = 27$

$\log 2^x = \log 27$

$x \log 2 = \log 27$

$x = \dfrac{\log 27}{\log 2}$

$x \approx \dfrac{1.4314}{0.3010}$

$x \approx 4.7549$

(b) $3^{x-1} = 16$

$\log 3^{x-1} = \log 16$

$(x - 1) \log 3 = \log 16$

$x - 1 = \dfrac{\log 16}{\log 3}$

$x = 1 + \dfrac{\log 16}{\log 3}$

$x \approx 1 + \dfrac{1.2041}{0.4771}$

$x \approx 1 + 2.5237$

$x \approx 3.5237$

(c) $3^{x+2} = 1$

$\log 3^{x+2} = \log 1$

$(x + 2) \log 3 = \log 1$

$x + 2 = \dfrac{\log 1}{\log 3}$

$x = -2 + \dfrac{\log 1}{\log 3}$

$x = -2 + \dfrac{0}{0.4771}$

$x = -2$

(d) $2^{3x} = 3$

$\log 2^{3x} = \log 3$

$3x \log 2 = \log 3$

$3x = \dfrac{\log 3}{\log 2}$

$x = \dfrac{\log 3}{3 \log 2}$

$x \approx \dfrac{0.4771}{3(0.3010)}$

$x \approx \dfrac{0.4771}{0.9030}$

$x \approx 0.5283$

(e) $4^{x^2} = 3$

$\log 4^{x^2} = \log 3$

$x^2 \log 4 = \log 3$

$x^2 = \dfrac{\log 3}{\log 4}$

$x = \pm \sqrt{\dfrac{\log 3}{\log 4}}$

$x \approx \pm \sqrt{\dfrac{0.4771}{0.6021}}$

$x \approx \pm \sqrt{0.7925}$

$x \approx \pm 0.8902$

(f) $e^{x+2} = 4$

$\ln e^{x+2} = \ln 4$

$(x + 2) \ln e = \ln 4$

$x + 2 = \dfrac{\ln 4}{\ln e}$

$x + 2 = \ln 4$

$x = -2 + \ln 4$

$x \approx -2 + 1.3863$

$x \approx -0.6137$

PROBLEM 15-20 Solve the following logarithmic equations:

(a) $\log x + \log x = 1$ **(b)** $\log (x^2 - 1) - \log (x + 1) = 2$ **(c)** $\log (x - 3) + \log x = 1$
(d) $\log (x - 4) + \log (x - 1) = 1$ **(e)** $\log_3 (x + 3) + \log_3 (x - 5) = 2$
(f) $\log_2 (x - 3) - \log_2 (2x + 1) = 1$

Solution: Recall that to solve a logarithmic equation, you obtain a single logarithm in each member, take the antilogarithm and solve [see Example 15-20].

(a) $\log x + \log x = 1$

$\log x \cdot x = \log 10$

$x^2 = 10$

$x = +\sqrt{10}$

(b) $\log (x^2 - 1) - \log (x + 1) = 2$

$\log \dfrac{x^2 - 1}{x + 1} = \log 100$

$\dfrac{(x + 1)(x - 1)}{x + 1} = 100$

$x - 1 = 100$

$x = 101$

(c) $\log (x - 3) + \log x = 1$

$\log (x - 3)x = \log 10$

$x^2 - 3x = 10$

$x^2 - 3x - 10 = 0$

$(x - 5)(x + 2) = 0$

$x = 5$

(d) $\log (x - 4) + \log (x - 1) = 1$

$\log (x - 4)(x - 1) = \log 10$

$x^2 - 5x + 4 = 10$

$x^2 - 5x - 6 = 0$

$(x - 6)(x + 1) = 0$

$x = 6$

(e) $\log_3 (x + 3) + \log_3 (x - 5) = 2$

$\log_3 (x + 3) + \log_3 (x - 5) = \log_3 9$

$\log_3 (x + 3)(x - 5) = \log_3 9$

$x^2 - 2x - 15 = 9$

$x^2 - 2x - 24 = 0$

$(x - 6)(x + 4) = 0$

$x = 6$

(f) $\log_2 (x - 3) - \log_2 (2x + 1) = 1$

$\log_2 \dfrac{x - 3}{2x + 1} = \log_2 2$

$\dfrac{x - 3}{2x + 1} = 2$

$x - 3 = 2(2x + 1)$

$x - 3 = 4x + 2$

$-5 = 3x$

No Solution.

PROBLEM 15-21 Solve each problem using an exponential formula:

(a) A savings account paying interest that is compounded continuously is opened with a deposit of $1000. At the end of 2 years, there is $1127.50 in the account. How much is in the account at the end of 5 years? (Assume no money is withdrawn from the account and no money, other than the original $1000, is deposited.)

(b) The atmospheric pressure at sea level is 14.7 pounds per square inch (psi). At an altitude of 2000 ft, the atmospheric pressure is 13.5 psi. What is the atmospheric pressure on the outside of an airliner flying at 36,000 ft, to the nearest tenth psi?

Solution: Recall that to solve a problem using an exponential formula, you:

1. *Read* the problem carefully, several times.
2. *Identify* the dependent quantity and independent quantity.
3. *Decide* how to represent the dependent and independent quantities using two different variables.
4. *Translate* to a general exponential formula of the form $y = y_0 e^{kx}$ or $y = y_0 e^{-kx}$.
5. *Replace* y_0 by the initial value of the dependent variable y when $x = 0$.
6. *Find k* by evaluating the exponential formula from step 5.
7. *Replace k* with the constant found in step 6 to find a specific exponential formula.
8. *Solve* the specific exponential formula for the unknown variable.
9. *Interpret* the solution from step 8 with respect to the unknown quantity to find the proposed solution of the original problem.
10. *Check* to see if the proposed solution satisfies all the conditions of the original problem [see Examples 15-21 and 15-22].

(a) *Identify:* The dependent quantity is the amount of money in the account at the end of a given time period.
The independent quantity is the amount of elapsed time from time zero (the opening of the account).

Decide: Let A = the amount of money in the account at the end of a given time period

and t = the amount of elapsed time from time zero.

Translate: $A = A_0 e^{kt}$ ⟵ general exponential growth formula

Replace A_0: $A = 1000 e^{kt}$ $A = 1000$ (dollars) when $t = 0$ (time zero) means $A_0 = 1000$.

Find k: $1127.50 = 1000 e^{k(2)}$ $A = 1127.50$ (dollars) at $t = 2$ (years).

$\ln 1127.50 = \ln 1000 e^{2k}$

$\ln 1127.50 = \ln 1000 + \ln e^{2k}$

$\ln 1127.50 = \ln 1000 + 2k$

$\ln 1127.50 - \ln 1000 = 2k$

$\ln \left(\dfrac{1127.50}{1000} \right) = 2k$

$\dfrac{\ln 1.1275}{2} = k$ ⟵ exponential growth constant (≈ 0.06)

Replace k: $A = 1000 e^{\left(\frac{\ln 1.1275}{2} \right) t}$ ⟵ specific exponential growth formula

Solve: $A = 1000 e^{\left(\frac{\ln 1.1275}{2} \right)(5)}$ Find A (the amount of money) when $t = 5$ (years).

$A \approx 1000 e^{0.300006981}$

$A \approx 1000(1.349868231)$

Interpret: $A = 1349.87$ [dollars] ⟵ solution (check as before)

(b) *Identify:* The dependent variable is the atmospheric pressure at a given altitude.
The independent variable is the altitude above sea level.

Decide: Let p = the atmospheric pressure at a given altitude

and a = the altitude above sea level.

Translate: $$p = p_0 e^{-ka} \longleftarrow \text{ general exponential decay formula}$$

Replace p_0: $$p = 14.7 e^{-ka}$$

$p = 14.7$ (psi) when $a = 0$ (feet above sea level) means $p_0 = 14.7$.

Find k: $$13.5 = 14.7 e^{-k(2000)}$$

$p = 13.5$ (psi) when $a = 2000$ (feet above sea level).

$$\ln 13.5 = \ln 14.7 e^{-2000k}$$

$$\ln 13.5 = \ln 14.7 + \ln e^{-2000k}$$

$$\ln 13.5 - \ln 14.7 = -2000k$$

$$\frac{\ln\left(\dfrac{13.5}{14.7}\right)}{-2000} = k \longleftarrow \text{ exponential decay constant } (\approx 0.00004)$$

Replace k: $$p = 14.7 e^{\left[\frac{\ln\left(\frac{13.5}{14.7}\right)}{-2000}\right]t} \longleftarrow \text{ specific exponential decay formula}$$

$$p = 14.7 e^{\left[\frac{\ln\left(\frac{13.5}{14.7}\right)}{2000}\right]t} \qquad \text{Simplify.}$$

Solve: $$p = 14.7 e^{\left[\frac{\ln\left(\frac{13.5}{14.7}\right)}{2000}\right](36,000)}$$

Find p (the atmospheric pressure in psi) when $a = 36,000$ (feet above sea level).

$$p \approx 14.7 e^{-1.53284055}$$

$$p \approx 14.7(0.21592146)$$

$$p \approx 3.17404546$$

Interpret: $p \approx 3.2$ (psi) $\longleftarrow$ solution (check as before)

PROBLEM 15-22 Solve this problem using a logarithmic formula:

The loudness of a sound is measured in decibels (db). The intensity of the sound is measured in watts per square meter (W/m^2). By international agreement, the intensity of sound at the threshold (beginning) of hearing is 10^{-12} W/m^2. The loudness of normal conversation is 60 db with an intensity of 10^{-6} W/m^2. Hearing damage can be caused by an intensity of sound of 1 W/m^2 or more. What is the decibel rating for sound that can cause hearing damage?

Solution: Recall that to solve a problem using a logarithmic formula:

1. *Read* the problem very carefully, several times.
2. *Identify* the dependent quantity and the independent quantity.
3. *Decide* how to represent the dependent and independent quantities using two different variables.
4. *Translate* to a general logarithmic formula of the form $y = k \log\left(\dfrac{x}{x_0}\right)$.
5. *Replace* x_0 by the initial value of the independent variable x when $y = 0$.
6. *Find k* by evaluating the logarithmic formula from step 5.
7. *Replace k* with the constant found in step 6 to find a specific logarithmic formula.
8. *Solve* the specific logarithmic formula for the unknown variable.
9. *Interpret* the solution from step 8 with respect to the unknown quantity to find the proposed solution of the original problem.
10. *Check* to see if the proposed solution satisfies all the conditions of the original problem [see Example 15-23].

Identify: The dependent quantity is the loudness of sound (db).
The independent quantity is the intensity of sound (W/m^2).

Decide: Let n = the loudness of sound (db)

and I = the intensity of sound (W/m^2).

Translate: $n = k \log \left(\dfrac{I}{I_0} \right)$ ⟵ general logarithmic formula

Replace I_0: $n = k \log \left(\dfrac{I}{10^{-12}} \right)$ $I_0 = 10^{-12}$ (W/m^2) by international agreement.

$n = k \log (10^{12} I)$ Simplify

Find k: $60 = k \log [10^{12}(10^{-6})]$ $n = 60$ (db) when $I = 10^{-6}$ (W/m^2).

$60 = k \log (10^6)$

$60 = k(6)$

$10 = k$ ⟵ logarithmic constant

Replace k: $n = 10 \log (10^{12} I)$ ⟵ specific logarithmic formula

Solve: $n = 10 \log [10^{12}(1)]$ Find n (the number of db) when $I = 1$ (W/m^2).

$n = 10 \log (10^{12})$

$n = 10(12)$

$n = 120$ (db or more) ⟵ solution (check as before)

Supplementary Exercises

PROBLEM 15-23 Evaluate the following expressions. Round to the nearest thousandth:

(a) $2^{1.4}$ (b) $2^{1.55}$ (c) $4^{0.25}$ (d) $4^{0.75}$ (e) $e^{1.1}$ (f) $e^{1.2}$

PROBLEM 15-24 Graph the following exponential functions:

(a) $f(x) = 4^x$ (b) $f(x) = \left(\frac{3}{4}\right)^x$ (c) $f(x) = 2^{x+2}$
(d) $f(x) = 2^{x-1}$ (e) $f(x) = 2^{2x+3}$ (f) $f(x) = 2^{2x-3}$

PROBLEM 15-25 Graph the following exponential functions:

(a) $x = 3^y$ (b) $x = \left(\frac{1}{3}\right)^y$ (c) $x = 2^{y+3}$ (d) $x = 2^{y-2}$ (e) $x = 2^{2y+3}$ (f) $x = 3^{y+2}$

PROBLEM 15-26 Write the following logarithmic equations in exponential form:

(a) $\log_{10} 10 = 1$ (b) $\log_2 8 = 3$ (c) $\log_5 625 = 4$
(d) $\log_{\sqrt{2}} 4 = 4$ (e) $\log_{16} 2 = \frac{1}{4}$ (f) $\log_e e = 1$

PROBLEM 15-27 Write the following exponential equations in logarithmic form:

(a) $10^2 = 100$ (b) $10^{-1} = 0.1$ (c) $3^5 = 243$
(d) $2^8 = 256$ (e) $8^{\frac{1}{2}} = 2\sqrt{2}$ (f) $e^0 = 1$

PROBLEM 15-28 Solve the following equations:

(a) $\log_{10} x = 2$ (b) $\log_3 x = 4$ (c) $\log_x 0.1 = -1$
(d) $\log_x 10{,}000 = 4$ (e) $\log_2 \frac{1}{2} = x$ (f) $\log_{\frac{1}{9}} \frac{1}{3} = x$

PROBLEM 15-29 Graph the following logarithmic equations:

(a) $y = \log_4 x$ (b) $y + \log_{\frac{1}{4}} x = 0$ (c) $y + \log_e x = 0$

PROBLEM 15-30 Express the following logarithms in terms of simpler logarithms:

(a) $\log_3 (9 \times 81)$ (b) $\log_4 \dfrac{64}{16}$ (c) $\log p^5 q^2 r^3$ (d) $\log \dfrac{x^3}{y^5 z^2}$ (e) $\log \dfrac{\sqrt[4]{x^3}}{\sqrt[3]{y^2}}$ (f) $\log \sqrt[4]{\dfrac{x^3}{y^2 z}}$

PROBLEM 15-31 Express the following logarithmic expressions as a single logarithm:

(a) $2 \log y + 5 \log z$ (b) $\frac{1}{3} \log w - \frac{2}{3} \log x$ (c) $\frac{1}{3} \log x + \frac{2}{3} \log y - \frac{1}{2} \log z$
(d) $\frac{1}{4} \log x - \frac{1}{2} \log y - \frac{3}{4} \log z$ (e) $\frac{1}{5} \log x^4 - \frac{2}{5} \log x - \frac{3}{5} \log x + \frac{4}{5} \log x$
(f) $\log_b (x^2 - y^2) - \log_b (x + y)$

PROBLEM 15-32 Approximate the following logarithms in terms of $\log_{10} 2 \approx 0.3010$, $\log_{10} 3 \approx 0.4771$ and $\log_{10} 5 \approx 0.6990$:

(a) $\log 100$ (b) $\log \sqrt{2}$ (c) $\log \sqrt[3]{10}$ (d) $\log 1.5$ (e) $\log 112.5$ (f) $\log 3.6$

PROBLEM 15-33 Find the common logarithm of the following numbers using the table in Appendix 1:

(a) 2.54 (b) 9.02 (c) 0.456 (d) 27.4 (e) 7240 (f) 0.00874

PROBLEM 15-34 Find the common antilogarithm of the following numbers using the table in Appendix 1:

(a) 0.7059 (b) 2.3075 (c) 3.9063 (d) 1.1761 (e) $0.9978 - 2$ (f) $0.6542 - 1$

PROBLEM 15-35 Find the common logarithm and the natural logarithm of the following numbers. Use a calculator and round each answer to the nearest thousandth.

(a) 727 (b) 35.3 (c) 5.07 (d) 0.572 (e) 0.0409 (f) 4170

PROBLEM 15-36 Find the common antilogarithm and the natural antilogarithm of the following numbers:

(a) 18 (b) 8.5 (c) 4.32 (d) 0.514 (e) 1.603 (f) 0.0381

PROBLEM 15-37 Find the logarithm of the following numbers using the change-of-base formula. Round each answer to the nearest thousandth.

(a) $\log_8 84$ (b) $\log_6 81$ (c) $\log_3 39$ (d) $\log_2 42$ (e) $\ln 27$ (f) $\ln 72$

PROBLEM 15-38 Use logarithms to compute the following expressions. Give your answers in 3 significant digits.

(a) 481^2 (b) 561×1.56 (c) 8410×6.34 (d) $\sqrt[3]{4.73}$ (e) $4^{2.3}$ (f) $\dfrac{141 \sqrt[4]{31.3}}{23.1^2}$

PROBLEM 15-39 Solve the following exponential equations:

(a) $2^x = 128$ (b) $3^{x-4} = 81$ (c) $9^x = 27$
(d) $2^{3-x} = 8$ (e) $3^{x^2+2} = 729$ (f) $2^{x^2+2} = 16$

PROBLEM 15-40 Solve the following exponential equations using the logarithms of both members:

(a) $3^x = 16$ (b) $2^{x-4} = 12$ (c) $2^{x+3} = 1$ (d) $3^{2x} = 4$ (e) $3^{x^2} = 6$ (f) $e^{x-2} = 4$

PROBLEM 15-41 Solve the following logarithmic equations:

(a) $2 \log x - \log x = 1$ (b) $\log (x^2 - 1) - \log (x - 1) = 2$ (c) $\log (x + 3) + \log x = 1$
(d) $\log (x + 4) + \log (x + 1) = 1$ (e) $\log_4 (x - 3) - \log_4 (x - 4) = 1$
(f) $\log_2 (2x - 3) - \log_2 (1 - 2x) = 1$

PROBLEM 15-42 Solve each problem using an exponential formula:

(a) The population of the U.S. was 204 million in 1970. In 1980, the population of the U.S. was 226.5 million. Project the population of the U.S. for the year 2000, to the nearest ten million. Project the year that the U.S. population will reach 300 million.

(b) The first year sales for a certain product were 50,000 units. Because of a lack of promotion, the second year sales fell to 45,242 units. Assuming a continued lack of promotion, find the third year sales. How long will it take for sales to drop to 40,000 units, to the nearest month, assuming no advertising takes place?

PROBLEM 15-43 Solve each problem using a logarithmic formula:

(a) The Alaska Good Friday earthquake of 1964 had an amplitude of 316,228 mm 100 km from the epicenter. What was the magnitude of the Alaska Good Friday earthquake?

(b) The Coalinga, California earthquake of 1983 had a magnitude of 6.5 on the Richter scale. What was the amplitude of the Coalinga earthquake 100 km from the epicenter?

(c) The intensity of sound from a normal whisper is 10^{-10} W/m². What is the decibel rating for a normal whisper?

(d) The decibel rating for a jet airplane at takeoff is 140 db. What is the intensity of sound for a jet at takeoff?

(e) The acidity or alkalinity of a solution is found by determining the hydrogen potential (pH) rating of the solution. The concentration of hydrogen ions (H^+) in the solution is measured in moles per liter [mol/L]. If a given solution has a pH rating less than 7, it is acidic. If its pH rating is greater than 7, the solution is alkalinic. If its rating is equal to 7, the solution is neutral. By international agreement, the initial value for concentration of hydrogen ions is 1 mol/L. Pure water is a neutral solution with a concentration of hydrogen ions of 10^{-7} mol/L. Find the specific logarithmic formula for pH ratings.

(f) The concentration of hydrogen ions in vinegar is 1.3×10^{-3} mol/L. The hydrogen potential rating of ammonia is 11.8. Find the hydrogen potential rating of vinegar. Find the concentration of hydrogen ions in ammonia. Is vinegar acidic or alkalinic? Is ammonia acidic or alkalinic? How many times greater is the concentration of hydrogen ions in vinegar than in ammonia?

Answers to Supplementary Exercises

(15-23) (a) 2.639 **(b)** 2.928 **(c)** 1.414 **(d)** 2.828 **(e)** 3.004 **(f)** 3.320

(15-24)

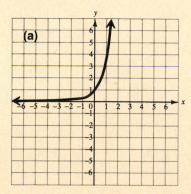
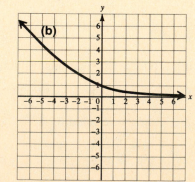
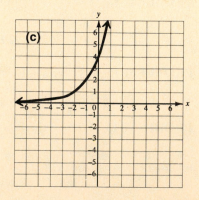

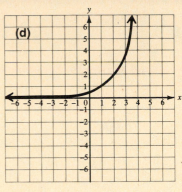 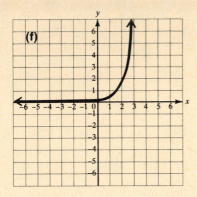 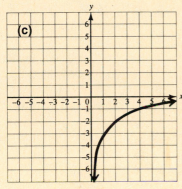

(15-25)

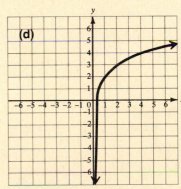 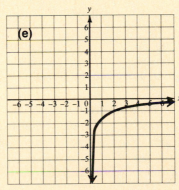 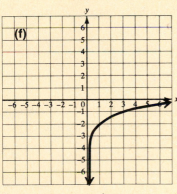

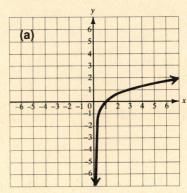 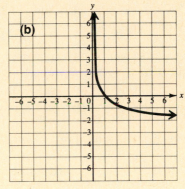 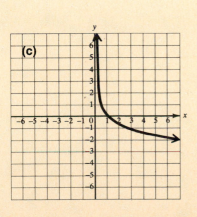

(15-26) **(a)** $10 = 10^1$ **(b)** $8 = 2^3$ **(c)** $625 = 5^4$ **(d)** $4 = (\sqrt{2})^4$ **(e)** $2 = 16^{\frac{1}{4}}$
(f) $e = e^1$

(15-27) **(a)** $\log_{10} 100 = 2$ **(b)** $\log_{10} 0.1 = -1$ **(c)** $\log_3 243 = 5$ **(d)** $\log_2 256 = 8$
(e) $\log_8 2\sqrt{2} = \frac{1}{2}$ **(f)** $\log_e 1 = 0$

(15-28) **(a)** 100 **(b)** 81 **(c)** 10 **(d)** 10 **(e)** -1 **(f)** $\frac{1}{2}$

(15-29)

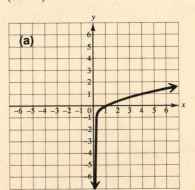 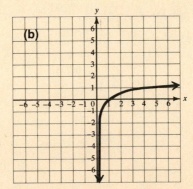 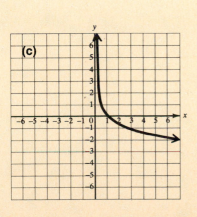

(15-30) (a) 6   (b) 1   (c) $5 \log p + 2 \log q + 3 \log r$ (d) $3 \log x - 5 \log y - 2 \log z$
(e) $\frac{3}{4} \log x - \frac{2}{3} \log y$ (f) $\frac{3}{4} \log x + \frac{1}{2} \log y - \frac{1}{4} \log z$

(15-31) (a) $\log y^2 z^5$ (b) $\log \sqrt[3]{\dfrac{w}{x^2}}$ (c) $\log \dfrac{\sqrt[3]{xy^2}}{\sqrt{z}}$ (d) $\log \sqrt[4]{\dfrac{x}{y^2 z^3}}$ (e) $\log \sqrt[5]{x^3}$
(f) $\log_b (x - y)$

(15-32) (a) 2   (b) 0.1505   (c) 0.3333   (d) 0.1761   (e) 2.0512   (f) 0.5562

(15-33) (a) 0.4048   (b) 0.9552   (c) $0.6590 - 1$ (d) 1.4378   (e) 3.8597   (f) $0.9415 - 3$

(15-34) (a) 5.08   (b) 203   (c) 8060   (d) 15.0   (e) 0.0995   (f) 0.451

(15-35) (a) 2.8615, 6.5889   (b) 1.5478, 3.5639   (c) 0.7050, 1.6233   (d) $-0.2426, -0.5586$
(e) $-1.3883, -3.1966$ (f) 3.6201, 8.3357

(15-36) (a) 1,000,000,000,000,000,000, 65,659,969   (b) 316,227,770, 4,914.7688
(c) 20,892.961, 75.1886   (d) 3.2659, 1.6720   (e) 40.0867, 4.9679   (f) 1.0917, 1.0388

(15-37) (a) 2.1308   (b) 2.4526   (c) 3.3347   (d) 5.3923   (e) 3.2958   (f) 4.2767

(15-38) (a) 231,000   (b) 875   (c) 53,300   (d) 1.68   (e) 24.3   (f) 0.625

(15-39) (a) 7   (b) 8   (c) $\frac{3}{2}$ (d) 0   (e) ± 2 (f) $\pm \sqrt{2}$

(15-40) (a) 2.5237   (b) 7.5850   (c) -3 (d) 0.6309   (e) ± 1.277 (f) 3.3863

(15-41) (a) 10   (b) 99   (c) 2   (d) 1   (e) $\frac{13}{3}$ (f) no solution

(15-42) (a) 279.2 million, 2006   (b) 40,937 units, 3 yr 3 mo

(15-43) (a) 8.5 on the Richter scale   (b) 3162 mm   (c) 20 db   (d) 100 W/m²
(e) $pH = -\log (H^+)$ (f) 2.9 rating, 1.6×10^{-12} mol/L, acidic, alkalinic, 8.125×10^8 times greater, or 812,500,000 times greater

16 SEQUENCES AND SERIES

THIS CHAPTER IS ABOUT

☑ Understanding Sequences and Series
☑ Arithmetic Sequences and Series
☑ Solving Problems Using Arithmetic Series
☑ Geometric Sequences and Series
☑ Solving Problems Using Geometric Series
☑ Binomial Expansions

16-1. Understanding Sequences and Series

A. Finding the terms of a sequence

A **sequence** is a function whose domain is the set of all or some consecutive portion of the natural numbers. If the domain is the set of all natural numbers, then the sequence is called an **infinite sequence.** If the domain is the set of some consecutive portion of the natural numbers (called the first n natural numbers), then the sequence is called a **finite sequence.** For example, $f(n) = 3n + 5$ for $n = 1, 2,$ and 3 is a finite sequence. Its domain is $\{1, 2, 3\}$, and its range is found by evaluating $f(n) = 3n + 5$ for $n = 1, 2,$ and 3:

$$f(1) = 3(1) + 5 = 8, f(2) = 3(2) + 5 = 11, f(3) = 3(3) + 5 = 14$$

In this sequence the range is $\{8, 11, 14\}$.

An infinite sequence is shown by $g(n) = 2n$ for $n \in N$. Its domain is the set of all natural numbers. Any range member k can be found by evaluating $g(n) = 2n$ for $n = k$:

$$g(1) = 2(1) = 2, g(2) = 2(2) = 4, g(3) = 2(3) = 6, \ldots, g(k) = 2k$$

In this sequence, the range is the set of all even natural numbers.

The range members of a sequence are called the **terms of the sequence.** The terms of the sequence a are denoted by using the subscript notation $a_1, a_2, a_3, \ldots, a_n, \ldots$ with $n \in N$. Subscript and functional notation are related as shown below:

$$a_1 = a(1), a_2 = a(2), a_3 = a(3), \ldots, a_n = a(n), \ldots$$

The "nth" term a_n is called the **general term of the sequence,** and it gives the formula that defines the sequence, or function, a. For instance, in the sequence a with terms $1, 6, 15, 28, \ldots, 2n^2 - n, \ldots$

the first term a_1 of the sequence is 1

the second term a_2 of the sequence is 6

the third term a_2 of the sequence is 15

$$\vdots$$

the nth term a_n of the sequence is $2n^2 - n$

Thus this sequence is defined by the function $a(n) = 2n^2 - n$.

To find the terms of a sequence whose general term is defined by a formula, we substitute the domain members for n and evaluate the formula.

EXAMPLE 16-1: Find the first two terms and the tenth term of each of the following sequences:
(a) $a_n = 8n - 1$ (b) $a_n = n^3 + n$ (c) $a_n = (-1)^n(2n + 3)$

Solution: Evaluate the formula for the general term for $n = 1, 2,$ and 10.

(a) $a_n = 8n - 1$ $a_n = 8n - 1$ $a_n = 8n - 1$
$\quad a_1 = 8(1) - 1$ $a_2 = 8(2) - 1$ $a_{10} = 8(10) - 1$
$\quad\quad = 7$ $\quad = 15$ $\quad\quad = 79$

(b) $a_n = n^3 + n$ $a_n = n^3 + n$ $a_n = n^3 + n$
$\quad a_1 = (1)^3 + (1)$ $a_2 = (2)^3 + (2)$ $a_{10} = (10)^3 + (10)$
$\quad\quad = 2$ $\quad\quad = 10$ $\quad\quad = 1010$

(c) $a_n = (-1)^n(2n + 3)$ $a_n = (-1)^n(2n + 3)$ $a_n = (-1)^n(2n + 3)$
$\quad a_1 = (-1)^1(2(1) + 3)$ $a_2 = (-1)^2(2(2) + 3)$ $a_{10} = (-1)^{10}(2(10) + 3)$
$\quad\quad = -5$ $\quad\quad = 7$ $\quad\quad = 23$

B. Finding the sum of a series

The indicated sum of the terms of a sequence is called a **series.** For example, $1 + 3 + 5 + 7$ is a series; it is the indicated sum of the terms in the sequence 1, 3, 5, 7. Although the domain members of a sequence must be positive, the range members may be negative. Therefore $1 - 2 + 3 - 4 + 5$ is a series also; it is the indicated sum of the terms in the sequence 1, -2, 3, -4, 5.

We often use the following special notation, called **summation notation,** to represent series. Given a sequence of terms $a_1, a_2, a_3, \ldots, a_n, \ldots$ the symbol $\sum_{k=1}^{n} a_k$ represents the indicated sum of the first n terms of the sequence.

$$\underbrace{\sum_{k=1}^{n} a_k}_{\substack{\text{summation} \\ \text{notation form}}} = \underbrace{a_1 + a_2 + a_3 + \cdots + a_n}_{\text{expanded form}}$$

In the summation above, k is called the **index of summation.** The terms in the expanded form are obtained from the left member by successively replacing k in a_k with the natural numbers starting with 1 and ending with n.

Note: We may use any letter to represent the index of summation, and these should not be confused with other numbers or usages, e.g., i with $\sqrt{-1}$ and k with variation constants.

To find the sum of a series written in summation notation, we first write the series in expanded form.

EXAMPLE 16-2: Find the sum of each series written in summation notation:

(a) $\sum_{k=1}^{3} (k - 2)$ (b) $\sum_{k=1}^{4} (k^2 - 3)$ (c) $\sum_{j=1}^{3} (-1)^j(2j)$

Solution: (a) $\sum_{k=1}^{3} (k - 2) = (1 - 2) + (2 - 2) + (3 - 2)$

$\quad\quad\quad\quad = (-1) + (0) + (1)$

$\quad\quad\quad\quad = 0$

(b) $\displaystyle\sum_{k=1}^{4} (k^2 - 3) = [(1)^2 - 3] + [(2)^2 - 3] + [(3)^2 - 3] + [(4)^2 - 3]$

$$= \quad -2 \quad + \quad 1 \quad + \quad 6 \quad + \quad 13$$

$$= 18$$

(c) $\displaystyle\sum_{j=1}^{3} (-1)^j(2j) = (-1)^1(2(1)) + (-1)^2(2(2)) + (-1)^3(2(3))$

$$= -2 + 4 + (-6)$$

$$= -4$$

If every term of a series is the same constant, then the series is called a **constant series.** For example, $3 + 3 + 3 + 3 + 3$ is a constant series with all 5 terms having a value of 3. To find the sum of a finite constant series, we multiply the number of terms n by the constant term c. For example, the series $3 + 3 + 3 + 3 + 3 = 5 \cdot 3 = 15$, and the series $7 + 7 + 7 + 7 + 7 + 7 = 6 \cdot 7 = 42$. We can write a finite constant series using summation notation as $\displaystyle\sum_{k=1}^{n} c$. For example, the series $2 + 2 + 2 = \displaystyle\sum_{k=1}^{3} 2$, and the series $0.4 + 0.4 + 0.4 + 0.4 + 0.4 = \displaystyle\sum_{k=1}^{5} 0.4$. Because the sum of a constant series with n terms can be computed by multiplying n times c (the value of each of the constant terms), it follows that $\displaystyle\sum_{k=1}^{n} c = nc$.

EXAMPLE 16-3: Find the sum of each of the following constant series:

(a) $\displaystyle\sum_{k=1}^{30} 8$ **(b)** $\displaystyle\sum_{k=1}^{300} (-5)$ **(c)** $\displaystyle\sum_{j=1}^{53} 20$

Solution: Use $\displaystyle\sum_{k=1}^{n} c = nc$.

(a) $\displaystyle\sum_{k=1}^{30} 8 = 30 \cdot 8 = 240$ **(b)** $\displaystyle\sum_{k=1}^{300} (-5) = 300(-5) = -1500$ **(c)** $\displaystyle\sum_{j=1}^{53} 20 = 53 \cdot 20 = 1060$

16-2. Arithmetic Sequences and Series

A. Finding terms of an arithmetic sequence

Some sequences have the property that every term after the first can be found by adding the same constant to the preceding term.

A sequence $a_1, a_2, a_3, \ldots, a_n, \ldots$ is called an **arithmetic sequence (arithmetic progression)** if there is a constant d, called the **common difference,** such that $a_n = a_{n-1} + d$ for every $n \geq 2$.

To find the terms of an arithmetic sequence, we use the formula $a_n = a_{n-1} + d$.

EXAMPLE 16-4: Find the first four terms of an arithmetic sequence with $a_1 = 8$ and $d = 7$.

Solution:

$$a_1 = 8$$

$$a_2 = a_1 + d = 8 + 7 = 15 \qquad \text{Add the common difference } d = 7.$$

$$a_3 = a_2 + d = 15 + 7 = 22 \qquad \text{Add 7 again.}$$

$$a_4 = a_3 + d = 22 + 7 = 29 \qquad \text{Add 7 again.}$$

The first four terms of the arithmetic sequence with $a_1 = 8$ and $d = 7$ are 8, 15, 22, and 29.

Every term of an arithmetic sequence can be written in terms of the first term a_1 and the common difference d. For example,

$$a_1 = 2, \qquad a_2 = 9, \qquad a_3 = 16, \qquad a_4 = 23$$

can be written as

$$a_1 = 2 + 0 \cdot 7, \quad a_2 = 2 + 1 \cdot 7, \quad a_3 = 2 + 2 \cdot 7, \quad a_4 = 2 + 3 \cdot 7$$

In each case, the subscript is one more than the coefficient of the common difference. The following formula is a generalization of this pattern.

From the pattern above we can see that the general term of an arithmetic sequence with first term a_1 and common difference d is given by the **nth-term formula**

$$a_n = a_1 + (n - 1)d$$

EXAMPLE 16-5: Given an arithmetic sequence with $a_1 = 8$ and $d = 4$, find a_{42}.

Solution:
$a_n = a_1 + (n - 1)d$ Write the nth-term formula.

$a_{42} = 8 + (42 - 1)8$ Evaluate a_{42} by letting $a_1 = 8$, $d = 4$, and $n = 42$.

$= 8 + (41)8$

$= 336$

Given any three of the four unknowns in $a_n = a_1 + (n - 1)d$, we can solve for the remaining unknown.

EXAMPLE 16-6: Given an arithmetic sequence with:

(a) $a_1 = 12$, and $a_8 = 61$, find d **(b)** $a_6 = 125$, and $d = -15$, find a_1
(c) $a_1 = -40$, $a_n = 260$, and $d = 12$, find n

Solution: **(a)** $a_n = a_1 + (n - 1)d$ Write the nth-term formula for an arithmetic sequence.

$61 = 12 + (8 - 1)d$ Evaluate d by letting $a_1 = 12$, $a_8 = 61$, and $n = 8$.

$61 = 12 + 7d$

$49 = 7d$

$7 = d$

(b) $a_n = a_1 + (n - 1)d$ **(c)** $a_n = a_1 + (n - 1)d$

$125 = a_1 + (6 - 1)(-15)$ $260 = -40 + (n - 1)(12)$

$125 = a_1 + 5(-15)$ $300 = (n - 1)(12)$

$125 = a_1 - 75$ $25 = n - 1$

$200 = a_1$ $26 = n$

B. Finding the sum of an arithmetic series

The indicated sum of the terms of an arithmetic sequence is called an **arithmetic series.** For example, $2 + 4 + 6 + 8$ is an arithmetic series associated with the arithmetic sequence 2, 4, 6, 8. To find the sum of a finite arithmetic series, we can simply add the terms, or use the summation formula $S_n = n\left(\dfrac{a_1 + a_n}{2}\right)$, where S_n denotes the sum of the first n terms of an arithmetic sequence with first term a_1 and nth-term a_n.

EXAMPLE 16-7: Find the sum of the first 21 terms of the arithmetic series with $a_1 = 8$ and $a_{21} = 128$.

Solution: $S_n = n\left(\dfrac{a_1 + a_n}{2}\right)$ Write the summation formula.

$S_{21} = 21\left(\dfrac{8 + 128}{2}\right)$ Evaluate S_{21} by letting $a_1 = 8$, $a_{21} = 128$, and $n = 21$.

$= 21(68)$

$= 1428$ ←—— the sum of the first 21 terms of the arithmetic series with $a_1 = 8$ and $a_{21} = 128$

Sometimes it is necessary to evaluate other unknowns before we can apply the summation formula.

EXAMPLE 16-8: Find the sum of all the even numbers from 40 to 76, inclusive.

Solution: $a_n = a_1 + (n - 1)d$ First, find the number of even numbers between 40 and 76 inclusive by using the nth-term formula to find n.

$76 = 40 + (n - 1)2$ Successive even numbers have a common difference of 2.

$36 = (n - 1)2$

$18 = n - 1$

$19 = n$ ←—— the number of even numbers between 40 and 76 inclusive

$S_n = n\left(\dfrac{a_1 + a_n}{2}\right)$ Write the summation formula.

$S_{19} = 19\left(\dfrac{40 + 76}{2}\right)$ Evaluate S_{19} by letting $a_1 = 40$, $a_{19} = 76$ and $n = 19$.

$= 1102$ ←—— the sum of all even numbers from 40 to 76 inclusive

16-3. Solving Problems Using Arithmetic Series

To **solve a problem using an arithmetic series,** we:

1. *Read* the problem carefully, several times.
2. *Identify* the unknown arithmetic series (or number).
3. *Decide* how to represent the first term of the arithmetic sequence, the common difference, and the number of terms.
4. *Translate* to the arithmetic series formula $S_n = n\left(\dfrac{a_1 + a_n}{2}\right)$.
5. *Evaluate* (or *solve*) the arithmetic series formula from step 4.
6. *Interpret* S_n (or n) with respect to the unknown arithmetic series (or number of terms) to find the proposed solution of the original problem.
7. *Check* to see if the proposed solution satisfies all the conditions of the original problem.

EXAMPLE 16-9: Solve this problem using an arithmetic series.

Read: In a certain city, the fine for a first parking offense is \$20. For each additional parking offense, a fine of \$8 is added. What is the total amount of money that must be paid for 10 parking offenses in this city?

Identify: The unknown arithmetic series is the sum of the first 10 parking offenses.

Decide: Let $a_1 = 20$ (dollars) ←—— first term of sequence

and $d = 8$ (dollars) ←—— common difference between terms

and $n = 10$ (parking offenses) ←—— number of terms in sequence

Translate: $a_n = a_1 + (n - 1)d$ ←—— nth-term formula

$a_{10} = 20 + (10 - 1)8$ Find the 10th term of the arithmetic sequence.

$a_{10} = 92$ ←—— the 10th parking offense costs \$92

$S_n = n\left(\dfrac{a_1 + a_n}{2}\right)$ ←—— arithmetic series formula

Evaluate: $S_{10} = 10\left(\dfrac{20 + 92}{2}\right)$ Find the sum of the first 10 terms of the arithmetic sequence.

$S_{10} = 560$

Interpret: $S_{10} = 560$ means the total that must be paid for 10 parking offenses is \$560.

Check: Do the first 10 parking offenses add up to be \$560?
Yes: $20 + 28 + 36 + 44 + 52 + 60 + 68 + 76 + 84 + 92 = 560$

16-4. Geometric Sequences and Series

A. Finding the terms of a geometric sequence

Some sequences have the property that every term after the first can be determined by multiplying the preceding term by the same constant.

A sequence $a_1, a_2, a_3, \ldots, a_n, \ldots$ $(a_1 \neq 0)$ is called a **geometric sequence (geometric progression)** if there is a nonzero constant r, called the **common ratio,** such that

$$a_n = ra_{n-1}$$

for every $n \geq 2$.

To find the terms of a geometric sequence, we use the formula $a_n = ra_{n-1}$.

EXAMPLE 16-10: Find the first four terms of the geometric sequence with $a_1 = 4$ and $r = 3$:

Solution: $a_1 = 4$

$a_2 = ra_1 = 3 \cdot 4\ = 12$ Multiply by the common ratio $r = 3$.

$a_3 = ra_2 = 3 \cdot 12 = 36$ Multiply by 3 again.

$a_4 = ra_3 = 3 \cdot 36 = 108$ Multiply by 3 again.

The first four terms of the geometric sequence with $a_1 = 4$ and $r = 3$ are 4, 12, 36, and 108.

Every term of a geometric sequence can be written in terms of the first term a_1 and the common ratio r. For instance, the geometric sequence with

$$a_1 = 4, \qquad a_2 = 12, \qquad a_3 = 36, \qquad a_4 = 108$$

can be written as

$$a_1 = 4 \cdot 3^0, \qquad a_2 = 4 \cdot 3^1, \qquad a_3 = 4 \cdot 3^2, \qquad a_4 = 4 \cdot 3^3$$

Note: In each case, the subscript is one more than the exponent of the common ratio 3.

From the pattern above we can see that the general term of a geometric sequence with first term a_1 and common ratio r is given by the nth-term formula

$$a_n = a_1 r^{n-1}$$

EXAMPLE 16-11: Given a geometric sequence with $a_1 = 3$ and $r = 2$, find a_7.

Solution: $a_n = a_1 r^{n-1}$ Write the nth-term formula for a geometric sequence.

$a_7 = 3(2)^{7-1}$

$= 3(2)^6$

$= 192$

Because $a_2 = a_1 \cdot r$ in any geometric sequence, you can compute r using $r = a_2/a_1$.

EXAMPLE 16-12: Find the seventh term in a geometric sequence with $a_1 = 256$ and $a_2 = 128$.

Solution: $r = \dfrac{a_2}{a_1}$ Evaluate r.

$$r = \frac{256}{128}$$

$$= \frac{1}{2}$$

$$a_n = a_1 r^{n-1} \qquad\qquad \text{Write the } n\text{th-term formula.}$$

$$a_n = 256\left(\frac{1}{2}\right)^{7-1} \qquad \text{Evaluate } a_7.$$

$$= 256\left(\frac{1}{2}\right)^6$$

$$= 256\left(\frac{1}{64}\right)$$

$$= 4 \leftarrow \text{the seventh term in a geometric sequence with } a_1 = 256 \text{ and } a_2 = 128.$$

B. Finding the sum of a finite geometric series

The indicated sum of the terms of a geometric sequence is called a **geometric series.** For example, $16 + 8 + 4 + 2$ is the geometric series associated with the geometric sequence 16, 8, 4, 2. To find the sum of a finite geometric series, we can simply add the terms. However, it is often more convenient to use the following summation formula.

The sum of the first n terms of a geometric series with ratio r is given by

$$S_n = \frac{a_1(1 - r^n)}{1 - r}, \qquad \text{where } r \neq 1$$

EXAMPLE 16-13: Find the sum of the first 10 terms of the geometric series with $a_1 = 16$ and $r = \dfrac{1}{2}$.

Solution: $S_n = \dfrac{a_1(1 - r^n)}{1 - r}$ Write the summation formula.

$$S_n = \frac{16\left[1 - \left(\frac{1}{2}\right)^{10}\right]}{1 - \frac{1}{2}} \qquad \text{Substitute.}$$

$$= \frac{16\left[1 - \frac{1}{1024}\right]}{\frac{1}{2}} \qquad \text{Simplify.}$$

$$= 16 \cdot \left[\frac{1023}{1024}\right] \div \frac{1}{2}$$

$$= \frac{1023}{32}$$

The sum of the first 10 terms of the geometric series with $a_1 = 16$ and $r = \dfrac{1}{2}$ is $\dfrac{1023}{32}$.

Caution: The formula $S_n = \dfrac{a_1(1 - r^n)}{1 - r}$ does not apply if $r = 1$, However, a finite series with a com-

mon ratio of 1 would be a constant series, and its sum could be found using $\sum\limits_{k=1}^{n} c = nc$.

C. Finding the sum of certain infinite geometric series

Consider the sum of the geometric series $S = \frac{1}{2} + \frac{1}{4} + \frac{1}{8} + \cdots + (\frac{1}{2})^n + \cdots$, which also has an infinite number of terms.

For $S = \underbrace{\frac{1}{2} + \frac{1}{4}} + \frac{1}{8} + \cdots + (\frac{1}{2})^n + \cdots$

$S_2 = \underbrace{\frac{3}{4}}$

$S_3 = \underbrace{\frac{7}{8}}$

$S_n = \dfrac{\frac{1}{2}(1 - (\frac{1}{2})^n)}{1 - \frac{1}{2}}$ $\longleftarrow$ summation formula for a finite geometric series

Observe that in this case the power of the ratio $(\frac{1}{2})^n$ can be made smaller by making n larger, and because $(\frac{1}{2})^n$ approaches 0 as n increases without bound, the sum

$$S_n = \frac{\frac{1}{2}[1 - (\frac{1}{2})^n]}{1 - \frac{1}{2}}$$

will approach

$$\frac{\frac{1}{2}[1 - 0]}{1 - \frac{1}{2}} = \frac{\frac{1}{2}}{1 - \frac{1}{2}} = 1$$

as n increases without bound. The number that S_n is approaching (1 in this case) is called a **limit**, and is defined as the sum of the infinite series. If S_n does not have a limit, then the series does not have a sum.

We can summarize this special case as follows: if an infinite geometric series has a ratio of r such that $|r| < 1$ (that is, $-1 < r < 1$), then r^n will approach 0 as n increases without bound. The sum

$$S_n = \frac{a_1(1 - r^n)}{1 - r}$$

must then approach

$$\frac{a_1(1 - 0)}{1 - r} = \frac{a_1}{1 - r}$$

as n increases without bound. Thus, the sum of an infinite geometric series with ratio r such that $|r| < 1$, is given by

$$S = \frac{a_1}{1 - r}$$

If $|r| \geq 1$, the infinite geometric series does not have a sum. The series $1 + 2 + 4 + 8 + 32 + \cdots$ has a ratio of 2; therefore it does not have a sum.

Caution: Before you use the formula $S = \dfrac{a_1}{1 - r}$, it is important to check that the series is an infinite geometric series with a ratio r such that $|r| < 1$.

EXAMPLE 16-14: Find the sum (provided there is a sum) of each of the following infinite geometric series: (a) $81 - 27 + 9 - 3 + 1 + \cdots$ (b) $2 + 2\sqrt{3} + 6 + 6\sqrt{3} + 18 + \cdots$

Solution: **(a)** $r = \dfrac{a_2}{a_1} = \dfrac{-27}{81} = -\dfrac{1}{3}$

$$S = \frac{a_1}{1-r} \qquad \text{Because } |r| = \frac{1}{3} < 1, S \text{ can be evaluated by using } S = \frac{a_1}{1-r}.$$

$$= \frac{81}{1 - \left(-\dfrac{1}{3}\right)}$$

$$= \frac{243}{4}$$

The sum of $81 - 27 + 9 - 31 + 1 + \cdots$ is $\dfrac{243}{4}$.

(b) $r = \dfrac{a_2}{a_1} = \dfrac{2\sqrt{3}}{2} = \sqrt{3}$

Because $|r| = \sqrt{3} > 1$, the sum $2 + 2\sqrt{3} + 6 + 6\sqrt{3} + 18 + \cdots$ does not exist.

To write a repeating decimal as the quotient of two integers, we write the repeating decimal as an infinite geometric series with $|r| < 1$, then find the sum of the series by using $S = \dfrac{a_1}{1-r}$.

EXAMPLE 16-15: Write $0.\overline{54}$ as the quotient of two integers.

Solution: $0.\overline{54} = 0.545454\cdots$

$$= 0.54 + 0.0054 + 0.000054 + \cdots$$

$$r = \frac{a_2}{a_1} = \frac{0.0054}{0.54} = 0.01$$

$$S = \frac{a_1}{1-r} \qquad \text{Because } |r| = 0.01 < 1, \text{ the sum may be evaluated using } S = \frac{a_1}{1-r}.$$

$$= \frac{0.54}{1 - 0.01}$$

$$= \frac{0.54}{0.99}$$

$$= \frac{54}{99}$$

$$0.\overline{54} = \frac{6}{11}$$

16-5. Solving Problems Using Geometric Series

To solve a problem using a geometric series, we:

1. *Read* the problem carefully, several times.
2. *Identify* the unknown geometric series (or numbers).
3. *Decide* how to represent the first term of the geometric sequence, the common ratio, and the number of terms.
4. *Translate* to the geometric series formula $S_n = \dfrac{a_1(1 - r^n)}{1 - r}$.
5. *Evaluate* (or *solve*) the geometric series formula from step 4.
6. *Interpret* S_n (or n) with respect to the unknown geometric series (or number of terms) to find the proposed solution of the original problem.
7. *Check* to see if the proposed solution satisfies all the conditions of the original problem.

EXAMPLE 16-16: Solve this problem using a geometric series:

Read: A standard checkerboard has 64 squares, 32 red squares and 32 black squares. If 1 penny is put on the first black square, 2 pennies on the second black squares, 4 pennies on the third black square, 8 pennies on the fourth black square, and so on, how much money will be on all 32 black squares?

Identify: The unknown geometric series is the sum of all the pennies.

Decide: Let $a_1 = 1$ (penny)

          and $r = 2$ (because $1 = 2(\frac{1}{2})$, $2 = 2(1)$, $4 = 2(2)$, $8 = 2(4)$, and so on)

          and $n = 32$ (black squares).

Translate: $S_n = \dfrac{a_1(1 - r^n)}{1 - r}$ $\longleftarrow$ finite geometric series formula

Evaluate: $S_{32} = \dfrac{1(1 - 2^{32})}{1 - 2}$ Find the sum of the first 32 terms of the geometric sequence.

           $S_{32} = 2^{32} - 1$

           $S_{32} = 4{,}294{,}967{,}295$

Interpret: $S_{32} = 4{,}294{,}967{,}295$ means the total number of pennies on the 32 black squares is 4,294,967,295 pennies, or \$42,949,672.95!

Check: Do the pennies on the 32 black squares add up to 4,294,967,295?
           Yes: Find the first 32 terms using $a_n = a_1 r^{n-1}$, then add to check.

16-6. Binomial Expansions

A. Expanding $(a + b)^n$

We can expand the the binomial expression $(a + b)^n$ by using repeated multiplication. However, a careful examination of the following expansions will help you discover patterns that can be used to write the expansion of $(a + b)^n$.

$$(a + b)^1 = a + b$$
$$(a + b)^2 = a^2 + 2ab + b^2$$
$$(a + b)^3 = a^3 + 3a^2b + 3ab^2 + b^3$$
$$(a + b)^4 = a^4 + 4a^3b + 6a^2b^2 + 4ab^3 + b^4$$

Patterns

1. The expansion of $(a + b)^n$ has $n + 1$ terms.
   Example: $(a + b)^4 = a^4 + 4a^3b + 6a^2b^2 + 4ab^3 + b^4$, which has $4 + 1 = 5$ terms.
2. The power of a is n in the first term, and it decreases by 1 for each successive term.
   Example: $(a + b)^4 = a^4 + 4a^3b + 6a^2b^2 + 4a^1b^3 + a^0b^4$
3. The power of b is 0 in the first term, and it increases by 1 for each successive term.
   Example: $(a + b)^4 = a^4b^0 + 4a^3b^1 + 6a^2b^2 + 4a^1b^3 + a^0b^4$
4. The sum of the exponents of each term is n.

$$4 + 0 = 4 \quad 3 + 1 = 4 \quad 2 + 2 = 4 \quad 1 + 3 = 4 \quad 0 + 4 = 4$$

   Example: $(a + b)^4 = a^4b^0 + 4a^3b^1 + 6a^2b^2 + 4a^1b^3 + a^0b^4$
5. The coefficients of the terms of $(a + b)^n$ are the numbers found in the nth row of the following triangular array, which is known as **Pascal's triangle.**

Examples: For $(a + b)^1$ use: 1   1
          For $(a + b)^2$ use: 1   2   1
          For $(a + b)^3$ use: 1   3   3   1
          For $(a + b)^4$ use: 1   4   6   4   1
          For $(a + b)^5$ use: 1   5   10   10   5   1
          For $(a + b)^6$ use: 1   6   15   20   15   6   1

Note: Each row of Pascal's triangle begins and ends with 1. The other entries in each row can be found by adding the two digits above that entry. Although only the first 6 rows are given here, Pascal's triangle could be extended to be used for any value of *n*.

EXAMPLE 16-17: Write the expansion of $(a + b)^6$.

Solution: Use the indicated patterns to first write the variable part of each term, then use Pascal's Triangle to write the coefficient of each term.

$$(a + b)^6 = ?a^6b^0 + ?a^5b^1 + ?a^4b^2 + ?a^3b^3 + ?a^2b^4 + ?a^1b^5 + ?a^0b^6$$

$$= 1a^6b^0 + 6a^5b^1 + 15a^4b^2 + 20a^3b^3 + 15a^2b^4 + 6a^1b^5 + 1a^0b^6$$

$$= a^6 + 6a^5b + 15a^4b^2 + 20a^3b^3 + 15a^2b^4 + 6ab^5 + b^6$$

To expand a binomial such as $(2x - 3y)^4$, we first expand $(a + b)^4$, then substitute $2x$ for a and $-3y$ for b.

EXAMPLE 16-18: Write the expansion of $(2x - 3y)^4$.

Solution:

$(a + b)^4 = a^4 + 4a^3b + 6a^2b^2 + 4ab^3 + b^4$ Expand $(a + b)^4$.

$(2x - 3y)^4 = (2x)^4 + 4(2x)^3(-3y) + 6(2x)^2(-3y)^2 + 4(2x)(-3y)^3 + (-3y)^4$ Substitute $a = 2x$ and $b = -3y$.

$$= 16x^4 + 4 \cdot 8x^3(-3y) + 6 \cdot 4x^2 \cdot 9y^2 + 4 \cdot 2x(-27)y^3 + 81y^4 \qquad \text{Simplify.}$$

$$= 16x^4 - 96x^3y + 216x^2y^2 - 216xy^3 + 81y^4$$

Note: The expansion of $(a - b)^n$ is the same as the expansion of $(a + b)^n$ except the *even-numbered* terms ·in the expansion of $(a - b)^n$ are preceded by a minus sign.

Although Pascal's triangle can be used to find the coefficients of the expanded form of $(a + b)^n$, this method is inconvenient when *n* is large. Another procedure for determining these coefficients is based on the following function.

The **factorial function** is denoted by *n*!. If *n* is a natural number greater than one, then *n*! (read as "*n* factorial") is the product of the first *n* natural numbers. This function is given by:

$$n! = \begin{cases} n(n - 1) \cdots (2)(1), & \text{for } n \in N \text{ and } n > 1 \\ 1 \text{ for } n = 1 \\ 1 \text{ for } n = 0 \end{cases}$$

For example, $6! = 6 \cdot 5 \cdot 4 \cdot 3 \cdot 2 \cdot 1 = 720$. Be sure to note that, by definition, $0! = 1$ and $1! = 1$.

EXAMPLE 16-19: Find each of the following: **(a)** 2!   **(b)** 3!   **(c)** 4!   **(d)** 8!

Solution: **(a)** $2! = 2 \cdot 1$ **(b)** $3! = 3 \cdot 2 \cdot 1$ **(c)** $4! = 4 \cdot 3 \cdot 2 \cdot 1$ **(d)** $8! = 8 \cdot 7 \cdot 6 \cdot 5 \cdot 4 \cdot 3 \cdot 2 \cdot 1$

                $= 2$ $= 6$ $= 24$ $= 40,320$

We could compute $\dfrac{7!}{5!}$ as $\dfrac{7 \cdot 6 \cdot 5 \cdot 4 \cdot 3 \cdot 2 \cdot 1}{5 \cdot 4 \cdot 3 \cdot 2 \cdot 1} = \dfrac{7 \cdot 6(5 \cdot 4 \cdot 3 \cdot 2 \cdot 1)}{(5 \cdot 4 \cdot 3 \cdot 2 \cdot 1)} = 7 \cdot 6 = 42$. However, if we write the numerator as a product that contains the same factorial as in the denominator, the factorials will divide, leaving us with a simple multiplication problem.

$$\frac{7!}{5!} = \frac{7 \cdot 6 \cdot 5!}{5!} = 7 \cdot 6 = 42$$

This method is much more compact.

The expression $\dbinom{n}{k}$, where $0 \leq k \leq n$, and n and $k \in W$ is called a **binomial coefficient,** and is defined by

$$\binom{n}{k} = \frac{n!}{k!(n - k)!}$$

To evaluate binomial coefficients, $\dbinom{n}{k}$, we substitute in the previous definition, then simplify.

EXAMPLE 16-20: Evaluate $\dbinom{9}{7}$.

Solution: $\dbinom{9}{7} = \dfrac{9!}{7!(9 - 7)!}$ Use the definition of $\dbinom{n}{k}$.

$$= \frac{9!}{7!2!}$$

$$= \frac{9 \cdot 8 \cdot 7!}{7!2!}$$ Write the numerator as a product that contains the largest factorial that is in the denominator.

$$= \frac{9 \cdot 8}{2!}$$ Reduce.

$$= \frac{9 \cdot 8}{2 \cdot 1}$$ Simplify.

$$\binom{9}{7} = 36$$

Because the coefficients of $(a + b)^n$ are the numbers $\dbinom{n}{k}$ for $k = 0, 1, 2, \ldots, n$, it is possible to write the expansion of $(a + b)^n$ by the following formula.

The Binomial Expansion Formula

$$(a + b)^n = \binom{n}{0}a^n + \binom{n}{1}a^{n-1}b + \binom{n}{2}a^{n-2}b^2 + \cdots + \binom{n}{k}a^{n-k}b^k + \cdots + \binom{n}{n}b^n$$

In summation notation the binomial expansion formula is

$$(a + b)^n = \sum_{k=0}^{n} a^{n-k}b^k$$

Note: The above formula is sometimes called the **binomial formula** or **binomial theorem.**

EXAMPLE 16-21: Use the binomial expansion formula to expand $(a + b)^5$.

Solution: $(a + b)^5 = \dbinom{5}{0}a^5 + \dbinom{5}{1}a^{5-1}b + \dbinom{5}{2}a^{5-2}b^2 + \dbinom{5}{3}a^{5-3}b^3 + \dbinom{5}{4}a^{5-4}b^4 + \dbinom{5}{5}b^5$

$$= \quad 1a^5 \quad + \quad 5a^4b \quad + \quad 10a^3b^2 \quad + \quad 10a^2b^3 \quad + \quad 5ab^4 \quad + \quad 1b^5$$

B. Finding the indicated term of a binomial expansion

To find a single term in the expansion of $(a + b)^n$, we use the following formula.

The General-Term Formula for the Expansion of $(a + b)^n$

The $(k + 1)$th term in the expansion of $(a + b)^n$ is given by $\binom{n}{k} a^{n-k} b^k$.

Therefore, if we are asked to find the fifth term of an expansion, we use $k = 4$, because $k + 1 = 5$, the term we wish to find.

EXAMPLE 16-22: Find the 7th term in the expansion of $(a + b)^9$.

Solution: The $(k + 1)$th term of $(a + b)^n = \binom{n}{k} a^{n-k} b^k$.

$$\text{The } (6 + 1)\text{th term of } (a + b)^2 = \binom{9}{6} a^{9-6} b^6. \qquad \text{Because } k + 1 = 7, k = 6.$$

$$= 84a^3 b^6$$

To find the kth term of the expansion of $(3p - q)^n$, we first find the kth term of $(a + b)^n$ and then substitute $3p$ for a and $-q$ for b.

EXAMPLE 16-23: Find the 4th term of $(3p - q)^7$.

Solution: The $(k + 1)$th term of $(a + b)^n = \binom{n}{k} a^{n-k} b^k$.

$$\text{The } (3 + 1)\text{th term of } (a + b)^7 = \binom{7}{3} a^{7-3} n^3.$$

$$= 35a^4 b^3$$

$$\text{The 4th term of } (3p - q)^7 = 35(3p)^4(-q)^3 \qquad \text{Substitute } a = 3p \text{ and } b = -q.$$

$$= 35 \cdot 81p^4(-1)q^3$$

$$= -2835p^4 q^3$$

RAISE YOUR GRADES

Can you . . . ?

☑ find the terms of a sequence using an nth-term formula
☑ find the sum of a series which is written in summation notation
☑ find the sum of a constant series using $\displaystyle\sum_{k=1}^{n} c = nc$
☑ find the terms of an arithmetic sequence using $a_n = a_{n-1} + d$ and $a_n = a_1 + (n-1)d$
☑ evaluate $a_n = a_1 + (n-1)d$ for the indicated unknown
☑ find the sum of an arithmetic series using $S_n = n\left(\dfrac{a_1 + a_n}{2}\right)$
☑ solve a problem using an arithmetic series
☑ find the terms of a geometric sequence using $a_n = ra_{n-1}$ and $a_n = a_1 r^{n-1}$
☑ find the sum of a finite geometric series using $S_n = \dfrac{a_1(1 - r^n)}{1 - r}$
☑ find the sum of an infinite geometric series with ratio r ($|r| < 1$), using $S = \dfrac{a_1}{1 - r}$
☑ write a repeating decimal as the quotient of two integers
☑ solve a problem using a geometric series.

☑ use the given patterns and Pascal's triangle to write the indicated expansion of $(a + b)^n$
☑ evaluate factorial expressions

☑ evaluate the binomial coefficient $\binom{n}{k}$, where $n \in W$, $k \in W$, and $0 \le k \le n$

☑ use the binomial expansion formula to write the indicated expansion of $(a + b)^n$
☑ use the general-term formula for the expansion of $(a + b)^n$ to find the indicated term of $(a + b)^n$

SOLVED PROBLEMS

PROBLEM 16-1 Find the first two terms and the tenth term of the sequences defined by the following nth-term formulas: **(a)** $a_n = n^2 - 5$ **(b)** $a_n = 3n^3 - 4n$ **(c)** $a_n = (-1)^{n+1}(n)$

Solution: Recall that to find the kth term of a sequence, you substitute k for n in the nth-term formula [see Example 16-1].

(a) $a_n = n^2 - 5$ $a_n = n^2 - 5$ $a_n = n^2 - 5$

$a_1 = (1)^2 - 5$ $a_2 = (2)^2 - 5$ $a_{10} = (10)^2 - 5$

$= -4$ $= -1$ $= 95$

(b) $a_n = 3n^3 - 4n$ $a_n = 3n^3 - 4n$ $a_n = 3n^3 - 4n$

$a_1 = 3(1)^3 - 4(1)$ $a_2 = 3(2)^3 - 4(2)$ $a_{10} = 3(10)^3 - 4(10)$

$= -1$ $= 16$ $= 2960$

(c) $a_n = (-1)^{n+1}(n)$ $a_n = (-1)^{n+1}(n)$ $a_n = (-1)^{n+1}(n)$

$a_1 = (-1)^{1+1}(1)$ $a_2 = (-1)^{2+1}(2)$ $a_{10} = (-1)^{10+1}(10)$

$= (-1)^2(1)$ $= (-1)^3(2)$ $= (-1)^{11}(10)$

$= 1$ $= -2$ $= -10$

PROBLEM 16-2 Find the sum of each of the following series which are written in summation notation:

(a) $\sum_{k=1}^{4} k^2$ **(b)** $\sum_{j=1}^{3} (-1)^j(2j + 5)$ **(c)** $\sum_{i=1}^{2} (i^2)(i + 3)$ **(d)** $\sum_{k=2}^{5} 7(k - 3)$

Solution: Recall that to find the sum of a series which is written in summation notation, you first write the series in expanded form and then compute the sum of the terms [see Example 16-2].

(a) $\sum_{k=1}^{4} k^2 = 1^2 + 2^2 + 3^2 + 4^2$

$= 1 + 4 + 9 + 16$

$= 30$

(b) $\sum_{j=1}^{4} (-1)^j(2j + 5) = (-1)^1(2(1) + 5) + (-1)^2(2(2) + 5) + (-1)^3(2(3) + 5)$

$= -7 + 9 - 11$

$= -9$

(c) $\sum_{i=1}^{2} (i^2)(i + 3) = (1^2)((1) + 3) + (2^2)((2) + 3)$

$= 4 + 20$

$= 24$

(d) $\displaystyle\sum_{k=2}^{5} 7(k-3) = 7(2-3) + 7(3-3) + 7(4-3) + 7(5-3)$

$$= -7 + 0 + 7 + 14$$

$$= 14$$

PROBLEM 16-3 Find the sum of each of the following constant series:

(a) $\displaystyle\sum_{k=1}^{50} 3$ (b) $\displaystyle\sum_{k=1}^{400} (-7)$ (c) $\displaystyle\sum_{j=1}^{25} 25$ (d) $\displaystyle\sum_{i=1}^{5000} 2$

Solution: Recall that for any constant c, $\displaystyle\sum_{k=1}^{n} c = nc$ [see Example 16-3].

(a) $\displaystyle\sum_{k=1}^{50} 3 = 50 \cdot 3 = 150$ (b) $\displaystyle\sum_{k=1}^{400} (-7) = 400(-7) = -2800$

(c) $\displaystyle\sum_{j=1}^{25} 25 = 25 \cdot 25 = 625$ (d) $\displaystyle\sum_{i=1}^{5000} 2 = 5000 \cdot 2 = 10,000$

PROBLEM 16-4 Find the first four terms of each of the following arithmetic series which have the indicated first term a_1 and common difference d:

(a) $a_1 = 6, d = 4$ (b) $a_1 = -3, d = -2$ (c) $a_1 = -20, d = 3$

Solution: Recall that any term of an arithmetic sequence can be found by adding the common difference d to the preceding term [see Example 16-4].

(a) $a_1 = 6$

$a_2 = 6 + 4 = 10$

$a_3 = 10 + 4 = 14$

$a_4 = 14 + 4 = 18$

(b) $a_1 = -3$

$a_2 = -3 + (-2) = -5$

$a_3 = -5 + (-2) = -7$

$a_4 = -7 + (-2) = -9$

(c) $a_1 = -20$

$a_2 = -20 + 3 = -17$

$a_3 = -17 + 3 = -14$

$a_4 = -14 + 3 = -11$

PROBLEM 16-5 Given an arithmetic sequence with:

(a) $a_1 = 6$, and $d = 3$, find a_8 (b) $a_{12} = -52$, and $d = -6$, find a_1
(c) $a_1 = 20$, $a_n = 380$, and $d = 18$, find n (d) $a_1 = 12$, and $a_{51} = -138$, find d.

Solution: Recall that the nth-term formula for an arithmetic sequence is $a_n = a_1 + (n-1)d$ [see Examples 16-5 and 16-6].

(a) $a_n = a_1 + (n-1)d$

$a_8 = 6 + (8-1)3$

$a_8 = 6 + 21$

$a_8 = 27$

(b) $a_n = a_1 + (n-1)d$

$-52 = a_1 + (12-1)(-6)$

$-52 = a_1 + (-66)$

$14 = a_1$

(c) $a_n = a_1 + (n-1)d$

$380 = 20 + (n-1)(18)$

$360 = (n-1)(18)$

$20 = n - 1$

$21 = n$

(d) $a_n = a_1 + (n-1)d$

$-138 = 12 + (51-1)d$

$-150 = 50d$

$-3 = d$

PROBLEM 16-6 Given an arithmetic sequence with:

(a) $a_1 = 10$, and $a_{20} = 124$, find S_{20} (b) $a_1 = 100$, and $a_{31} = -260$, find S_{31}

(c) $a_3 = 15$, and $a_{50} = 203$, find S_{50}

Solution: Recall that the sum of the first n terms of an arithmetic series is given by $S_n = n\left(\dfrac{a_1 + a_n}{2}\right)$ [see

Examples 16-7 and 16-8].

(a) $S_n = n\left(\dfrac{a_1 + a_n}{2}\right)$ (b) $S_n = n\left(\dfrac{a_1 + a_n}{2}\right)$

$S_{20} = 20\left(\dfrac{10 + 124}{2}\right)$ $S_{31} = 31\left(\dfrac{100 + (-260)}{2}\right)$

$= 20(67)$ $= 31(-80)$

$= 1340$ $= -2480$

(c) $a_3 = 15$ means $15 = a_1 + (3 - 1)d$ or $15 = a_1 + 2d$

$a_{50} = 203$ means $203 = a_1 + (50 - 1)d$ or $203 = a_1 + 49d$

Solving the system $\begin{cases} 15 = a_1 + 2d \\ 203 = a_1 + 49d \end{cases}$ produces $a_1 = 7$ and $d = 4$.

$S_n = n\left(\dfrac{a_1 + a_n}{2}\right)$

$S_{50} = 50\left(\dfrac{7 + 203}{2}\right)$

$= 50(105)$

$= 5250$

PROBLEM 16-7 Solve this problem using an arithmetic sequence and/or series:

A free-falling object will fall from rest 16 ft in the 1st second, 48 ft during the 2nd second, 80 ft during the 3rd second and so on. How long will it take a free-falling object to fall from 1600 ft above the earth's surface?

Solution: Recall that to solve a problem using an arithmetic sequence or series, you:

1. *Read* the problem carefully, several times.
2. *Identify* the unknown arithmetic series (or number).
3. *Decide* how to represent the first term of the arithmetic sequence, the common difference, and the number of terms.
4. *Translate* to the arithmetic series formula: $S_n = n\left(\dfrac{a_1 + a_n}{2}\right)$.
5. *Evaluate* (or *solve*) the arithmetic series formula from step 4.
6. *Interpret* S_n (or n) with respect to the unknown arithmetic series (or number of terms) to find the proposed solution of the original problem [see Example 16-9].

Identify: The unknown number is the number of seconds needed to fall 1600 ft.

Decide: Let $a_1 = 16$ (ft)

and $d = 32$ (ft) (because $48 - 16 = 32$, $80 - 48 = 32$, and so on)

and $n =$ the unknown number of seconds to fall 1600 ft.

Translate: $a_n = a_1 + (n - 1)d$ ⟵ *n*th-term formula

$a_n = 16 + (n - 1)32$ Represent a_n for $a_1 = 16$ and $d = 32$.

$S_n = n\left(\dfrac{a_1 + a_n}{2}\right)$ ⟵ arithmetic series formula

Solve:

$$1600 = n\left(\frac{16 + [16 + (n-1)32]}{2}\right)$$

Find n (the number of seconds to fall 1600 ft) from $S_n = 1600$, $a_1 = 16$, $d = 32$, and $a_n = 16 + (n-1)32$.

$$1600 = n\left(\frac{16 + 16 + 32n - 32}{2}\right)$$

$$1600 = n(16n)$$

$$1600 = 16n^2$$

$$0 = 16n^2 - 1600$$

$$0 = n^2 - 100$$

$$0 = (n + 10)(n - 10)$$

$$n + 10 = 0 \quad \text{or} \quad n - 10 = 0$$

Interpret: $\quad n = -10$ or $\quad n = 10$ *(seconds)* ⟵ solution (check S_n for $n = 10$)

PROBLEM 16-8 Find the first four terms of each of the following geometric series which have the indicated first term a_1 and common ratio r:

(a) $a_1 = 10$, $r = 3$ (b) $a_1 = -64$, $r = -\frac{1}{2}$ (c) $a_1 = 5\sqrt{2}$, $r = \sqrt{2}$

Solution: Recall that any term of a geometric sequence can be found by multiplying the common ratio r times the preceding term [see Example 16-10].

(a) $a_1 = 10$

$a_2 = 10 \cdot 3 = 30$

$a_3 = 30 \cdot 3 = 90$

$a_4 = 90 \cdot 3 = 270$

(b) $a_1 = -64$

$a_2 = -64 \cdot (-\frac{1}{2}) = 32$

$a_3 = \quad 32 \cdot (-\frac{1}{2}) = -16$

$a_4 = -16 \cdot (-\frac{1}{2}) = 8$

(c) $a_1 = 5\sqrt{2}$

$a_2 = 5\sqrt{2}(\sqrt{2}) = 10$

$a_3 = 10(\sqrt{2}) = 10\sqrt{2}$

$a_4 = 10\sqrt{2}(\sqrt{2}) = 20$

PROBLEM 16-9 Given a geometric sequence with:

(a) $a_1 = 2$, and $r = 3$, find a_4 (b) $a_1 = 128$ and $r = -\frac{1}{2}$, find a_5
(c) $a_1 = 17$, and $r = \frac{1}{2}$, find a_{10} (d) $a_1 = 4$ and $a_2 = -4$, find a_{100}

Solution: Recall that the nth-term formula for a geometric sequence is $a_n = a_1 r^{n-1}$ [see Examples 16-11 and 16-12].

(a) $a_n = a_1 r^{n-1}$

$a_4 = 2 \cdot 3^{4-1}$

$\quad = 2 \cdot 3^3$

$\quad = 2 \cdot 27$

$\quad = 54$

(b) $a_n = a_1 r^{n-1}$

$a_5 = 128(-\frac{1}{2})^{5-1}$

$\quad = 128(-\frac{1}{2})^4$

$\quad = 128(\frac{1}{16})$

$\quad = 8$

(c) $\quad a_n = a_1 r^{n-1}$

$a_{10} = 17(\frac{1}{2})^{10-1}$

$\quad = 17(\frac{1}{2})^9$

$\quad = 17(\frac{1}{512})$

$\quad = \frac{17}{512}$

(d) $r = \dfrac{a_2}{a_1} = \dfrac{-4}{4} = -1$ First find r.

$a_n = a_1 r^{n-1}$

$a_{100} = 4(-1)^{100-1}$

$\quad = 4(-1)^{99}$

$\quad = -4$

PROBLEM 16-10 Given a geometric sequence with: (a) $a_1 = 16$ and $r = \frac{1}{2}$, find S_4
(b) $a_1 = 1$ and $r = 2$, find S_6 (c) $a_1 = 243$ and $r = -\frac{1}{3}$, find S_5

Solution: Recall that the sum of the first n terms of a geometric series is given by $S_n = \dfrac{a_1(1 - r^n)}{1 - r}$ [see Example 16-13].

(a) $S_n = \dfrac{a_1(1 - r^n)}{1 - r}$ **(b)** $S_n = \dfrac{a_1(1 - r^n)}{1 - r}$ **(c)** $S_n = \dfrac{a_1(1 - r^n)}{1 - r}$

$S_4 = \dfrac{16[1 - (\frac{1}{2})^4]}{1 - \frac{1}{2}}$ $S_6 = \dfrac{1(1 - 2^6)}{1 - 2}$ $S_5 = \dfrac{243[1 - (-\frac{1}{3})^5]}{1 - (-\frac{1}{3})}$

$= \dfrac{16[\frac{15}{16}]}{\frac{1}{2}}$ $= \dfrac{-63}{-1}$ $= \dfrac{243(\frac{244}{243})}{\frac{4}{3}}$

$= 30$ $= 63$ $= 183$

PROBLEM 16-11 Find the sum (if it exists) of each of the following infinite geometric series:

(a) $1 + \frac{1}{2} + \frac{1}{4} + \frac{1}{8} + \frac{1}{16} + \cdots$ **(b)** $81 + 27 + 9 + 3 + 1 + \cdots$
(c) $2 + 4 + 8 + 16 + 32 + 64 + \cdots$ **(d)** $16 - 8 + 4 - 2 + 1 - \frac{1}{2} + \cdots$
(e) $32 + 16\sqrt{2} + 16 + 8\sqrt{2} + 8 + \cdots$

Solution: Recall that for an infinite geometric series with a ratio r such that $|r| < 1$, $S = \dfrac{a_1}{1 - r}$ [see Example 16-14].

(a) $r = \dfrac{a_2}{a_1} = \dfrac{\frac{1}{2}}{1} = \dfrac{1}{2}$ **(b)** $r = \dfrac{a_2}{a_1} = \dfrac{27}{81} = \dfrac{1}{3}$

$S = \dfrac{1}{1 - \frac{1}{2}} = \dfrac{1}{\frac{1}{2}} = 2$ $S = \dfrac{81}{1 - \frac{1}{3}} = \dfrac{81}{\frac{2}{3}} = 81 \cdot \dfrac{3}{2} = \dfrac{243}{2}$

(c) $r = \dfrac{a_2}{a_1} = \dfrac{4}{2} = 2$ Because $|r| = 2 > 1$, the sum $2 + 4 + 8 + 16 + 32 + 64 + \cdots$ does not exist.

(d) $r = \dfrac{-8}{16} = -\dfrac{1}{2}$

$S = \dfrac{16}{1 - (-\frac{1}{2})} = \dfrac{16}{\frac{3}{2}} = \dfrac{32}{3}$

(e) $r = \dfrac{16\sqrt{2}}{32} = \dfrac{\sqrt{2}}{2}$

$S = \dfrac{32}{1 - \dfrac{\sqrt{2}}{2}} = \dfrac{32}{\dfrac{2 - \sqrt{2}}{2}} = 32 \cdot \dfrac{2}{2 - \sqrt{2}} = \dfrac{64}{2 - \sqrt{2}} = 64 + 32\sqrt{2}$

PROBLEM 16-12 Write each of the following repeating decimals as the quotient of two integers:

(a) $0.\overline{4}$ **(b)** $0.1\overline{6}$ **(c)** $0.\overline{27}$ **(d)** $0.\overline{9}$

Solution: Recall that to write a repeating decimal as the quotient of two integers, you write the repeating decimal as a infinite geometric series and then find the sum of the series by using $S = \dfrac{a_1}{1 - r}$ [see Example 16-15].

(a) $0.\overline{4} = 0.44444 \cdots$

$= 0.4 + 0.04 + 0.004 + 0.0004 + 0.00004 + \cdots$

$r = \dfrac{a_2}{a_1} = \dfrac{0.04}{0.4} = 0.1$

$S = \dfrac{a_1}{1 - r} = \dfrac{0.4}{1 - 0.1} = \dfrac{0.4}{0.9} = \dfrac{4}{9}$

(b) $0.1\overline{6} = 0.16666$

$$= 0.1 + [0.06 + 0.006 + 0.0006 + 0.00006 + \cdots]$$

$$r = \frac{a_2}{a_1} = \frac{0.006}{0.06} = 0.1$$

$$S = \frac{a_1}{1 - r} = \frac{0.06}{1 - 0.1} = \frac{0.06}{0.9} = \frac{6}{90} = \frac{1}{15}$$

$$0.16 = 0.1 + \frac{1}{15} = \frac{1}{10} + \frac{1}{15} = \frac{3}{30} + \frac{2}{30} = \frac{5}{30} = \frac{1}{6}$$

(c) $0.\overline{27} = 0.272727\cdots$

$$= 0.27 + 0.0027 + 0.000027 + 0.00000027 + \cdots$$

$$r = \frac{a_2}{a_1} = \frac{0.0027}{0.27} = 0.01$$

$$S = \frac{a_1}{1 - r} = \frac{0.27}{1 - 0.01} = \frac{0.27}{0.99} = \frac{27}{99} = \frac{3}{11}$$

(d) $0.\overline{9} = 0.99999\cdots$

$$= 0.9 + 0.09 + 0.009 + 0.0009 + 0.00009 + \cdots$$

$$r = \frac{a_2}{a_1} = \frac{0.09}{0.9} = 0.1$$

$$S = \frac{a_1}{1 - r} = \frac{0.9}{1 - 0.1} = \frac{0.9}{0.9} = 1$$

PROBLEM 16-13 Solve this problem using a geometric series:
A ball dropped from 10 ft rebounds $\frac{1}{2}$ its previous height on each bounce. How far does the ball travel by the time it comes to rest?

Solution: Recall that to solve a problem using a geometric series, you:

1. *Read* the problem carefully, several times.
2. *Identify* the unknown geometric series (or number).
3. *Decide* how to represent the first term of the geometric sequence, the common ratio, and the number of terms.
4. *Translate* to the geometric series formula: $S_n = \dfrac{a_1(1 - r^n)}{1 - r}$ or $S_n = \dfrac{a_1}{1 - r}$.
5. *Evaluate* (or solve) the geometric series formula from step 4.
6. *Interpret* S_n (or n) with respect to the unknown geometric series (or number of terms) to find the proposed solution of the original problem.
7. *Check* to see if the proposed solution satisfies all the conditions of the original problem [see Example 16-16].

Identify: The unknown geometric series is the sum of the infinite number of rebounds.

Decide: Let $a_1 = 5$ (ft) $\longleftarrow$ [$\frac{1}{2}(10$ ft) or $\frac{10}{2}$ ft]

and $r = \frac{1}{2}$. $\longleftarrow$ the common ratio

Translate: $S = \dfrac{a_1}{1 - r}$ $\longleftarrow$ infinite geometric series formula

Evaluate: $S = \dfrac{5}{1 - \frac{1}{2}}$ Find the sum of the terms of the infinite geometric sequence $[5 + \frac{5}{2} + \frac{5}{4} + \frac{5}{8} + \cdots]$.

$$S = \frac{5}{\frac{1}{2}}$$

$$S = 10$$

Interpret: $S = 10$ means the sum of all the rebounds is 10 ft.

$10 + 2S = 10 + 2(10) = 30$ means the total distance that the ball travels before coming to rest (the original 10 ft the ball was dropped, plus the sum of the rebounds, plus the sum of all the falls after the rebounds).

PROBLEM 16-14 Write the expansion of each of the following:

(a) $(a + b)^7$ **(b)** $(x + 3y)^4$ **(c)** $(2u - v)^6$

Solution: Recall that you can expand a binomial by using the patterns (see Section 16-6) to first write the variable part of each term and then use Pascal's triangle to write the coefficient of each term [see Examples 16-17 and 16-18].

(a) $(a + b)^7 = ?a^7b^0 + ?a^6b^1 + ?a^5b^2 + ?a^4b^3 + ?a^3b^4 + ?a^2b^5 + ?a^1b^6 + ?a^0b^7$

$$= 1a^7b^0 + 7a^6b^1 + 21a^5b^2 + 35a^4b^3 + 35a^3b^4 + 21a^2b^5 + 7a^1b^6 + 1^0b^7$$

$$= a^7 + 7a^6b + 21a^5b^2 + 35a^4b^3 + 35a^3b^4 + 21a^2b^5 + 7ab^6 + b^7$$

(b) $(x + 3y)^4 = ?x^4 + ?x^3(3y) + ?x^2(3y)^2 + ?x(3y)^3 + ?(3y)^4$

$$= 1x^4 + 4x^3(3y) + 6x^2(3y)^2 + 4x(3y)^3 + 1(3y)^4$$

$$= x^4 + 12x^3y + 54x^2y^2 + 108xy^3 + 81y^4$$

(c) $(2u - v)^6 = ?(2u)^6 + ?(2u)^5(-v) + ?(2u)^4(-v)^2 + ?(2u)^3(-v)^3 + ?(2u)^2(-v)^4 + ?(2u)(-v)^5 + ?(-v)^6$

$$= 1(2u)^6 + 6(2u)^5(-v) + 15(2u)^4(-v)^2 + 20(2u)^3(-v)^3 + 15(2u)^2(-v)^4 + 6(2u)(-v)^5 + 1(-v)^6$$

$$= 64u^6 - 192u^5v + 240u^4v^2 - 160u^3v^3 + 60u^2v^4 - 12uv^5 + v^6$$

PROBLEM 16-15 Evaluate each of the following factorials:

(a) 5!   **(b)** 7!   **(c)** 10!   **(d)** 0!   **(e)** 1!

Solution: Recall that the factorial function is defined by

$$n! = \begin{cases} n(n - 1) \cdots (2)(1) \text{ for } n \in N \text{ and } n > 1 \\ 1 \text{ for } n = 1 \\ 1 \text{ for } n = 0 \end{cases}$$

[see Example 16-19].

(a) $5! = 5 \cdot 4 \cdot 3 \cdot 2 \cdot 1 = 120$ **(b)** $7! = 7 \cdot 6 \cdot 5 \cdot 4 \cdot 3 \cdot 2 \cdot 1 = 5040$
(c) $10! = 10 \cdot 9 \cdot 8 \cdot 7 \cdot 6 \cdot 5 \cdot 4 \cdot 3 \cdot 2 \cdot 1 = 3{,}628{,}800$ **(d)** $0! = 1$ (by definition)
(e) $1! = 1$ (by definition)

PROBLEM 16-16 Evaluate the following binomial coefficients:

(a) $\begin{pmatrix} 6 \\ 2 \end{pmatrix}$ **(b)** $\begin{pmatrix} 5 \\ 1 \end{pmatrix}$ **(c)** $\begin{pmatrix} 3 \\ 0 \end{pmatrix}$ **(d)** $\begin{pmatrix} 10 \\ 8 \end{pmatrix}$ **(e)** $\begin{pmatrix} 7 \\ 7 \end{pmatrix}$

Solution: Recall that $\begin{pmatrix} n \\ k \end{pmatrix}$, where $0 \le k \le n$, n and $k \in W$, is defined by $\begin{pmatrix} n \\ k \end{pmatrix} = \dfrac{n!}{k!(n - k)!}$ [see Example 16-20].

(a) $\begin{pmatrix} 6 \\ 2 \end{pmatrix} = \dfrac{6!}{2!(6 - 2)!} = \dfrac{6!}{2!4!} = \dfrac{6 \cdot 5 \cdot 4!}{2!4!} = \dfrac{6 \cdot 5}{2!} = \dfrac{6 \cdot 5}{2} = 15$

(b) $\begin{pmatrix} 5 \\ 1 \end{pmatrix} = \dfrac{5!}{1!(5 - 1)!} = \dfrac{5!}{1!4!} = \dfrac{5 \cdot 4!}{1!4!} = \dfrac{5}{1!} = \dfrac{5}{1} = 5$

(c) $\dbinom{3}{0} = \dfrac{3!}{0!(3-0)!} = \dfrac{3!}{0!3!} = \dfrac{1}{0!} = \dfrac{1}{1} = 1$

(d) $\dbinom{10}{8} = \dfrac{10!}{8!(10-8)!} = \dfrac{10!}{8!2!} = \dfrac{10\cdot 9\cdot 8!}{8!2!} = \dfrac{10\cdot 9}{2!} = \dfrac{10\cdot 9}{2} = 45$

(e) $\dbinom{7}{7} = \dfrac{7!}{7!(7-7)!} = \dfrac{7!}{7!0!} = \dfrac{1}{0!} = \dfrac{1}{1} = 1$

PROBLEM 16-17 Use the binomial expansion formula to expand **(a)** $(a+b)^4$ and **(b)** $(2p-3q)^5$.

Solution: Recall the binomial expansion formula:

$$(a+b)^n = \binom{n}{0}a^n + \binom{n}{1}a^{n-1}b + \binom{n}{2}a^{n-2}b^2 + \cdots + \binom{n}{k}a^{n-k}b^k + \cdots + \binom{n}{n}b^n$$

[see Example 16-21].

(a) $(a+b)^4 = \binom{4}{0}a^4 + \binom{4}{1}a^{4-1}b + \binom{4}{2}a^{4-2}b^2 + \binom{4}{3}a^{4-3}b^3 + \binom{4}{4}b^4$

$= 1a^4 + 4a^3b + 6a^2b^2 + 4ab^3 + 1b^4$

(b) $(2p-3q)^5 = \binom{5}{0}(2p)^5 + \binom{5}{1}(2p)^{5-1}(-3q) + \binom{5}{2}(2p)^{5-2}(-3q)^2 + \binom{5}{3}(2p)^{5-3}(-3q)^3$

$\quad + \binom{5}{4}(2p)^{5-4}(-3q)^4 + \binom{5}{5}(-3q)^5$

$= 1(2p)^5 + 5(2p)^4(-3q) + 10(2p)^3(-3q)^2 + 10(2p)^2(-3q)^3$

$\quad + 5(2p)(-3q)^4 + 1(-3q)^5$

$= 32p^5 - 240p^4q + 720p^3q^2 - 1080p^2q^3 + 810pq^4 - 243q^5$

PROBLEM 16-18 Find the indicated term in each of the following expansions:

(a) the 5th term of $(a+b)^8$ **(b)** the 8th term of $(2x-y)^{10}$ **(c)** the 3rd term of $(3x-5y)^5$

Solution: Recall that the $(k+1)$th term in the expansion of $(a+b)^n$ is $\binom{n}{k}a^{n-k}b^k$ [see Examples 16-22 and 16-23].

(a) The $(4+1)$th term of $(a+b)^8 = \binom{8}{4}a^{8-4}b^4$

$= 70a^4b^4$

(b) The $(7+1)$th term of $(2x-y)^{10} = \binom{10}{7}(2x)^{10-7}(-y)^7$

$= 120(2x)^3(-y)^7$

$= -960x^3y^7$

(c) The $(2+1)$th term of $(3x-5y)^5 = \binom{5}{2}(3x)^{5-2}(-5y)^2$

$= 10(3x)^3(-5y)^2$

$= 10(27x^3)(25y^2)$

$= 6750x^3y^2$

Supplementary Exercises

PROBLEM 16-19 Find the first three terms and the tenth term of the sequences defined by the following *n*th-term formulas:

(a) $a_n = n^4 - 1$ **(b)** $a_n = 2n - 8$ **(c)** $a_n = (-1)^n(5n - 3)$ **(d)** $a_n = (-1)^{n+1}$

(e) $a_n = |1 - 5n|$ **(f)** $a_n = 3n^2 + 2n$ **(g)** $a_n = \dfrac{1}{n+1}$ **(h)** $a_n = \dfrac{1}{2}n$

PROBLEM 16-20 Find the sum of each of the following series, which are written in summation notation:

(a) $\displaystyle\sum_{k=1}^{6} k$ **(b)** $\displaystyle\sum_{k=1}^{5} 3k$ **(c)** $\displaystyle\sum_{k=1}^{4} (k^2 + 1)$ **(d)** $\displaystyle\sum_{k=1}^{7} (2k - 5)$ **(e)** $\displaystyle\sum_{k=1}^{6} (-1)^k(4k)$

(f) $\displaystyle\sum_{k=1}^{10} 5$ **(g)** $\displaystyle\sum_{k=1}^{2000} (-6)$ **(h)** $\displaystyle\sum_{j=1}^{27} 3$ **(i)** $\displaystyle\sum_{i=1}^{51} 15$ **(j)** $\displaystyle\sum_{i=1}^{600} 40$

PROBLEM 16-21 Find the first five terms of the following arithmetic series, which have the indicated first term a_1 and common difference d:

(a) $a_1 = 6, d = 10$ **(b)** $a_1 = -20, d = -4$ **(c)** $a_1 = -50, d = 15$ **(d)** $a_1 = 80, d = -3$

(e) $a_1 = 2000, d = 25$ **(f)** $a_1 = 0, d = -100$ **(g)** $a_1 = \frac{1}{2}, d = \frac{5}{2}$ **(h)** $a_1 = 4.75, d = 6.25$

PROBLEM 16-22 Given an arithmetic sequence with:

(a) $a_1 = 30$, and $d = 6$, find a_{21} 　　　　　 **(b)** $a_1 = 140$, and $d = -25$, find a_{36}

(c) $a_{14} = -80$, and $d = 12$, find a_1 　　　　 **(d)** $a_{10} = 20$, and $d = -8$, find a_1

(e) $a_1 = 4$, and $a_{16} = 109$, find d 　　　　 **(f)** $a_1 = -32$, and $a_{211} = -662$, find d

(g) $a_1 = 10$, and $a_n = 335$, and $d = 13$, find n 　　 **(h)** $a_1 = -42$, and $a_n = 2582$, and $d = 8$, find n

PROBLEM 16-23 Solve each problem using an arithmetic sequence and series:

(a) A certain object will roll down an inclined plane 3 ft in the first second, 6 ft in the second second, 9 ft in the third second, and so on. How far will the object roll in 20 seconds? How long will it take the object to roll 18 ft?

(b) The top row of a stack of logs contains 3 logs, the next row down 5 logs, the next row down 7 logs, and so on. If there are 20 rows, then how many logs are in the stack? How many complete rows are needed to stack more than 100 logs?

PROBLEM 16-24 Find the first four terms on the following geometric sequences, which have the indicated first term a_1 and common ratio r:

(a) $a_1 = 10, r = \frac{1}{2}$ **(b)** $a_1 = 6, r = 2$ **(c)** $a_1 = -4, r = -3$ **(d)** $a_1 = 0.2, r = 0.4$
(e) $a_1 = \frac{5}{3}, r = -\frac{1}{3}$ **(f)** $a_1 = 24, r = \frac{1}{6}$ **(g)** $a_1 = 100, r = -\frac{1}{10}$ **(h)** $a_1 = 0.4, r = 0.1$

PROBLEM 16-25 Given a geometric sequence with:

(a) $a_1 = 4$, and $r = 2$, find a_6 　　　　　　 **(b)** $a_1 = 512$, and $r = \frac{1}{2}$, find a_7
(c) $a_1 = 10$, and $r = -\frac{1}{2}$, find a_8 　　　 **(d)** $a_1 = -30$, and $r = \frac{1}{3}$, find a_5
(e) $a_1 = 6000$, and $r = 0.1$, find a_9 　　　 **(f)** $a_1 = -\frac{1}{32}$, and $r = -2$, find a_{10}

PROBLEM 16-26 Given a geometric sequence with:

(a) $a_1 = 20$, and $r = 2$, find S_4 　　　　　 **(b)** $a_1 = 2$, and $r = \frac{1}{2}$, find S_6
(c) $a_1 = 1024$, and $r = -\frac{1}{4}$, find S_3 　 **(d)** $a_1 = 1$, and $r = -2$, find S_5
(e) $a_1 = 7000$, and $r = 0.1$, find S_3 　　 **(f)** $a_1 = 0.0004$, and $r = 3$, find S_4

PROBLEM 16-27 Find the sum (if it exists) of each of the following infinite geometric series:

(a) $100 + 50 + 25 + \frac{25}{2} + \frac{25}{4} + \cdots$

(b) $60 - 20 + \frac{20}{3} - \frac{20}{9} + \frac{20}{27} - \cdots$

(c) $12 + 24 + 48 + 96 + 192 + \cdots$

(d) $0.5 + 0.05 + 0.005 + 0.0005 + 0.00005 + \cdots$

(e) $5 + 5\sqrt{2} + 10 + 10\sqrt{2} + 20 + 20\sqrt{2} + \cdots$

(f) $4 + 2\sqrt{2} + 2 + \sqrt{2} + 1 + \frac{\sqrt{2}}{2} + \cdots$

PROBLEM 16-28 Write each of the following repeating decimals as the quotient of two integers:

(a) $0.\overline{6}$ (b) $0.\overline{23}$ (c) $0.5\overline{1}$ (d) $0.00\overline{2}$ (e) $0.\overline{18}$ (f) $-1.\overline{27}$

PROBLEM 16-29 Solve each problem using a geometric series:

(a) Assuming no duplications of ancestors, how many direct ancestors has each of us had over the last 10 generations? How many generations would we have to go back to have at least 100 direct ancestors?

(b) A ball is dropped from 20 m and rebounds 75% of its previous height on each bounce. How far does the ball travel by the time it comes to rest? How many rebounds are necessary for the ball to travel at least 100 m?

PROBLEM 16-30 Write the expansion of each of the following:

(a) $(x + y)^8$ (b) $(2w + 1)^4$ (c) $(p - 3)^5$ (d) $(2p - 1)^7$ (e) $(1 - 4x)^3$ (f) $(3a + 4b)^6$

PROBLEM 16-31 Find the indicated term in each of the following expansions:

(a) the 6th term of $(a + b)^9$

(b) the 4th term of $(3x - 2y)^6$

(c) the 10th term of $(x + 1)^{12}$

(d) the 2nd term of $\left(x - \dfrac{y}{2}\right)^{10}$

(e) the middle term of $(x + 2y)^6$

(f) the last term of $(u - v)^{15}$

Answers to Supplementary Exercises

(16-19) (a) $a_1 = 0, a_2 = 15, a_3 = 80, a_{10} = 9999$ (b) $a_1 = -6, a_2 = -4, a_3 = -2, a_{10} = 12$
(c) $a_1 = -2, a_2 = 7, a_3 = -12, a_{10} = 47$ (d) $a_1 = 1, a_2 = -1, a_3 = 1, a_{10} = -1$
(e) $a_1 = 4, a_2 = 9, a_3 = 14, a_{10} = 49$ (f) $a_1 = 5, a_2 = 16, a_3 = 33, a_{10} = 320$
(g) $a_1 = \frac{1}{2}, a_2 = \frac{1}{3}, a_3 = \frac{1}{4}, a_{10} = \frac{1}{11}$ (h) $a_1 = \frac{1}{2}, a_2 = 1, a_3 = \frac{3}{2}, a_{10} = 5$

(16-20) (a) 21   (b) 45   (c) 34   (d) 21   (e) 12   (f) 50   (g) $-12,000$ (h) 81
(i) 765   (j) 24,000

(16-21) (a) $a_1 = 6, a_2 = 16, a_3 = 26, a_4 = 36, a_5 = 46$
(b) $a_1 = -20, a_2 = -24, a_3 = -28, a_4 = -32, a_5 = -36$
(c) $a_1 = -50, a_2 = -35, a_3 = -20, a_4 = -5, a_5 = 10$
(d) $a_1 = 80, a_2 = 77, a_3 = 74, a_4 = 71, a_5 = 68$
(e) $a_1 = 2000, a_2 = 2025, a_3 = 2050, a_4 = 2075, a_5 = 2100$
(f) $a_1 = 0, a_2 = -100, a_3 = -200, a_4 = -300, a_5 = -400$
(g) $a_1 = \frac{1}{2}, a_2 = 3, a_3 = \frac{11}{2}, a_4 = 8, a_5 = \frac{21}{2}$
(h) $a_1 = 4.75, a_2 = 11, a_3 = 17.25, a_4 = 23.5, a_5 = 29.75$

(16-22) (a) 150   (b) -735 (c) -236 (d) 92   (e) 7   (f) -3 (g) 26   (h) 329

(16-23) (a) 630 ft, 3 sec   (b) 440 logs, 10 complete rows

(16-24) **(a)** $a_1 = 10$, $a_2 = 5$, $a_3 = \frac{5}{2}$, $a_4 = \frac{5}{4}$
(b) $a_1 = 6$, $a_2 = 12$, $a_3 = 24$, $a_4 = 48$
(c) $a_1 = -4$, $a_2 = 12$, $a_3 = -36$, $a_4 = 108$
(d) $a_1 = 0.2$, $a_2 = 0.08$, $a_3 = 0.032$, $a_4 = 0.0128$
(e) $a_1 = \frac{5}{3}$, $a_2 = -\frac{5}{9}$, $a_3 = \frac{5}{27}$, $a_4 = -\frac{5}{81}$
(f) $a_1 = 24$, $a_2 = 4$, $a_3 = \frac{2}{3}$, $a_4 = \frac{1}{9}$
(g) $a_1 = 100$, $a_2 = -10$, $a_3 = 1$, $a_4 = -\frac{1}{10}$
(h) $a_1 = 0.4$, $a_2 = 0.04$, $a_3 = 0.004$, $a_4 = 0.0004$

(16-25) **(a)** 128   **(b)** 8   **(c)** $-\frac{5}{64}$ **(d)** $-\frac{10}{27}$ **(e)** 0.00006   **(f)** 16

(16-26) **(a)** 300   **(b)** $\frac{63}{16}$ **(c)** 832   **(d)** 11   **(e)** 7770   **(f)** 0.016

(16-27) **(a)** 200   **(b)** 45   **(c)** does not exist   **(d)** $\frac{5}{9}$ **(e)** does not exist   **(f)** $8 + 4\sqrt{2}$

(16-28) **(a)** $\frac{2}{3}$ **(b)** $\frac{23}{99}$ **(c)** $\frac{17}{33}$ **(d)** $\frac{1}{450}$ **(e)** $\frac{2}{11}$ **(f)** $-\frac{14}{11}$

(16-29) **(a)** 2046 direct ancestors, 7 generations   **(b)** 140 m, 4 rebounds

(16-30) **(a)** $x^8 + 8x^7y + 28x^6y^2 + 56x^5y^3 + 70x^4y^4 + 56x^3y^5 + 28x^2y^6 + 8xy^7 + y^8$
(b) $16w^4 + 32w^3 + 24w^2 + 8w + 1$
(c) $p^5 - 15p^4 + 90p^3 - 270p^2 + 405p - 243$
(d) $128p^7 - 448p^6 + 672p^5 - 560p^4 + 280p^3 - 84p^2 + 14p - 1$
(e) $1 - 12x + 48x^2 - 64x^3$
(f) $729a^6 + 5832a^5b + 19{,}440a^4b^2 + 34{,}560a^3b^3 + 34{,}560a^2b^4 + 18{,}432ab^5 + 4096b^6$

(16-31) **(a)** $126a^4b^5$ **(b)** $-4320x^3y^3$ **(c)** $220x^3$ **(d)** $-5x^9y$ **(e)** $160x^3y^3$ **(f)** $-v^{15}$

FINAL EXAMINATION

Chapters 9–16

Part 1: Skills and Concepts (40 questions)

1. Evaluate $16^{0.75}$:
   (a) 8   **(b)** 4   **(c)** 12   **(d)** 1   **(e)** none of these

2. Simplify $\left(\dfrac{12a^{-3}b^{-4}}{6ab^{-2}}\right)^{-2}$:

   (a) $\dfrac{2}{a^4b^2}$ **(b)** $\dfrac{4}{a^8b^4}$ **(c)** $\dfrac{a^8b^4}{4}$ **(d)** $-\dfrac{a^8b^4}{4}$ **(e)** none of these

3. Write $\sqrt{\dfrac{49b^3}{28a}}$ in simplest radical form:

   (a) $\dfrac{7b\sqrt{b}}{2\sqrt{7a}}$ **(b)** $\dfrac{b}{2}\sqrt{\dfrac{7b}{a}}$ **(c)** $\dfrac{b\sqrt{7ab}}{2a}$ **(d)** $\dfrac{2b\sqrt{7b}}{a}$ **(e)** none of these

4. Simplify $\sqrt{8a} + 5\sqrt{2a} - \sqrt{2a}$:
   (a) $\sqrt{8a} + 4\sqrt{2a}$ **(b)** $\sqrt{8a} + 5$ **(c)** $7\sqrt{2a}$ **(d)** $6\sqrt{2a}$ **(e)** none of these

5. Solve $\sqrt{3y-3} - \sqrt{2y-4} = 1$:
   (a) 4   **(b)** 2   **(c)** 3   **(d)** 0   **(e)** none of these

6. Write $(4 - 7i) - (2 + 11i)$ in $a + bi$ form:
   (a) $2 - 18i$ **(b)** $2 + 4i$ **(c)** $-2 - 18i$ **(d)** $2 - 4i$ **(e)** none of these

7. Write $(3 - 2i)(4 - 5i)$ in $a + bi$ form:
   (a) $22 - 23i$ **(b)** $12 + 10i$ **(c)** $12 - 10i$ **(d)** $2 - 23i$ **(e)** none of these

8. Simplify i^{322}:
   (a) 1   **(b)** -1 **(c)** i **(d)** $-i$ **(e)** none of these

9. Write the conjugate of $2 + 7i$:
   (a) $2 - 7i$ **(b)** $-2 + 7i$ **(c)** $-2 - 7i$ **(d)** $7 + 2i$ **(e)** none of these

10. Write $\dfrac{4 + 3i}{2 - 5i}$ in $a + bi$ form:

    (a) $-\dfrac{7}{29} - \dfrac{26}{29}i$ **(b)** $\dfrac{23}{29} - \dfrac{14}{29}i$ **(c)** $-\dfrac{23}{29} - \dfrac{14}{29}i$ **(d)** $-\dfrac{7}{29} + \dfrac{26}{29}i$ **(e)** none of these

11. Determine the domain of the function specified by $\{(1, 5), (-2, 7), (3, 8)\}$:
    (a) $\{5, 7, 8\}$ **(b)** $\{-2, 1, 3\}$ **(c)** $\{-2, 1, 3, 5, 7, 8\}$ **(d)** $\{1, 3\}$ **(e)** none of these

12. Evaluate $f(x) = 3x^2 - 2x + 4$ for $x = -6$:
    (a) 100   **(b)** 28   **(c)** 124   **(d)** 89   **(e)** none of these

13. Find the inverse of the function specified by $f(x) = \dfrac{2x}{x - 4}$:

    (a) $f^{-1}(x) = \dfrac{x - 4}{2x}$ **(b)** $f^{-1}(x) = \dfrac{4x}{x - 2}$ **(c)** $f^{-1}(x) = \dfrac{2x}{x + 4}$ **(d)** $f^{-1}(x) = \dfrac{4x}{x + 2}$

    (e) none of these

14. If y varies directly as x, and $y = 12$ when $x = 2\frac{1}{2}$, find y when $x = 3$:
(a) $\frac{72}{5}$ (b) 90 (c) 14 (d) $\frac{74}{5}$ (e) none of these

15. w varies directly as u and inversely as the square of v. If $w = 10$ when $u = 8$, and $v = 4$, find w when $u = 12$ and $v = 10$:
(a) 24 (b) 2 (c) $\frac{1}{2}$ (d) $\frac{12}{5}$ (e) none of these

16. The solution(s) of $3x^2 = 6x$ are:
(a) 2 (b) 0 (c) 0, 2 (d) ± 2 (e) none of these

17. The solution(s) of $6x^2 - 7x - 3 = 0$ are
(a) $-\frac{3}{2}, -\frac{1}{3}$ (b) $\frac{3}{2}, \frac{1}{3}$ (c) $\frac{1}{3}, -\frac{3}{2}$ (d) $-\frac{1}{3}, \frac{3}{2}$ (e) none of these

18. The solution(s) of $3x^2 + 2x - 3 = 0$ are:

(a) $\dfrac{-2 \pm \sqrt{10}}{3}$ (b) $\dfrac{-1 \pm \sqrt{10}}{6}$ (c) $\dfrac{1 \pm \sqrt{10}}{3}$ (d) $\dfrac{-1 \pm \sqrt{10}}{3}$ (e) none of these

19. The solution(s) of $3x + 1 - x + 4 = 1$ are:
(a) 5 (b) 0, 5 (c) 0 (d) no solution (e) none of these

20. The solution set of $6x^2 < 11x - 3$ is:
(a) $\{x | \frac{1}{3} < x < \frac{3}{2}\}$ (b) $\{x | x < \frac{1}{3} \text{ or } x > \frac{3}{2}\}$ (c) $\{x | x < -\frac{1}{3} \text{ or } x > \frac{3}{2}\}$
(d) $\{x | x < -\frac{2}{3} \text{ or } x > \frac{1}{3}\}$ (e) none of these

21. The graph of $y = 3x^2 - x - 2$ is:
(a)   (b)   (c)

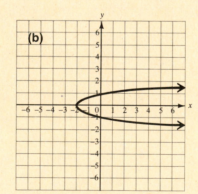
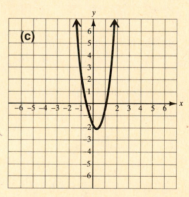

(d)   (e) none of these

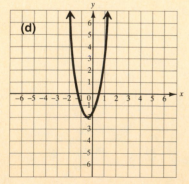

22. The graph of $x^2 + y^2 + 6x - 8y + 16 = 0$ is:

(a)   **(b)**   **(c)**

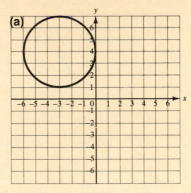   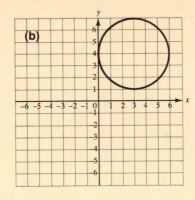   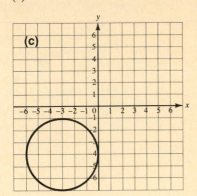

(d)   **(e)** none of these

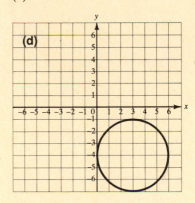

23. The graph of $9x^2 + 16y^2 = 144$ is:

(a)   **(b)**   **(c)**

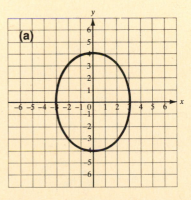   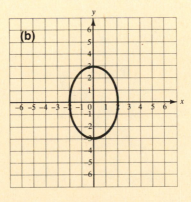   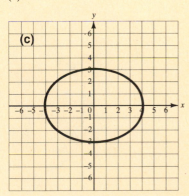

(d)   **(e)** none of these

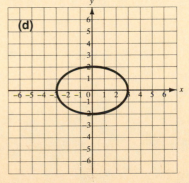

24. The graph of $4x^2 - 9y^2 = 36$ is:

(a)   **(b)**   **(c)**

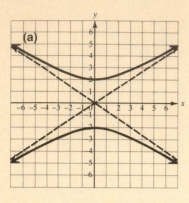   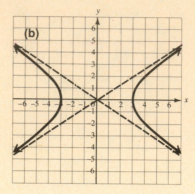   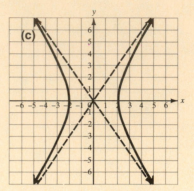

(d)   **(e)** none of these

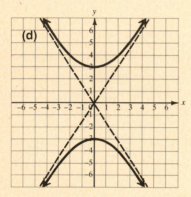

25. The graph of $9y^2 - 16x^2 \leq 144$ is:

(a)   **(b)**   **(c)**

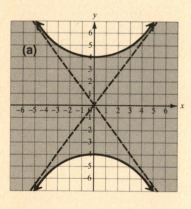   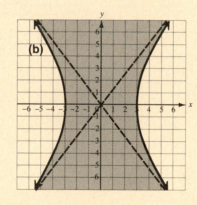   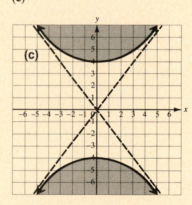

(d)   **(e)** none of these

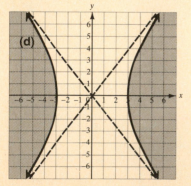

26. The solution(s) of $\left\{\begin{array}{c} x^2 + y^2 = 25 \\ x + y = 7 \end{array}\right\}$ are:

    (a) $(4, 3), (3, 4)$ **(b)** $(-4, 3), (4, -3)$ **(c)** $(-4, -3), (4, 3)$ **(d)** $(-4, -3), (4, -3)$
    (e) none of these

27. The solution(s) of $\left\{\begin{array}{c} x^2 + 6x - y + 6 = 0 \\ x + y = 0 \end{array}\right\}$ are:

    (a) $(-6, 6), (1, -1)$ **(b)** $(-6, 6), (-1, 1)$ **(c)** $(6, -6) (1, -1)$ **(d)** $(6, -6), (-1, 1)$
    (e) none of these

28. The solution(s) of $\left\{\begin{array}{c} xy - 3x = 6 \\ xy = -6 \end{array}\right\}$ are:

    (a) $(4, \frac{3}{2})$ **(b)** $(\frac{3}{2}, 4)$ **(c)** $(-4, \frac{3}{2})$ **(d)** $(4, -\frac{3}{2})$ **(e)** none of these

29. The solution(s) of $\left\{\begin{array}{c} x^2 - 4x - y^2 = -4 \\ x^2 - y^2 = 8 \end{array}\right\}$ are:

    (a) $(-3, 1), (3, -1)$ **(b)** $(-3, -1), (3, 1)$ **(c)** $(3, -1), (-3, 1)$ **(d)** $(-1, -3), (1, 3)$
    (e) none of these

30. The graph of $\left\{\begin{array}{c} x^2 + y^2 \le 16 \\ x + y \ge 0 \end{array}\right\}$ is:

(a)   **(b)**   **(c)**

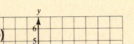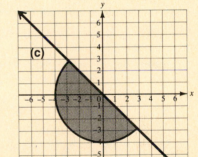

(d)   **(e)** none of these

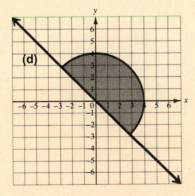

31. The exponential form of $\log_6 216 = 3$ is:
    (a) $3 = 216^6$ **(b)** $216 = 3^6$ **(c)** $216 = 6^3$ **(d)** $6 = 216^3$ **(e)** none of these

32. The common logarithm (to the nearest ten thousandth) of 43.5 is:
    (a) 0.6385   **(b)** 1.6385   **(c)** 0.6395   **(d)** 1.6375   **(e)** none of these

33. The solution(s) of $2^{x^2} = 16$ are:
    (a) 4   (b) 2   (c) ± 4 (d) ± 2 (e) none of these

34. The solution (to the nearest ten thousandth) of $2^{3x} = 3$ is:
    (a) 1.585   (b) 0.6309   (c) 0.5283   (d) 0.2103   (e) none of these

35. The solution(s) of $\log(x - 4) + \log(x - 1) = 1$ are:
    (a) 6   (b) $-6, 1$ (c) $6, -1$ (d) -6 (e) none of these

36. Find $\sum_{k=1}^{4} (k^2 - 1)$:

    (a) 14   (b) 16   (c) 20   (d) 26   (e) none of these

37. Given an arithmetic sequence with $a_1 = 12$ and $d = 6$, find a_{32}:
    (a) 196   (b) 204   (c) 208   (d) 188   (e) none of these

38. Find the sum of the first 31 terms of the arithmetic series with $a_1 = 6$ and $a_{31} = 126$:
    (a) 2046   (b) 1980   (c) 4092   (d) 132   (e) none of these

39. Find the sum of the infinite geometric series $9 + 3 + 1 + \frac{1}{3} + \frac{1}{9} \cdots$:
    (a) 13   (b) $\frac{27}{2}$ (c) 18   (d) $\frac{29}{2}$ (e) none of these

40. Find the fourth term of $(2x - 1)^7$:
    (a) $-35x^4$ (b) $-560x^4$ (c) $560x^4$ (d) $-280x^4$ (e) none of these

Part 2: Problem Solving (10 problems)

41. *Period of a Pendulum:* The period of a pendulum (the time it takes a pendulum to make one complete swing from left to right and then back again) varies directly as the square root of the length and inversely as the square root of the gravitational pull. The gravitational pull near the earth's surface is about 32 feet per second per second [32 ft/s²]. The period of an 8-foot pendulum on earth is π seconds. What is the period of a 2-foot pendulum on earth?

42. Working alone, Paul can complete a certain job in 4 hours less time than Ernie. Working together, they can complete the same job in 3 hours. Exactly how long would it take each to complete the job alone?

43. The sum of a number and its reciprocal is 4. Find the two possible solutions.

44. Two cars start at the same place and travel on roads that are at right angles to each other. The difference between their constant rate is 35 mph. At the end of 4 hours, the cars are 260 miles apart. What is the speed of the slower car? How far did the faster car travel?

45. The perimeter of a right triangle is 12 m with a hypotenuse of 5 m. Find the length of each leg so that the area is maximized.

46. The area of a rectangle is 8 m. The perimeter of the rectangle is 11.4 m. Find the length of a diagonal to the nearest centimeter.

47. The number of bacteria present in a culture initially is 200. At the end of 3 hours, there are 300 bacteria in the culture. How many bacteria will be in the culture after 5 hours, to the nearest whole bacterium?

48. The *intensity of an earthquake* is the ratio a/a_0. Find the intensity of the San Francisco earthquake of 1906 [see Example 15-26].

49. During a free-fall, a parachutist fell 12 ft in the first second, 37 ft during the second second, 62 ft during the third second, and so on. How far did she fall in 14 seconds?

50. A ball dropped from 6 m up rebounds 75% of its previous height on each bounce. How far does the ball travel before it comes to rest?

Final Examination Answers

Part 1

1. (a)	9. (a)	17. (d)	25. (a)	33. (d)
2. (c)	10. (d)	18. (d)	26. (a)	34. (c)
3. (c)	11. (b)	19. (e)	27. (b)	35. (a)
4. (d)	12. (c)	20. (a)	28. (c)	36. (d)
5. (a)	13. (b)	21. (c)	29. (e)	37. (e)
6. (a)	14. (a)	22. (a)	30. (d)	38. (a)
7. (d)	15. (d)	23. (c)	31. (c)	39. (b)
8. (b)	16. (c)	24. (b)	32. (b)	40. (b)

Part 2

41. $\frac{\pi}{2}$ seconds

42. $(1 + \sqrt{13})$ hr, $(5 + \sqrt{13})$ hr

43. $2 + \sqrt{3}, 2 - \sqrt{3}$

44. 25 mph, 240 mi

45. each leg is 3.5 m, maximum area is 12.25 m^2

46. 4.06 m or 406 cm

47. 393 bacteria

48. $10^{8.3}$ [$\approx 1.995 \times 10^8$]

49. 2443 ft

50. 42 m

x	0	1	2	3	4	5	6	7	8	9
1.0	.0000	.0043	.0086	.0128	.0170	.0212	.0253	.0294	.0334	.0374
1.1	.0414	.0453	.0492	.0531	.0569	.0607	.0645	.0682	.0719	.0755
1.2	.0792	.0828	.0864	.0899	.0934	.0969	.1004	.1038	.1072	.1106
1.3	.1139	.1173	.1206	.1239	.1271	.1303	.1335	.1367	.1399	.1430
1.4	.1461	.1492	.1523	.1553	.1584	.1614	.1644	.1673	.1703	.1732
1.5	.1761	.1790	.1818	.1847	.1875	.1903	.1931	.1959	.1987	.2014
1.6	.2041	.2068	.2095	.2122	.2148	.2175	.2201	.2227	.2253	.2279
1.7	.2304	.2330	.2355	.2380	.2405	.2430	.2455	.2480	.2504	.2529
1.8	.2553	.2577	.2601	.2625	.2648	.2672	.2695	.2718	.2742	.2765
1.9	.2788	.2810	.2833	.2856	.2878	.2900	.2923	.2945	.2967	.2989
2.0	.3010	.3032	.3054	.3075	.3096	.3118	.3139	.3160	.3181	.3201
2.1	.3222	.3243	.3263	.3284	.3304	.3324	.3345	.3365	.3385	.3404
2.2	.3424	.3444	.3464	.3483	.3502	.3522	.3541	.3560	.3579	.3598
2.3	.3617	.3636	.3655	.3674	.3692	.3711	.3729	.3747	.3766	.3784
2.4	.3802	.3820	.3838	.3856	.3874	.3892	.3909	.3927	.3945	.3962
2.5	.3979	.3997	.4014	.4031	.4048	.4065	.4082	.4099	.4116	.4133
2.6	.4150	.4166	.4183	.4200	.4216	.4232	.4249	.4265	.4281	.4298
2.7	.4314	.4330	.4346	.4362	.4378	.4393	.4409	.4425	.4440	.4456
2.8	.4472	.4487	.4502	.4518	.4533	.4548	.4564	.4579	.4594	.4609
2.9	.4624	.4639	.4654	.4669	.4683	.4698	.4713	.4728	.4742	.4757
3.0	.4771	.4786	.4800	.4814	.4829	.4843	.4857	.4871	.4886	.4900
3.1	.4914	.4928	.4942	.4955	.4969	.4983	.4997	.5011	.5024	.5038
3.3	.5051	.5065	.5079	.5092	.5105	.5119	.5132	.5145	.5159	.5172
2.3	.5185	.5198	.5211	.5224	.5237	.5250	.5263	.5276	.5289	.5302
3.4	.5315	.5328	.5340	.5353	.5366	.5378	.5391	.5403	.5416	.5428
3.5	.5441	.5453	.5465	.5478	.5490	.5502	.5514	.5527	.5539	.5551
3.6	.5563	.5575	.5587	.5599	.5611	.5623	.5635	.5647	.5658	.5670
3.7	.5682	.5694	.5705	.5717	.5729	.5740	.5752	.5763	.5775	.5786
3.8	.5798	.5809	.5821	.5832	.5843	.5855	.5866	.5877	.5888	.5899
3.9	.5911	.5922	.5933	.5944	.5955	.5966	.5977	.5988	.5999	.6010
4.0	.6021	.6031	.6042	.6053	.6064	.6075	.6085	.6096	.6107	.6117
4.1	.6128	.6138	.6149	.6160	.6170	.6180	.6191	.6201	.6212	.6222
4.2	.6232	.6243	.6253	.6263	.6274	.6284	.6294	.6304	.6314	.6325
4.3	.6335	.6345	.6355	.6365	.6375	.6385	.6395	.6405	.6415	.6425
4.4	.6435	.6444	.6454	.6464	.6474	.6484	.6493	.6503	.6513	.6522
4.5	.6532	.6542	.6551	.6561	.6571	.6580	.6590	.6599	.6609	.6618
4.6	.6628	.6637	.6646	.6656	.6665	.6675	.6684	.6693	.6702	.6712
4.7	.6721	.6730	.6739	.6749	.6758	.6767	.6776	.6785	.6794	.6803
4.8	.6812	.6821	.6830	.6839	.6848	.6857	.6866	.6875	.6884	.6893
4.9	.6902	.6911	.6920	.6928	.6937	.6946	.6955	.6964	.6972	.6981
5.0	.6990	.6998	.7007	.7016	.7024	.7033	.7042	.7050	.7059	.7067
5.1	.7076	.7084	.7093	.7101	.7110	.7118	.7126	.7135	.7143	.7152
5.2	.7160	.7168	.7177	.7185	.7193	.7202	.7210	.7218	.7226	.7235
5.3	.7243	.7251	.7259	.7267	.7275	.7284	.7292	.7300	.7308	.7316
5.4	.7324	.7332	.7340	.7348	.7356	.7364	.7372	.7380	.7388	.7396

x	0	1	2	3	4	5	6	7	8	9

APPENDIX 1: *Common Logarithms*
(continued)

x	0	1	2	3	4	5	6	7	8	9
5.5	.7404	.7412	.7419	.7427	.7435	.7443	.7451	.7459	.7466	.7474
5.6	.7482	.7490	.7497	.7505	.7513	.7520	.7528	.7536	.7543	.7551
5.7	.7559	.7566	.7574	.7582	.7589	.7597	.7604	.7612	.7619	.7627
5.8	.7634	.7642	.7649	.7657	.7664	.7672	.7679	.7686	.7694	.7701
5.9	.7709	.7716	.7723	.7731	.7738	.7745	.7752	.7760	.7767	.7774
6.0	.7782	.7789	.7796	.7803	.7810	.7818	.7825	.7832	.7839	.7846
6.1	.7853	.7860	.7868	.7875	.7882	.7889	.7896	.7903	.7910	.7917
6.2	.7924	.7931	.7938	.7945	.7952	.7959	.7966	.7973	.7980	.7987
6.3	.7993	.8000	.8007	.8014	.8021	.8028	.8035	.8041	.8048	.8055
6.4	.8062	.8069	.8075	.8082	.8089	.8096	.8102	.8109	.8116	.8122
6.5	.8129	.8136	.8142	.8149	.8156	.8162	.8169	.8176	.8182	.8189
6.6	.8195	.8202	.8209	.8215	.8222	.8228	.8235	.8241	.8248	.8254
6.7	.8261	.8267	.8274	.8280	.8287	.8293	.8299	.8306	.8312	.8319
6.8	.8325	.8331	.8338	.8344	.8351	.8357	.8363	.8370	.8376	.8382
6.9	.8388	.8395	.8401	.8407	.8414	.8420	.8426	.8432	.8439	.8445
7.0	.8451	.8457	.8463	.8470	.8476	.8482	.8488	.8494	.8500	.8506
7.1	.8513	.8519	.8525	.8531	.8537	.8543	.8549	.8555	.8561	.8567
7.2	.8573	.8579	.8585	.8591	.8597	.8603	.8609	.8615	.8621	.8627
7.3	.8633	.8639	.8645	.8651	.8657	.8663	.8669	.8675	.8681	.8686
7.4	.8692	.8698	.8704	.8710	.8716	.8722	.8727	.8733	.8739	.8745
7.5	.8751	.8756	.8762	.8768	.8774	.8779	.8785	.8791	.8797	.8802
7.6	.8808	.8814	.8820	.8825	.8831	.8837	.8842	.8848	.8854	.8859
7.7	.8865	.8871	.8876	.8882	.8887	.8893	.8899	.8904	.8910	.8915
7.8	.8921	.8927	.8932	.8938	.8943	.8949	.8954	.8960	.8965	.8971
7.9	.8976	.8982	.8987	.8993	.8998	.9004	.9009	.9015	.9020	.9025
8.0	.9031	.9036	.9042	.9047	.9053	.9058	.9063	.9069	.9074	.9079
8.1	.9085	.9090	.9096	.9101	.9106	.9112	.9117	.9122	.9128	.9133
8.2	.9138	.9143	.9149	.9154	.9159	.9165	.9170	.9175	.9180	.9186
8.3	.9191	.9196	.9201	.9206	.9212	.9217	.9222	.9227	.9232	.9238
8.4	.9243	.9248	.9253	.9258	.9263	.9269	.9274	.9279	.9284	.9289
8.5	.9294	.9299	.9304	.9309	.9315	.9320	.9325	.9330	.9335	.9340
8.6	.9345	.9350	.9355	.9360	.9365	.9370	.9375	.9380	.9385	.9390
8.7	.9395	.9400	.9405	.9410	.9415	.9420	.9425	.9430	.9435	.9440
8.8	.9445	.9450	.9455	.9460	.9465	.9469	.9474	.9479	.9484	.9489
8.9	.9494	.9499	.9504	.9509	.9513	.9518	.9523	.9528	.9533	.9538
9.0	.9542	.9547	.9552	.9557	.9562	.9566	.9571	.9576	.9581	.9586
9.1	.9590	.9595	.9600	.9605	.9609	.9614	.9619	.9624	.9628	.9633
9.2	.9638	.9643	.9647	.9652	.9657	.9661	.9666	.9671	.9675	.9680
9.3	.9685	.9689	.9694	.9699	.9703	.9708	.9713	.9717	.9722	.9727
9.4	.9731	.9736	.9741	.9745	.9750	.9754	.9759	.9763	.9768	.9773
9.5	.9777	.9782	.9786	.9791	.9795	.9800	.9805	.9809	.9814	.9818
9.6	.9823	.9827	.9832	.9836	.9841	.9845	.9850	.9854	.9859	.9863
9.7	.9868	.9872	.9877	.9881	.9886	.9890	.9894	.9899	.9903	.9908
9.8	.9912	.9917	.9921	.9926	.9930	.9934	.9939	.9943	.9948	.9952
9.9	.9956	.9961	.9965	.9969	.9974	.9978	.9983	.9987	.9991	.9996
x	0	1	2	3	4	5	6	7	8	9

APPENDIX 2: *Powers and Roots*

n	n^2	$\sqrt{n}$	n^3	$\sqrt[3]{n}$	n	n^2	$\sqrt{n}$	n^3	$\sqrt[3]{n}$
1	1	1.000	1	1.000	51	2,601	7.141	132,651	3.708
2	4	1.414	8	1.260	52	2,704	7.211	140,608	3.733
3	9	1.732	27	1.442	53	2,809	7.280	148.877	3.756
4	16	2.000	64	1.587	54	2,916	7.348	157,464	3.780
5	25	2.236	125	1.710	55	3,025	7.416	166,375	3.803
6	36	2.449	216	1.817	56	3,136	7.483	175,616	3.826
7	49	2.646	343	1.913	57	3,249	7.550	185,193	3.849
8	64	2.828	512	2.000	58	3,364	7.616	195,112	3.871
9	81	3.000	729	2.080	59	3,481	7.681	205,379	3.893
10	100	3.162	1,000	2.154	60	3,600	7.746	216,000	3.915
11	121	3.317	1,331	2.224	61	3,721	7.810	226,981	3.936
12	144	3.464	1,728	2.289	62	3,844	7.874	238,328	3.958
13	169	3.606	2,197	2.351	63	3,969	7.937	250,047	3.979
14	196	3.742	2,744	2.410	64	4,096	8.000	262,144	4.000
15	225	3.873	3,375	2.466	65	4,225	8.062	274,625	4.021
16	256	4.000	4,096	2.520	66	4,356	8.124	287,496	4.041
17	289	4.123	4,913	2.571	67	4,489	8.185	300,763	4.062
18	324	4.243	5,832	2.621	68	4,624	8.246	314,432	4.082
19	361	4.359	6,859	2.668	69	4,761	8.307	328,509	4.102
20	400	4.472	8,000	2.714	70	4,900	8.367	343,000	4.121
21	441	4.583	9,261	2.759	71	5,041	8.426	357,911	4.141
22	484	4.690	10,648	2.802	72	5,184	8.485	373,248	4.160
23	529	4.796	12,167	2.844	73	5,329	8.544	389,017	4.179
24	576	4.899	13,824	2.884	74	5,476	8.602	405,224	4.198
25	625	5.000	15,625	2.924	75	5,625	8.660	421,875	4.217
26	676	5.099	17,576	2.962	76	5,776	8.718	438,976	4.236
27	729	5.196	19,683	3.000	77	5,929	8.775	456,533	4.254
28	784	5.292	21,952	3.037	78	6,084	8.832	474,552	4.273
29	841	5.385	24,389	3.072	79	6,241	8.888	493,039	4.291
30	900	5.477	27,000	3.107	80	6,400	8.944	512,000	4.309
31	961	5.568	29,791	3.141	81	6,561	9.000	531,441	4.327
32	1,024	5.657	32,768	3.175	82	6,724	9.055	551,368	4.344
33	1,089	5.745	35,937	3.208	83	6,889	9.110	571,787	4.362
34	1,156	5.831	39,304	3.240	84	7,056	9.165	592,704	4.380
35	1,225	5.916	42,875	3.271	85	7,225	9.220	614,125	4.397
36	1,296	6.000	46,656	3.302	86	7,396	9.274	636,056	4.414
37	1,369	6.083	50,653	3.332	87	7,569	9.327	658,503	4.431
38	1,444	6.164	54,872	3.362	88	7,744	9.381	681,472	4.448
39	1,521	6.245	59,319	3.391	89	7,921	9.434	704,969	4.465
40	1,600	6.325	64,000	3.420	90	8,100	9.487	729,000	4.481
41	1,681	6.403	68,921	3.448	91	8,281	9.539	753,571	4.498
42	1,764	6.481	74,088	3.476	92	8,464	9.592	778,688	4.514
43	1,849	6.557	79,507	3.503	93	8,649	9.644	804,357	4.531
44	1,936	6.633	85,184	3.530	94	8,836	9.695	830,584	4.547
45	2,025	6.708	91,125	3.557	95	9,025	9.747	857,375	4.563
46	2,116	6.782	97,336	3.583	96	9,216	9.798	884,736	4.579
47	2,209	6.856	103,823	3.609	97	9,409	9.849	912,673	4.595
48	2,304	6.928	110,592	3.634	98	9,604	9.899	941,192	4.610
49	2,401	7.000	117,649	3.659	99	9,801	9.950	970,299	4.626
50	2,500	7.071	125,000	3.684	100	10,000	10.000	1,000,000	4.642

APPENDIX 3: *Exponential Functions e^x and e^{-x}*

x	e^x	e^{-x}	x	e^x	e^{-x}
0.00	1.0000	1.0000	1.5	4.4817	0.2231
0.01	1.0101	0.9901	1.6	4.9530	0.2019
0.02	1.0202	0.9802	1.7	5.4739	0.1827
0.03	1.0305	0.9702	1.8	6.0496	0.1653
0.04	1.0408	0.9608	1.9	6.6859	0.1496
0.05	1.0513	0.9512	2.0	7.3891	0.1353
0.06	1.0618	0.9418	2.1	8.1662	0.1225
0.07	1.0725	0.9324	2.2	9.0250	0.1108
0.08	1.0833	0.9231	2.3	9.9742	0.1003
0.09	1.0942	0.9139	2.4	11.023	0.0907
0.10	1.1052	0.9048	2.5	12.182	0.0821
0.11	1.1163	0.8958	2.6	13.464	0.0743
0.12	1.1275	0.8869	2.7	14.880	0.0672
0.13	1.1388	0.8781	2.8	16.445	0.0608
0.14	1.1503	0.8694	2.9	18.174	0.0550
0.15	1.1618	0.8607	3.0	20.086	0.0498
0.16	1.1735	0.8521	3.1	22.198	0.0450
0.17	1.1853	0.8437	3.2	24.533	0.0408
0.18	1.1972	0.8353	3.3	27.113	0.0369
0.19	1.2092	0.8270	3.4	29.964	0.0334
0.20	1.2214	0.8187	3.5	33.115	0.0302
0.21	1.2337	0.8106	3.6	36.598	0.0273
0.22	1.2461	0.8025	3.7	40.447	0.0247
0.23	1.2586	0.7945	3.8	44.701	0.0224
0.24	1.2712	0.7866	3.9	49.402	0.0202
0.25	1.2840	0.7788	4.0	54.598	0.0183
0.30	1.3499	0.7408	4.1	60.340	0.0166
0.35	1.4191	0.7047	4.2	66.686	0.0150
0.40	1.4918	0.6703	4.3	73.700	0.0136
0.45	1.5683	0.6376	4.4	81.451	0.0123
0.50	1.6487	0.6065	4.5	90.017	0.0111
0.55	1.7333	0.5769	4.6	99.484	0.0101
0.60	1.8221	0.5488	4.7	109.95	0.0091
0.65	1.9155	0.5220	4.8	121.51	0.0082
0.70	2.0138	0.4966	4.9	134.29	0.0074
0.75	2.1170	0.4724	5.0	148.41	0.0067
0.80	2.2255	0.4493	5.5	244.69	0.0041
0.85	2.3396	0.4274	6.0	403.43	0.0025
0.90	2.4596	0.4066	6.5	665.14	0.0015
0.95	2.5857	0.3867	7.0	1096.6	0.0009
1.0	2.7183	0.3679	7.5	1808.0	0.0006
1.1	3.0042	0.3329	8.0	2981.0	0.0003
1.2	3.3201	0.3012	8.5	4914.8	0.0002
1.3	3.6693	0.2725	9.0	8103.1	0.0001
1.4	4.0552	0.2466	10.0	22,026	0.00005

APPENDIX 4: *Natural Logarithms (base e)*

n	$\log_e n$	n	$\log_e n$	n	$\log_e n$
		4.5	1.5041	9.0	2.1972
0.1	-2.0326	4.6	1.5261	9.1	2.2083
0.2	-1.6094	4.7	1.5476	9.2	2.2192
0.3	-1.2040	4.8	1.5686	9.3	2.2300
0.4	-0.9163	4.9	1.5892	9.4	2.2407
0.5	-0.6931	5.0	1.6094	9.5	2.2513
0.6	-0.5108	5.1	1.6292	9.6	2.2618
0.7	-0.3567	5.2	1.6487	9.7	2.2721
0.8	-0.2231	5.3	1.6677	9.8	2.2824
0.9	-0.1054	5.4	1.6864	9.9	2.2925
1.0	0.0000	5.5	1.7047	10	2.3026
1.1	0.0953	5.6	1.7228	11	2.3979
1.2	0.1823	5.7	1.7405	12	2.4849
1.3	0.2624	5.8	1.7579	13	2.5649
1.4	0.3365	5.9	1.7750	14	2.6391
1.5	0.4055	6.0	1.7918	15	2.7081
1.6	0.4700	6.1	1.8083	16	2.7726
1.7	0.5306	6.2	1.8245	17	2.8332
1.8	0.5878	6.3	1.8405	18	2.8904
1.9	0.6419	6.4	1.8563	19	2.9444
2.0	0.6931	6.5	1.8718	20	2.9957
2.1	0.7419	6.6	1.8871	25	3.2189
2.2	0.7885	6.7	1.9021	30	3.4012
2.3	0.8329	6.8	1.9169	35	3.5553
2.4	0.8755	6.9	1.9315	40	3.6889
2.5	0.9163	7.0	1.9459	45	3.8067
2.6	0.9555	7.1	1.9601	50	3.9120
2.7	0.9933	7.2	1.9741	55	4.0073
2.8	1.0296	7.3	1.9879	60	4.0943
2.9	1.0647	7.4	2.0015	65	4.1744
3.0	1.0986	7.5	2.0149	70	4.2485
3.1	1.1314	7.6	2.0281	75	4.3175
3.2	1.1632	7.7	2.0412	80	4.3820
3.3	1.1939	7.8	2.0541	85	4.4427
3.4	1.2238	7.9	2.0669	90	4.4998
3.5	1.2528	8.0	2.0794	100	4.6052
3.6	1.2809	8.1	2.0919	110	4.7005
3.7	1.3083	8.2	2.1041	120	4.7875
3.8	1.3350	8.3	2.1163	130	4.8676
3.9	1.3610	8.4	2.1282	140	4.9416
4.0	1.3863	8.5	2.1401	150	5.0106
4.1	1.4110	8.6	2.1518	160	5.0752
4.2	1.4351	8.7	2.1633	170	5.1358
4.3	1.4586	8.8	2.1748	180	5.1930
4.4	1.4816	8.9	2.1861	190	5.2470

	Figure	Perimeter (P)	Area (A)
Square		$P = 4s$	$A = s^2$
Rectangle		$P = 2(l + w)$	$A = lw$
Parallelogram		$P = 2(a + b)$	$A = bh$
Triangle		$P = a + b + c$	$A = \frac{1}{2}bh$

	Figure	Circumference (C)	Area (A)
Circle		$C = \pi d$ $C = 2\pi r$	$A = \pi r^2$

	Figure	Volume (V)	Surface Area (SA)
Cube		$V = e^3$	$SA = 6e^2$
Rectangular Prism (box)		$V = lwh$	$SA = 2(lw + lh + wh)$
Cylinder		$V = \pi r^2 h$	$SA = 2\pi r(r + h)$
Sphere		$V = \frac{4}{3}\pi r^3$	$SA = 4\pi r^2$

INDEX

B
C
D 3
E 4
F 5
G 6
H 7
I 8
J 9